NAVAIR 01-F14AAA-1

F-14 TOMCAT
PILOT'S FLIGHT
OPERATING MANUAL

ISSUED BY AUTHORITY OF THE CHIEF OF NAVAL OPERATIONS
AND UNDER THE DIRECTION OF THE COMMANDER,
NAVAL AIR SYSTEMS COMMAND

Reprint ©2009 Periscope Film LLC
All Rights Reserved
www.PeriscopeFilm.com

ISBN # 978-1-935327-72-1   1-935327-72-0

SECTION V — EMERGENCY PROCEDURES

TABLE OF CONTENTS

Introduction

This section covers the recommended procedures for coping with emergencies and malfunctions that may be encountered during aircraft operations. Knowledge of the aircraft systems and emergency procedures must be reviewed on a regular basis to ensure that the flightcrew will take the correct course of action under adverse conditions.

Each emergency presents a different problem that requires positive, specific, remedial action in accordance with recommended procedures and good airmanship. Judgment, precision, and teamwork are essential during emergencies. The flightcrew must weigh all the factors of a given situation and then take appropriate action. This section discusses the preplanned, likely courses of action and

recommended procedures for certain emergencies. References to emergency procedures promulgated in NWP-41 is also required. As soon as possible, the pilot should notify the RIO, flight leader, flight, and ground station in as much detail as possible of the existing emergency and of the intended action.

Critical Procedures

Procedures marked with asterisks (*) are considered critical. Flightcrew members should be able to accomplish asterisked procedures without reference to the checklist. Critical procedures are presented in an abbreviated, easy to remember form in NAVAIR 01-F14AAA-1B.

PART 1 — GROUND EMERGENCIES

ENGINE FIRE ON THE GROUND

PILOT

*1. Both FUEL SHUT OFF handles . . . . PULL

*2. Both throttles . . . . OFF

3. If conditions permit . . . . . . WINDMILL ENGINE

If Fire Light And/Or Other Secondary Indications:

4. Fire extinguisher pushbutton
   (affected engine) . . . . .DEPRESS

5. EGRESS

RIO

1. Notify ground and/or tower

2. EGRESS

ABNORMAL START

1. Throttle (affected engine) . . . . OFF

Note

- If hot start, windmill engine until TIT is below 400°C.

- If wet start, continue cranking until tailpipe is clear of fuel.

UNCOMMANDED THROTTLE(S)—ENGINE ACCELERATION ON THE GROUND

*1. Paddle switch . . . . DEPRESS AND HOLD

*2. THROTTLE MODE switch . . . . MAN

*3. Throttle(s) . . . . .AS DESIRED

If Throttle(s) Still Uncommanded and Aircraft Is Not In Catapult Tension:

4. Throttle(s) . . . . OFF

5. FUEL SHUTOFF handle(s) . . . . PULL

Note

- Uncommanded throttles can result in increased or decreased throttle settings depending on type of failure.

- Approximately fifty pounds of force must be applied to the throttles to override the boost system to ensure disengagement of APC BIT self-test.

- The quickest and most reliable method to secure uncommanded throttles is to revert the throttle system to the manual mode and secure the throttle(s). Since manual is, by design, a backup mode the throttle rigging may not be the same as the boost mode. It may take a hard snapping motion to position the throttle into idle cutoff. If throttle(s) are mis-rigged in manual mode the idle cutoff position may not secure fuel flow to the engine.

- Both throttles cannot be secured simul-taneously, however, reverting to manual mode will allow both throttles to be repositioned to IDLE simultaneously.

GROUND EGRESS WITHOUT PARACHUTE AND SURVIVAL KIT

Note

If possible, kneel the aircraft for easier ground egress.

*1. Canopy . . . . .OPEN OR JETTISON

*2. Parking brake . . . . .PULL

*3. Ejection seat . . . . .SAFE
   (Safe by pulling the Emergency Restraint Release Handle and with the Face Curtain Locking Tab)

*4. All fittings . . . . RELEASE
   (Restraint fittings and Oxygen hose)

Note

To retain survival kit, do not release lap
belt restraint fittings.

EMERGENCY ENTRANCE

See figure 5-1 for procedures for entering the cockpit for
emergency rescue.

DITCHING

*1. Canopy . . . . JETTISON BEFORE IMPACT

WEIGHT ON-OFF WHEELS SWITCH MALFUNCTION

There are weight on-off wheels switches on the left and
right main gear that interact with many aircraft subsystems
to provide safety interlocks. The interlocks prevent operation of various components or systems on deck or in flight,
as appropriate.

```
CAUTION
```

Failure of the left or right weight on-off
wheels switches to the in-flight mode
can cause loss of engine ejector air to the
engine oil and IDGs causing thermal disconnect and/or heat damage to the
generators.

Failure Of Weight On-Off Wheels Switch To In-flight Mode

INDICATIONS:

- WOW acronym displayed

- Approach indexers illuminated

- Nozzles closed at idle RPM

- Nosewheel steering inoperative

- Launch bar light illuminated
  (if nose gear turned $> 10°$)

- Ground roll braking inoperative

- Wing sweep MASTER TEST disabled

- Oversweep disabled

- Outboard spoiler module on with FLAP
  handle UP

- Aircraft will not kneel

If two or more of the preceding anomalies are detected,
the following action should be taken:

PILOT

1. Clear runway (if applicable)

2. Generators . . . . OFF

3. Throttles . . . . OFF
                    (after down locks are in place)

RIO

1. WCS switch . . . . OFF

```
WARNING
```

With failure of the weight on-off wheels
switch to the in-flight mode, the following functions are enabled:

- AWG-9 can scan and radiate.

- ALQ-100 or ALQ-126 can transmit.

- Probe heaters will be on in AUTO.

EMERGENCY ENTRANCE

1. PUSHBUTTON TO OPEN DOOR.
2. SQUEEZE T-HANDLE AND PULL TO JETTISON CANOPY.

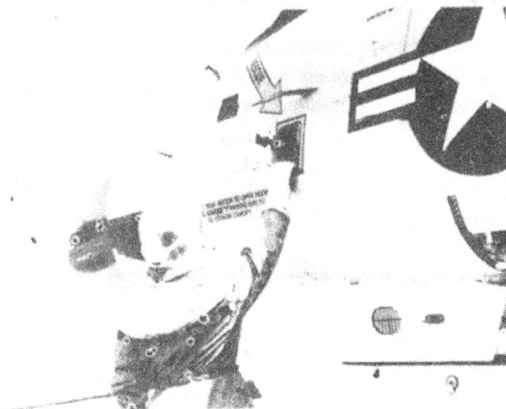

NORMAL COCKPIT ENTRANCE

IT IS PREFERABLE TO USE
THE NORMAL COCKPIT
ENTRANCE PROCEDURE.
HOWEVER, IF IT IS
INOPERATIVE OR
TIME IS CRITICAL — — — JETTISON.

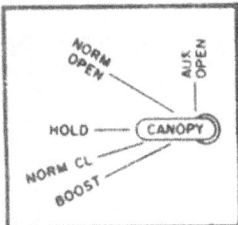

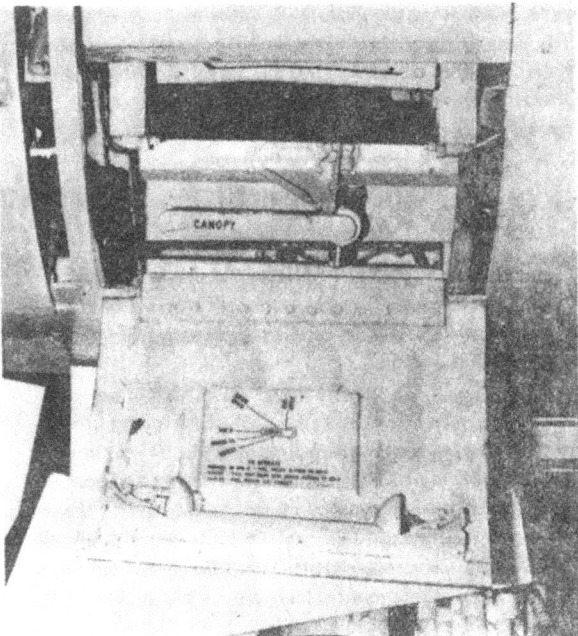

Figure 5-1. Emergency Entrance

F-F050-004
106-0

5-5

.PART 2 — TAKEOFF EMERGENCIES

ABORTED TAKEOFF

Emergencies during takeoff are extremely critical and require fast analysis and quick decision by the pilot. The decision to abort should not be delayed just because emergency arresting gear is available at the end of the runway. Whether to abort or continue the takeoff depends on the length of runway remaining, refusal speed, best single-engine climb speed, and the arresting gear available. Failure of either engine, a fire warning light, or a blown tire during takeoff dictates an immediate abort if enough runway is available. The ejection seats will provide safe escape at ground level and low airspeeds. Therefore, if a safe aborted takeoff cannot be performed and takeoff is impossible, — EJECT.

In an aborted takeoff, aerodynamic ground roll braking is assisted by simultaneous deflection of all spoilers (flaps down) or inboard spoilers only (flaps up) to 55° when both throttles are retarded to idle. When securing the starboard engine, use caution to prevent inadvertent shutdown of both engines. If both engines are shut down, hydraulic pressure is lost, along with antiskid, nosewheel steering, spoiler braking, and normal braking. Full aft stick is used to augment aerodynamic braking. Care should be taken while positioning the stick aft to avoid any nose rotation. The aircraft's tendency to rotate is accentuated with the flaps up due to increased longitudinal control effectiveness, and aft stick must be applied at a slower rate to avoid rotation.

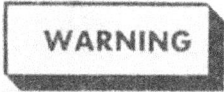

WARNING

Maximum braking effort in aborts initiated near rotation speed at takeoff gross weights may result in blown tires even with antiskid engaged.

If arresting gear is available, use it to avoid rolling off the runway. Always inform the control tower of your intention to abort the takeoff and engage the arresting gear, so that aircraft landing behind you can be waved off. Lower the hook in sufficient time for it to fully extend (normally 1,000 feet before engagement). Use nosewheel steering to maintain directional control and aim for the center of the runway. At night, use the taxi light to see the arresting gear. If off center just before engaging the arresting gear,

do not turn the aircraft but continue straight ahead, parallel to the centerline. After engagement, secure the engines if no directional control problems exist.

If aborting with a blown nosewheel tire, it is likely that either or both engines have FOD. The flaps should be left in the down position until they also can be inspected for FOD.

Aircraft control following loss of an engine during the takeoff phase is adequate up to 15 units AOA with thrust asymmetries to maximum after-burner thrust on the operating engine.

Aborted Takeoff

*1. Throttles . . . . IDLE

*2. Speed brakes . . . . EXT

*3. Stick. . . . . AFT

Note

The stick should be positioned fully aft at a rate that will not cause any nose rotation.

*4. Hook . . . . DN
   (1,000 feet before wire)

*5. Brakes. . . . . AS REQUIRED

*6. Right engine. . . . . OFF
   (if required)

SINGLE ENGINE FAILURE — FIELD/ CATAPULT LAUNCH

Loss of an engine during takeoff is a critical emergency situation. Until VMc and VMcg (minimum control speed and minimum control speed ground) have been determined, no specific procedures can be established to counter the effects of an engine failure during field takeoff.

Pending completion of testing, the following guidelines apply:

- If loss of an engine during takeoff occurs before liftoff, the takeoff should be aborted.

- If loss of an engine during takeoff occurs after liftoff, continue the takeoff and adhere to procedures under Single Engine Failure Field/Catapult.

- Do not rotate the aircraft below 130 KIAS in any configuration. Refer to NAVAIR 01-F14AAA-1.1, Section XI, Part 2 for higher rotation speeds and no flap configurations.

CATAPULT LAUNCH. A single engine failure during catapult launch is controllable if the flightcrew uses correct procedures and includes preflight planning to select the appropriate launch thrust setting to maintain a 200 foot per minute single-engine rate of climb. See figure 5-2. Controllability is not affected by gross weight or external stores configurations. Rudder control power to offset yawing moment due to thrust asymmetry is a function of airspeed. (Refer to Section IV, Rudder Effectiveness).

F-14A TOMCAT

THIS PAGE INTENTIONALY LEFT BLANK.

Rudder is the primary control for countering yaw due to asymmetric thrust since lateral stick inputs alone will induce adverse yaw in an already critical flight regime. Compounding the situation, visual cues for ascertaining yaw excursions may be absent at night. While roll due to yaw will always be apparent, yaw excursions during night/IFR conditions may be first indicated by the turn and slip indicator and heading indicator if in near wings level flight. At the first indication of an engine failure, the pilot should be prepared to apply rudder as the primary means of maintaining aircraft control. The pilot should not hesitate to apply up to full rudder to counter roll and yaw.

ANGLE-OF-ATTACK/PITCH ATTITUDE CONSIDERATIONS. Failure to limit pitch attitude will place the aircraft in a regime of reduced directional stability, rudder control and rate of climb. The aircraft may be uncontrollable at angles of attack above 20 units. A smooth rotation to 10° pitch attitude (approximately 14 units indicated AOA) will enhance controllability, and provide a good initial flyaway attitude, ensure single engine acceleration and generate adequate rate of climb performance.

RATE OF CLIMB CONSIDERATIONS. Stabilized zone 5 afterburner launch shall be performed when required to ensure a positive single engine rate of climb. (Refer to NAVAIR 01-F14AAA-1.1, Section XI, Part 2.) Figure 5-2 summarizes single engine flyaway capabilities as a function of ambient temperature, aircraft gross weight and angle of attack. The chart is useable for all combinations of external stores. Prior to launch the aircrew must take into account the ambient temperature and aircraft gross weight to determine climb capability. If a turn is required, airspeed control and angle of bank will greatly affect rate of climb performance. (Refer to NAVAIR 01-F14AAA-1.1, Section XI, Part 2.)

ENGINE CONSIDERATIONS. A properly trimmed engine exhibits essentially the same stall margin at stabilized zone 5 as at stabilized military power. Throttle/afterburner transients cause a significant reduction in engine stall margin and should be avoided whenever possible. This is best accomplished by reaching a stabilized zone 5 or military power condition prior to catapult release. Mid-zone afterburner launches may result in inadvertent afterburner staging during catapult stroke and increased likelihood of afterburner blowout/engine stall. Aircrews should realize that pilot reaction time and afterburner lighting characteristics may not provide adequate safety margin in countering aircraft deceleration/rate of climb problems associated with heavy gross weight single engine military power catapult launches.

STORES JETTISON CONSIDERATIONS. If a positive rate of climb cannot be maintained or deceleration cannot be countered by thrust alone, emergency jettison should be selected without hesitation. The benefits of an instantly lighter aircraft and lower drag configuration always produce positive effects on performance.

AIRCREW COORDINATION. Of paramount importance is a common and knowledgeable understanding by both pilot and RIO of what to expect when confronted with engine failure during launch. Both must have already determined during a preflight briefing the points to be considered, i.e. controllability, AOA/pitch, rate of climb, engine and jettison considerations. The pilot will probably be the only one to know if an engine fails during launch. The RIO will probably be the only one in a position to successfully initiate ejection prior to departing the successful ejection envelope. Each launch must be made with the aircrew prepared for the worst case situation.

SINGLE ENGINE FAILURE FIELD/CATAPULT

*1. Establish 10° pitch attitude (not to exceed 14 units AOA).

*2. Counter yaw/roll with rudder supplemented by lateral stick.

*3. Throttles (both engines) as required to maintain a positive rate of climb.

*4. EMERG STORES JETT pushbutton . . . . . DEPRESS
(if required)

5. Landing gear. . . . . .UP

6. If unable to control aircraft. . . . . EJECT

7. Establish 10 unit AOA climb.

8. Climb to safe altitude.

9. Flaps. . . . . .UP

BLOWN TIRE DURING TAKEOFF

If a tire blows during the takeoff and an abort is impossible, do not raise the landing gear or flaps. Leave the landing gear down to avoid fouling the blown tire in the wheel well.

SINGLE ENGINE MILITARY POWER FLYAWAY CAPABILITIES DATA FOR 15 KNOT EXCESS SPEED LAUNCHES

AIRCRAFT CONFIGURATION:
(6) PHOENIX MISSLES,
+ (2) 280 GALLON FUEL TANKS,
35° FLAPS, GEAR DOWN,
DLC STOWED

DATE: 1 AUGUST 1981
DATA BASIS: ESTIMATED BASED
ON FLIGHT TEST

REMARKS
ENGINE(S): (2) TF30-P-414

FUEL GRADE: JP-5 (JP-8, JP-4)
FUEL DENSITY: 6.8 (6.7, 6.5) lb/gal

INOPERATIVE ENGINE: LOCKED ROTOR
(N1 = N2 = 0 RPM)

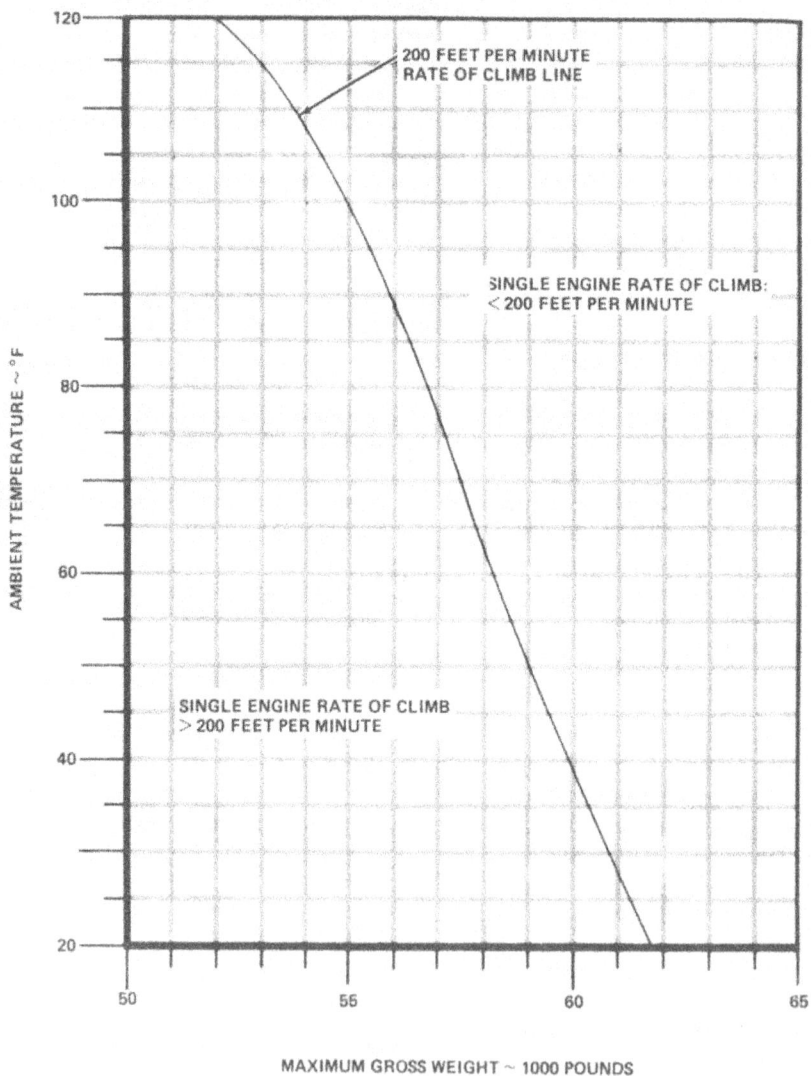

Figure 5-2. Single Engine Military Power Flyaway Capabilities Data for 15-Knot Excess Launches

3-F50-234-0

Leave the flaps down; they may be damaged by pieces of ruptured tire. Also, climbing with the gear and flaps down is an optimum flight attitude for emergency fuel dumping.

If a Tire Blows During Takeoff:

TAKEOFF NOT CONTINUED

*1. ABORT

*2. Nosewheel steering . . . . ENGAGED

*3. ANTI SKID SPOILER BK switch . . . . SPOILER BK

CAUTION

Engines should not be secured until crash crew is in position to extinguish a possible fire created from drain fuel contacting hot wheel assembly.

Note

Antiskid will sense a constant release on a dragging blown tire.

TAKEOFF CONTINUED

*1. Throttles . . . . MAX AB

*2. Landing gear and flaps . . . . LEAVE AS
SET FOR TAKEOFF

CAUTION

Blown nose tire can cause engine FOD.

3. Refer to Blown Tire Landing Procedures. (Refer to page 5-42.)

LAUNCH BAR

Launch Bar Light Illuminated

1. Landing gear. LEAVE DN

2. Obtain visual inspection.

If Launch Bar Is Down and Visual Inspection is Not Available:

3. Remove arresting cables for field landing.

4. Remove crossdeck pendants 1 and 4 for CV landings.

PART 3 — IN-FLIGHT EMERGENCIES

COMMUNICATIONS FAILURE

1. Check mikes and earphone plugs.

2. Check oxygen mask connection and oxygen hose disconnect.

3. RIO check console connector adjacent to shoulder harness control lever. Pilot check console connector aft of G valve.

4. Increase ICS volume, and attempt B/U and EMERG positions.

5. Attempt intercommunications with UHF transceiver.

6. If cockpit altitude is safe, oxygen mask can be removed so that when helmet earmuff is held open, verbal communications can be maintained.

FLIGHTCREW ATTENTION SIGNALS

When no other method of communicating exists, the following signals should be used:

1. Pilot will attract RIO by rocking of wings.

2. RIO will attract pilot by shouting . . . . !

3. Acknowledgement will be thumbs-up, high on left-hand side of cockpit, and future communications will be conducted by visual hand signals using HEFOE code.

COMM-NAV EMERGENCY PROCEDURES

Lost (Without Navigation Aids or Radio Receiver)

1. Pilot select running lights on FLASH.

2. RIO squawk mode 3 Code 7700 for 1 minute and then mode 3 code 7600.

3. Attempt home base location by radar mapping, or DR to best known position. Attempt marshal pattern location by APX-76 interrogation.

4. Drop four bundles of chaff at 2-mile intervals, then complete series of four standard left-hand 360° turns at 20-second intervals.

5. If no chaff, fly minimum of two triangular patterns to left with 1-minute legs.

6. Repeat patterns at 20-minute intervals.

7. Conserve fuel throughout and facilitate radar pickup by maintaining highest feasible altitude consistent with situation.

8. Be alert for aircraft attempting to join.

9. After joining, communicate with appropriate hand or light signals.

Lost (Without Navigation Aids but With Radio Receiver)

1. Same as without radio, but make turns to right.

No Radio (With Navigation Aids)

1. Proceed to alternate marshal.

2. Energize IP function at least once each minute.

3. Commence penetration or jetdown at Expected Approach Clearance time (EAC). If not given EAC, commence approach at ETA.

4. Be alert for aircraft vectored to join.

PITOT-STATIC SYSTEM FAILURES

If the altimeter and Mach airspeed indicators are erroneous, pitot pressure, static pressure, and total temperature inputs to the CADC may also be inaccurate. Placing the ENG/PROBES ANTI-ICE switch in the ORIDE or AUTO position may restore operation if the malfunction was caused by icing.

Note

Pitot-static system failures due to icing may input an erroneous Mach number to the AICS programmer, which will result in the ramps being in the wrong position for the actual Mach number (engine stall may result).

If it is apparent that icing is not the problem, use the angle-of-attack indicator in place of airspeed for flight conditions as shown in figure 5-3. Descend to an altitude below 43,000 feet. When cabin altitude stabilizes at 8,000 feet, aircraft altitude will be approximately 23,000 feet. Below 23,000 feet, aircraft altitude can be determined by dumping cabin pressure and using the cabin altitude indicator above 5,000 feet. Below 5,000 feet, use the radar altimeter.

Reduce airspeed and set wing sweep to 20° using the emergency wing sweep mode. The landing should be without the auto throttle engaged. If the AWG-9 computer computations are affected, the RIO can manually enter estimated wind direction and velocity through the computer address panel.

AIRSPEED INDICATOR FAILURE

FLIGHT CONDITION		ANGLE-OF-ATTACK UNITS
CATAPULT		
Transition From Catapult		15.0
MILITARY POWER CLIMB		
Drag Index = 8	Sea Level	6.0
	Combat Ceiling	9.5
Drag Index = 100	Sea Level	6.0
	Combat Ceiling	9.5
MAXIMUM POWER CLIMB		
All Drag Indexes	(DI=8) Sea Level	5.0
	(DI=8) Combat Ceiling	8.0
CRUISE AT ALTITUDES BELOW 20,000 ft (All Gross Weights)		
Drag Index = 8		8.0
Drag Index = 100		9.0
CRUISE AT OPTIMUM ALTITUDE		
Drag Index = 8		8.0
Drag Index = 100		
ENDURANCE AT OPTIMUM ALTITUDE		
Drag Index = 8		9.0
Drag Index = 100		10.0
DESCENTS (Low to Medium Gross Weights)		
250 KCAS, IDLE Power		9.0
GEAR AND FLAPS EXTENSION		
Safe Gear Extension (Flaps UP) at 280 KIAS		6.5
Safe Flap Extension (Gear DN) at 225 KIAS		8.0
APPROACH		
CCA/GCA Pattern, 220 KCAS, Gear UP, Flaps UP; 57,000 pounds.		9.0
Final ON-SPEED Approach (Gear DN):		
Two Engines (Flap Configurations)		15.0
Single Engine (Flap Configurations)		14.0

NOTES		
DRAG INDEX	CONFIGURATION	
8	(4)	AIM-7F
100	(6)	AIM-54 PLUS
	(2)	267 gallon external tanks

Figure 5-3. Airspeed Indicator Failure

EMERGENCY JETTISON

All stores including external fuel tanks (stations 2 and 7), except Sidewinder missiles (AIM-9), are jettisoned in a fixed interval between sequenced stations to avoid store-to-aircraft collision. See figure 5-4 for external stores jettison table.

WARNING

With landing flaps and slats down, do not fire Sidewinder missiles.

Note

Sidewinder missiles cannot be jettisoned from aircraft with AAC 688 incorporated.

1. EMERG STORES
   JETT pushbutton . . . . DEPRESS

Note

- The EMERG STORES JETT pushbutton-indicator remains illuminated until the pushbutton is depressed again.

- A weight-off-wheels signal from the left or right main wheel is sufficient to enable emergency jettison.

- A complete emergency store jettison sequence can take 1.7 seconds.

- In aircraft BUNO 161282 and subsequent and aircraft incorporating AFC 628, the MASTER CAUTION light and the EMERG JETT caution light illuminates when the EMERG STORES JETT pushbutton is activated.

If Step 1 Fails, Proceed With ACM Jettison:

ACM jettison will release all stores selected except Sidewinder missiles.

1. LDG GEAR handle . . . . UP

2. Station select switches . . . . AS REQUIRED

3. ACM . . . . ON (cover up)

4. ACM JETT. . . . . DEPRESS AND HOLD
   (at least 2 seconds)

EXTERNAL STORES JETTISON

NOTE

- WHEN JETTISONING MER/TER WITH OR WITHOUT STORES, JETT OPTIONS SWITCH MUST BE IN MER/TER POSITION, UNLESS IT IS DESIRED TO JETTISON INDIVIDUAL STORES.

- FOR AUXILIARY, SELECT ONLY ONE STATION AT A TIME.

- FUZING SAFED IN ALL JETTISON MODES. (DOES NOT PRECLUDE INADVERTENT ARMING OF MECHANICAL FUZES.)

- AAC 688 PRECLUDES SIDEWINDER JETTISON.

JETTISON MODE	TYPE OF STORES					REMARKS
	EXTERNAL TANKS	PHOENIX	SPARROW	SIDE-WINDER	AIR TO GROUND	
EMERGENCY (PILOT)	✓	✓	✓	—	✓	(✱) VERIFY SWITCH OFF DURING LTS CHECK PRESTART – PILOT ❶
ACM (PILOT)	✓	✓	✓	—	✓	(✱) SEQUENCE JETTISON SELECTED STATIONS ❷ ❹ ❺
SELECTIVE (RIO)	✓	✓	✓	—	✓	❷ ❸ ❺
AUXILIARY (RIO)	—	—	—	—	✓	GRAVITY DROP SELECTED STATION RELEASES IN SALVO. ❷ ❸ ❺

INTERLOCKS

❶ WEIGHT OFF WHEELS (EITHER RIGHT OR LEFT MAIN GEAR)

❷ LANDING GEAR HANDLE UP

❸ MASTER ARM SWITCH ON

❹ ACM COVER UP

❺ STATION SELECT

(✱) JETTISON SEQUENCE

STATIONS 1B, 8B, 2, 7, 4D, 5D, 4A, 5A, 4C, 5C, 4B, 5B, 3D, 6D, 3A, 6A, 3C, 6C, 3B, 6B

NOTE

- The time interval between stations indicated by (-) is 100 ms.
- Substations A, B, C, and D of rail are numbered clockwise, looking down at rail with A the left rear station on each rail.
- Stations 1B, 8B, 2, and 7 are jettisoned simultaneously.

2 1 050 004
1.34.0

Figure 5-4. External Stores Jettisoning

Note

- ACM jettison follows the same sequence as emergency jettisoning, but requires individual selection of stations to be released. Station not selected is skipped.

- When jettisoning bombs from stations 3, 4, 5, and 6 the interval between sequenced stations is automatically designated at 100 milliseconds to avoid store-to-store and store-to-aircraft collision.

ENGINE FAILURES AND MALFUNCTIONS

L OR R FIRE LIGHT AND/OR FIRE IN FLIGHT

Fire may be accompanied by other indications such as explosion, vibration, smoke, or fumes in the cockpit, trailing smoke, or abnormal engine instrument indications.

A fire in flight precipitated by a failure in the engine can be catastrophic in an extremely short period of time. The shrapnel generated by the engine can rupture fuel and/or hydraulic lines, resulting in a raging fire. The sequence of events for the failure could include all or some of the following.

- A low-amplitude vibration and noise

- Intermittent bursts of white sparks in the vicinity of aft edge of the overwing fairing (OWF)

- Sparks turning to flames

- Continuous yellow sparks in an area of increasing size

- Flames and/or smoke spreading forward to wing pivot point and encompassing the area of the Over Wing Fairing

- Flames, smoke, and/or heat crossing the centerline of aircraft and exiting in the other Over Wing Fairing area

These indications may or may not be accompanied by a FIRE light. This midship passage of heat and flames could be through the area containing the flight control system control rods, which run fore and aft through the back of the aircraft. Heat and flames progressing through this area would impinge on the longitudinal and lateral directional control rods causing possible distortion or

failure. Loss of aircraft may follow. The flightcrew faced with this type of fire in flight must react immediately.

*1. Throttle (affected engine) . . . . IDLE

*2. AIR SOURCE pushbutton . . . . OFF

If Light Goes Off

Note

Fire detection test is not available on the emergency generator.

*3. MASTER TEST switch . . . . FIRE DET test

If Light Remains Illuminated, FIRE DET Test Fails, or Other Secondary Indications:

*4. FUEL SHUTOFF handle (affected engine) . . PULL

*5. Throttle (affected engine) . . . . OFF

*6. Climb and decelerate.

*7. Fire extinguisher pushbutton . . . . DEPRESS

8. Land as soon as possible.

9. If fire persists . . . . EJECT

OIL SYSTEM MALFUNCTION

Malfunctions in the oil system are indicated by an L or R OIL HOT light, OIL PRESS light, or by oil pressure below or above normal.

For all normal in-flight engine power settings, oil pressure outside the range of 40 to 50 psi indicates a system or indicator malfunction. If oil pressure is over psi, retard power until pressure is within the normal range. If pressure cannot be reduced, the engine should be shut down to avoid rupturing oil lines. If oil pressure is less than 40 psi, bearing wear can be minimized by maintaining a constant throttle setting and avoiding unnecessary aircraft maneuvers. Bearing failure is normally characterized by vibration, increasing in intensity with bearing deterioration. When vibration becomes moderate to heavy, engine seizure is imminent if engine is not shut down. Continued operation of an engine with oil pressure less than 30 psi is likely to result in illumination of OIL HOT light or an engine seizure. If conditions permit, it is advisable to shut down the engine to reduce damage and to save it for emergency use

OIL PRESS Light And/Or Abnormal Oil Pressure

1. Throttle (affected engine) . . . . IDLE

   If oil pressure above 50 psi or below 30 psi, or engine vibration:

2. Throttle (affected engine) . . . . OFF

   If shutdown not feasible:

3. RPM . . . . SET 78% RPM

4. Avoid high g or large throttle movements.

5. Land as soon as practicable.

L or R OIL HOT Light

1. Oil pressure. CHECK

2. Throttle (affected engine) . . . . AS HIGH FUEL FLOW AS PRACTICABLE

3. Induce slight left sideslip condition (right rudder).

4. If, after 1 minute, light is still illuminated. . . . . THROTTLE OFF

5. Land as soon as practicable.

6. Relight engine for landing if necessary.

FUEL SYSTEM MALFUNCTIONS

L or R FUEL PRESS Light

ONE LIGHT

1. No afterburner above 15,000 feet.

2. Fuel distribution . . . . MONITOR
   (balance if required)

3. Land as soon as practicable.

BOTH LIGHTS

1. Descend to below 25,000 feet.

2. Maintain cruise power settings or less.

3. Land as soon as possible.

WARNING

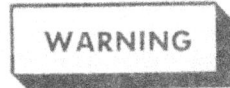

- Illumination of both lights may be indicative of a total motive flow failure. In this event, wing fuel will not be available. Zero- or negative-g flight should be avoided.

- Complete loss of motive flow will result in the feed tank interconnect and the engine crossfeed valve remaining in the closed position, thus isolating the forward and aft systems. Consequently, single-engine operation will cause fuel on the opposite side to be unavailable.

L or R FUEL LOW Light

1. DUMP switch . . . . OFF

2. Fuel distribution . . . . CHECK
   (balance if required)

If Wing And/or External Fuel Remaining:

3. WING/EXT TRANS switch . . . . ORIDE

Fuel Transfer or Fuel Dump Failures

If Wings Do Not Transfer

1. WING/EXT TRANS switch . . . . ORIDE

If External Tanks Fail to Transfer or Transfer Slowly:

1. WING/EXT TRANS switch . . . . ORIDE

If Fuel Continues to Transfer Improperly or Does Not Transfer:

2. MASTER TEST switch. . . . . FLT GR UP

Note

If GO light illuminates, transfer air
pressure is available. If transfer still
does not occur, a failure exists down-
stream in the tanks or air transfer lines.

3. Apply cyclic positive or negative g's.

If Problem Not Corrected or Transfer Complete:

4. MASTER TEST switch . . . . OFF

5. WING/EXT TRANS switch . . . . AUTO

If Fuel Dumps With Dump Switch In Off Position:

1. FUEL FEED/DUMP cb . . . . PULL (RD1)

Note

If pulling the FUEL FEED/DUMP
cb does not stop fuel dumping, re-
engage cb at 4,000 pounds total fuel
to allow fuel feed interconnect valve
to operate.

If Wings Do Not Accept Fuel With Switch In All EXTD Position:

1. REFUEL PROBE switch . . . . FUS EXTD

2. WING/EXT TRANS switch . . . . OFF

If Wings Accept Fuel With Switch In FUS EXTD Position:

1. WING/EXT TRANS switch . . . . ORIDE

Note

With AIR SOURCE OFF pushbutton
selected, external fuel tanks will not
transfer.

Fuel Leak

In the absence of actual visual detection, a fuel leak resulting from a malfunction or failure of a fuel system component will usually result in a split in the fuel quantity tapes
or feeds. The flightcrew must determine from available
instruments (fuel flow and total fuel quantity) whether
the aircraft is losing more fuel than the engines indicate

they are losing. Corrective steps are based on confirmation of the leak. Upon confirmation of abnormal decrease
in fuel quantity:

1. Land as soon as possible.

CAUTION

Use of afterburner with fuel leak should
be limited to emergency use only.

2. WING/EXT TRANS switch . . . . OFF

If Abnormal Fuel Quantity Decrease Ceases, Fuel Leak Is In Wing/Wing Pivot or Attachment Points For Auxiliary Tanks:

3. If leak is not stopped, it is in engine/nacelle area.
Proceed immediately with next step.

4. FUEL FEED/DUMP cb . . . . PULL (RD1)

Note

Enough time should be allowed for quantity tapes/feeds to develop split so that
leak can be isolated to left or right feed
group. Affected side will be low side.

5. Throttle (affected side) . . . . OFF

6. Conditions permitting, allow RPM to decelerate to
windmill RPM.

7. FUEL SHUT OFF handle . . . . PULL

Setting the WING/EXT TRANS switch to OFF stops
motive flow to the wings and inhibits external tank transfer
and fuselage tank pressurization. Pulling the FUEL FEED/
DUMP circuit breaker (RD1) isolates the right and forward
system and the left and aft fuel system. This aids in determining the location of the leak and prevents loss of fuel
from the good side via the fuel system interconnects. The
circuit breaker also deactivates the function of the FEED
switch, the automatic balance functions, and the fuel dump
system. Securing the engine and, if necessary, pulling the
FUEL SHUT OFF handle should stop most engine- or
afterburner-related leaks. This ensures that operable nozzles and full military thrust are available should a subsequent emergency require a restart of the secured engine to
save the aircraft.

EXHAUST NOZZLE FAILED OPEN

Nozzle position is controlled by high-pressure fuel from afterburner pump. A rupture in this fuel line could spill fuel into engine cavity.

*1. FUEL SHUT OFF
handle (affected engine) . . . . .PULL

*2. Throttle (affected engine) . . . . OFF

3. Land as soon as practicable.

EXHAUST NOZZLE FAILED CLOSED

Exhaust nozzle failed closed will result in afterburner mis-light, afterburner blowout, and compressor stalls if afterburner is selected. Excess residual thrust will be present on landing rollout.

1. Throttles . . . . .BASIC ENGINE ONLY

2. Monitor engine instruments.

3. Land as soon as practicable.

COMPRESSOR STALL

A compressor stall is an aerodynamic disruption of the airflow through the compressor. Compressor stalls may occur at any altitude/airspeed combination, including supersonic, and can be identified by any one or a combination of the following indications.

- Loud bangs or vibrations
- Rapid yaw or nose slice
- Increasing TIT
- RPM rollback and/or thrust loss
- Lack of throttle response
- Inlet buzz (supersonic only)
- Fireball emanating from the exhaust and/or intake.

Compressor Stall

*1. Unload aircraft (0.5 to 1.0 g)

*2. Throttles . . . . . . RETARD SMOOTHLY TO IDLE
(conditions permitting) (If step 1 cannot be accomplished immediately, DO NOT delay step 2.)

If TIT Above 1,000°C And/Or Engine Response Abnormal:

*3. Throttle (stalled engine) . . . . OFF (if less than 1.2 IMN)

CAUTION

If both engines are stalled, at least one engine must be secured immediately to prevent turbine damage and provide maximum potential for a successful airstart.

Note

Airspeed and altitude will determine whether both engines can be safely shut down (with

dual compressor stalls), or whether one should be secured and relit prior to shutting down the other. If airspeed is insufficient to provide windmill RPM for hydraulic pressure, one engine should be left in hung stall.

There is a threefold danger present when one engine has experienced a compressor stall. The most serious danger manifests itself at slow airspeeds and high power settings, where the sudden thrust asymmetry (a stalled engine yields negligible thrust) will induce or aggravate a departure, and may produce sufficient yaw rate to cause a flat spin.

WARNING

There is no known technique that will consistently recover the aircraft from a fully developed flap spin. Compressor stalls have been a contributing factor in departure/spin aircraft losses; any hesitation in following prescribed emergency procedures could lead rapidly to a flat spin.

The other two dangers from a compressor stall are that the stalled engine may suffer overtemperature damage, and that the good engine might also stall. Although the emergency procedures are designed to address all three dangers, the pilot must understand that aircraft controllability takes priority over engine considerations, and involves both throttle position and flight controls. Reference to the TIT gage will probably be required to determine the stalled engine, but this should not be attempted until the pilot has ensured that thrust asymmetry has been minimized and that yaw rate and AOA are under control. The rationale for each individual step in the emergency procedure is as follows:

Step 1: Unload The Aircraft (0.5 to 1.0 g) - Unloading the aircraft reduces the likelihood of a departure, while providing a more normal engine inlet airflow. It is not intended that the pilot push full forward stick or induce negative g, but merely that any g-load on the aircraft be reduced to as near 1.0 g as possible. In the nose high, slow airspeed case, the pilot may temporarily lose control effectiveness. This should not be cause for alarm and the pilot should be able to expeditiously establish a wings level nose low attitude as long as step 2 is followed immediately.

Step 2: Throttles - Retard Smoothly To IDLE (conditions permitting) - This action minimizes the asymmetric thrust. In the case of a violent slicing departure involving asymmetric thrust, reduction of throttles to IDLE is the most critical step and must be done immediately, regardless of whether the aircraft is unloaded. Pilot reaction times, aircraft pitch, yaw and roll rates at the time of the compressor stall, and aircraft configuration could additionally compound the situation.

Minimizing asymmetric thrust at high AOA and low airspeed shall be accomplished whenever possible. The probability of stalling the good engine is reduced if throttles are smoothly retarded to IDLE. Obviously, there are situations (landing pattern, catapult launch, low altitude, and airspeed) where idle power is unacceptable, and emergency procedures must be tempered by pilot judgment.

Step 3: Stalled Engine - Throttle OFF - When an engine stalls, the combustor flame does not extinguish. Airflow through the engine and cooling flow to the turbine blades are severely reduced, and the turbine blades may suffer overtemperature damage. Shutting the stalled engine throttle off extinguishes the combustor flame, thereby reducing the turbine-blade temperature. Additionally, the throttle must be shut off to allow the stall to clear. Successful airstarts can only be accomplished after the stall has cleared.

> **CAUTION**
>
> Engine turbine damage may result from prolonged stalls (greater than 20 seconds) with TIT stabilized as low as 1,000°C, or from short duration stalls (less than 5 seconds) with TIT stabilized above 1,300°C.

Subsonic compressor stalls are characterized by a thump or bang, RPM decrease, and an increase in TIT. At low airspeed or high AOA, rapid yaw rate or nose slice may be the first indication. Some stalls are quiet (usually stalls that occur at low power or high altitude) and cannot be heard by the aircrew. If the stall recovers immediately, the engine will return to normal operation, and the pilot may not see any movement in the engine instruments. This type of stall does not damage the engine and poses no problem with aircraft controllability. If the stall hangs or stagnates, the TIT will increase and generally stabilize within 2 to 3 seconds at 200 to 250°C above the normal operating temperature at which the stall occurred. Some low-power stalls may not stabilize and will continue to climb as high as 500° over a 30-second period unless the throttle is placed in the OFF position. The RPM will rapidly decelerate to below 60% RPM and possibly decelerate further at a rate dependent on airspeed. The engine will not respond to throttle inputs, but does not flameout (as some other engines do), and this accounts for the increase in TIT. The pilot's first consideration should be to maintain airplane control. He must also react quickly and set the stalled engine's throttle in OFF to extinguish the

combustor flame and prevent turbine damage. The TF30-P-414 engine turbine will run without damage continuously at temperatures as high as 1,175°C when the engine is operating normally. In hung stall, turbine cooling is reduced, and the turbine can be catastrophically damaged (blades melted) at temperatures less than 1,175°C if subjected to the conditions for more than 20 seconds or if subjected to temperatures above 1,300°C for any period of time.

> **CAUTION**
>
> ● Above 40,000 feet and below 0.9 M, the minimum RPM should be 90% to prevent quiet stalls that may be undetected by the pilot.
>
> ● Immediately after missile firing, a check should be made for indications of compressor stall.

Supersonic compressor stalls are characterized by a loud bang, slight increase in TIT, and slow RPM decay. If the stall hangs (which they normally do), the inlet will buzz. Inlet buzz is an unstable flow in the inlet duct and is characterized by pressure oscillations and an oscillatory movement of the inlet shock wave system. This results in a rough, bumpy ride (+2.5 to -1 g at 6-cycles per second). The proper technique to recover from a supersonic compressor stall is to smoothly retard throttles to IDLE, keep feet on the deck, and control any wing drop tendencies with lateral stick. The inlet buzz will cease at approximately 1.2 IMN, and the engine will generally return to normal operation at this point. TIT does not increase significantly during a supersonic compressor stall, and turbine cooling remains at an acceptable level. Therefore, an engine experiencing a supersonic compressor stall does not require immediate shutdown and should not be shut down unless the stall persists below 1.2 IMN. NATOPS procedures for supersonic compressor stalls are no different than for subsonic compressor stalls except that step 3 should not be performed until the aircraft has decelerated to below 1.2 IMN as there is a high probability of stall recovery passing 1.2 IMN.

Dual Engine Compressor Stall

Dual supersonic compressor stalls pose no problems of aircraft control or engine operation but become a concern if the engines do not recover at 1.2 IMN. While supersonic, all normal hydraulic and electrical power will be available. Dual subsonic compressor stalls can become a very serious problem unless NATOPS procedures are strictly followed. A compressor stall that occurs at 100% RPM may decelerate through 40% RPM in as little as 15 seconds. Lower air-speed, lower thrust setting at stall, and greater hydraulic demands after stall will shorten engine deceleration times; dual compressor stall will also hasten engine deceleration and probably result in lower RPM than single-engine stalls. When both engines decelerate through 40% RPM, electrical power will be momentarily lost until the emergency generator comes up to speed. Power will be restored (after approximately 1 second) but may be immediately lost again as the rotors continue to decelerate. Windmilling engine speeds of 15% to 20% (approximately 320 KIAS) will maintain electrical power to the number 1 and number 2 essential buses, but if flight control inputs are made, the number 2 essential bus (therefore, the engine instruments) may drop off, and the emergency generator switch must be recycled.

WARNING

During recovery from a dual engine compressor stall at low RPM, flight control inputs may cause temporary loss of combined hydraulic system pressure, and the emergency generator may drop the number 2 essential bus. In order to restore power to the number 2 essential bus and restore engine instruments required for monitoring engine starting, the emergency generator switch must be cycled.

If both engines are stalled after retarding throttles to IDLE, at least one engine must be immediately secured to prevent turbine damage and provide maximum potential for an airstart. If possible, secure the engine that did not initiate the event (the second engine to stall). The cause of the first engine stall may not be known at this point; however, it is highly probable that the second stall will have been induced during the throttle transient to IDLE. Leaving one engine in hung stall minimizes the likelihood of total loss of hydraulic and electrical power (emergency generator). The method of relight to be used (windmill or spooldown) depends on the situation. If sufficient altitude, airspeed, and hydraulic pressure exist, a windmill airstart should be attempted. After a low-airspeed compressor stall and subsequent engine spooldown, there may be considerable time delay after accelerating to more than 300 KIAS before achieving sufficient windmill RPM for relight and regaining emergency electrical power, if lost. Should flight conditions (such as low altitude) require a rapid airstart, perform a spooldown airstart as soon as TIT drops below 400°C. Although dual engine shutdowns are not generally recommended, this procedure is acceptable when the aircraft is nose down and accelerating and the flight path is under control.

WARNING

Leaving one engine in hung stall may catastrophically damage the turbine. It is, therefore, imperative that the pilot expeditiously secure and relight one engine to prevent turbine damage. Attention should be given to the remaining stalled engine as soon as possible.

L OR R OVSP/VALVE LIGHT

If Not After Start Attempt

*1. Throttle (affected engine) . . . . IDLE

*2. Nozzle position . . . . CHECK

  3. Land as soon as practicable.

Note

Refrain from high power settings.

If After Start Attempt Airborne

*1. AIR SOURCE pushbutton . . . . OFF

  2. Ensure ENG CRANK switch . . . . OFF

If OVSP/VALVE Light Does Not Go Off:

  3. Land as soon as practicable.

If OVSP/VALVE Light Goes Off

  3. AIR SOURCE pushbutton . . . . BOTH ENG

  4. Land as soon as practicable.

CAUTION

With AIR SOURCE OFF pushbutton selected, remain below 300 KIAS/0.8 IMN.

If After Start Attempt on Deck

  1. AIR SOURCE pushbutton . . . . OFF

  2. Throttle (affected engine) . . . . OFF

MACH LEVER HIGH OR LOW IDLE RPM

Indications

- Engine stalls greater than 1.7 IMN
- Low idle RPM greater than 1.5 IMN
- High residual thrust less than 0.5 IMN

1. Do not exceed 1.5 IMN.
2. MACH LVR/L or R FUEL
   CONTR cb . . . . CYCLE
   (LF1 - left or LG1 - right)

WARNING

Do not use APC if high idle rpm still exists.

If Conditions Still Exist After Lowering Gear:

3. MACH LVR/L or R FUEL
   CONTR cb. . . . . .PULL
   (LF1 or LG1)
4. Throttle (affected engine) . . . . OFF
   (if required for safe landing)
5. HYD TRANSFER PUMP SWITCH . . . . CHECK
   NORMAL
6. Make arrested landing (if necessary).

SINGLE-ENGINE FLIGHT CHARACTERISTICS

Single-engine flight characteristics are dependent on gross weight, configuration, angle of attack, wing sweep, and maneuvering requirements. In the cruise configuration, with one engine inoperative at military power settings, a slight rudder deflection and/or trim is required to prevent yaw toward the failed engine. However, single-engine performance capabilities can be significantly restricted by adverse flying qualities in approach power configuration, particularly at high gross weights in turning flight, due to the effects of thrust asymmetry at normal approach speed. This degrades with turns into the failed engine such that rudder requirements to maintain level flight can exceed available rudder control. Flight in this configuration should be planned to avoid turns into the failed engine with bank angles limited to 20° maximum and AOA limited to 12 units. The aircraft design is such that no one system (flight control, pneumatic, electrical, etc) depends on a specific engine. Therefore, loss of an engine does not result in loss of any complete system as long as the HYD TRANSFER PUMP is operative. Refer to section XI for single-engine performance data.

Single-Engine Failure During Flight

It is not uncommon to encounter high power compression stalls that require immediate engine shutdown to avoid engine over-temperature. Lower power stalls that do not self-clear also require shutdown. These engine shutdowns would require an airstart to regain operation. Occasionally, mechanical failure of TF30 engine components results in engine failure. These failures may be obvious as when accompanied by severe engine vibration, or may be subtle as indicated by a lack of engine response to throttle changes. Turbine failure for example, may appear only as an apparent loss of thrust and/or the inability to obtain a successful airstart. For confirmed mechanical failures the engine should be secured and the FUEL SHUT OFF handle pulled.

WARNING

If an engine fails or a mechanical malfunction has been determined, the respective FUEL SHUT OFF handle shall be pulled immediately after engine shutdown to reduce the possibility of fire.

AIRSTART

Engine ignition is available when engine RPM is more than 8% regardless of electrical power distribution. Each engine has an automatic restart feature that instantly actuates the ignition system for approximately 30 seconds when there is a rapid decrease in combusion chamber pressure below a preset value, such as during a compressor stall. A relight is automatic when the flameout is produced by something other than fuel starvation or a fuel control malfunction. Even though this automatic ignition feature is available, placing the AIR START switch to ON ensures positive engine ignition, and manual airstart procedures should not be delayed.

Note

When an engine is shut down from a high operating RPM or at a high engine temperature and a reduction in airspeed results in zero percent RPM, a restart should be attempted within 5 minutes to prevent rotor lock. Delaying for 5 to 12 minutes before attempting a windmill or cross-bleed airstart can result in a rotor lock. Maintaining an indicated airspeed of 300 knots or greater will prevent a rotor lock. If the speed remains at zero RPM, allow 10 to 20 minutes for the engine temperature to decrease, which will release the locked rotor. Monitor RPM for a windmilling indication before attempting the airstart.

Three types of airstarts are available: windmill, cross-bleed, and spooldown. If sufficient altitude, airspeed, and hydraulic pressure exist, a windmill aircraft with RPM stabilized at 15% or greater, should be used. See figure 5-5 for windmill envelope.

WINDMILL AIRSTART ENVELOPE

TF30-P-414 ENGINE
DATE: 15 MAY 1976
DATA BASIS: FLIGHT TEST

NOTES
- ZERO RPM INDICATES AIRSPEED AT WHICH ENGINE RPM GOES TO ZERO NECESSITATING ACCELERATING TO 320 KIAS TO EFFECT A WINDMILLING START.
- THE RPM MUST BE STABILIZED AT 15% OR GREATER TO EFFECT AN AIRSTART.
- CROSSBLEED STARTS PROHIBITED WITH RPM GREATER THAN 5%.
- EXCEPT IN EMERGENCY, INTENTIONAL SHUTDOWN PROHIBITED ABOVE 1.18 IMN.

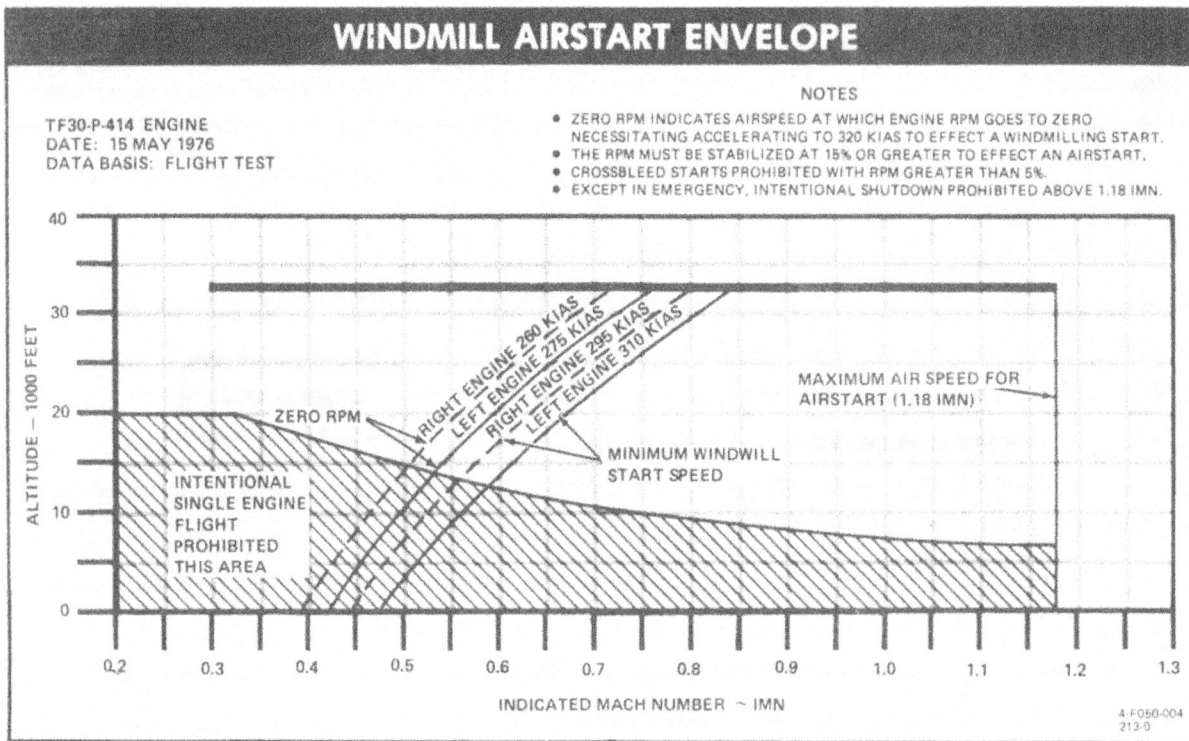

Figure 5-5. Windmill Airstart Envelope

Note

When an airstart is attempted at the edge of the windmill airstart envelope, the possibility of a starting sub-idle engine stall exists. The problem is evidenced by a low engine idle RPM, a lack of engine response to throttle advancement and a slowly increasing TIT. The sub-idle stall may be cleared by cycling the throttle to off and immediately returning it to idle.

The highest probability of achieving a successful airstart is realized if the TIT is below 400°C. Testing has shown that after an engine has experienced a hung compressor stall, the throttle may have to be held in the OFF position for as long as 10 to 15 seconds to clear the stall and up to an additional 15 seconds to reduce TIT to less than 400°C.

Note

- Airstarts generally result in TIT increase of 300°C to 400°C. Therefore, attempts to start with TIT above 450°C may result in a hung and/or hot start (870°C). If hot start occurs, reattempt start with AIR SOURCE selected to good engine.

- TIT during an airstart will peak 8 to 10 seconds after idle power is selected. Regardless of throttle setting/movement, the TIT then comes back down for about two seconds before it responds to throttle commands. RPM and fuel flow should be monitored to confirm normal engine start.

With RPM below 5%, the other engine running, and acceleration to windmill airspeeds (greater than 300 KIAS) not practicable, a crossbleed airstart may be used.

WARNING

If RPM is between 5 and 15%, do not attempt an airstart.

Spooldown airstarts should be attempted only when both engines are stalled or flamed out and altitude is below 10,000 feet AGL.

Note

The probability of achieving a successful spooldown airstart is enhanced by high airspeed, high RPM, and low TIT. An unsuccessful spooldown start will result in a hung and/or hot start and may result in turbine damage.

WARNING

If both engines fail below 1,500 feet AGL
and 250 KIAS — EJECT.

AIRSTART

Windmill (RPM stabilized greater than or equal to 15%)/Spooldown

*1. Throttle. . . . . OFF

*2. Airstart switch . . . . ON

When TIT is Less Than 400°C:

*3. Throttle. . . . . IDLE

4. Engine Instruments, fuel feeds . . . . MONITOR

5. Airstart switch . . . . NORMAL

Note

If airstart is unsuccessful, reattempt start
with AIR SOURCE selected to good
engine.

Crossbleed (RPM less than 5%)

1. Throttle (bad engine). . . . . OFF

2. Throttle (good engine) . . . . . . 85% RPM (minimum)

3. Crank switch . . . . BAD ENGINE

4. Throttle (bad engine) . . . . IDLE at 22%

5. Engine instruments, fuel feeds . . . . MONITOR

6. Crank switch . . . . ENGINE OFF AT 55% RPM

Windmill Airstart

*1. Fuel quantity . . . . CHECK

*2. Throttle. . . . . OFF

If RPM Stabilized at 15% or More and TIT Less Than 400°C:

*3. AIR START switch . . . . ON

*4. Throttle. . . . . IDLE

5. Engine instruments . . . . MONITOR

6. AIR START switch . . . . OFF (at 45%)

7. AIR SOURCE pushbutton . . . . BOTH ENG
(both engines at IDLE
or above)

AICS MALFUNCTION

If INLET and/or RAMPS Lights Illuminated:

1. Decelerate to below 0.9 IMN.

2. Avoid abrupt throttle movements.

3. Affected INLET RAMPS switch . . . . STOW

4. Remain below 0.9 IMN.

Note

Do not return to AUTO with INLET
light illuminated.

If RAMPS Light Remains Illuminated:

5. Throttle (bad engine) . . . . 80% OR LESS

6. Land as soon as practicable.

INLET Light Illuminated Only - Attempt AICS Programmer Reset

7. Decelerate to 0.5 IMN.

8. Affected L or R AICS cb . . . . CYCLE
(LF2, left or LG2, right)

WARNING

If WING SWEEP advisory light is illuminated, cycling R AICS cb may cause unintentional wing sweep unless WING
SWEEP DRIVE NO. 1 and WG SWP DR
NO. 2/MANUV FLAP cb's are pulled.

If INLET Light Goes Off:

9. Affected INLET RAMPS switch . . . . AUTO

If INLET Light Remains Illuminated:

10. Affected INLET RAMPS
switch . . . . REMAIN IN STOW

11. Remain below 0.9 IMN

INLET ICE LIGHT

1. ENG/PROBE ANTI-ICE switch . . . . ORIDE

When Clear of Known Icing Conditions:

2. ENG/PROBE ANTI-ICE switch . . . . AUTO

ELECTRICAL FAILURES OR MALFUNCTIONS

GENERATOR FAILURE

A mechanical generator failure or an overheating automatically causes the CSD unit of the generator transmission to decouple from the engine. Once disengaged,
the CSD cannot be reconnected in flight.

Either generator by itself is capable of supplying the electrical requirements of the aircraft. Even double generator failure will not cause total loss of electrical power: the 5-kVA emergency generator will automatically pick up the load for the essential ac and dc buses no. 1 and no. 2.

If the bidirectional pump is operating and pressure drops to between 2,000 and 1,100 psi (dependent upon the load placed on the generator), the emergency generator will automatically shift to the 1-kVA mode and power only the essential ac and dc no. 1 buses. If combined system hydraulic pressure subsequently recovers, the emergency generator switch must be cycled through OFF/RESET to NORM to regain the essential no. 2 ac and dc buses. Figure 5-6 lists the equipment available with only the emergency generator operating.

With both engines inoperative, windmilling engines provides hydraulic pressure for both the flight controls and the emergency generator. However, the flight controls have first priority and may cause the emergency generator to loiter when low airspeeds reduce engine windmilling rpm. Approximately 290 KIAS must be maintained to ensure adequate engine windmilling rpm for hydraulic pressure.

L or R GEN Light

1. Generator (affected gen) . . . . OFF/RESET, THEN NORM

Note

If the generator fault is corrected, the generator will be reconnected and the caution light will go off.

If Generator Does Not Reset:

2. Generator (affected gen) . . . . TEST

If the light goes off with the switch in TEST position, the fault is in the respective electrical distribution system. If light remains illuminated, the generator has been disconnected automatically and the fault is in the IDG or generator control unit.

L or R GEN and TRANS/RECT Lights

1. Generator (affected gen) . . . . OFF/RESET, THEN NORM

2. If both lights remain illuminated, select EMERG GEN on MASTER TEST panel.

3. Land as soon as practicable.

DOUBLE GENERATOR FAILURE

1. Generators . . . . CYCLE

If Operating on Emergency Generator, Following Important Systems Are Inoperative

● EMERGENCY FLIGHT HYDRAULICS

● OUTBOARD SPOILER MODULE AND EMERGENCY FLAP ACTIVATION

If Temporary Loss of Combined System Pressure Causes Emergency Generator to Drop No. 2 Essential Bus

2. EMERG generator switch . . . . CYCLE

3. Land as soon as practicable.

DOUBLE TRANSFORMER-RECTIFIER FAILURE

The 5-kVA emergency generator will automatically activate and power the essential ac and dc no. 1 and no. 2 and AFCS buses. See figure 5-7 for listing of inoperable dc-powered equipment.

TRANS/RECT Light

The TRANS/RECT light will illuminate if either or both T/R malfunction. If one transformer-rectifier (T/R) fails, the operating T/R will assume the dc load. If both T/R's fail, the emergency generator will come on the line and tie to essential dc buses no. 1 and no. 2.

Land as soon as practicable.

Note

Do not reset circuit breaker more than three times.

The loss of one generator and failure to tie the ac main buses will illuminate the affected GEN light. The TRANS/RECT light will also illuminate because the affected generator's associated T/R is not receiving ac power to convert. Upon observing a TRANS/RECT light, the pilot can check

EMERGENCY GENERATOR DISTRIBUTION

ESSENTIAL BUSES NO. 1

ACM Control
AICS Ramp Stow
*Airspeed Indication
Altitude Low Warning
*Angle-of-Attack Indication
AWG-15
*Barometric Altimeter
ECM Destruct
*Engine Stall Tone
*Fire Detection
Fire Extinguishing
*Flap/Slat Position Indicators
*Fuel Flow Indicators
*Fuel Quantity Indicators
*Hydraulic Pressure Indication
    (Flight and Combined)

ICS
IFF
Jettison (Emergency)
Master Arm
*Mid Compression Bypass
Missile Release
Radar Altimeter
Red Floodlights
*RPM Indicators
Standby Altitude Indicator
*TIT Indicators
UHF (Main Receiver-Transmitter
    Antenna Selection, Remote
    Indicator)
*Wing Position Indications

*In aircraft BUNO 161168 and earlier not incorporating AFC 610, located on essential buses no. 2.

ESSENTIAL BUSES NO. 2

ACM Panel Lights
ADF
Advisory Lights
AFCS/SAS
AHRS
AICS
Air Conditioning
Air Source Control
Antiskid Control
Approach Lights
Approach Power Control
Arresting Hook Control
ARI AOA System
Auxiliary Flap Control
BDHI
Beacon Augmenter
Cabin Pressure Dump
CADC
Caution Lights
Cockpit Temperature Control
Control Surface Position Indicator
    (Tail, Rudder, and Spoilers)
CSDC
ECM (Navigational Aids, IFF
    Interrogator, and Transmitter)
Eject Command Indicator
Electronic Cooling
Emergency Generator Test
Engine:
    Anti-Ice
    Nozzle position Indicator
    Oil Cooling (GRD only)
    Oil Pressure Indicators
    Overspeed Indicator
    Starting (Ground Use or Crank)
    Fuel Control

Fuel Control Indicators
Flight Control Trim
Fuel Feed/Dump/Transfer
Fuel Management Panel
Ground Roll Braking
HUD
Hydraulic Valve Control
Ice Detection
ILS
In-Flight Refueling Probe
In-Flight Refueling Probe Light
KY-28
Launch Bar
Mach Lever
Mach Trim
MDIG (Display Panel)
Nose Landing Gear Kneeling
Nosewheel Steering
Oxygen Quantity Indicators
Probe Anti-Ice (Angle-of-Attack)
    Total Temperature, and Pitot)
Radar Beacon
Rain Repellant
Servopneumatic Altimeter
Speed Brakes
Spoiler Control
TACAN
UHF (No. 2 Receiver-Transmitter,
    Auxiliary Receiver, Cryptographics)
Utility/Map Lights
VDIG
Warning Lights
Wheels and Speed Brakes Position
    Indicator
Windshield Anti-Ice
Wing Sweep

Figure 5-6. Emergency Generator Distribution

FAILURE OF BOTH TRANSFORMER-RECTIFIERS EQUIPMENT INOPERATIVE LIST

AIM-7 MOTOR FIRE	ELECTRONIC COOLING	MACH LEVER BIT
AIM-54A MISSILE	EMERGENCY GENERATOR	MASTER TEST
ALE-29 CHAFF/FLARE	CONTROL	MECHANICAL FUZING STA 5 AND 6
DISPENSER		MECHANICAL FUZING STA 3 AND 4
ALE-39 CHAFF/FLARE	FUZE FUNCTION CONTROL	MECHANICAL FUZING STA 1 AND 2
DISPENSER		MID COMPRESSION BYPASS
AMC BIT	GROUND POWER	POWER
AMCS ENABLE	GROUND TEST	MISSILE AUXILIARY
ANTENNA LOCK EXCITER	GUN ARMED POWER	SUBSYSTEMS
APX-72 TEST SET	GUN CLEAR POWER	MONITOR BUS CONTROL
AUTO THROTTLE	GUN CONTROL POWER	
	GYRO POWER	OUTBOARD SPOILER CONTROL
CONTROL DISPLAY		
SUBSYSTEMS	IFF AIR TO AIR	RADAR SUBSYSTEMS
COOLING INTERLOCK	INTEGRATED TRIM	RIGHT DC TEST
COUNTING ACCELEROMETER	INTERRUPTION FREE	
	DC BUS FDR NO. 1	STA 1A, 1B, 8A, 8B AIM-9
	AND 2	POWER
DATA LINK		STA 1, 8 AIM-9 COOLING
DDI/ANNUNCIATOR PANEL	L AND R AIM-7 BATTERY	POWER
DIM CONTROL	ARM	
DEHYDRATOR UNIT	LIQUID COOLING CONTROL	WINDSHIELD DEFOG CONTROL
DIGITAL DATA INDICATOR		

Figure 5-7. Failure of Both Transformer-Rectifiers Equipment Inoperative List

that he is actually experiencing a T/R failure and not a bus tie failure. If the seat adjust, white floods, or instrument lights are still operative with the R GEN light illuminated, or if the throttles are still boosted with the L GEN light illuminated, the bus is tied.

If the hydraulic transfer pump is operating and pressure drops to between 2,000 and 1,100 psi (dependent upon the load placed on the generator), the emergency generator will automatically shift to the 1-kVA mode and power only the essential ac and dc no. 1 buses. If combined hydraulic pressure subsequently recovers, the EMERG generator switch must be cycled through OFF/RESET to NORM to regain the essential ac and dc no. 2 and AFCS buses.

ELECTRICAL FIRE

Electrical fires may be indicated by visual or audible arcing or an ozone odor in the cockpit and popping circuit breakers. Electrical fires produced by 400°F air leaks can result in any one or combination of the following:

- Pinballing caution/advisory lights and instrument indications.

- CADC associated caution/advisory lights.

- Uncommanded movement of electrically controlled components (SAS, spoilers, wingsweep, throttles).

- Complete electrical failure.

- Smoke, fumes, and/or heat in the cockpit.

The most effective method to extinguish an electrical fire is to secure all electrical power. However, some conditions may not permit securing the emergency generator after both main generators are secured. Night/IFR flight or ECS duct leak induced electrical fires are cases where securing all electrical power is not feasible.

*1.   L and R generators . . . . OFF

If Uncommanded SAS or Spoiler Inputs Present:

*2.   ROLL AND PITCH CMPTR AC cb's . . . . .PULL
(LA1, LB1)

*3.   YAW SAS switch. . . . . OFF

If Conditions Permit:

4.   EMERG generator switch . . . . OFF

Note

Prior to securing the EMERG generator, inform the RIO of intentions, ICS will be lost and further communications should be accomplished by whatever means are deemed possible.

If Cause of Fire Can Be Isolated:

5. Pull cb's of affected equipment.

6. All generators . . . . NORM

If Cause of Fire Cannot Be Isolated:

5. Secure all unnecessary equipment.

6. EMERG generator switch . . . . NORM

7. Land as soon as practicable.

Note

Securing all electrical power while airborne causes the ECS to go to full cold.

CAUTION

Do not operate engines on the ground without electrical power. Ground cooling fans are shut off, causing hot bleed air to cook off oil and hydrocarbons in the ECS ducting, resulting in smoke in the cockpit and possible damage to the ECS turbine compressor.

COMPLETE ELECTRICAL SYSTEM FAILURE

The conditions at the time of the electrical failure will, in part, dictate the corrective action that must be taken by the flightcrew if the aircraft loses complete electrical inputs. However, the flightcrew has certain instruments that can aid in turning a precarious situation into a relatively uneventful landing.

If the electrical system aboard the aircraft fails, the following systems are still available:

- AIRSPEED INDICATOR
- ALTIMETER (STBY mode)
- VERTICAL VELOCITY INDICATOR

- ARRESTING HOOK
- STBY ATTITUDE GYRO (for 9 minutes after complete loss of power and accurate ±6° in all axes)
- EMERGENCY WING SWEEP
- FLAPS (EMER UP or DN)
- LANDING GEAR EXTENSION
- STBY COMPASS

All other normal system and cockpit cues are not available.

When all electrical power is shut off, the cockpit dump valve closes, and the environmental control system supplies only cold air to the cockpit and forced air-cooled avionics. Pressurization will slowly bleed off. If flight conditions permit, descend to a lower altitude. If the system failure occurs in the day or night VFR environment, immediate return to base and an emergency landing shall be accomplished. In the day or night IFR environment, ascend or descend to known VFR conditions. (Extreme care should be exercised due to partial panel environment.) Reduce power setting to maximum endurance. Contact nearest ground facility by hand-held survival radio. Once positive radar identification is made, follow controllers' directions to landing (make barricade arrestment aboard ship at night to avoid partial-panel bolter).

Note

If possible, section IFR descent should be conducted to VFR conditions for landing.

ENVIRONMENTAL CONTROL SYSTEM FAILURES OR MALFUNCTIONS

ECS DUCT LEAK/BLEED DUCT LIGHT/ELIMINATION OF SMOKE OR FUMES

A bleed air leak between the engine firewall forward of the primary heat exchanger discharge and the 400°F shutoff/modulating valve (575°F minimum) should be indicated by illumination of a BLEED DUCT light. Bleed air leaks that cannot be detected by the bleed air detection system have no single positive failure indication. The first symptom of an undetected bleed air leak (no BLEED AIR light) may be an audible pop or squeal that may be associated with some changes in ECS air flow volume, pressure, or temperature. An airflow volume decrease is the most likely symptom to be observed. ECS duct failures may eventually lead to smoke and fumes in the cockpit if early recognition of the failure was not made and appropriate action (that is, AIR SOURCE to OFF) taken. The RIO may feel heat below and behind the starboard side of the cockpit in the

F-14A TOMCAT

THIS PAGE INTENTIONALY LEFT BLANK.

vicinity of number eight circuit breaker panel. Other ancillary indications that may confirm a bleed air leak are the illumination of the AWG-9 COND or MSL COND advisory lights.

When an ECS duct leak is suspected, AIR SOURCE should be immediately selected OFF. Bleed duct leaks may melt wiring splice junctions and create conditions that may induce an electrical fire. If an associated electrical fire occurs, smoke, fumes, heat, and damage to the surrounding aircraft structure may intensify. Since electrical fire procedures are not compatible with measures to eliminate smoke and fumes, canopy jettison may become necessary as a last ditch procedure.

WARNING

Selection of AIR SOURCE to RAM allows bleed air to circulate throughout the 400°F manifold system.

Note

Selecting AIR SOURCE OFF eliminates pressurization to the service system (canopy, G-suit, external fuel tanks, pressure/ventilation, and air bag seals). Rain removal, defog, and heating systems are also eliminated. Judicious reselection of AIR SOURCE to BOTH or RAM in order to regain critical support/service systems is predicted on severity of ECS malfunction and operational requirements.

*1. AIR SOURCE pushbutton . . . . OFF

*2. If smoke or fumes present:

    a. Altitude . . . . .BELOW 35,000 FEET

    b. CABIN PRESS switch . . . . DUMP

*3. RAM AIR switch . . . . .INCR

4. Airspeed . . . . .BELOW 300 KIAS/0.8 IMN

5. Non essential electrical equipment . . . . SECURE

6. CANOPY DEFOG/CABIN
    AIR lever . . . . CANOPY DEFOG

7. Land as soon as possible.

If Electrical Fire:

8. Follow electrical fire procedures (page 5-24).

CAUTION

The emergency generator switch should be left in the NORM position unless there are overriding considerations which mandate turning the emergency generator off.

Note

- Securing all electrical power while airborne causes the cockpit dump valve to close, cabin hot air valve to close, bleed air shutoff valves to open, dual pressure regulator to open, and the ram air door to remain at its last commanded position (ram air door takes up to 50 seconds to open). This results in full cold air to the cockpit, uncontrolled bleed air to circulate, and the loss of normal cabin dump capability. Additionally, low speed (less than 0.25 IMN) and ground operations should be minimized as the heat exchanger cooling fan will be inoperative and ECS overheat condition will result.

- Elimination of smoke or fumes without any electrical power may be accomplished by ECS air flow. In order to obtain maximum smoke/fume removal capability under this condition, fly below 8,000 feet MSL and set the throttle to maximum practical position. This will open the cabin regulator valve to the maximum position and permit maximum ECS air flow. If smoke or fumes are not eliminated, it is most probable that smoke/fumes are being regenerated by an ECS air leak. As a last resort, jettison the canopy.

ECS COMPONENT FAILURE

ECS failure other than ECS duct leaks can include malfunctions of ECS components such as the turbine compressor assembly, valves, controllers, or sensors. Indications of an ECS component malfunction can be the illumination of a COOLING AIR and/or MSL COND light, or erratic temperature control (that is, full hot or full cold). If the aircrew cannot discern between an ECS duct leak and an ECS component failure, ECS duct leak procedures take precedence over ECS component failure procedures.

An ECS turbine failure may be indicated by a continuous low pitch tone emanating from the right side of the aircraft aft of the RIO cockpit and may result in smoke in the cockpit. Other ancillary symptoms that reflect a bleed duct failure/bleed air leak are usually not present.

1. AIR SOURCE pushbutton . . . . RAM

2. Airspeed . . . . SLOW TO LESS THAN
350 KIAS OR 1.5 IMN

Note

After ram airflow is stabilized, airspeed may be increased as required for flightcrew comfort or to increase flow to electronic equipment.

3. RAM AIR switch . . . . INCR

4. Non-essential equipment. . . . . SECURE

5. Check g-suit or vent air operation.

If G-Suit or Vent Air Not Operating:

6. AIR SOURCE pushbutton . . . . OFF

7. WCS switch . . . . OFF

8. Remain below 300 KIAS or 0.8 IMN, and land as soon as practicable.

If G-Suit or Vent Air Operating:

9. AIR SOURCE pushbutton . . . . BOTH ENG
(engine ECS airflow could take more than 90 seconds to recover)

If ECS Failure Recurs or Fails To Reset, Perform Steps 1 through 3, 7, and 8.

Note

- The ram-air door takes approximately 50 seconds to open fully.

- AIR SOURCE selection to RAM or OFF inhibits gun firing.

AWG-9 COND LIGHT ILLUMINATED AND/ OR PUMP PHASE CIRCUIT BREAKERS POPPED

1. LIQ COOLING switch . . . . OFF

2. WCS switch . . . . OFF

3. AN/AWG-9 PUMP
PH A, B, C cb. . . . . .PULL
(2G3, 2G6, 2G7)

4. Land as soon as practicable.

AWG-9 PM ACRONYM

1. AWG-9 LIQ COOLING switch . . . . OFF

2. WCS switch . . . . OFF

3. AN/AWG-9 PUMP
PH A, B, C cb. . . . . .PULL
(2G3, 2G6, 2G7)

MSL COND LIGHT (AIM-54A ABOARD)

1. LIQ COOLING switch . . . . CHECK

IF LIQ COOLING Switch Is In AWG-9/AIM-54 Position:

2. LIQ COOLING switch AWG-9

3. STA 3/6 AIM-7/AIM-54
PUMP PH A, B, C cb . . . . .PULL
(1D1, 1D3, 1D7)

4. MSL HTR PH A, B, C cb. . . . . .PULL
(1C2, 1C5, 1C6)

Note

In aircraft BUNO 161168 and subsequent MSL HTR PH A, B, C cb's are 2G2, 2G5, and 2G8.

5. Land as soon as practicable.

Note

With aircraft configured for AIM-54
missiles, a MSL COND light is normal
if the LIQ COOLING switch is set to
the AWG-9 position. There is no reason
to secure the AWG-9 with a MSL COND
light since the AWG-9 and AIM-54
cooling loops are separate.

COCKPIT TEMPERATURE CONTROL MALFUNCTION

1. TEMP mode selector switch . . . . MAN

2. TEMP thumbwheel control . . . AS DESIRED

If Temperature Control Not Regained:

3. VENT AIRFLOW thumbwheel . . . . OFF

CAUTION

Reduce airspeed to 350 KIAS or 1.5 IMN,
whichever is lower, to prevent ram air
at temperature above 110°F from entering
aircraft. After ram air flow is stabilized,
airspeed may be increased as required for
flightcrew comfort or to increase flow to
electronic equipment.

4. AIR SOURCE pushbutton . . . . RAM
(below 35,000 feet)

5. RAM AIR switch. . . . . .INCR
(Select amount
of ram air desired for
flightcrew comfort.)

Note

In this configuration, normal operation
of all systems can be maintained.

CAUTION

High cockpit temperature and smoke
during ground operation indicates ECS
cooling fan shutdown. This will occur
with an external air source (start cart)
without electric power on the aircraft.
This results in an overtemperature condition due to operating without ground
cooling fans.

WSHLD HOT LIGHT

1. WSHLD switch . . . . OFF

2. AIR SOURCE
pushbutton . . . . OFF
(below 35,000 feet)

Note

If light remains illuminated after air
source is off, the indication is faulty.
Turn ECS on and land as soon as
practicable.

3. RAM AIR switch. .INCR

4. Reduce airspeed to less than 300 KIAS or
0.8 IMN.

5. Land as soon as practicable.

OXYGEN SYSTEM FAILURE

1. Oxygen quantity . . . . CHECK

2. If possible, descent to cabin altitude
of less than 10,000 feet.

3. If not possible to descend, actuate
emergency oxygen.

WARNING

Once emergency oxygen is actuated, the oxygen flows until the emergency bottle is depleted (approximately 10 minutes).

Note

Normally, it is desirable to conserve the emergency oxygen for landing.

OXY LOW LIGHT

1. Oxygen quantity . . . . LESS THAN 2 LITERS

2. Cabin altitude . . . . LESS THAN 10,000 FEET

3. Oxygen masks . . . . RELEASE ONE SIDE

4. OXYGEN switch . . . . OFF

5. Before landing, OXYGEN switch and masks . . . . ON

CANOPY LIGHT AND/OR LOSS OF CANOPY

In the event of canopy loss in flight, the pilot will be adequately shielded by the forward windscreen to maintain control of the airplane. Vision may be impaired briefly by dust in the cockpit, and moderate head buffet may occur which can be alleviated by lowering the seat and/or leaning forward. The RIO will be exposed to a significantly more hazardous and disorientating environment, which will include vision impairment, loss of communications, windblast injury and breathing difficulties. The degree to which these will be experienced is directly related to airspeed and seat height. In addition, the possibility of helmet loss becomes greater as airspeed increases above 300 KIAS. ICS and RIO UHF communications will probably be impossible above 200 KIAS, although the pilot will be able to effectively utilize UHF at airspeeds up to approximately 400 KIAS. After lowering his seat, the RIO should lean forward to take advantage of the windblast protection provided by the DDD and instrument panel, while the pilot decelerates the airplane by utilizing idle power, speed brakes, and moderate g. The RIO should deselect HOT MIC ICS to prevent interference with UHF communications due to windblast across his oxygen mask microphone. Helmet loss will result in severe disorientation due to total loss of communications and vision impairment due to windblast.

If canopy loss is experienced at high speed, or if helmet loss appears to be possible due to windblast or buffeting, retain the helmet by pulling down on the visor cover (keeping arms close to the body).

*1. Canopy . . . . BOOST CLOSE

*2. EJECT CMD lever . . . . PILOT

3. Airspeed and altitude . . . . BELOW 200 KIAS/ 15,000 FEET

4. Seats and Visors . . . . DOWN

5. Perform slow flight check. (If canopy lost)

6. Land as soon as possible.

HYDRAULIC SYSTEM MALFUNCTIONS

HYDRAULIC PRESSURE LOW

Combined Pressure Approximately 2,400 to 2,600 PSI

1. HYD ISOL switch . . . . FLT

Note

Monitor AUX BRAKE PRESSURE gage. Tap wheel brakes to seat priority valve if pressure is decreasing.

2. Wing sweep . . . . SET AT 20°

3. L INLET RAMP switch . . . . STOW (less than 0.9 IMN)

4. Equipment inoperative . . . . DLC

5. EMERG FLT HYD switch . . . . HIGH (just before dirty-up)

6. Land as soon as practicable.

Flight Pressure Approximately 2,400 to 2,600 PSI

1. Wing sweep . . . . SET AT 20°

2. R INLET RAMP switch . . . . STOW (less than 0.9 IMN)

3. EMERG FLT HYD switch . . . . HIGH (just before dirty-up)

FOLLOWING IMPORTANT EQUIPMENT IS INOPERATIVE

    NORMAL HOOK - Restored by weight on wheels

Note

    Arrested landing will require emergency hook extension.

4. Land as soon as practicable.

Combined Pressure Zero

1. HYD ISOL switch . . . . FLT

2. HYD TRANSFER PUMP switch . . . . SHUTOFF

3. Wing sweep . . . . SET AT 20°

FOLLOWING IMPORTANT EQUIPMENT IS INOPERATIVE

- L AICS
- L GLOVE VANE
- NOSEWHEEL STEERING
- GUN DRIVE
- INBOARD SPOILERS
- HOOK EXTEND
- FLAPS AND SLATS
- LANDING GEAR   Emergency Actuation Available
- WHEEL BRAKES
- REFUEL PROBE
- EMERG GEN
- AUX FLAPS
- DLC
- SPEED BRAKES
- NORMAL HOOK

4. HOOK . . . . EMERG DN
(restow handle)

5. EMERG FLT HYD switch . . . . HIGH
(just before dirty-up)

6. LDG GEAR handle . . . . EMERG DN

7. Flaps (no aux flaps) . . . . DN

8. Brake accumulator (handpump) . . . . CHECK

9. ANTI SKID SPOILER BK switch . . . . SPOILER BK
(OFF for CVA)

10. Make arrested landing as soon as practicable.

AFTER LANDING:

11. Do not taxi out of arresting gear.

12. Engines . . . . OFF

Flight Pressure Zero

1. HYD TRANSFER PUMP switch . . . . SHUTOFF

2. Wing sweep . . . . SET AT 20°

3. EMERG FLT HYD switch . . . . HIGH
(just before dirty-up)

FOLLOWING IMPORTANT EQUIPMENT IS INOPERATIVE:

- R GLOVE VANE
- ACLS
- R AICS
- NORMAL HOOK - Restored by weight on wheels

4. Land as soon as practicable.

Note

    Arrested landing will require emergency hook extension.

Both Combined and Flight Pressure Zero

1. EMERG FLT HYD
   switch . . . . LOW (clean) -
   HIGH (just before dirty-up)

2. External stores . . . . JETTISON

FOLLOWING IMPORTANT EQUIPMENT IS OPERATIVE IN FLIGHT:

- HORIZONTAL TAILS ⎫
- RUDDERS ⎬ Reduced Rate
- MAIN FLAPS AND SLATS ⎭
- OUTBOARD SPOILERS
- HYDRAULIC HAND PUMP
- LANDING GEAR ⎫
- HOOK EXTEND ⎬ Emergency
- REFUELING PROBE ⎬ Actuation Available
- WHEEL BRAKES ⎭

3. LDG GEAR handle. . . . . EMERG DN

4. Flaps (no aux flaps) . . . . DN

5. HOOK. . . . . EMERG DN

6. Brake accumulator . . . . CHECK

7. Make arrested landing as soon as practicable.

8. Fly straight-in approach.

After Landing:

9. Do not taxi out of arresting gear.

10. Engines . . . . OFF

BACKUP FLIGHT MODULE MALFUNCTION

1. FLT HYD BACKUP
   PH A, B, C, cb . . . . IN
   (2A1, 2C1, 2E1)

2. Land as soon as practicable.

OUTBOARD SPOILER MODULE MALFUNCTION

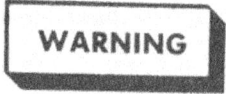

WARNING

If outboard spoilers fail with wing sweep angles less than 57°, limit lateral stick deflection to 1/2 pilot authority.

Note

In aircraft BUNO 160887 and subsequent and aircraft incorporating AFC 523, OUTBD SPOILER PUMP PH A, B, C cb are replaced by one circuit breaker (2B2)

1. OUTBD SPOILER PUMP
   PH A, B, C cb . . . . CHECK
   (2A2, 2B2, 2C2)

   a. If OUT . . . . ATTEMPT RESET

   b. If IN and outboard spoiler module
      flag indicates OFF . . . . PULL
      (2A2, 2B2, 2C2)

Following Important Equipment Is Inoperative

- OUTBOARD SPOILERS
- FLAP AND SLAT BACKUP
- DLC AND ACL

2. Evaluate flaps-down lateral control characteristics at safe altitude.

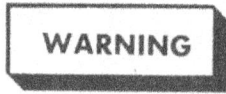

WARNING

An outboard spoiler module failure with flaps extended, below 180 KIAS, and with a combined hydraulic failure rendering the inboard spoilers inoperative, can result in assymetric spoiler float such that the aircraft may not be flyable at normal approach airspeeds.

If Unacceptable:

3. Make flaps-up landing

LOW BRAKE ACCUMULATOR PRESSURE

In Flight

1. HYD ISOL switch . . . . T.O./LDG

If Pressure Does Not Recover:

2. LDG GEAR handle . . . . DN

3. HYD HAND PUMP . . . . RECHARGE
ACCUMULATOR

Note

Monitor AUX BRAKE PRESSURE gage.
Tap wheel brakes to seat priority valve if
pressure is decreasing.

If Accumulator Cannot Be Recharged:

4. Make arrested landing as soon as practicable.

5. Parking brake . . . . PULL
(to lock wheels)

FLIGHT CONTROL FAILURES OR MALFUNCTIONS

UNCOMMANDED ROLL AND/OR YAW

Note

- If uncommanded roll and/or yaw
occurs during high AOA maneuvering
(above 15 units), assume departure
from controlled flight and apply
appropriate departure and/or spin
recovery procedures. Otherwise
perform appropriate procedures
below.

- Failures that may cause uncommanded roll and/or yaw include, but
are not limited to:

  - Stuck up spoilers

  - Asymmetric flaps and slats

- Uncommanded differential stabilizer
and/or rudder automatic flight control system inputs due to abnormal
power transients.

- Rudder hardover from yaw SAS (19°).

*1. If flap transition:
FLAP handle . . . . PREVIOUS POSITION

*2. Rudder and stick. . . . . OPPOSITE ROLL/YAW

*3. AOA. . . . . BELOW 12 UNITS

*4. Downwing engine . . . . MAX THRUST
(if required)

5. Roll SAS . . . . ON

6. Roll trim . . . . OPPOSITE STICK
(if required)

7. Out of control below 10,000 feet. . . . . EJECT

8. Control regained - climb and investigate for:
   - Flap and slat asymmetry
   - SAS malfunction
   - Spoiler malfunction
   - Hardover rudder
   - Structural damage

9. Stow flight aircraft to determine controllability at
10,000 feet AGL minimum.

STABILITY AUGMENTATION FAILURE

Yaw Channel Failure

Failure of a single yaw channel does not affect yaw stabilization since the system is triple redundant. Single channel
failure is indicated when the YAW STAB OP light illuminates while the STAB AUG switch remains ON. A second
yaw channel failure is indicated when the YAW STAB OUT
light illuminates. The YAW STAB switch is not automatically positioned to OFF. CV landings with total YAW
SAS failure require increased attention to yaw oscillations
in turbulence and/or adverse yaw during lineup corrections.

YAW STAB OP Light

No corrective action is required - gain and authority will be
normal.

1. MASTER RESET pushbutton . . . . DEPRESS

2. If light remains illuminated . . . . STAY BELOW
0.93 IMN

YAW STAB OUT Light

1. YAW STAB switch .... OFF

2. MASTER RESET pushbutton .... DEPRESS

3. Recheck light:

   a. If YAW STAB OUT
     light off .... RESET YAW STAB
     AUG SWITCH

   b. If YAW STAB OUT
     light illuminated .... REMAIN BELOW
     0.93 IMN

Pitch or Roll Channel Failure

Failure of a single pitch or roll channel damper is indicated by that channel STAB caution light. The remaining roll channel provides 50% of the scheduled gain and damping authority. The remaining pitch channel provides 25% authority flaps up and 50% flaps down. The affected STAB AUG switch remains in the ON position if only one channel fails.

Failure of two channels illuminates both STAB lights of the affected system. The affected STAB AUG switch is automatically positioned to OFF. If the stability augmentation system self-test circuit can isolate the faulty channel, the remaining channel light will go out; however, the STAB AUG switch must be reset to the ON position. CV landings with either or both PITCH and ROLL SAS failures can be accomplished with little degradation in normal approach handling qualities.

Single PITCH or ROLL STAB Light

1. Reset appropriate STAB AUG switch - no corrective action or limitations.

Both PITCH or Both ROLL STAB Lights

1. Airspeed .... REDUCE TO
     STAB LIMITS

2. Pitch .... NOT RESTRICTED

3. Roll .... 0.93 IMN

4. Wait 10 seconds for self-test.

5. Recheck lights:

   a. If one STAB
     light off .... RESET STAB AUG
     SWITCH, NO LIMITS

   b. If both lights remain
     illuminated ....LEAVE APPROPRIATE
     STAB AUG SWITCHES
     OFF, STAY BELOW
     STAB LIMITS

WARNING

Do not engage ACLS or DLC.

STAB AUG Transients

1. Paddle switch .... DEPRESS

2. Airspeed .... DECELERATE TO
     0.93 IMN

3. STAB AUG switches .... ALL OFF

4. MASTER RESET pushbutton .... DEPRESS

5. STAB AUG switches .... RESET
     (Reset individually
     to isolate failure.)

Note

Pitch SAS loss may result in loss of outboard spoilers. Roll SAS loss may result in loss of inboard spoilers.

GLOVE VANE LIGHT

1. Hydraulic pressure .... CHECK

2. Maneuvering devices .... RETRACT

3. MASTER RESET pushbutton .... DEPRESS

Note

Above 1.4 IMN, glove vane is not resettable.

RUDDER AUTHORITY FAILURE

Scheduling of allowable rudder deflection is computed in the CADC as a function of dynamic pressure. If the command signals and position feedback do not agree, power is removed, stopping further movement and the RUDDER AUTH light illuminates. Directional authority is never less than 9.5° of rudder.

RUDDER AUTH Light

1. MASTER RESET pushbutton. . . . . DEPRESS
(10 seconds)

2. If light remains
illuminated . . . . ABOVE 250 KIAS,
RESTRICT RUDDER
INPUTS TO LESS
THAN 10°

```
CAUTION
```

- With rudder authority stops failed open, excess rudder authority is available and could result in structural damage above 250 KIAS.

- After landing, nosewheel steering authority may be restricted to 10° (with neutral directional trim) and differential braking is required coming out of the arresting gear.

RUNAWAY STABILIZER TRIM

A runaway trim failure is sensed by the pilot by both uncommanded stick motion and by changes in aircraft pitch and normal acceleration. This failure state causes the horizontal tail to move along the normal stick-to-tail gearing curve for the hands-off condition. Aircraft response to a runaway stabilizer trim, even in the high-speed configuration, is slow enough (about 1° per second stabilizer change) to be recovered from safely.

The most critical steady-state trim conditions are those for which the greatest stick force is required. A field or carrier landing with either a full noseup or nosedown runaway stabilizer trim requires an average stick force of 14 to 19 pounds to maintain longitudinal control. If pilot fatigue becomes a factor with full nose up trim, stick forces may be significantly reduced by placing the wings aft of 21° and lowering the FLAP handle causing the main flaps to extend while the auxiliary flaps remain retracted. This overrides

the wing sweep 21° interlock and the FLAP light will be illuminated. This configuration is not recommended for landing. At approach speed the worst nosedown trim condition requires a maximum stick pull of 27 pounds without DLC engaged and approximately 24 pounds with DLC engaged. A full noseup runaway trim requires a maximum of 17 pounds stick push without DLC engaged and 23 pounds with DLC engaged.

Note

With abnormal stabilizer trim response, continuing to trim may preclude ability to retrim to a neutral position.

1. SPD BK/P-ROLL TRIM
ENABLE cb . . . . PULL (RA2)

2. Decelerate to below 300 KIAS.

3. Use AFCS, if available, in cruise configuration to reduce pilot workload.

4. Minimum stick forces are achieved under the following conditions:

- Runaway nosedown - flaps up

- Runaway noseup - flaps down

5. Straight-in approach

Note

Force required (push or pull) may be as much as 30 pounds.

HORIZONTAL TAIL AUTHORITY FAILURE

Lateral stick inputs are limited by control authority stops scheduled by the CADC as a function of dynamic pressure. Failure of the lateral stick stops is indicated by the HZ TAIL AUTH caution light. Failure of the stops in the full-closed position does limit low-speed rolling performance but ample roll control is available for all landing conditions and configurations. Failure in the open condition, with SAS on, requires the pilot to manually limit stick deflection to prevent exceeding fuselage torsional load limits.

HZ TAIL AUTH Light

1. MASTER RESET pushbutton . . . . DEPRESS
(10 seconds)

2. ROLL STAB AUG switch . . . . OFF

3. Restrict lateral control inputs above
   400 KIAS/0.9 IMN to 1/4 throw.

4. ROLL STAB AUG switch . . . . ON FOR LANDING

Note

At low airspeeds, lateral control effectiveness may be reduced.

5. Do not select OV SW after landing.

SPOILER MALFUNCTION

The inboard and outboard spoilers are powered and controlled by separate hydraulic and electrical command systems and are protected by separate spoiler failure detection
circuits. The pitch computer and outboard spoiler module
control the outboard spoilers, and the roll computer and
the combined hydraulic system control the inboard spoilers.
It is highly unlikely that both spoiler pairs on one wing
would fail up/float, but if this situation occurred, control-
ability would be marginal to impossible.

The severity of a spoiler failure is influenced by spoiler
position and deflection, aircraft configuration, and flight
conditions. The rolling moment generated from a deflected
spoiler varies from the inboard to most outboard spoiler.
The moment generated by the no. 4 spoiler is approximately twice as great as that of the no. 1 spoiler. Sweeping the wings aft reduces the effect of a deflected spoiler
by decreasing the moment arm of the spoiler and its aerodynamic effectiveness. A fully deployed no. 4 spoiler at
200 KIAS and 20,000 feet in the cruise configuration
generates 25° per second roll rate with the wings at 22°,
and a 4° per second roll rate with the wings at 62°.

In the cruise configuration, the increase in rolling moment
is essentially linear with increasing spoiler deflection angle,
whereas in the landing configuration, the increase in rolling
moment is nonlinear with increasing spoiler deflection
angle (approximately 50% of total rolling moment is
generated in the first 10° of spoiler deflection). Flap position has a very pronounced effect on rolling moment. With
the flaps down in the high-lift (landing) configuration, a
deflected spoiler reduces lift considerably more than with
the flaps up. Consequently, a failed up spoiler causes a
significantly higher rolling moment with the flaps down
than with flaps up. Selecting flaps up with a failed up
spoiler greatly reduces the amount of lateral stick required
to maintain a wings level attitude. The same lateral stick

position of 6° differential stabilizer will balance 55° of
no. 4 spoiler deflection with flaps up, and only 8° of
no. 4 spoiler deflection with flaps down. The use of
lateral trim to reduce stick forces may result in a reduction of available opposite wing spoiler deflection. Trimming in the direction of stick forces (normal pilot reaction) reduces effective spoiler authority from 55° to
approximately 30° at full lateral trim. Wing and external
stores have little effect on rolling moments under normal
conditions, but with failed up or floating spoilers, asymmetric weapon loading may mean the difference between control and out of control.

The control capability remaining with a failed up spoiler is
influenced by flap poisition, roll SAS operation, and availability of the remaining spoiler set. Any single, fully deflected, failed up spoiler is controllable even with flaps
down and roll SAS off if the remaining spoiler set is operating. With a second failure, airplane configuration is the
critical factor, and with flaps down, roll control using lateral stick alone is impossible. However, with flaps up, adequate roll control to wings level flight is available with use
of lateral stick alone. If a second failure disables the remaining spoiler set, the flaps should remain up.

Spoilers Light/Malfunction

If associated with Abnormal Roll and/or Yaw:

1. Counter roll with at least 1 inch of lateral stick.

WARNING

Do not depress MASTER RESET pushbutton. Stick deflection of over 1 inch
enables spoiler asymmetry logic and
electrically fails all affected spoilers
down. SPOILERS light should illuminate.

2. Visually check spoiler position/operation

If Spoilers Failed Down:

3. Failed spoiler cb . . . . PULL (INBD 7G9)
   (OTBD 8C5)

4. Land as soon as practicable.

If Spoilers Fail Up or Float:

3. Failed spoiler cb . . . . PULL (INBD 7G9)
   (OTBD 8C5)

Note

- Outboard spoiler position indicators will indicate down with cb 8C5 pulled.

- Allow up to 60 seconds for spoilers to return to trail.

If AFC 573 Not Incorporated: (SPOILER FLR ORIDE)

4. Slow-fly aircraft at safe altitude to determine minimum control speed.

5. Verify that full spoiler authority is available by trimming laterally opposite to stick deflection.

6. Fly straight-in, flaps-up approach above minimum control speed (Refer to page 5-42.)

If AFC 573 is Incorporated: (SPOILER FLR ORIDE)

4. Affected SPOILER FLR ORIDE switch . . . . .ORIDE

5. MASTER RESET pushbutton . . . . .DEPRESS

6. Spoiler control cb (affected spoiler) . . . . . . .RESET

7. Slow-fly aircraft at safe altitude to determine minimum control speed.

8. Verify full spoiler authority is available by trimming laterally opposite to stick deflection.

9. Fly straight-in flaps up approach above minimum control speed. (Refer to page 5-42.)

If Not Associated With Abnormal Roll and/or Yaw:

1. Lateral control . . . . NEUTRALIZE

2. Climb to safe altitude.

3. MASTER RESET pushbutton . . . . .DEPRESS

If Light Remains Illuminated:

4. Avoid abrupt lateral control movements and high roll rates.

WARNING

With wings forward of 57°, excessive horizontal tail differential may cause severe structural damage.

5. Land as soon as practicable

FLAP LIGHT

If Not After Landing/Takeoff Flap Transition:

1. Airspeed . . . . BELOW 225 KIAS

2. FLAP handle . . . . ENSURE FULL UP

3. MASTER RESET pushbutton . . . . .DEPRESS

4. While holding MASTER RESET pushbutton depressed, maneuver flap thumbwheel . . . . FULL FORWARD

5. Check FLAP light out.
(Light can take up to 10 seconds to re-illuminate.)

If After Landing/Takeoff Flap Transition, or Re-illumination After Above Procedures:

1. MASTER RESET pushbutton . . . . .DEPRESS

2. If light still illuminated, check FLAP handle and indicator position, then proceed with appropriate steps below.

FLAP Handle Up and Flaps Not Fully Retracted

1. FLAP handle . . . . EMER UP

If FLAP handle or flaps will not respond or FLAP light remains illuminated, refer to Flap Asymmetry Lockout Procedures.

FLAP Handle Up and Flaps Indicating Full Up

1. Flaps. . . . . CYCLE

If FLAP handle or flaps will not respond or FLAP light remains illuminated, refer to Flap Asymmetry Lockout Procedures.

FLAP Handle Down and Flaps Not Fully Extended

1. Wing sweep . . . . ENSURE AT 20°

2. FLAP handle . . . . EMER DN

If FLAP handle or flaps will not respond or FLAP light remains illuminated, refer to Flap Asymmetry Lockout Procedures.

FLAP Handle Down and Flaps Down

1. Wing sweep . . . . ENSURE AT 20°

2. MASTER RESET pushbutton . . . . .DEPRESS
(Allow 10 seconds for auxiliary flaps to extend.)

If FLAP handle or flaps will not respond or FLAP light remains illuminated, refer to Flap Asymmetry Lockout Procedures.

FLAP AND SLAT ASYMMETRY LOCKOUT

If the flaps fail to respond to FLAP handle movement, asymmetry flap and slat lockout has occurred within tolerances and no roll will be experienced. Under these conditions, the auxiliary flaps will retract, and the FLAP light will remain illuminated.

1. Flaps. . . . . MATCH HANDLE
WITH POSITION

2. Obtain visual check, if possible, to ascertain position of all flap and slat surfaces.

3. Slow-fly aircraft in existing configuration to determine approach characteristics.

4. Land as soon as practicable.

FLAP AND SLAT ASYMMETRY WITHOUT LOCKOUT

CAUTION

These procedures do not apply if a proper flap and slat asymmetry lockout has occurred.

Actuator flap and slat asymmetry can occur with failure of an asymmetry sensor and subsequent failure of the flap and slat drive mechanism for one wing. The pilot's only indication will be an uncommanded roll followed by a FLAP light approximately 10 seconds later. The flap indicator does not indicate actual flap position, but the position to which the flap and slat control box has been driven. The slat indicator shows up, down, or transition (barber pole) for the starboard slat only. The port slat position is not monitored. Asymmetric flaps cause an immediate roll. Asymmetric slats may not be apparent until just before wing stall. Asymmetric slats can cause rapid roll-off above 15 units AOA. Slat position must be monitored by the RIO during transition.

WARNING

The use of lateral trim to reduce stick force will reduce spoiler control significantly. An uncontrollable situation can develop if lateral trim is out of neutral before flap and slat asymmetry or if the pilot trims laterally in the neutral direction (opposite

the roll) during flap and slat transition. This situation will be aggravated and recovery may not be possible with roll SAS off due to reduced differential tail authority. Once asymmetry occurs, do not trim out stick forces. If lateral control is marginal, trim opposite to the natural direction until full spoiler deflection is available. For example; stick to the right, trim left.

If a roll is encountered during flap and slat transition or if the RIO notes asymmetric slat extension or retraction:

1. Perform uncommanded roll/yaw procedure.

CAUTION

If asymmetric condition is eliminated, do not reposition FLAP handle.

2. If flaps and slats do not respond to command and asymmetry exists, perform following steps to clear or minimize asymmetric condition.

a. Climb above 10,000 feet AGL.

b. AUX FLAP/FLAP
CONTR cb. . . . . PULL (7G3)

WARNING

Failure to complete step 2b before the subsequent steps can result in large uncommanded pitch trim changes due to auxiliary flap movement.

c. FLAP handle . . . . CORRESPONDING
TO INDICATOR

d. FLAP/SLAT CONTR
SHUTOFF cb. . . . . PULL (RE2)

e. Slowly move FLAP handle in direction to minimize asymmetry and/or lateral control requirements.

f. Stop flap and slat travel before reaching full up or full down.

g. FLAP/SLAT CONTR
SHUTOFF cb. . . . . RESET (RE2)

h. If asymmetry has been corrected, land using 15 units AOA.

i. If asymmetry has not been corrected, flaps and slats did not respond to above procedure, or lateral control problems exist, use 10 to 12 units AOA if landing is elected.

VERTICAL RECOVERY

1. Above 100 KIAS, use longitudinal stick to pitch the nose down. At extreme nose-high attitudes, aft stick facilitates recovery time and will avoid prolonged engine operation with zero oil pressure.

2. Below 100 KIAS, release controls and wait for aircraft to pitch nose down. This prevents depletion of hydraulic pressure in the event both engines are lost and provides quickest recovery.

3. If roll and/or yaw develop, wait until aircraft is in a nosedown attitude and accelerating before correcting with rudder or lateral stick.

4. Use longitudinal control as necessary to keep nose down and accelerating.

5. Above 100 KIAS, pull out, using 17 units AOA.

6. Recovery to level flight from point of pitchover can normally be completed in less than 10,000 feet.

UPRIGHT DEPARTURE/FLAT SPIN

*1. Stick . . . . . . FORWARD/NEUTRAL LATERAL, HARNESS-LOCK

*2. Rudder . . . . . OPPOSITE TURN NEEDLE/YAW

If No Recovery:

*3. Stick . . . . INTO TURN NEEDLE

*4. If engine stalls. . . . . . BOTH THROTTLES IDLE

If Recovery Indicated:

*5. Controls . . . . NEUTRALIZE

*6. Recover at 17 units AOA.

If Flat Spin Verified By Flat Attitude, Increasing Yaw Rate, Increasing Eyeball, Out g, and Lack of Pitch and Roll Rates:

*7. Canopy . . . . JETTISON

*8. EJECT (RIO COMMAND EJECT)

WARNING

Ejection guidelines are not meant to prohibit earlier canopy jettison and/or ejection. If insufficient altitude exists to recover from departed flight the flightcrew should not hesitate to eject.

Note

At high yaw rates where eyeball out g is sensed, aft stick and full lateral stick into the turn needle may arrest the yaw rate and increase the possibility of recovery. At these yaw rates, the additional differential tail provided by roll SAS on will also increase the possibility of recovery.

INVERTED DEPARTURE/SPIN

*1. Stick . . . . FULL AFT/NEUTRAL LATERAL, HARNESS − LOCK

*2. Rudder . . . . OPPOSITE TURN NEEDLE/YAW

*3. If engine stalls. . . . . BOTH THROTTLES IDLE

If Recovery Indicated:

*4. Controls . . . . NEUTRALIZE

*5. Recover at 17 units AOA

If Spinning Below 10,000 Feet AGL:

*6. EJECT (RIO COMMAND EJECT)

CAUTION

Dual compressor stalls/overtemperatures should be expected in an inverted spin.

Note

If pedal adjustment and/or pilot positioning (due to negative g forces) is such that full rudder pedal travel cannot be obtained, full lateral control opposite the turn needle/yaw may provide an alternate recovery method. Aft longitudinal stick should be relaxed enough to allow full lateral stick application.

WING SWEEP LIGHTS

Advisory Light Only—No Loss of Normal Control

1. MASTER RESET pushbutton . . . . .DEPRESS

Advisory and Warning Lights—No Auto or Manual Control

1. Airspeed . . . . .DECELERATE TO
0.9 IMN OR LESS

2. Check spider detent engaged.

3. MASTER RESET pushbutton . . . . . . .DEPRESS
(Wait 15 seconds to determine system status.)

If Advisory and Warning Lights Illuminate Again:

4. WING SWEEP DRIVE NO. 1
and WG SWP DR NO. 2/
MANUV FLAP cb . . . . .PULL
(LE1, LE2)

5. Emergency WING SWEEP handle -
comply with below schedule:

- <0.4 IMN . . . . 20°

- <0.7 IMN . . . . 25°

- <0.8 IMN . . . . 50°

- <0.9 IMN . . . . 60°

- >0.9 IMN . . . . 68°

```
CAUTION
```

Avoid ACM and aerobatics.

Note

Before landing, the FLAP handle EMER DN
position must be used to ensure that flaps
remain fully extended.

UNSCHEDULED WING SWEEP

1. Emergency WING
SWEEP handle . . . . .POSITION AND HOLD

- If speed greater than 0.6 IMN . . . . FULL AFT

- If speed less than 0.6 IMN . . . . FULL FWD

2. Airspeed . . . . DECELERATE TO
0.6 IMN OR LESS

3. WING SWEEP DRIVE NO. 1
and WG SWP DR NO. 2/
MANUV FLAP cb . . . . PULL
(LE1, LE2)

4. Remain in emergency wing sweep and
comply with emergency wing sweep
schedule.

5. Land as soon as practicable.

```
CAUTION
```

Unscheduled wing sweep at supersonic
speed may cause structural damage.

Note

- Before landing, the FLAP handle
EMER DN position must be used to
ensure that flaps remain fully
extended.

- After a wing sweep malfunction,
the WING SWEEP warning light
may take 15 seconds to illuminate.

- FLAP light will be illuminated with
cb LE2 pulled.

CADC LIGHT

1. MASTER RESET pushbutton . . . . .DEPRESS

If Light Remains Illuminated:

2. Remain below 1.5 IMN.

One or more of following systems may
be affected by CADC malfunction that
illuminates CADC light only.

- MAXIMUM SAFE MACH

- AUTO PILOT

- MACH LEVER

- WING SWEEP INDICATOR

- COCKPIT COOLING LESS THAN MACH 0.25

- HUD DISPLAY (TAKEOFF AND LANDING)

- SERVO ALTIMETER

Note

If illumination of the CADC light is accompanied by other caution or advisory light(s), refer to the appropriate procedure that will dictate the most restrictive limitation.

LADDER LIGHT

1. Airspeed . . . . . .   MINIMUM SAFE OPERATING

2. Obtain in-flight visual check if possible.

3. Land as soon as practicable.

WEIGHT ON-OFF WHEELS SWITCH MALFUNCTION

For most systems, failure of both the left and right weight-on-wheels switches is required to cause the systems to revert to the on-deck mode. Should such failures occur, the following anomalies can result:

- Ground-roll braking

- Nozzles full open (with throttles at IDLE)

- Approach indexers inoperative

- Autopilot cannot be engaged.

- Outboard spoiler module off (flaps up)

- APC will not engage.

- At high altitude, ground cooling fans may overspeed and shut down, causing smoke in cockpit.

- Radar will not scan.

If two or more of the above anomalies are detected, the following action should be taken:

Pilot

1. ANTI SKID SPOILER BK switch . . . . OFF

2. Land as soon as practicable.

WARNING

Do not move both throttles to IDLE unless ANTI SKID SPOILER BK switch is set to OFF if weight on-off wheels switch is suspected, due to loss of thrust and lift caused by nozzles opening and spoilers deploying.

RIO

1. MLG SAFETY RLY NO. 1 and
   NO. 2 cb . . . . PULL
   (6F5, 6F4)

Note

Circuit breakers can be reset after touch-down on field landings to enable ground roll braking and antiskid.

PART 4 – LANDING EMERGENCIES

SINGLE-ENGINE LANDING

At normal landing weights, a single engine landing is essentially the same as a normal landing except the pattern should be flown to avoid turns in excess of 20° and turns into the failed engine. At high gross weights, immediately following takeoff, flight in the power approach configuration is critical. In this situation, the pattern should be planned to turn away from the failed engine using bank angles that do not exceed 20°. Remain below 12 units AOA until established on final approach. Flap and slats should be fully extended as in a normal two-engine landing; however, speed brakes should remain retracted to reduce thrust required and eliminate inadvertent retraction. Single-engine landing, bolter, or waveoff may be accomplished safely up to the gross weight limits for two-engine operation.

During waveoffs or bolters, select MIL thrust, rotate not to exceed 14 units AOA as for normal two-engine takeoff, and climb out at the airspeed indicated in Climb Performance After Takeoff (single-engine) charts in part 2 of section XI. External fuel tanks have a negligible effect on thrust required and need be dropped only if necessary for gross weight considerations.

For single-engine landing, proceed as follows:

1. Fuselage fuel . . . . DUMP AS REQUIRED

Note

Fuel dumping stops automatically at approximately 4,000 pounds remaining.

2. Refer to appropriate hydraulic system failure.

3. Speed brakes . . . . RET

4. Check afterburner operation and full rudder authority. (RUDDER AUTH light out.)

5. EMERG FLT HYD switch. . . . . HIGH
   (just before dirty-up)

6. LDG GEAR handle . . . . DN

7. Flaps. . . . . DN

8. For landing pattern use 12 units AOA for pattern airspeed and do not attempt turns greater than 20° angle of bank.

9. APC . . . . DO NOT ENGAGE

10. Final approach airspeed . . . . 14 units AOA

WARNING

- Ensure that full rudder throw is available before selecting afterburner in landing configuration due to excessive yaw caused by asymmetric thrust. Check afterburner if possible.

- Military power climb performance during heavy gross weight waveoffs may not adequately arrest high sink rate conditions. Use of full afterburner provides a significant increase in climb performance. Up to full rudder may be required to counter afterburner asymmetric thrust yawing moment during waveoff or bolter. Do not exceed 14 units AOA during waveoff/bolter.

CAUTION

- Rudder should be simultaneously applied with afterburner. Lateral stick should not be used alone to counter yawing moment.

- In the event of a bolter, hold slight aft control stick until the desired flyaway attitude is established. Application of full aft stick will partially stall the tail surface, delaying aircraft rotation and causing the aircraft to settle off the angle.

Landing Gear Emergency Lowering

Use emergency lowering of the landing gear only as a last resort. Once this system is used, the gear cannot be retracted; therefore, the landing must be made in whatever configuration you have at that time. If a long flight is necessary to make a field landing, it will have to be made with the gear down. (See figure 5-8.)

1. Airspeed . . . . LESS THAN
   280 KIAS

LANDING GEAR MALFUNCTION EMERGENCY LANDING GUIDE

FINAL CONFIG- URATION	CARRIER LANDINGS	NOTES	FIELD LANDING			
			ARRESTING GEAR AVAILABLE	NOTES	NO ARRESTING GEAR AVAILABLE	NOTES
COCKED NOSE GEAR	LAND	1,8,11	ARRESTED LANDING	6,8,9,11, 12	LAND	6,9,11
SIDE BRACE NOT IN PLACE	LAND	1,2,8,11	NO ARRESTED LANDING	3,6,7,8,11	LAND	3,6,7,8, 11
NOSE GEAR DOWN UNSAFE INDICA- TION	LAND	1,8,11	NO ARRESTED LANDING	4,6,8,9, 10,11	LAND	6,8,9, 10,11
NOSE GEAR UP	LAND	1,2,4 8,11	NO ARRESTED LANDING	4,6,8,9 10,11	LAND	6,8,9 10,11
STUB NOSE GEAR	LAND	1,2,4 8,11	NO ARRESTED LANDING	4,6,8,9, 10,11	LAND	6,8,9, 10,11
NOSE GEAR UP, ONE MAIN UP	EJECT PILOT OPTION TO LAND IF TANKS INSTALLED	1,2,4 8,11	PILOT OPTION EJECT OR ARREST	6,8,10, 11,12	EJECT	—
ONE MAIN UP UNSAFE	LAND	1,2,4, 8,11	ARRESTED LANDING	6,8,10, 11,12	PILOT OPTION EJECT OR LAND	5,6,8 10,11
BOTH MAIN UP UNSAFE	EJECT PILOT OPTION TO EJECT IF TANKS NOT IN- STALLED	1,2,4, 8,11	PILOT OPTION EJECT OR ARREST	6,8,10, 11,12	PILOT OPTION EJECT OR LAND	6,8,10, 11
ONE OR BOTH STUB MAINS	LAND	1,2,4 8,11	NO ARRESTED LANDING	4,5,6 8,11	LAND	5,6,8,11
ALL GEAR UP	EJECT PILOT OPTION TO LAND IF TANKS INSTALLED	1,2,4 8,11	PILOT OPTION EJECT OR NO ARRESTED LANDING	4,6,8, 10,11	PILOT OPTION EJECT OR LAND	6,8,10, 11

1. DIVERT IF POSSIBLE
2. HOOK DOWN BARRICADE ENGAGEMENT
3. MINIMIZE SKID AND DRIFT ROLLOUT
4. REMOVE ALL ARRESTING GEAR
5. LAND OFF-CENTER TO GEAR DOWN SIDE
6. MINIMUM RATE OF DESCENT LANDING

7. GRADUAL SYMMETRICAL BRAKING
8. RETAIN EMPTY DROP TANKS
9. LOWER NOSE GENTLY PRIOR TO FALL THROUGH
10. SECURE ENGINES AT AIR FRAME CONTACT
11. JETTISON EXTERNAL ORDNANCE
12. HOLD DAMAGED GEAR OFF DECK UNTIL PENDANT ENGAGEMENT

7-F52-005-0

Figure 5-8. Landing Gear Malfunction
Emergency Landing Guide

2. LDG GEAR handle . . . . DN

3. Push LDG GEAR handle in hard, turn it 90°
   clockwise, pull, and hold.

4. Gear position indication . . . . CHECK
   (12 seconds)

The LDG GEAR handle must be held in
the fully extended emergency position for
a minimum of 1 second to ensure complete
actuation of the air release valve. Approximately 55-pound pull force is required to
fully discharge the emergency nitrogen
bottle.

5. Make arrested landing if available.

Note

- The nose gear cannot be confirmed as
  locked by visual observation. If both
  the indicator and transition light indicate unsafe, assume that the downlock
  is not in place.

- If there is disagreement between the
  indicator and light and the gear appears
  down, the malfunction may be due to a
  faulty contact on the nose gear downlock microswitch.

- Use of emergency gear extension results
  in loss of nosewheel steering.

- To facilitate in-flight refueling probe
  extension when the gear has been blown
  down, raise the LDG GEAR handle to
  give priority to the refueling probe
  system.

If Any Gear Does Not Come Down:

6. Increase airspeed to at least 280 KIAS.

7. Apply positive and negative g to force gear down.

8. Obtain visual in-flight check if possible.

LANDING GEAR MALFUNCTION

1. Remain below 280 KIAS.

2. Combined hydraulic pressure . . . . CHECK

3. If less than 3,000 psi, refer to combined hydraulic
   failure procedures (page 5-26).

Landing Gear Indicates Unsafe Gear Up, or Transition Light Illuminated

1. LDG GEAR handle . . . . DOWN

If Safe Gear Down Indication Obtained and Transition Light Out:

2. LDG GEAR . . . . LEAVE DOWN

3. Land as soon as practicable

> **CAUTION**
>
> If landing gear indicates unsafe after retraction and a down-and-locked indication can be obtained, the brake pedals should be depressed for 60 seconds to ascertain whether brake hydraulic lines have been severed. If brake hydraulic lines are severed and a combined hydraulic failure occurs, refer to combined hydraulic system failure procedures (page 5-27).

Landing Gear Indicates Unsafe Gear Down, Transition Light Out

This indication means a failure in one of the dual-pole down-lock microswitches.

1. Transition light bulb — CHECK (LTS TEST)
2. Landing gear. . . . . CYCLE

If Condition Still Exists:

3. Obtain visual check if possible.
4. Make normal landing.

Landing Gear Indicates Unsafe, Gear Down, Transition Light Illuminated

Nose gear unsafe indicates that the downlock pin through the drag brace is not in place. Visual determination of nose gear unlocked status is assisted by red stripes on both ends of the nose gear brace oleo. However, a positive check for locked nose gear is not possible visually. Main gear unsafe should be verified by visual inspection. If the drag brace is fully extended, the main gear should be down and locked.

1. LDG GEAR handle . . . . CYCLE

If Still Unsafe:

2. Increase airspeed to 280 KIAS, pull positive g's and yaw aircraft.

If Still Unsafe:

3. LDG GEAR handle . . . . EMERG DN
(Refer to landing gear emergency lowering.)

Note

Use of the emergency gear lowering procedure will result in loss of nosewheel steering.

4. Obtain visual check if possible.

If Still Unsafe and Visually Confirmed Unsafe, Or Gear Position Cannot be Confirmed:

5. Refer to Landing Gear Malfunctions Emergency Landing Guide.

Landing Gear Indicates Safe Gear Down, Transition Light Illuminated

This indication can be caused by a malfunction of:

- Half of the dual-pole micro in the nose gear downlock.

- Half of the dual-pole micros in either of the main gear downlocks.

- The proximity micros in the side braces.

- Failure of the LDG GEAR handle position micro.

- If a visual check confirms the gear extended, and both side braces in place, a malfunction of one of the transition light micros is indicated.

1. LDG GEAR handle . . . . CYCLE

If Transition Light Remains On:

2. Obtain visual check.

3. Gear/Sidebraces appear
in place . . . . NORMAL LANDING

Sidebraces Confirmed Not in Place:

4. Refer to Landing Gear Malfunction Emergency Landing Guide.

Cocked Nose Gear In Flight

1. Airspeed . . . . LESS THAN 280 KIAS

2. Landing gear . . . . LEAVE DN

3. Obtain in-flight visual check if possible.

4. Refer to Landing Gear Malfunction Emergency
   Landing Guide.

WHEEL BRAKE EMERGENCY OPERATION

1. ANTI SKID SPOILER
   BK switch . . . . SPOILER BK

2. HYD HAND PUMP . . . . RECHARGE
   BRAKE ACCUMULATOR
   (LDG GEAR handle DN)

Note

Approximately 6 to 8 differential pedal
applications of auxiliary brakes are
available.

If Toe Brake Pedals Are Ineffective:

3. Parking brake . . . . PULL
   (After auxiliary brake depletion,
   parking brake will lock
   both main wheels.)

```
CAUTION
```

Maximum airspeed for wheel brake application is 165 KIAS at a gross weight of
46,000 pounds and 145 KIAS at 51,000
pounds.

BLOWN-TIRE LANDING

Blown-tire landings should be performed into arresting
gear whenever possible. MAX AB should be selected
immediately upon touchdown in anticipation of a possible bolter. Rollout is extremely rough on blown tires
and MIL acceleration is minimal. Do not apply full aft
stick in attempt to rotate the aircraft before reaching
flying speed. The drag from full-up deflection of the

stabilizers is large and significantly delays acceleration.
Blown-tire landings will frequently result in damaged
main landing gear hydraulic lines. Anticipate possible
combined hydraulic system failure and attendant
commital to gear down bingo following blown tire.

```
CAUTION
```

Engines should not be secured until crash
crew is in position to extinguish a possible
fire created from drain fuel contacting
hot wheel assembly.

1. Obtain in-flight visual check if possible.

2. ANTI SKID SPOILER
   BK switch . . . . SPOILER BK
   (OFF for CV)

3. HOOK . . . . DN

4. Make carrier or short-field fly-in arrested landing
   as soon as practicable. (MAX AB at touchdown
   will be required for adequate bolter performance.)

If Arresting Gear Is Not Available:

5. Land on centerline.

6. Nosewheel steering . . . . ENGAGED

EMERGENCY FLAP AND SLAT EXTENSION

If the flaps and slats fail to fully extend or retract with
normal flap handle movement and either the combined
hydraulic system or the outboard spoiler module is pressurized, proceed as follows:

1. Airspeed . . . . LESS THAN 225 KIAS

2. Flaps . . . . EMER DN

3. Flap and slat position
   indicator . . . . CHECK
   (45 seconds)

NO FLAPS AND NO SLATS LANDING

A no flaps and no slats landing is basically the same as a normal landing, except that the pattern is extended and the
approach speed is slightly higher than normal. Field arresting gear should be used if necessary. CV arrestments are
permitted. Consult applicable recovery bulletins for WOD
requirements.

1. Gross weight .... REDUCE
   (Weight consistent with existing
   runway length and conditions.)

2. Flaps .... UP

Note

If outboard spoilers are needed for ground
roll braking, FLAP handle must be set to
DN.

3. Fly landing pattern slightly wider than normal or
   make straight-in approach at 15 units AOA (14
   units AOA if AFB 198 has not been complied
   with).

4. Use normal braking technique.

┌─────────────┐
│   CAUTION   │
└─────────────┘

● Maximum airspeed for wheel brake
  application is 165 KIAS at a gross
  weight of 46,000 pounds and 145
  KIAS at 51,000 pounds.

● Use of full aft stick during landing in
  this configuration can result in tailpipe ground contact.

● Avoid slow approaches. Wing drop
  and increased sink rate may occur
  at 16.5 to 17.5 units AOA. If AFB
  198 has not been complied with, wing
  drop and increased sink rate may
  occur at 0.5 to 1 unit lower AOA.

AUXILIARY FLAP FAILURE

A no auxiliary flaps landing is basically the same as a
normal landing, except that the approach speed is slightly
higher than normal and the longitudinal stick position
during the approach is more aft than normal. CV arrestments are permitted; consult applicable recovery bulletin
for WOD requirements.

1. Wing sweep .... ENSURE AT 20°

2. AUX FLAP/FLAP CONTR cb .... PULL
                                        (7G3)

3. Approach .... 15 UNITS AOA

Note

With AUX FLAP/FLAP CONTR cb
pulled, wings will not sweep aft.

AFT WING SWEEP LANDINGS

Aft wing sweep landings should be conducted utilizing on-
speed angle of attack (15 units). Main flaps and slats
should be utilized to reduce approach speed with aft wing
sweeps up to 50°. Maneuver flaps may be utilized should
main flaps and slats fail to extend.

Note

● If maneuver flaps are used, avoid inadvertent actuation of maneuver thumbwheel at approach airspeeds.

● Due to additive approach speed increases and improper pitch trim compensation, do not utilize direct lift
  control during aft wing sweep landings.

APC performance is degraded (particularly turning performance) at aft wing sweeps and should not be utilized. CVA
arrestments are permitted up to 40° sweep, and emergency
barricade engagements are permitted up to 35° sweep.
Waveoff performance is slightly degraded. Consult applicable recovery bulletins for WOD requirements.

FIELD ARRESTMENTS

Field Arresting Gear

The types of field arresting gear in use include the anchor
chain cable, water squeezer, and Morest-type equipment.
All require engagement of the arresting hook in a cable
pendant rigged across the runway. Location of the
pendant in relation to the runway will classify the gear
as follows.

MIDFIELD GEAR - Located near the halfway point of the
runway. Usually requires prior notification in order to rig
for arrestment in the direction desired.

ABORT GEAR - Located 1,500 to 2,500 feet short of the
upwind end of the duty runway and usually rigged for
immediate use.

OVERRUN GEAR - Located shortly past the upwind end
of the duty runway. Usually rigged for immediate use.

Some fields will have all types of gear, others none. For
this reason, it is imperative that all pilots be aware of the
type, location, and compatibility of gear in use with the
aircraft, and the policy of the local air station with regard
to which gear is rigged for use and when.

As various modifications to the basic types of arresting
gear are made, exact speeds will vary accordingly. Certain
aircraft service changes may also affect engaging speed and
weight limitations.

An engagement in the wrong direction into
chain gear can severely damage the aircraft.

In general, arresting gear is engaged on the centerline at as
slow a speed as possible. Burn down to 2,000 pounds or
less fuel remaining. While burning down, make practice
passes to accurately locate the arresting gear. Engagement
should be made with feet off the brakes, shoulder harness
locked, and with the aircraft in a three-point attitude.
After engaging the gear, good common sense and existing
conditions dictate whether to keep the engines running or
to shut down and abandon the aircraft.

In an emergency situation, first determine the extent of
the emergency by whatever means are available (instruments, other aircraft, LSO, RDO, tower or other ground
personnel). Next, determine the most advantageous
arresting gear available and the type of arrestment to be
made under the conditions. Whenever deliberate field
arrestment is intended, notify control tower personnel as
much in advance as possible and state estimated landing
time in minutes.

If gear is not rigged, it will probably require 10 to 20
minutes to prepare. If foaming of the runway or area of
arrestment is required or desired, it should be requested by
the pilot at this time.

If fuel is streaming from the bottom of the aircraft, a field
arrested landing is not recommended due to the high
probability of sparks and heat from the arresting hook
igniting the streaming fuel and air mixture. If an arrested
landing is mandated due to the lack of adequate braking or
runway conditions, an effort should be made to foam the
runway in the runout area of the arresting gear.

Short-Field Arrestment

If at any time before landing a directional control
problem exists or a minimum rollout is desired, a short-
field arrestment should be made and the assistance of an
LSO requested. He should be stationed near the touchdown point and equipped with a radio. Inform the LSO
of the desired touchdown point. A constant glide slope
approach to touchdown is permitted (mirror or Fresnel
Lens Landing Aid) with touchdown on centerline at or
just before the arresting wire with the hook extended.
The hook should be lowered while airborne and a positive
hookdown check should be made. Use midfield gear or
Morest-type, whenever available. If neither is available,
use abort gear. Use an approach speed commensurate
with the emergency experienced. Landing approach

power will be maintained until arrestment is assured or
a waveoff is taken. Be prepared for a waveoff if the gear
is missed. After engaging the gear, retard the throttles to
IDLE or secure engines and abandon aircraft, depending
on existing conditions.

Long-Field Arrestment

The long-field arrestment is used when a stopping problem
exists with insufficient runway remaining (that is, aborted
takeoffs, icy or wet runways, loss of brakes after touchdown, etc). Lower the hook, allowing sufficient time for
it to extend fully before engagement (normally 1,000 feet
before reaching the arresting gear). Do not lower the
hook too early and weaken the hook point. Line up the
aircraft on the runway centerline. Inform the control
tower of your intentions to engage the arresting gear, so
that aircraft landing behind you may be waved off. If no
directional control problem exists (crosswind, brakes out,
etc), secure the engines.

Do not attempt an arrested landing with full
or partially full external tanks. If external
tanks are not empty, jettison.

Engaging Speeds

The maximum permissible engaging speed, gross weight,
and off-center engagement distance for field arrestment
are listed in figure 5-9. The data in the long-field landing
columns may be used for light-weight aborted takeoff
where applicable; data in the aborted takeoff columns
may be used for heavy gross-weight landings.

As various modifications to the basic types of arresting
gear are incorporated, engaging speeds or gross-weight
limitations may change. For this reason and for more
detailed information, the applicable Aircraft Recovery
Bulletin should be consulted.

BARRICADE ARRESTMENT

1. External stores . . . . JETTISON
   (Except AIM-7 or AIM-54A on fuselage
   stations if wing is at full forward
   sweep.)

2. External tanks . . . . JETTISON
   (Retain empty for landing gear   (unless empty)
   malfunction.)

3. Fuel . . . . .DUMP OR BURN
(Reduce to 2,000 pounds.)

4. HOOK. . . . . DN
(Lower to permit engagement of a
cross-deck pendant, which will minimize barricade engagement speed
and damage to aircraft.

5. Fly normal pattern and approach,
on-speed, angle of attack, centerline,
and meatball.

Note

Anticipate loss of meatball for a short
period of time during the approach.
Barricade stanchions may obscure the
meatball.

Upon Engaging the Barricade:

6. Throttles . . . . OFF

7. Evacuate aircraft as soon as practical.

WARNING

Weight Limits for Barricade Engagement
are:

Wing λ 20° . . . . . . 52,000 lbs. (max)

Wing λ > 20°, < 35°. 46,000 lbs. (max)

Wing λ > 35° . . . . not permitted

ARRESTING HOOK EMERGENCY DOWN

1. HOOK handle. . . . . DN

2. HOOK handle. . . . . PULL, THEN
ROTATE

Note

Pull handle aft approximately 4 inches
and turn counterclockwise. This will
mechanically release the uplatch mechanism and allow hook to extend.

3. HOOK handle. . . . . RESTOW TO
THE NORMAL POSITION

4. Hook transition light . . . . .CHECK OFF

If Light Illuminated:

5. HK CONTR/RAIN RPL/ . . . . .PULL
ANTI-ICE CONTR CB   (7C2)

Note

CB 7C2 also controls rain repellent and
anti-ice.

FORCED LANDING

Landing the aircraft on unprepared surfaces is not recommended. If it is necessary to do so, landing with the landing gear down, regardless of the terrain, will assist in absorb-
ing the shock of ground impact and reduce possibility of
flightcrew injuries. External stores should be jettisoned in
a safe area prior to touchdown. External tanks should be
jettisoned if they contain fuel, but retained to absorb landing shock if they are empty. If time permits, dump fuel
to allow touchdown at the slowest possible speed with full
flaps.

EMERGENCY FIELD ARRESTMENT GUIDE

TYPE OF ARRESTING GEAR	MAXIMUM ENGAGING SPEED KNOTS (a)										MAXIMUM OFF CENTER ENGAGEMENT FEET
	GROSS WEIGHT X 1000 POUNDS										
	40	44	48	51.8	54	57	60	64	68	69.8	
E-27	146	142	138	134	132	129	126	122	118	116	35
E-15 200-FOOT SPAN	180	180	177	175	173	171	168	164	159	156	35
E-15 300-FOOT SPAN	180	180	180	180	179	177	174	171	167	166	50
M-21	130	130	130	130	125	125	120	115	115	115	10
E-28	176(f)	180	179	178	177	176	175	174	172	172	40
E-28 (c)	176(f)	176	160	160	160	160	156	145	145	145	40
E-5 STD CHAIN (b)	150	145	140	136	134	131	128	125	122	121	(a)
E-5-1 STD CHAIN (b)	150	145	140	136	134	131	128	125	122	121	(a)
E-5 HEAVY CHAIN (b)	150	150	150	150	150	150	150	150	147	145	(a)
E-5-1 HEAVY CHAIN (b)	165	165	165	165	165	160	157	152	147	145	(a)
BAK-9	160	160	160	155	150	144	138	131	124	122	30
BAK-12 (d)	160	160	159	146	137	118	(g)	(g)	(g)	(g)	50
DUAL BAK-12 (e)	160	160	160	160	160	160	160	160	160	160	30

- ALL ENGAGING SPEEDS LIMITED BY ARRESTING GEAR CAPACITY EXCEPT AS NOTED.

(a) OFF-CENTER ENGAGEMENT MAY NOT EXCEED 25% OF THE RUNWAY SPAN.

(b) BEFORE MAKING AN ARRESTMENT, THE PILOT MUST CHECK WITH THE AIR STATION TO CONFIRM MAXIMUM ENGAGING SPEED BECAUSE OF A POSSIBLE INSTALLATION WITH LESS THAN MINIMUM REQUIRED RATED CHAIN LENGTH.

(c) ONLY FOR THE E-28 SYSTEMS AT KEFLAVIK, BERMUDA, AND WALLOPS FLIGHT CENTER WITH 920-FOOT TAPES.

(d) STANDARD BAK-12 LIMITS ARE BASED ON 150-FOOT SPAN, 1-INCH CROSS-DECK PENDANT, 40,000-POUND WEIGHT SETTING AND 950 FOOT RUNOUT. NO INFORMATION AVAILABLE REGARDING APPLICABILITY TO OTHER CONFIGURATIONS.

(e) DUAL BAK-12 LIMITS ARE BASED ON 225-FOOT SPAN, 1-1/4-INCH CROSS-DECK PENDANT, 50,000-POUND WEIGHT SETTING AND 1,200-FOOT RUNOUT. NO INFORMATION AVAILABLE REGARDING APPLICABILITY TO OTHER CONFIGURATIONS.

(f) MAXIMUM ENGAGING SPEED LIMITED BY AIRCRAFT LIMIT HORIZONTAL DRAG LOAD FACTOR (MASS ITEM LIMIT "G").

(g) ENGAGING SPEED LIMIT IS 96 KNOTS AT 59,000 POUNDS. DUE TO RUNOUT LIMITATIONS IT IS RECOMMENDED THIS GEAR NOT BE ENGAGED AT WEIGHTS GREATER THAN 59,000 POUNDS.

2-F50-164-0

Figure 5-9. Emergency Field Arrestment Guide

PART 5 — EJECTION AND BAILOUT

EJECTION

The escape system will function up to 0.9 IMN or 450 KIAS, whichever is greater; however, human limitations are more restrictive as indicated below:

- 0 to 350   SAFE EJECTION
                          (injury improbable)

- 350 to 450 KIAS   HAZARDOUS EJECTION
                          (injury may be sustained)

- above 450 KIAS   EXTREMELY HAZARDOUS
                          (serious injury highly probable)

The decision to eject under controlled conditions should not be delayed lower than 2,000 feet AGL, and in uncontrolled flight, lower than 10,000 feet AGL.

The ejection envelope curves shown in figure 5-10 represent the ejection seat capability with respect to airspeed limitations and minimum altitudes. For all ejections it is recommended that airspeed be reduced as slow as practicable; however, in uncontrolled situations, do not delay ejection because the aircraft is not within the published safe escape envelope. For ejection at low altitude, it is recommended that a climb be initiated to convert excess airspeed into altitude. Although the escape system is capable of zero-zero ejection, it should be borne in mind that a combination of low airspeed and high rate of descent at low altitude can present a condition more severe than zero-zero. Ejection sequence is shown on page FO-18.

CONTROLLED EJECTION

Time permitting, perform all or as much as possible of the following:

1. Place aircraft in safe envelope and attitude for ejection.

2. IFF/SIF EMERG/7700
                                                                    (RIO)

3. Position report . . . . .TRANSMIT

4. Shoulder harness . . . . LOCK

5. Helmet visor . . . . DOWN

6. Sit erect, head back, feet on pedals.

7. Grasp ejection handle with both hands.

8. Pull smartly with elbows in and seated erect.

If Seat Fails to Separate

1. Emergency restraint
   release handle . . . . PULL

2. Pitch forward, kicking clear of seat.

When Clear Of Seat:

3. Parachute ripcord . . . . PULL

4. Helmet visor . . . . RAISE
   (Survey landing site. If landing in
   trees seems likely, lower visor to
   protect face.)

5. Oxygen mask . . . . RELEASE ONE SIDE
   (Prevents suffocation if
   disabled when emergency
   O_2 runs out.)

MANUAL BAILOUT

Manual bailout is extremely hazardous, even under the most favorable conditions (level flight, slow airspeed, and optimum altitude) and is considered only as a last chance method of escape. If the canopy has been jettisoned, and the seat cannot be ejected, and manual bailout is elected, the following bailout procedures are suggested:

WARNING

Manual bailout below 2,000 feet AGL
(minimum sink rate) may not allow
sufficient altitude for parachute
deployment.

Seat Fails to Eject and Canopy is Off

1. Place aircraft in safe envelope.

2. Ensure leg-restraint cords clear.

3. Pitch trim . . . . .FULL NOSE DOWN

4. Emergency restraint release . . . . .PULL

5. Release stick.

6. Roll forward and push free of aircraft.

When Clear of Aircraft:

7. Parachute ripcord . . . . .PULL

CANOPY JETTISON

The canopy may be jettisoned by pulling the CANOPY JETTISON handle in either cockpit. The handle must be squeezed and held before and while pulling. On pulling, the handle disconnects from the instrument panel. Jettisoning the canopy before initiation of ejection sequence is not recommended because of the windblast on the RIO. If normal ejection sequence has been attempted, but the canopy has not jettisoned, jettisoning the canopy could initiate ejection. If the face curtain has been pulled, it should be grasped and held before jettisoning to prevent drogue chute entanglement.

SURVIVAL/POST-EJECTION PROCEDURES

Figure 5-11, sheets 1 through 14 describe step-by-step procedures for inflation of the life preserver assembly (LPA) configured with beaded handles, the 35-grams CO_2 cylinder, and using the SKU-2 (single handle) or RSSK-7 (dual handle) survival kit. The ejection situation is below-barostat, high-altitude and over water in which seat/man separation and parachute deployment was accomplished automatically.

EJECTION ENVELOPE CURVES

COMMAND DUAL EJECTION

NOTE

THESE CURVES REPRESENT THE EJECTION SEAT CAPABILITY
AND ARE ABSOLUTE MINIMUM ALTITUDES AT WHICH
EJECTION MUST HAVE BEEN INITIATED.

MINIMUM SAFE EJECTION ALTITUDES

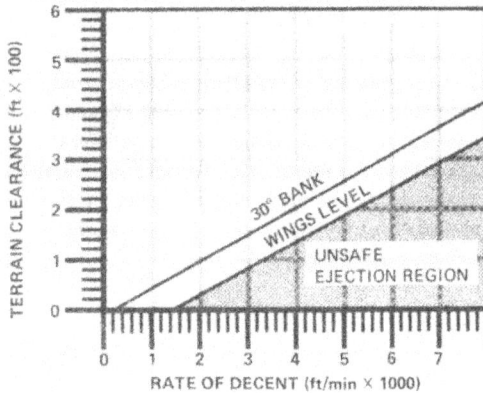

NOTE

- THESE CURVES DO NOT
  INCLUDE REACTION TIME

- FOR INVERTED FLIGHT ADD 400 ft TO TERRAIN
  CLEARANCE REQUIRED FOR WINGS LEVEL.

EJECTION SEAT - HUMAN FACTORS LIMITATIONS

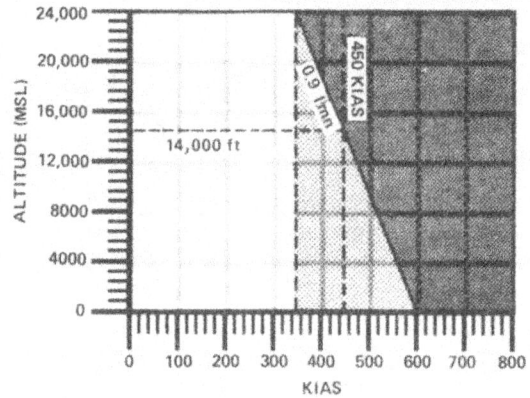

WHITE – SAFE (INJURY IMPROBABLE)
LIGHT GRAY – HAZARDOUS (INJURY MAY BE SUSTAINED)
DARK GRAY – EXTREMELY HAZARDOUS (SERIOUS
INJURY HIGHLY PROBABLE)

NOTE

- SOLID LINE IS SEAT LIMITATIONS
  DASHED LINES/SHADED AREAS ARE HUMAN LIMITATIONS

VELOCITY vs DIVE ANGLE

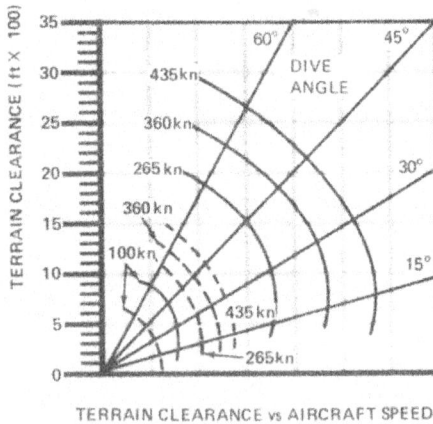

TERRAIN CLEARANCE vs AIRCRAFT SPEED

NOTE

- CURVES ARE BASED ON WINGS LEVEL, BANK
  ATTITUDE, AND APPROPRIATE ANGLE OF ATTACK

- ———— 2 SECONDS PILOT REACTION TIME
  - - - - 0 SECONDS PILOT REACTION TIME

- 600 kn DATA NOT AVAILABLE AT TIME OF CHARTS

TERRAIN CLEARANCE vs AIRCRAFT SPEED

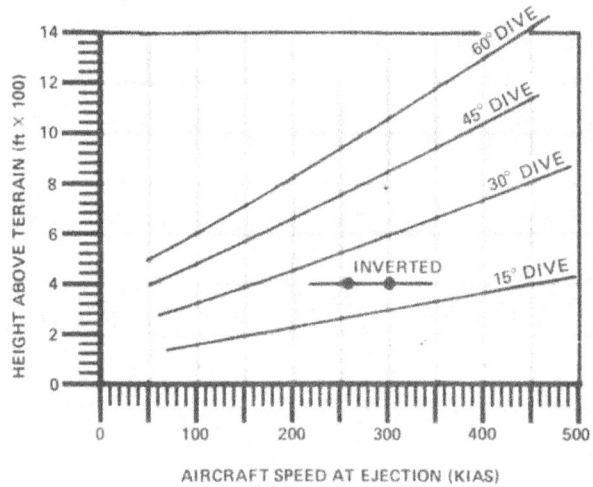

AIRCRAFT SPEED AT EJECTION (KIAS)

NOTE

- INVERTED AND BANK PLOTS
  AIRCRAFT AT CONSTANT ALTITUDE

- THESE CURVES DO NOT INCLUDE REACTION TIME

3-F050-004
074-0

Figure 5-10. Ejection Envelope Curves

POST EJECTION/BAILOUT PROCEDURES

LPA/LPU INFLATION

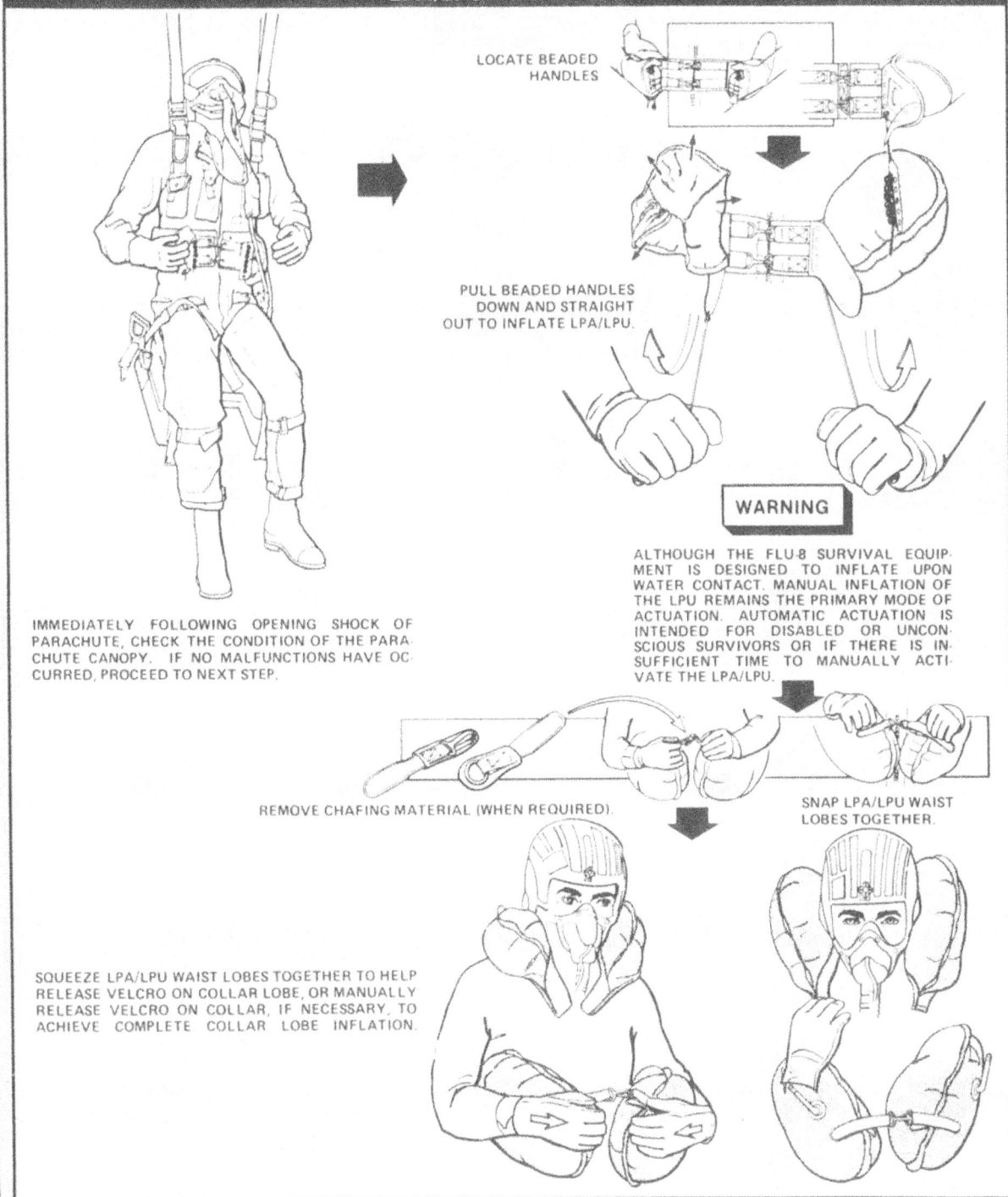

LOCATE BEADED HANDLES

PULL BEADED HANDLES DOWN AND STRAIGHT OUT TO INFLATE LPA/LPU.

IMMEDIATELY FOLLOWING OPENING SHOCK OF PARACHUTE, CHECK THE CONDITION OF THE PARACHUTE CANOPY. IF NO MALFUNCTIONS HAVE OCCURRED, PROCEED TO NEXT STEP.

WARNING

ALTHOUGH THE FLU-8 SURVIVAL EQUIPMENT IS DESIGNED TO INFLATE UPON WATER CONTACT, MANUAL INFLATION OF THE LPU REMAINS THE PRIMARY MODE OF ACTUATION. AUTOMATIC ACTUATION IS INTENDED FOR DISABLED OR UNCONSCIOUS SURVIVORS OR IF THERE IS INSUFFICIENT TIME TO MANUALLY ACTIVATE THE LPA/LPU.

REMOVE CHAFING MATERIAL (WHEN REQUIRED).

SNAP LPA/LPU WAIST LOBES TOGETHER.

SQUEEZE LPA/LPU WAIST LOBES TOGETHER TO HELP RELEASE VELCRO ON COLLAR LOBE, OR MANUALLY RELEASE VELCRO ON COLLAR, IF NECESSARY, TO ACHIEVE COMPLETE COLLAR LOBE INFLATION.

0 F50-262 1

Figure 5-11. Survival/Post Ejection Procedures (Sheet 1 of 14)

DEPLOY LIFERAFT

AIRCREWMAN UNDER CANOPY WITH LPA/LPU IN-FLATED PREPARING TO ACTIVATE KIT RELEASE HANDLE.

LEFT OR RIGHT

FRONT VIEW

SKU-2

RSSK-7

WITH THE RIGHT HAND, LOCATE THE SINGLE RAFT RELEASE HANDLE ON THE RIGHT SIDE OF THE SKU-2.

LOCATE AND TIGHTLY SQUEEZE EITHER RAFT RELEASE HANDLE (ON EITHER SIDE OF KIT).

NOTE

TRIGGER MECHANISM MUST BE SQUEEZED.

FIRMLY PULL UP ON THE RAFT RELEASE HANDLE UNTIL FREE OF KIT.

FIRMLY SQUEEZE AND PULL UP ON EITHER HANDLE UNTIL CABLE HAS TRAVELED ITS FULL DISTANCE AND KIT HAS RELEASED.

0-F50-262-2

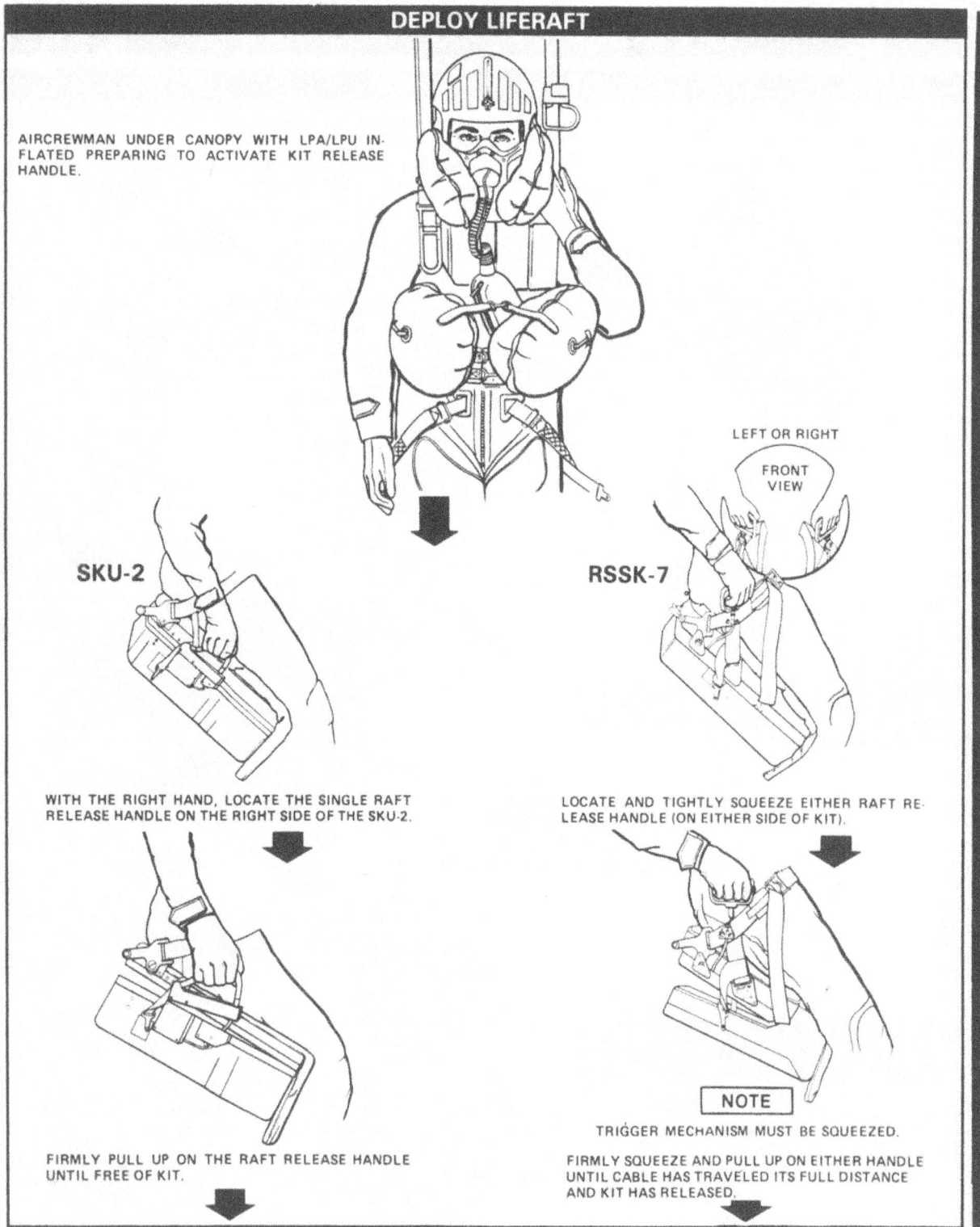

Figure 5-11. Survival/Post Ejection Procedures (Sheet 2 of 14)

DEPLOY LIFERAFT — CONT

SKU-2

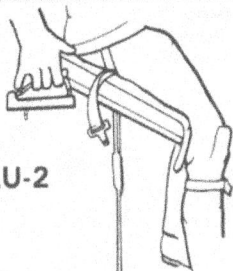

PULLING RAFT RELEASE HANDLE UNLOCKS CON-
TAINER; THE LOWER HALF FALLS AWAY BUT RE-
MAINS ATTACHED BY DROPLINE.

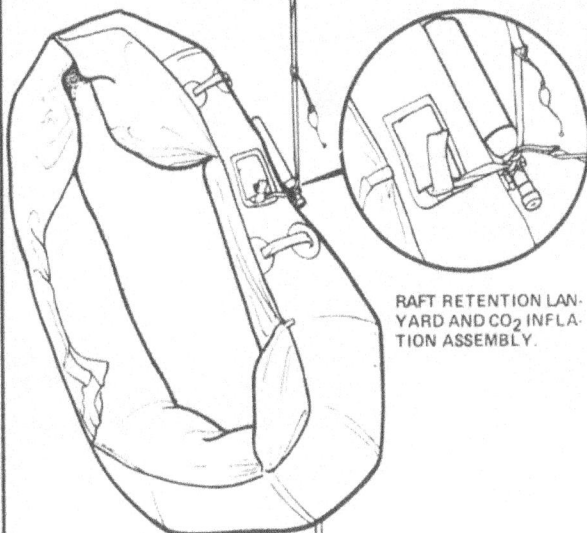

RAFT RETENTION LAN-
YARD AND CO_2 INFLA-
TION ASSEMBLY.

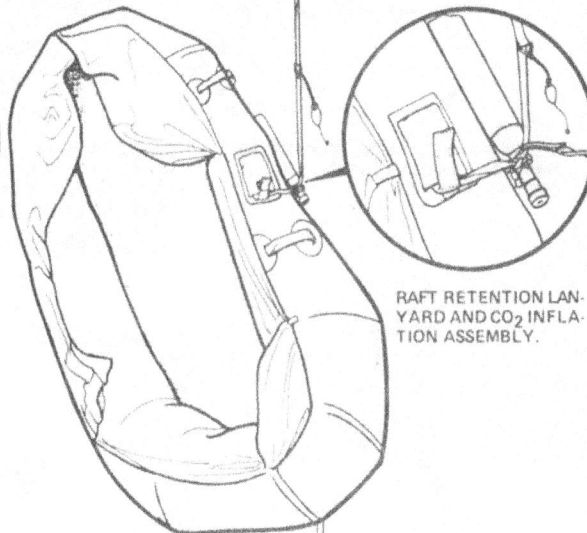

LIFERAFT FULLY INFLATED APPROXIMATELY 17
FEET BELOW UPPER HALF OF SEAT KIT CONTAINER.

RSSK-7

PULLING RAFT RELEASE HANDLE UNLOCKS CON-
TAINER; LOWER HALF FALLS AWAY BUT REMAINS
ATTACHED BY DROPLINE.

RAFT RETENTION LAN-
YARD AND CO_2 INFLA-
TION ASSEMBLY.

LIFERAFT FULLY INFLATED APPROXIMATELY 17
FEET BELOW UPPER HALF OF SEAT KIT CONTAINER.

0 F50 262 3

Figure 5-11. Survival/Post Ejection Procedures (Sheet 3 of 14)

OPTIONS

IF TIME AND ALTITUDE PERMIT OR RESCUE IS NOT IMMINENT,
THE FOLLOWING OPTIONS FOR OXYGEN MASK, VISOR, GLOVES,
AND FOUR-LINE RELEASE SYSTEM MAY BE CONSIDERED.

REMOVE OXYGEN MASK. OXYGEN HOSE MAY BE DISCONNECTED AND DISCARDED IF DESIRED.

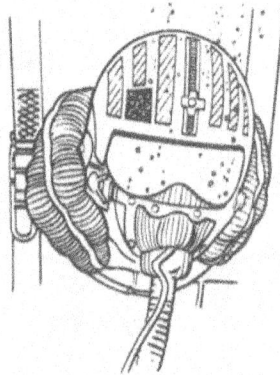

THE A-13 OR MBU-14 OXYGEN MASK AND MINIATURE REGULATOR PROVIDE UNDERWATER BREATHING CAPABILITIES AND WHEN FUNCTIONING PROPERLY SHOULD BE RETAINED IN LOW LEVEL OVERWATER EJECTIONS.

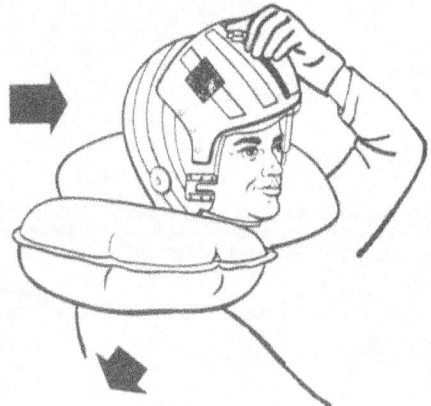

RAISE VISOR

REMOVE GLOVES

NOTE

STOW GLOVES IN A SAFE PLACE TO PREVENT LOSS. REMOVAL OF GLOVES MAY FACILITATE SUBSEQUENT RELEASE OF CANOPY RELEASE FITTINGS.

FOUR LINE RELEASE SYSTEM SHOULD BE UTILIZED TO REDUCE OSCILLATIONS OF THE 28 FOOT FLAT CIRCULAR CANOPY AND TO MAKE THE CANOPY STEERABLE BY PROVIDING A CONTROLLED CHANNEL OF ESCAPE FOR THE AIR.

WARNING

- CAREFULLY INSPECT THE CANOPY AND SUSPENSION LINES PRIOR TO USING THE FOUR LINE SYSTEM.

- IF ANY CANOPY DAMAGE IS EVIDENT OR IF THERE ARE BROKEN SUSPENSION LINES, DO NOT USE THE FOUR LINE RELEASE SYSTEM. THE SYSTEM ALSO SHOULD NOT BE USED AT NIGHT SINCE CANOPY DAMAGE MAY BE DIFFICULT TO DETERMINE.

TO OPERATE FOUR LINE RELEASE, LOCATE LANYARD PULL LOOPS ON THE INSIDE OF REAR RISERS.

0-F50 262 4

Figure 5-11. Survival/Post Ejection Procedures (Sheet 4 of 14)

OPTIONS — CONT

PULL LANYARD LOOPS DOWNWARD WITH A SHARP TUG.

NOTE

APPROXIMATELY 20 POUNDS OF PULL FORCE IS NEEDED TO BREAK THE TACKINGS AND FREE THE DAISY CHAIN COUPLINGS.

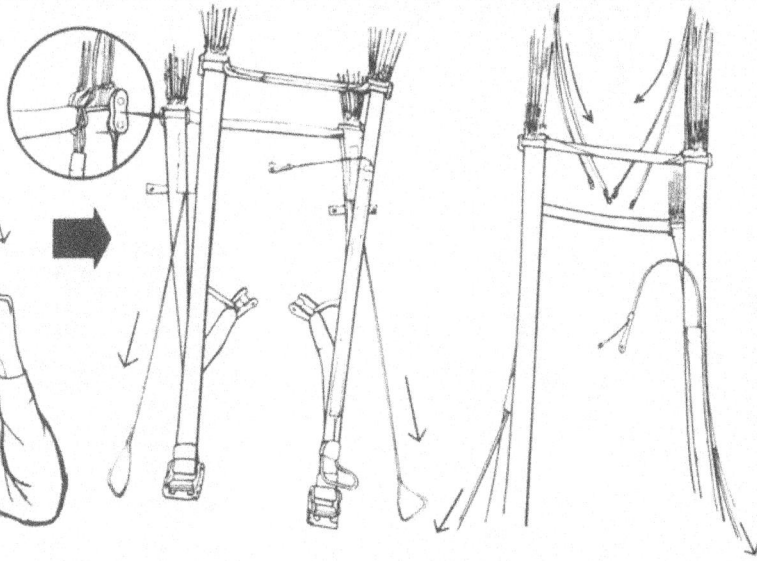

FOUR SUSPENSION LINES ARE RELEASED FROM THE CONNECTOR LINKS AND A LOBE FORMS AT THE CENTER REAR OF THE CANOPY CREATING A CONTROLED CHANNEL FOR AIR ESCAPE.

PULL DOWN ON LEFT LANYARD TO STEER LEFT.

PULL DOWN ON RIGHT LANYARD TO STEER RIGHT.

NOTE

A 180 DEGREE TURN CAN BE ACCOMPLISHED IN ABOUT 20 SECONDS.

0-F50 262-5

Figure 5-11. Survival/Post Ejection Procedures (Sheet 5 of 14)

LANDING PREPARATION

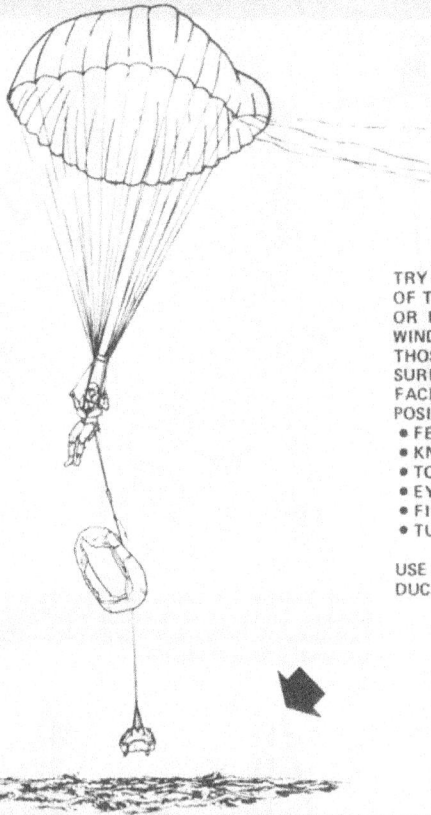

TRY TO DETERMINE THE WIND DIRECTION AT THE SURFACE
OF THE WATER USING WHITECAPS, SMOKE FROM WRECKAGE,
OR KNOWN SURFACE WINDS IN THE VICINITY. NOTE THAT
WINDS AT THE SURFACE MAY BE QUITE DIFFERENT FROM
THOSE ENCOUNTERED AT ALTITUDE. WHEN NEARING THE
SURFACE MANEUVER THE PARACHUTE SO THAT YOU ARE
FACING INTO THE WIND THEN ASSUME THE PROPER BODY
POSITION FOR LANDING:

- FEET TOGETHER
- KNEES BENT
- TOES POINTED SLIGHTLY DOWNWARD
- EYES ON THE HORIZON
- FIRMLY GRASP CANOPY RELEASE FITTINGS
- TUCK ELBOWS IN PRIOR TO WATER ENTRY

USE OF THE FOUR LINE RELEASE SYSTEM SIGNIFICANTLY RE-
DUCES THE CHANCES OF PARACHUTE OSCILLATION.

KOCH FITTING RELEASE

AS THE FEET MAKE CONTACT WITH THE WATER; RELEASE
PARACHUTE FITTINGS.

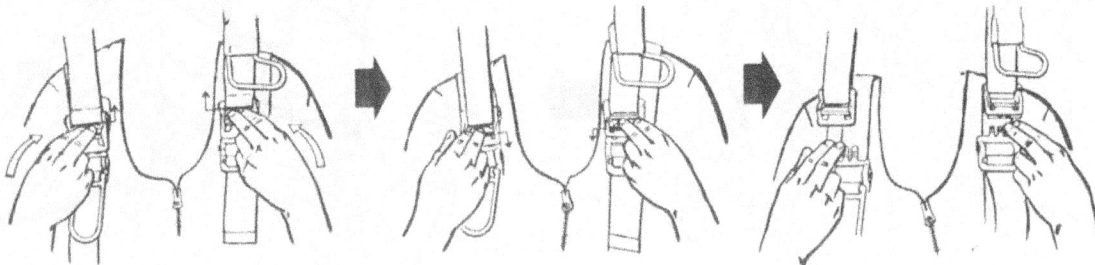

PUSH UP ON CANOPY RELEASE LOCKING
LEVER COVER.

PULL DOWN ON RELEASE LEVER.

CANOPY RELEASED

NOTE

THE PROCEDURES OUTLINED THUS FAR
APPLY AS WELL TO OVERLAND EJECTION. IN
SOME INSTANCES HOWEVER, INFLATION OF
THE LPA/LPU MAY BE UNDESIRABLE OVER
LAND. DEPLOYMENT OF THE RIGID SEAT
SURVIVAL KIT IS NOT RECOMMENDED FOR
OVERLAND EJECTIONS.

0-F50-262-6

Figure 5-11. Survival/Post Ejection Procedures (Sheet 6 of 14)

RAFT BOARDING

WHEN CLEAR OF THE PARACHUTE CANOPY, RETRIEVE THE LR-1 LIFERAFT BY LOCATING THE DROPLINE AND PULLING THE RAFT TO YOU.

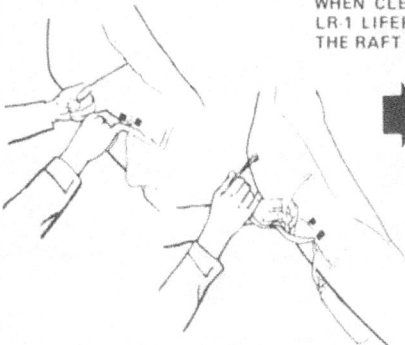

ATTACH SNAPHOOK TO GATED HELO LIFT D-RING.

> **WARNING**
>
> INSURE THAT RAFT RETENTION LANYARD IS SECURELY ATTACHED AND O_2 HOSE HAS BEEN DISCONNECTED FROM SEAT PAN (IF NOT PREVIOUSLY ACCOMPLISHED) BEFORE RELEASING UPPER HALF OF SEAT PAN.

LOCATE RAFT RETENTION LANYARD POCKET JUST ABOVE CO_2 CYLINDER; THEN REMOVE RAFT RETENTION LANYARD.

LOCATE MINI-KOCH FITTINGS; RELEASE UPPER HALF OF KIT.

BRING RAFT AROUND FOR ENTRY INTO SMALLER END (STERN).

GRASP STERN AND FORCIBLY PUSH UNDER LPA/LPU WAIST LOBES.

USING BOARDING HANDLES, PULL INTO RAFT AND TURN TOWARD A SEATED POSITION.

MOVE INTO A COMFORTABLE AND WELL-BALANCED POSITION.

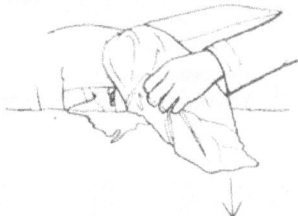

IF SEA ANCHOR HAS NOT DEPLOYED LOCATE IT ON LEFT SIDE OF RAFT AND...

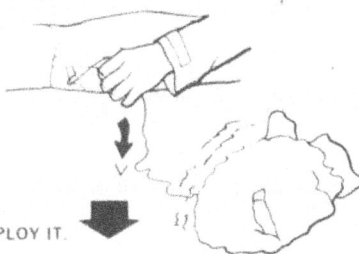

DEPLOY IT.

0 F50 262 7

Figure 5-11. Survival/Post Ejection Procedures (Sheet 7 of 14)

RAFT BOARDING — CONT

RETRIEVE LOWER HALF OF SEAT PAN.

SKU-2

RSSK-7

NOTE

INSURE THAT NO SHARP EDGES CONTACT THE LIFERAFT.

SKU-2

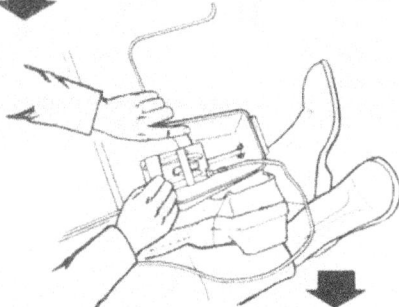

RSSK-7

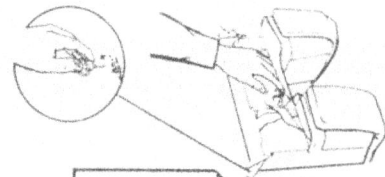

LOCATE URT-33 (RIGHT LEG OF LOWER CONTAINER) AND RELEASE VELCRO STRAPS.

SKU-2

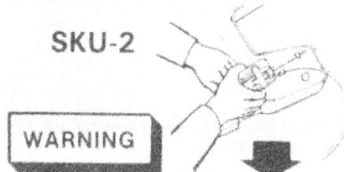

WARNING

URT-33 IS NOT TIED AND ONCE REMOVED FROM ITS BRACKET CARE MUST BE TAKEN TO PREVENT ITS LOSS.

URT-33 HAS RETRIEVAL LANYARD SECURED TO RADIO WITH RUBBER BANDS. SECURE LANYARD TO SUITABLE PLACE ON SURVIVAL EQUIPMENT, THEN REMOVE URT-33 FROM ITS BRACKET.

NOTE

URT-33 MUST BE TURNED OFF WHEN USING PRC-90 ON 243.0 MHz TO PREVENT INTERFERENCE FROM URT-33.

IMMEDIATELY SECURE EQUIPMENT CONTAINERS TO GATED HELO HOIST D-RING. THE SKU-2 EQUIPMENT CONTAINER CAN BE REMOVED FROM DROPLINE BY SLIPPING IT BACK THROUGH THE LARKSHEAD KNOT. THEN SECURE TO GATED HELO HOIST D-RING.

WARNING

ONCE SAFETY PIN IS RELEASED, EQUIPMENT CONTAINERS WILL NO LONGER BE SECURED TO LOWER HALF OF SEAT PAN AND CARE MUST BE TAKEN TO PREVENT ITS LOSS.

LOCATE EQUIPMENT CONTAINER SAFETY PIN IN CENTER OF LOWER HALF OF SEAT PAN, DEPRESS BUTTON AND REMOVE PIN.

RSSK-7

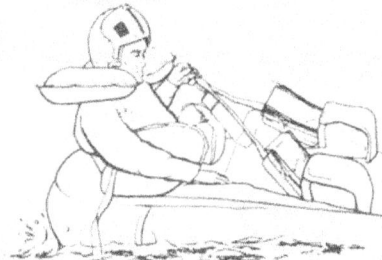

REMOVE EQUIPMENT CONTAINERS FROM LOWER HALF OF SEAT PAN.

RSSK-7

WARNING

ITEMS IN EQUIPMENT CONTAINERS ARE NOT TIED IN AND ONCE CONTAINERS ARE OPENED CARE MUST BE TAKEN TO AVOID THEIR LOSS.

LOCATE LARGE CONTAINER (WITH URT-33 AUTOMATIC ACTUATION ASSEMBLY PROTRUDING) THAT WOULD HAVE BEEN IN RH LEG OF SEAT PAN. OPEN CONTAINER, RELEASE VELCRO TAPE. (ONE PACKET OF SRU-31P KIT WILL BE ON TOP OF URT-33.)

G F50-262 8

Figure 5-11. Survival/Post Ejection Procedures (Sheet 8 of 14)

RAFT BOARDING — CONT

WARNING

URT-33 IS NOT TIED AND ONCE REMOVED FROM ITS BRACKET CARE MUST BE TAKEN TO PREVENT ITS LOSS.

RSSK-7

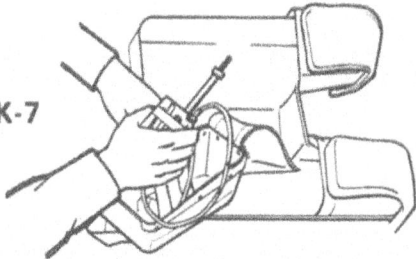

URT-33 HAS A RETRIEVAL LANYARD SECURED TO RADIO WITH RUBBER BANDS. SECURE LANYARD TO SUITABLE PLACE ON SURVIVAL EQUIPMENT, THEN REMOVE URT-33 FROM ITS BRACKET.

NOTE
- URT-33 MUST BE TURNED OFF WHEN USING PRC-90 ON 243.0 MHz TO PREVENT INTER- FERENCE FROM URT-33.

- DEPENDING ON SEA STATE, DECIDE WHETHER TO RETAIN THE SEAT PAN OR DISCARD IT BY CUTTING THE DROPLINE NEAR THE CO_2 CYLINDER. INSURE THAT THE EQUIPMENT CONTAINER(S) HAVE BEEN REMOVED FROM THE LOWER HALF OF THE SEAT PAN AND THAT THEY ARE SECURELY FASTENED TO THE D-RING.

SIGNALING DEVICES

THE FOLLOWING INFORMATION DESCRIBES THE USE OF SIG- NALING DEVICES WHILE IN THE LIFERAFT, AND IS NOT IN- TENDED TO PRESCRIBE ANY GIVEN ORDER OF PRIORITY WHICH WOULD BE DICTATED BY THE IMMEDIATE SITUATION OF THE SURVIVOR.

TRIGGER SCREW

SIGNAL PROJECTOR   ANGULAR SLOT

LANYARD

PLASTIC BANDOLIER   HAND FIRED SIGNAL

THE MK-79 MOD 0 ILLUMINATION SIGNAL KIT USES A PENCIL-TYPE LAUNCHER AND CARTRIDGE FLARE TO ATTRACT ATTENTION OF SAR.

WARNING

ENSURE PENCIL-TYPE LAUNCHER IS IN COCKED POSITION.

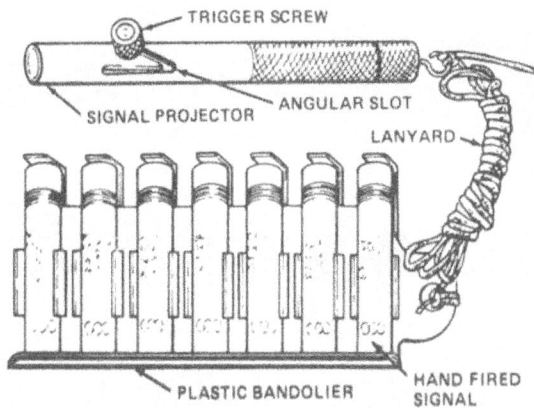

SCREW CARTRIDGE FLARE INTO LAUNCHER WHILE KEEPING FLARE POINTED IN A SAFE DIRECTION.

HOLD LAUNCHER FROM ABOUT A 45-DEGREE ANGLE TO SLIGHTLY OVERHEAD, PULL BACK ON TRIGGER AND RELEASE. CARTRIDGE FLARE HAS A MINIMUM 4-1/2-SECOND DURATION AND CAN BE LAUNCHED TO ABOUT 200 FEET.

0-F50-262-9

Figure 5-11. Survival/Post Ejection Procedures (Sheet 9 of 14)

SIGNALING DEVICES — CONT

SDU-5/E STROBE LIGHT CAN BE ATTACHED TO HELMET BY MATING HOOK AND PILE (VELCRO) TAPE. THIS FREES HANDS FOR USING OTHER SIGNALING DEVICES WHILE ALLOWING LIGHT TO FLASH UP INTO SKY AND REFLECT OFF HELMET.

MIRROR SIGNALING

1. REFLECT SUNLIGHT ONTO NEARBY SURFACE (RAFT, HAND, ETC.).
2. SLOWLY BRING MIRROR UP TO EYE LEVEL AND LOOK THROUGH SIGHTING HOLE. YOU WILL SEE A BRIGHT SPOT: THE AIM INDICATOR.
3. HOLD MIRROR CLOSE TO EYE, SLOWLY TURN AND MANIPULATE SO THAT BRIGHT LIGHT IS ON TARGET.
4. EVEN IF NO AIRCRAFT OR SHIPS ARE IN SIGHT, CONTINUE TO SWEEP HORIZON. MIRROR FLASHES CAN BE SEEN FOR MANY MILES, EVEN IN HAZY WEATHER.

MK-13 MOD 0 MARINE SMOKE AND ILLUMINATION SIGNAL (USED TO ATTRACT ATTENTION OF SAR AIRCRAFT AND TO GIVE WIND DRIFT DIRECTION).

IDENTIFICATION:

NIGHT
- RED CAP
- PROTRUSIONS ON CAP
- PROTRUSIONS ON CASE
- METAL WASHER ATTACHED TO LANYARD

NOTE

FLARE INCORPORATING PAPER VICE PLASTIC END CAPS HAS NO PROTRUSIONS ON CAP.

DAY
- ORANGE CAP
- NO PROTRUSIONS ON CAP OR CASE

BURNS APPROXIMATELY 20 SECONDS WITH APPROXIMATELY 3,000 CANDLE-POWER.

REMOVE CAP FROM DESIRED END.

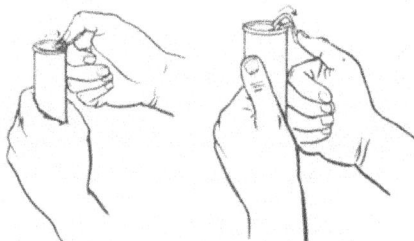

PULL FLIP RING OVER SIGNAL RIM TO BREAK LEAD SEAL. IF SEAL DOESN'T BREAK, PUSH RING UNTIL IT BENDS AGAINST CASE.

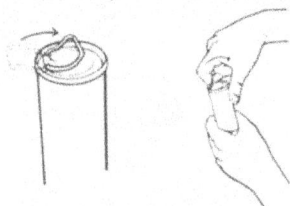

FLIP BENT RING BACK TO ORIGINAL POSITION AND USE AS A LEVER TO BREAK SEAL.

WARNING

MK-13 MOD 0 SIGNAL MAY REACH A TEMPERATURE THAT IS UNCOMFORTABLE TO HAND AFTER IGNITION. USE OF GLOVES IS SUGGESTED.

IGNITE SIGNAL BY QUICK PULL ON RING.

IGNITED MK-13 MOD 0 MUST BE HELD AT ARMS LENGTH, DOWN WIND, NO MORE THAN SHOULDER HIGH, AND OVER THE SIDE OF LIFERAFT TO PREVENT DAMAGE TO RAFT FROM HOT RESIDUE.

0 F50-262 10

Figure 5-11. Survival/Post Ejection Procedures (Sheet 10 of 14)

SIGNALING DEVICES — CONT

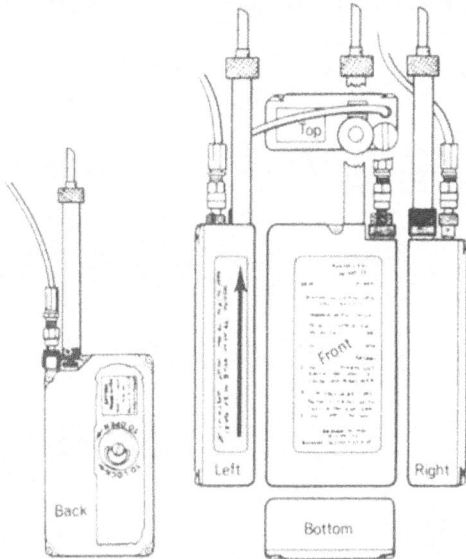

THE AN/URT-33 RADIO AUTOMATICALLY TRANSMITS A SWEPT TONE SIGNAL ON 243.0 MHz WHEN THE EJECTION SEAT LEAVES THE FLOOR OF THE AIR-CRAFT.

NOTE

- URT-33 MUST BE TURNED OFF WHEN USING PRC-90 ON 243.0 MHz TO PREVENT INTERFERENCE FROM URT-33.
- DO NOT POINT ANTENNA DIRECTLY AT RECEIVING AIRCRAFT.
- WHEN WORD ON IS VISIBLE, RADIO IS ON.

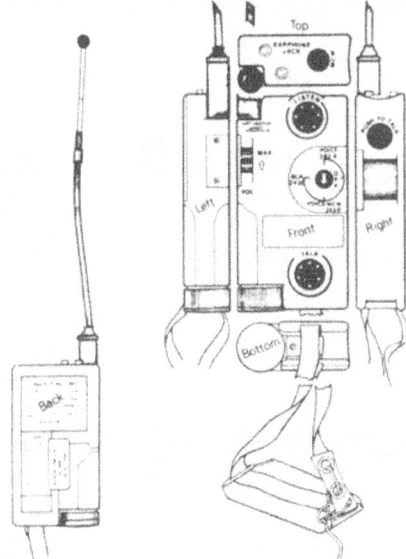

THE AN/PRC-90 IS A DUAL CHANNEL TRANSMITTER/ RECEIVER SURVIVAL RADIO CAPABLE OF TRANS-MITTING (VOICE MODE) UP TO 60 NMI (LINE OF SIGHT, DEPENDING ON RECEIVING AIRCRAFT ALTITUDE). IT OPERATES ON GUARD (243.0 MHz) OR SAR . PRI-MARY OPERATING FREQUENCY (282.8 MHz) WITH A MODE FOR SWEPT TONE SIGNAL ON 243.0 MHZ ONLY. TRANSMISSION OF THE BEACON OR CODE CAN BE UP TO 80 NMI.

RESCUE

WHEN RECOVERY AIRCRAFT IS READY TO EFFECT RESCUE:

STOW OR DISCARD LOOSE GEAR, ROLL OUT OF RAFT ON RIGHT SIDE (SIDE WITH CO_2 CYLINDER).

SWIM AWAY FROM RAFT. ENSURE THAT HELMET VISOR HAS BEEN LOWERED.

REMOVE RAFT ,RETENTION LANYARD AFTER RESCUE DEVICE HAS BEEN LOWERED.

WARNING

TO ALLOW DISCHARGE OF STATIC ELECTRICITY AND PRE-VENT ELECTRICAL SHOCK, DO NOT TOUCH HELO HOIST CABLE OR RESCUE DEVICE UNTIL IT HAS MADE CONTACT WITH WATER/GROUND.

0-F50-262-11

Figure 5-11. Survival/Post Ejection Procedures (Sheet 11 of 14)

RESCUE — CONT

PROCEDURES FOR D-RING WITH GATE.

ATTACH LARGE HOOK TO HELO LIFT D-RING.

NOTE

THE RESCUE HOOK HAS A SMALL AND LARGE HOOK. THE LARGE HOOK IS THE PRIMARY HOOK FOR HOISTING PERSONNEL.

WARNING

TO ALLOW DISCHARGE OF STATIC ELECTRICITY AND PREVENT ELECTRICAL SHOCK, AVOID TOUCHING RESCUE HOOK UNTIL IT HAS MADE CONTACT WITH WATER/GROUND.

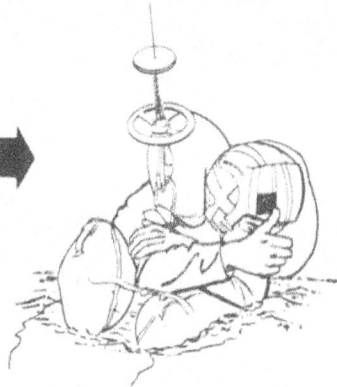

CROSS ARMS IN FRONT OF CHEST AND PLACE HEAD DOWN AND TO LEFT. GIVE THUMBS UP SIGNAL TO HELO HOIST OPERATOR.

POSITION OF AIRCREWMAN DURING HELO HOIST USING D-RING WITH GATE.

PROCEDURES FOR SURVIVOR'S SLING/HORSE COLLAR (DESIGNED TO ACCOMMODATE ONE SURVIVOR AT A TIME).

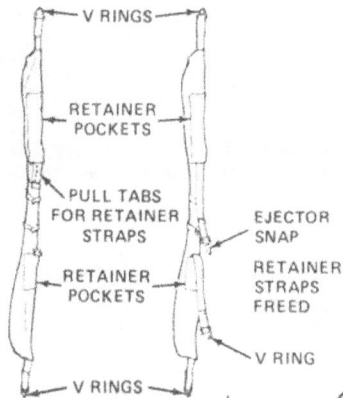

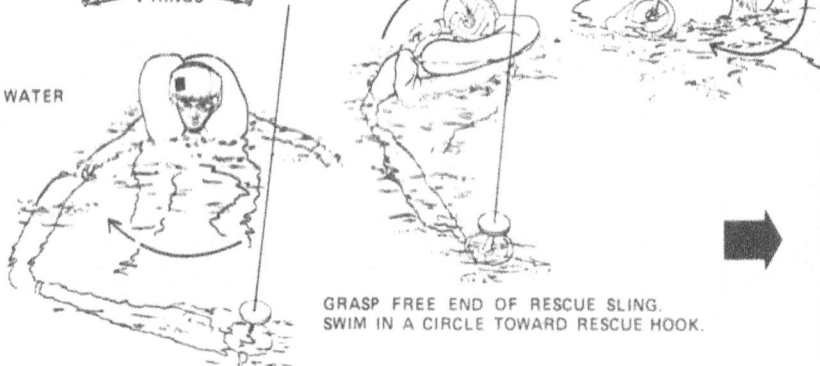

V RINGS

RETAINER POCKETS

PULL TABS FOR RETAINER STRAPS

EJECTOR SNAP

RETAINER POCKETS

RETAINER STRAPS FREED

V RING

V RINGS

UPON CLEARING WATER CROSS FEET.

GRASP FREE END OF RESCUE SLING. SWIM IN A CIRCLE TOWARD RESCUE HOOK.

0-F50-262-12

Figure 5-11. Survival/Post Ejection Procedures (Sheet 12 of 14)

RESCUE — CONT

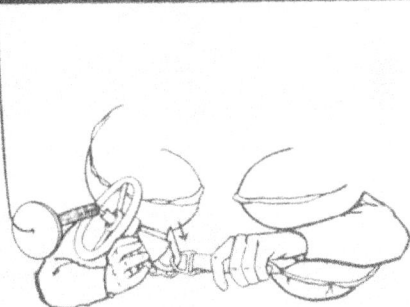

ATTACH FREE END OF SLING TO LARGE
HOOK.

PULL BOTH RETAINER STRAPS FREE
AND CONNECT EJECTOR SNAP TO V-RING
OF OTHER RETAINER STRAP. PULL
TIGHT.

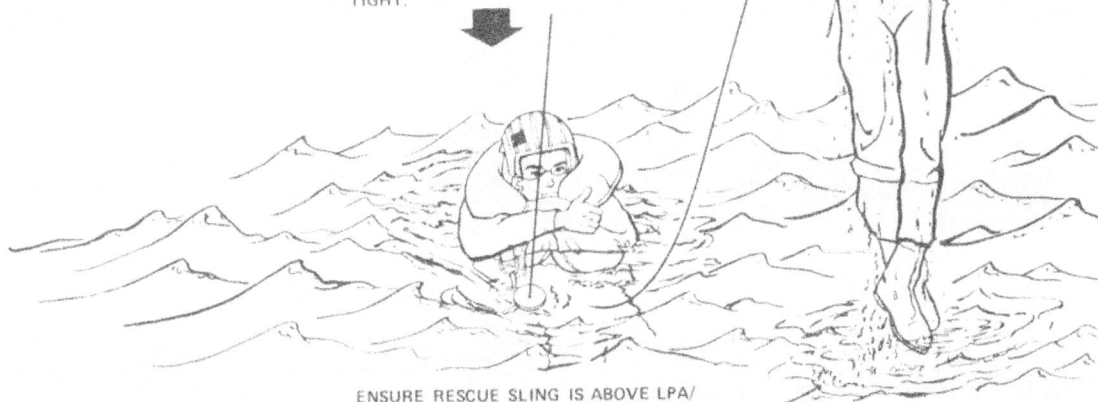

ENSURE RESCUE SLING IS ABOVE LPA/
LPU WAIST LOBES AND HIGH ON BACK.
WRAP ARMS AROUND SLING, KEEP HEAD
DOWN, CROSS LEGS, AND GIVE THUMBS
UP SIGNAL TO HELO HOIST OPERATOR.

POSITION OF AIRCREWMAN DURING
HOIST. UPON CLEARING WATER CROSS
FEET.

PROCEDURES FOR USE OF FOREST PENETRATOR

FOREST PENETRATOR WITH FLOTATION
COLLAR AND SEATS RETRACTED.

(SAFETY STRAPS OMITTED TO SHOW
CONNECTION OF RESCUE HOOK TO EYE-
BOLT.)

FOREST PENETRATOR WITH FLOTA-
TION COLLAR AND SEATS EXTENDED.

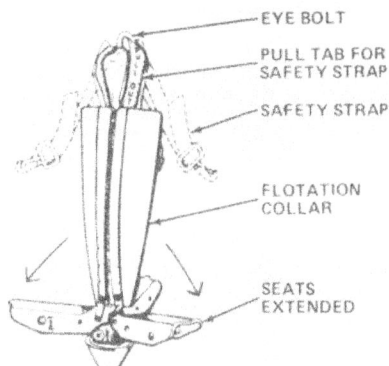

EYE BOLT

PULL TAB FOR
SAFETY STRAP

SAFETY STRAP

FLOTATION
COLLAR

SEATS
EXTENDED

WARNING

TO ALLOW DISCHARGE OF STA-
TIC ELECTRICITY AND PRE-
VENT ELECTRICAL SHOCK,
AVOID TOUCHING RESCUE
HOOK/FOREST PENETRATOR
UNTIL IT HAS MADE CONTACT
WITH WATER GROUND.

0-F-50-262-13

Figure 5-11. Survival/Post Ejection Procedures (Sheet 13 of 14)

RESCUE — CONT

UNSNAP LPA/LPU WAIST LOBES.

EXTEND ONLY ONE SEAT ON FOREST
PENETRATOR.

SIT ON SEAT FACING FLOTATION COL-
LAR. USING ELBOWS, SEPARATE LPA/
LPU WAIST LOBES AND PULL SHAFT OF
PENETRATOR CLOSE TO CHEST.

TURN HEAD DOWN AND TO THE LEFT.
GIVE THUMBS UP SIGNAL TO HELO
HOIST OPERATOR.

PASS SAFETY STRAP UNDER ARM, AROUND BACK,
AND UNDER OTHER ARM. CONNECT SAFETY STRAP
AND TIGHTEN.

WARNING

UNDER NO CIRCUMSTANCES SHOULD SUR-
VIVORS ATTEMPT TO ASSIST THEIR ENTRANCE
INTO HELICOPTER OR MOVE FROM RESCUE
DEVICE UNTIL HELICOPTER AIRCREWMAN
ASSISTS THEM TO A SEAT IN THE AIRCRAFT.

POSITION OF AIRCREWMAN
DURING HOIST WITH FOREST
PENETRATOR.

UPON CLEARING WATER
CROSS FEET.

0 F50 262 14

Figure 5-11. Survival/Post Ejection Procedures (Sheet 14 of 14)

SECTION VI — ALL-WEATHER OPERATIONS

TABLE OF CONTENTS

PART 1 — INSTRUMENT PROCEDURES

AUTOMATIC CARRIER LANDING SYSTEM

Automatic carrier landing system (ACLS) approaches apply to properly configured aircraft utilizing carrier or shorebased AN/SPN-10, or AN/SPN-42 ACLS radar facilities. Three primary modes of operation and two submodes are available.

- Mode I — approach automatically controlled to touchdown.

- Mode IA — approach automatically controlled to a minimum of 200 feet and 1/2-mile; manual control remainder of approach.

- Mode II — approach manually controlled using AN/SPN-41 or AN/SPN-42 VDI and/or HUD presentation for glide slope and lineup information.

- Mode IIT — approach manually controlled using AN/SPN-41 or AN/SPN-42 VDI and/or HUD presentation for glide slope and lineup information as well as information from the CCA controller.

- Mode III — approach manually controlled using only CCA-controller-supplied information.

Mode I

Mode I provides a fully automatic, hands-off landing capability, called automatic carrier landing (ACL) or all-weather landing (AWL). The landing radar system (AN/SPN-42) tracks the aircraft and compares its position with the desired position. The aircraft position is corrected to fly the desired glidepath by commands from the Navy Tactical Data System (NTDS) using the radar computer. These commands are transmitted over the UHF data link to the aircraft, where the automatic flight control system (AFCS) executes the pitch and bank commands. Additional ramp input commands tailored to each specific ship or field are applied at the proper time to assist the aircraft through the burble. In addition to control of the aircraft, discrete words and glide slope error signals are transmitted for cockpit displays to show the pilot where his aircraft is in relation to the desired glide slope. Independent glideslope error signals from the AN/SPN-41 instrument landing system may also be displayed. The pilot may take control at any time and continue the landing to Mode II. An approach that is automatically controlled to a minimum of 200 feet and 1/2-mile, then manually controlled to a landing, is called a Mode IA approach.

Mode II

The control of the aircraft remains with the pilot along the entire glide slope to touchdown. Glideslope error signals are transmitted to the aircraft for cockpit displays from the AN/SPN-41 or the AN/SPN-42. The pilot flies the aircraft to null the error and to keep the vertical and lateral crosshairs centered. During a Mode II T approach, the final controller provides a Mode III type talk down to assist the pilot in flying his needles.

Mode III

The pilot flies the aircraft in response to voice radio commands from the final controller to keep the aircraft on the proper glide slope. From the radar azimuth and elevation displays, the final controller determines the aircraft position with respect to the desired glidepath and gives guidance to the pilot.

AIRCRAFT SUBSYSTEMS

Mode I (automatic) landings are possible only if the ACLS installation, including data link, AFCS, radar beacon augmentor, INS, and ACLS displays (VDIG and/or HUD) are all fully operational. The approach power compensator (APC) should be used during the coupled portion of the approach. Mode II (manual) landings can be made using displayed cross-pointer information from either the data link or the AN/ARA-63A receiver decoder, or both (providing dual displays).

Data Link

Data link (link 4A) messages are received and transmitted by a UHF frequency-shift-key-modulated radio link. Data link receives control messages in serial form from the NTDS and processes each message as necessary. For ACL the position error information is furnished to the VDIG and/or HUD needles, discrete messages appear on the DDI panel, and control information is provided for the AFCS. Reply messages are transmitted to the NTDS with detailed information on aircraft heading, speed, altitude, fuel quantity, weapons, stores, and autopilot status.

The shipboard data link continuously transmits a universal test message (UTM) and a monitor control message (MCM). When in operation, the UTM or MCM is used by the aircraft as a self-test feature. The aircraft data link system self-test is performed by selecting AWL on the VDI and/or HUD (declutter off) and AWL/PCD steering on the pilot display control panel. For aircraft configured with AN/ASW-25B control boxes, set AN/ASW-25B control panel to the AN/SPN-42 data link frequency and select ON. When the TILT discrete appears, select and hold TEST on the data link mode switch to initiate test with the UTM. This will result in the precision course vector needles cycling every 6 seconds from fly up and right to fly down and left. The NO MSG discrete will illuminate at fly up and right, and go off at fly down and left; the WAVEOFF discrete will illuminate. When the system cycles back to fly up and right these lights will go off and the NO MSG

discrete will go on again. This cycle will continue as long as the data link mode switch is held in the TEST position.

For aircraft configured with the integrated communications control panel, select the AN/SPN-42 data link frequency and select power ON. Selection of D/L RAD on the pilot's master test function panel will initiate test with the MCM. See figure 6-1 (sheets 1 and 2) for correct display indications.

Note

In aircraft BUNO 159637 and subsequent and those aircraft incorporating CADC 1166 A/A, the command airspeed/Mach numeric is not displayed on the VDI. However, command airspeed error from data link is displayed.

Automatic Flight Control System (AFCS)

The AFCS performs two functions: stability augmentation and autopilot.

Stability augmentation (STAB AUG) provides added stability to the aircraft and is, in general, necessary for effective aircraft control.

The autopilot ACL mode can be engaged only after engaging all STAB AUG axes and then by placing the AUTO PILOT ENGAGE switch in the ON position. Selection of ACL on the AFCS control panel arms the mode and illuminates the A/P REF light on the pilot's DDI panel. The pilot engages ACL through the reference engage switch on the stick grip, at which time the A/P REF light is extinguished.

Note

If a pitch parallel actuator disconnects during an ACL, the A/P CPLR warning light will go out when ACL is selected on the AFCS control panel, but the aircraft will not respond to SPN 42 commands.

Following AFCS engagement to data link commands the pilot can take control of the aircraft by simply overriding the data link commands with his control stick. This condition is referred to as control stick steering and, when used with ACL engaged, causes immediate disengagement, and the AFCS will again revert to STAB AUG. A/P REF indicates that an AFCS pilot relief mode has been selected (in this case, ACL), but not engaged.

D/L DISPLAYS FOR D/L RAD

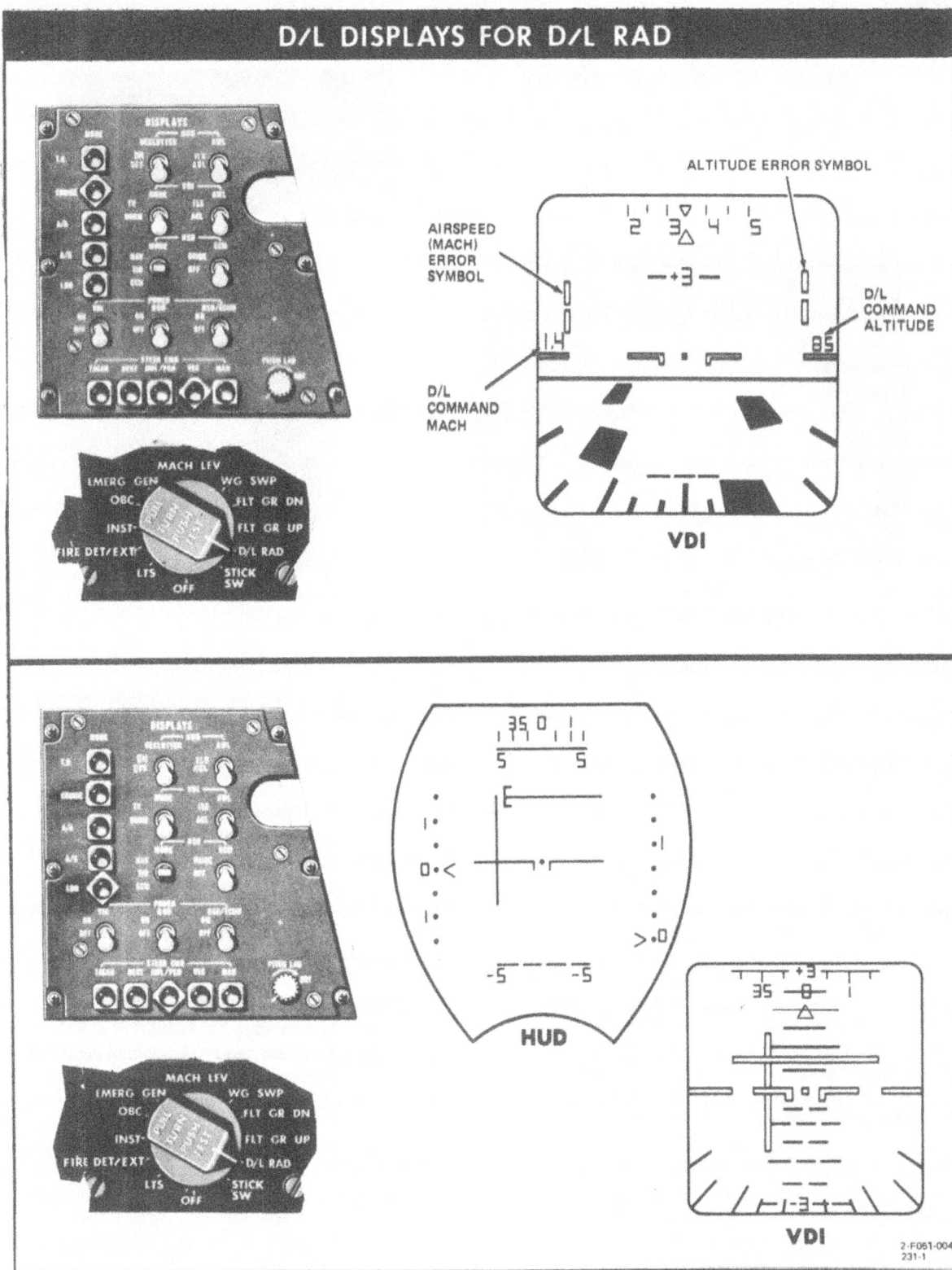

Figure 6-1. D/L Displays for D/L RAD (Sheet 1 of 2)

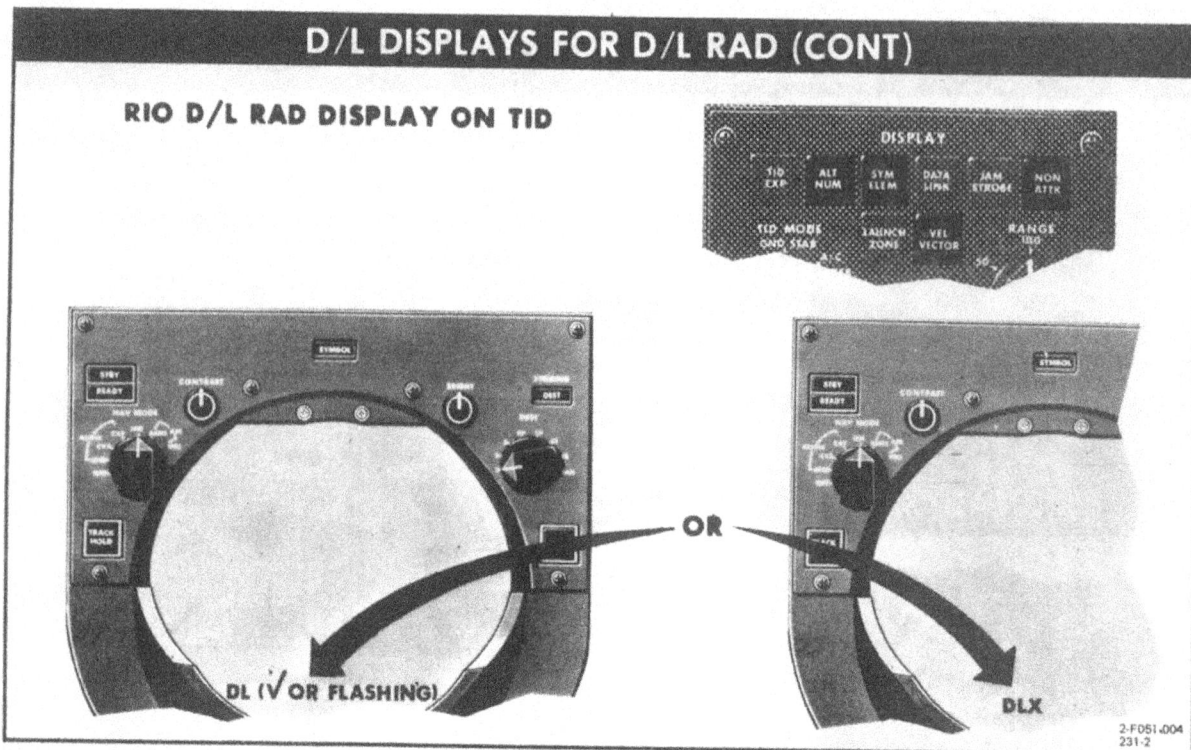

Figure 6-1. D/L Displays for D/L RAD (Sheet 2 of 2)

Radar Beacon Augmentor (AN/APN-154)

The radar beacon enhances aircraft tracking (range and accuracy) by ship and/or ground-based X-band radars for precision vectoring. Pulsed (coded) X-band signals transmitted by the surface radar station are received by the beacon and decoded; if they match the mode (six available) selected by the flight officer, the beacon responds with a return pulse to the radar site. The reply signal, considerably stronger than a normal radar echo, enhances the radar acquisition and tracking capability of the surface station.

The antenna on the IR fairing is a horizontally polarized antenna. The radar beacon set is a Ka-band crossband radar augmentor receiver that extends the tracking capability of the AN/SPN-42 shipboard radar. These two units give the AN/APN-154 radar beacon the capability of operating with either or both channels of the AN/SPN-42 without interference.

The beacon augmentor eliminates radar scintillation by providing a large source of reply energy from one point on the aircraft. The beacon augmentor receives interrogations from the AN/SPN-42 carrier-based radar in the Ka-band at 33.0 to 33.4 GHz, processes them,

and retransmits modulated X-band pulses at 8.8 to 9.6 GHz to the AN/SPN-42, which has an X-band receiving system mounted contiguous with the basic Ka-band radar transmitting antenna. The unique feature of the augmentor is that it uses the AN/APN-154 beacon as its X-band transmitter. This is accomplished by coupling the output of the augmentor to the AN/APN-154 and triggering its modulator and transmitter. During the landing phase, it is necessary to manually place the AN/APN-154 into its ACL position. In this mode the AN/APN-154 receiver is disabled to ensure that X-band signals in the area will not trigger the AN/APN-154 transmitter during landing.

The radar beacon control panel (figure 6-2) is on the flight officer's right console. The POWER or STBY position can be used for radar beacon warmup; to preclude response to a premature or unintentional interrogation, the STBY (ACLS not selected) position should be used.

There are no cockpit displays for the beacon, although the ACLS TEST button will be illuminated if the beacon is responding during an ACLS approach. A self-check of the beacon ACLS mode is accomplished by depressing the ACLS TEST or performing an OBC. Either of these two use the receiver video processing circuits of the

RADAR BEACON PANEL

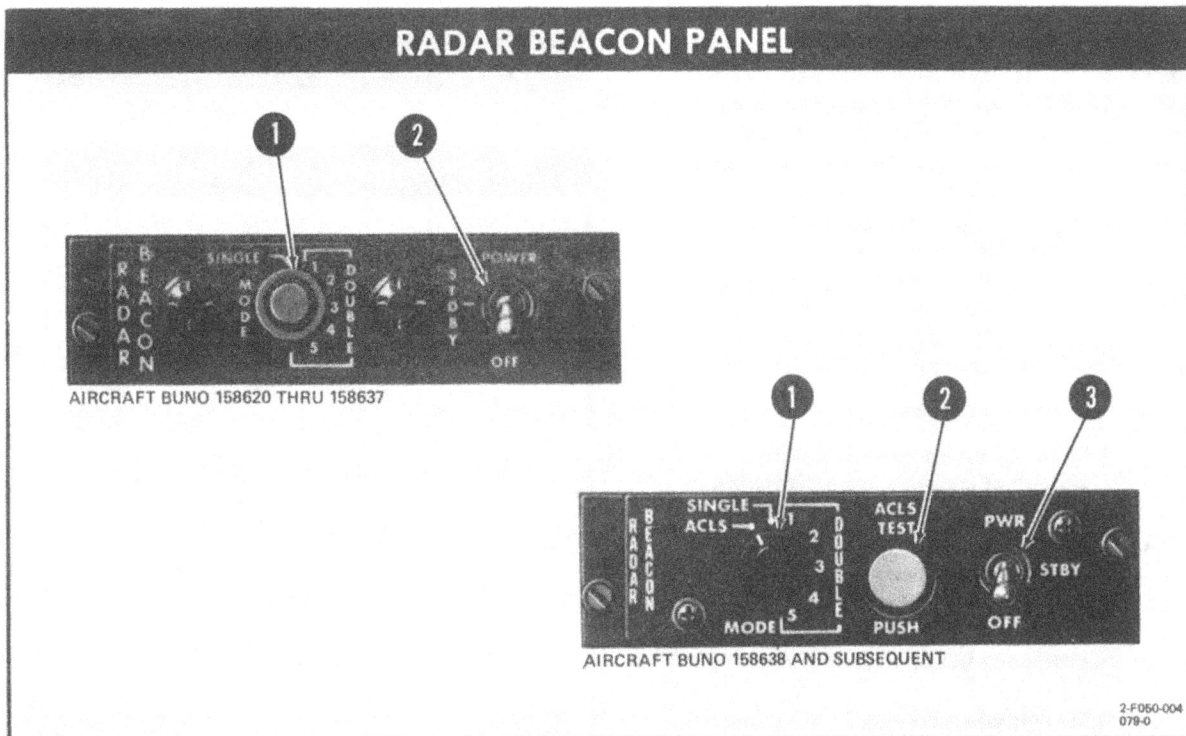

AIRCRAFT BUNO 158620 THRU 158637

AIRCRAFT BUNO 158638 AND SUBSEQUENT

2-F050-004
079-0

NOMENCLATURE	FUNCTION
① MODE switch	SINGLE — Limits beacon response to single pulse of any code group received. DOUBLE — Beacon response to one of five double-pulse interrogations. ACLS — Enables augmentor operation.
② ACLS TEST light/ pushbutton	On (green) — Indicates a AN/SPN-42 lockon in ACLS mode; when pressed with radar beacon mode selector in ACLS, indicates a satisfactory self-test of ACLS mode only. Flashing — Indicates AN/SPN-42 is sweeping through aircraft but has not locked on. Intermittent (or no light) — During self-test indicates a fault in the ACLS mode only.
③ PWR switch	PWR — With radar beacon mode selector in ACLS, enables X-band replies to Ka-band interrogations. STBY — With radar beacon mode selector in ACLS, enables X-band replies to Ka-band interrogations; used for warmup with radar beacon mode selector in SINGLE or DOUBLE. **Note** The beacon will warm up with the switch in either STBY or PWR position. To prohibit response to premature or unintentional interrogations, warmup should be accomplished in the STBY position. For optimum performance allow 5-minute warmup. OFF — Turns off all power to radar beacon.

Figure 6-2. Radar Beacon Panel

augmentor in the same manner as a Ka-band input from
the AN/SPN-42. If operation of the receiver is normal,
the ACLS TEST button on the radar beacon control
panel will illuminate. A BAG acronym will be displayed
when performing an OBC and in the event of a beacon
augmentor failure. The radar beacon has a minimum
warmup time of 5 minutes. During this time, failure
indications will be displayed and self-test results should
be regarded as inconclusive. A NO GO light during
OBC should be verified by depressing the ACLS TEST
button. If the ACLS test light illuminates the system is
functioning regardless of the NO GO light indication.

Note

- If the aircraft is parked on the flight
  deck aft of the island, the AN/APN-154
  should not be in the STBY or POWER
  positions, as stray energy from the
  AN/SPN-42 system can trigger beacon
  response and may degrade performance
  or preclude lock-on of aircraft
  attempting ACLS approaches.

- Do not depress the ACLS TEST button
  after coupling on a MODE I approach
  as it will cause the ground station to
  break lock.

- Do not attempt MODE I approaches
  while on MODE IA without beacon
  tracking.

ACLS Displays (VDIG And HUD)

Automatic carrier landing system (ACLS) and instrument
landing system (ILS) steering information can be
displayed on the VDIG and the HUD (figure 6-3). When
the AWL/PCD pushbutton is depressed, final determination of the display submode is governed by the AWL
switches on the pilot's display control panel, which
provides for separate ILS and ACLS selection for both
the HUD and VDI. This enables any mix of ILS
(AN/SPN-41/AN/ARA-63) or ACL (AN/SPN-42/data
link) displays at the pilot's option. Basically the two
displays are the same except for the driving source; ILS
or ACL. The one difference is that in the ACL mode,
the VDI displays a command heading marker. This
marker, during AN/SPN-42 approaches, indicates carrier
deck heading.

The ILS steering displays glide slope information in the
form of precision course vectors. A vertical vector is
used for azimuth steering while the horizontal vector is
for elevation. The pair form a crosspointer and are
displayed on the HUD and VDI simultaneously. On the

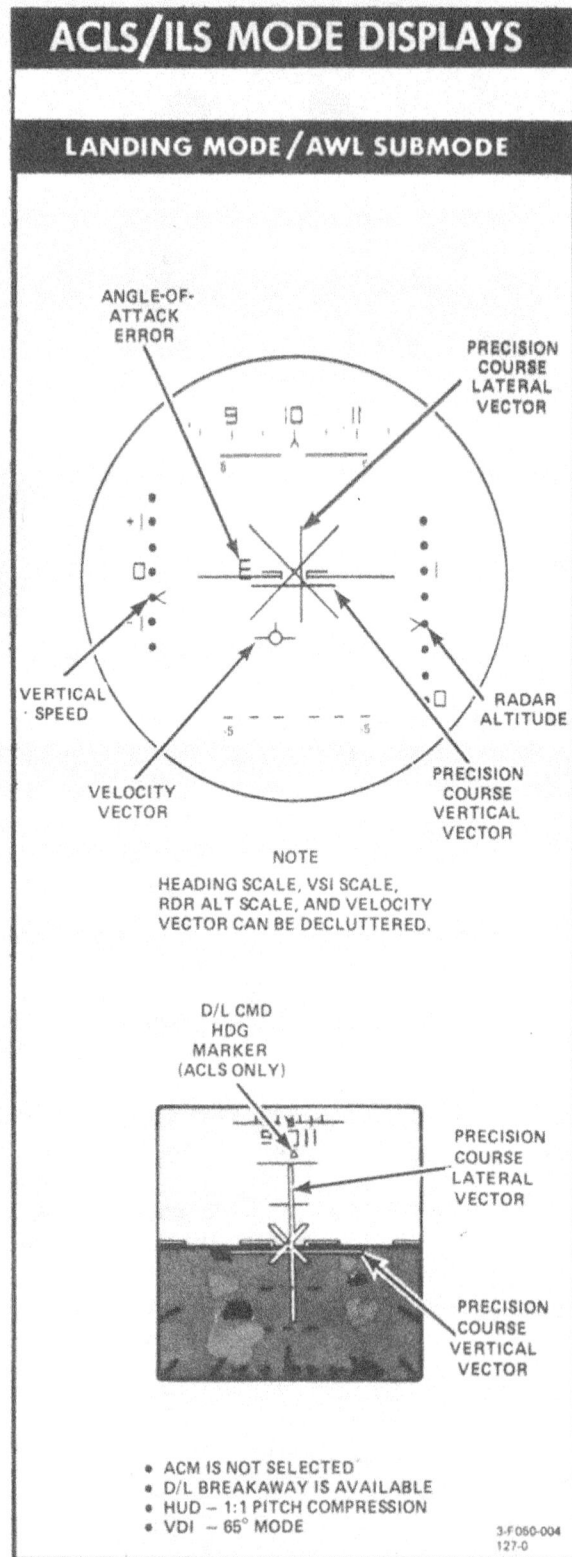

Figure 6-3. ACLS/ILS Mode Displays

HUD, full scale deflection will cause the vectors to deflect 2°. The vectors are limited to this deflection in order to insure that the displayed symbol will always have an intersection. The VDI vector symbols have a 1.5 inch full scale deflection.

The ACL submode also uses the pair of precision course vectors but they are driven by the data link instead of the AN/ARA-63A receiver decoder. Normal operating procedures for mode selection is to select ILS display for the HUD and ACL display for the VDI.

Additionally, certain ACLS commands which are uplinked to aircraft via the data link system are displayed to the R10 on the digital data indicator (DDI) and the pilot's repeater (figure 6-4).

Note

For more detailed information on the DDI refer to NAVAIR 01-F14AAA-1A.

Instrument Landing System (ILS) (AN/ARA-63A)

The aircraft instrument landing system (ILS) uses the AN/ARA-63A receiver decoder to process AN/SPN-41 information. This system is used for manual instrument landing approaches or as an independent monitor during final approach with the automatic carrier landing system (ACLS). The AN/ARA-63A control panel (figure 6-5) is located on the pilot's right side outboard console.

The aircraft system receives and decodes glideslope azimuth and elevation signals, which are converted into command fly-to indications in the CSDC and displayed on the VDI and/or HUD in the landing mode (figure 6-3). During ACL, the pilot normally displays ILS information on the HUD and selects automatic carrier landing (ACL) symbology on the VDI. If the ILS or ACL landing submodes selected on the pilot's display control panel becomes invalid, the VDIG will revert to a basic landing mode. The basic landing mode does not display the vertical and lateral cross-bars for glide-slope error and does not provide a waveoff signal.

Note

The ILS has a minimum warmup time of 1 minute. During this time, a failure indication should be disregarded.

The ILS performs a self-test when the BIT pushbutton on the AN/ARA-63 control panel is depressed and held. Response to the ILS self-test is displayed, providing the

HUD and VDI MODE switches are set to ILS. The correct ILS landing mode display on the HUD and VDI during system checkout shows the vertical precision course vector symbol slowly oscillating on the right side of the display, then on the left side. The horizontal precision course vector symbol remains stationary in the center of the display.

SURFACE SUBSYSTEMS

Automatic Landing System (AN/SPN-42)

The AN/SPN-42 radar uses a conically scanning antenna beam of Ka-bank energy, which is received at the aircraft in direct proportion to its position within the antenna coverage area. This microwave energy is received as amplitude modulation (AM) of the pulsed carrier and, by means of the beacon augmentor, the AM is put on the X-band beacon for retransmission back to the ship as an active radar signal. The AM on this retransmitted signal is therefore identical to the AM received at the aircraft. By relating the amplitude of the returned signal to the AN/SPN-42 antenna position within its conical scanning area, the system knows the exact location of the aircraft in relation to the axis of the conical scan, which is the desired glidepath. From this information, the system can generate corrections to bring the aircraft to the desired glidepath. Additional ramp input pitch commands, tailored to each specific ship or field by the Naval Air Test Center during Mode I certification, and applied at the proper time to assist the aircraft through the burble.

To satisfy the system capability and landing-rate requirements, the shipboard subsystem landing control central AN/SPN-42 has a dual-channel configuration. This provides increased system reliability through redundancy. At full operational capability, both channels are in use, controlling two aircraft on the glide slope at the same time. Two aircraft are normally spaced approximately 60-seconds apart along the glide slope. In addition, the three operating modes act as backups for each other should partial system failure occur.

Instrument Landing System (AN/SPN-41)

The aircraft instrument landing system (ILS) uses carrier or shore-based AN/SPN-41 (C-scan) transmitters. The system operates in the Ku-band, between 15.4 and 15.7 GHz, on any of 20 channels. The transmitted azimuth signal produces a 2° beam, which is scanned +20° from the deck centerline. The transmitted elevation signal produces a 1.3° beam with a scan pattern from 0° to 10°, locking up. A proportional azimuth angle for

ACL ADVISORY LIGHTS

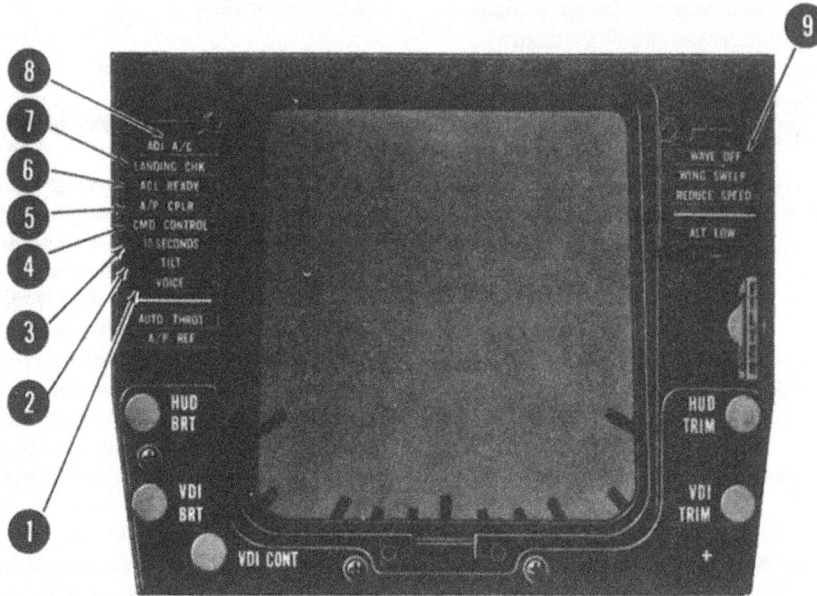

PILOT'S DDI REPEATER

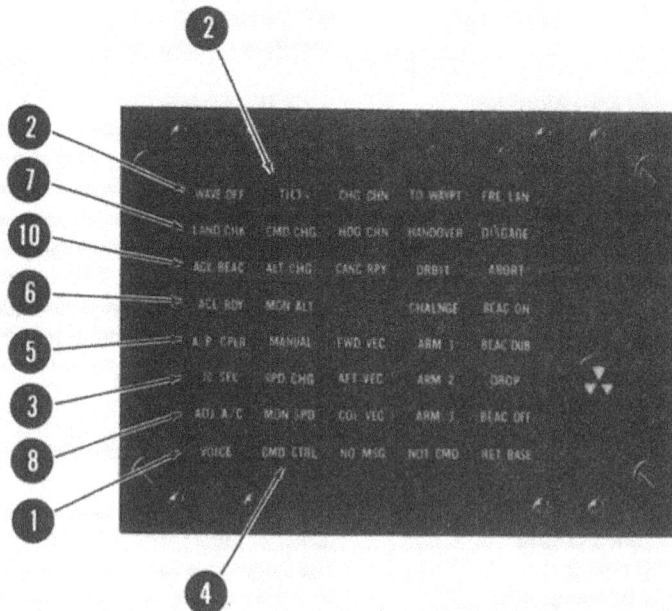

ADI A/C
LANDING CHK
ACL READY
A/P CPLR
CMD CONTROL
10 SECONDS
TILT
VOICE
AUTO THROT
A/P REF
HUD BRT
VDI BRT
VDI CONT
HUD TRIM
VDI TRIM

WAVE OFF
WING SWEEP
REDUCE SPEED
ALT LOW

RIO'S DIGITAL DATA INDICATOR (DDI)

WAVE OFF	TILT	CHG CHN	TO WAYPT	FRE LSN
LAND CHK	CMD CHG	HDG CHN	HANDOVER	DISEAGE
ACL BEAC	ALT CHG	CANC RPY	ORBIT	ABORT
ACL RDY	MON ALT		CHALNGE	BEAC ON
A/P CPLR	MANUAL	FWD VEC	ARM 1	BEAC DUB
10 SEC	SPD CHG	AFT VEC	ARM 2	DROP
ADI A/C	MON SPD	COI VEC	ARM 3	BEAC OFF
VOICE	CMD CTRL	NO MSG	NOT CMD	RET BASE

Figure 6-4. ACL Advisory Lights (Sheet 1 of 2)

1-F050-004
214-0

NOMENCLATURE	FUNCTION
① VOICE	Indicates CATCC (AN/SPN-42) not available for ACL. Standard voice commands to be used, and select ILS or TACAN steering for display.
② TILT	Indicates no D/L update for last 2 seconds during ACL mode. Disengages (automatically) the AFCS. For modes other than ACL, it indicates no message received during last 10 seconds, therefore the flight officer should change antenna position.
③ 10 SECONDS	During ACL, indicates ships motion is added to glidepath information and D/L commands. For other modes indicates 10 seconds or less before any action indicates for an arrival at a specific point in the approach path.
④ CMD CONTROL	Indicates aircraft is under D/L control for landing.
⑤ AP CPLR	Indicates CATCC is ready to control aircraft, autopilot should now be engaged.
⑥ ACL READY	Indicates aircraft has been acquired by CATCC and that glidepath information is being transmitted to aircraft for zero pitch and zero bank. Error symbols are now available on the VDI.
⑦ LANDING CHK	Indicates CATCC has a channel available for ACL and that aircraft should be prepared for carrier landing. Complete appropriate landing checklist. Positive D/L contact has been established.
⑧ ADJ A/C	Indicates that an aircraft is in (or close to) own aircraft's traffic pattern.
⑨ WAVE OFF	Indicates unsafe condition for landing: execute briefed wave-off procedures. Automatic AFCS disengagement and aircraft reverts to manual control.
⑩ ACL BEAC (RIO only)	Informs flight officer that CATCC requests that AN/APN-154 beacon should be turned on.

Figure 6-4. ACL Advisory Lights (Sheet 2 of 2)

AN/ARA-63A DECODER PANEL°

3-F050-004
080-0

NOMENCLATURE	FUNCTION
① CHANNEL selector	Twenty possible channel selections by rotation of selector knob.
② BIT PRESS-to-test button	Depressing button activates BIT test circuitry. Landing symbols available on VDIG if LDG MODE and AWL/PCD steering are selected on PDCP and ILS is selected on the HUD MODE or VDI MODE switches.
③ POWER switch (lock-lever)	ON — Activates receiver decoder for all-weather carrier landing. OFF — Turns system off. Lock-lever switch must be lifted to OFF.
④ Indicator light	Lights when AN/ARA-63A is on.

Figure 6-5. AN/ARA-63A Decoder Panel

steering is 6° right or left of centerline; proportional elevation angle for steering is 1.4° from the reference glide slope (above or below). Operating range is approximately 20 nautical miles. The signal is transmitted in J-band on a carrier frequency of 15.4 to 15.7 GHz.

The AN/SPN-41 can be used to guide the pilot to the window of the AN/SPN-42 radar for an ACL MODE I approach and as an independent monitor glide slope display during a MODE I approach. Should the AN/SPN-42 radar system fail, the AN/SPN-41 can be used for MODE II approaches.

LANDING PROCEDURE

The landing sequence can be divided into two distinct phases, approach and landing. Flight from the marshalling point to the radar acquisition window is referred to as the approach phase; radar acquisition to touchdown is defined as the landing phase.

Approach Phase

In ACL, the purpose of the approach phase is to get the aircraft to the acquisition window (figure 6-6). At the marshalling area, some 20-miles astern of the carrier, the aircraft about to land are stacked according to fuel status and other relevant parameters that determine landing priority, the ILS (AN/ARA-63) system is energized, and the proper channel and displays are selected. The pilot, in concurrence with the controller, has the option of choosing from three display submodes to aid him in reaching the radar acquisition window:

- Data Link vector

- TACAN

- AWL/PCD-ILS.

All are directly selectable on the pilot's display control panel. Switching between submodes requires a choice between VEC, TACAN, and AWL/PCD steering. If a submode selected becomes invalid, the steering information will cease. The pilot has the option of reselecting another landing display submode.

During the letdown from marshalling, an AN/SPN-42 channel is assigned to the aircraft and a computer program of aircraft control parameters are selected. A data link discrete message, (the first of a series to be transmitted) LANDING CHECK, is sent to the aircraft to initiate data link communications with Carrier Air

Traffic Control Center (CATCC) and to indicate to the pilot that an AN/SPN-42 channel is available. The aircraft will usually already be in a landing configuration upon receipt of LANDING CHECK.

DATA LINK VECTOR APPROACH

When VEC is selected, the data link vector display is added to the basic landing display (BLS). Command heading relative to the heading tape is added to the VDI, along with altitude and airspeed error symbols on the right and left side, respectively. When the double bar is above the reference bar, it means the aircraft is below commanded altitude or commanded speed. Data link vector information is available only for the approach phase; that is, to the radar acquisition window. When the aircraft is vectored (D/L vector commands) to the acquisition window, the pilot has to make a new submode selection for the descent phase. This is not the case with the TACAN and ILS submodes, as both TACAN and ILS information are available throughout landing, from marshalling to touchdown.

Note

In aircraft BUNO 159637 and subsequent and those aircraft incorporating CADC 1166 A/A, the command airspeed/March numeric is not displayed on the VDI. However, command airspeed error from data link is displayed.

TACAN APPROACH

The vertical precision course vectoring symbols (CVS) bar is used for TACAN deviation, along with carrier deck heading, through the data link, if available, on both the HUD and VDI. The receipt of the data link waveoff in any landing submode causes the breakaway symbol to be displayed.

AWL/PCD-ILS-APPROACH

ILS information from the AN/SPN-41 is available during both the approach and descent phase. Selection of AWL/PCD on the pilot's display panel enables vertical and lateral glide-slope error display. Final determination of the AWL/PCD mode is governed by the ILS/ACL selection, which provides for separate HUD and VDI selection.

If the pilot intends to make a MODE I landing he will normally display ILS (AN/SPN-41) information on the HUD (HUD selection switch to ILS position) and prepare the VDI for automatic carrier landing AN/SPN-42

ACLS MODE I AND II APPROACHES

BEFORE REACHING 6-MILE DME FIX

MARSHAL

1. ARA-63 . . . . SET
2. PILOT DISPLAY CONTROL PANEL
   a. MODE . . . . LDG
   b. HUD DECLUTTER . . . . OFF
   c. HUD AWL . . . . ILS
   d. VDI MODE . . . . NORM
   e. VDI MODE . . . . ACL
   f. HSD MODE . . . . NAV
   g. HSD ECM . . . . OFF
   h. STEER CMD . . . . TACAN OR VEC
3. APN-154
   a. MODE SELECTOR . . . . ACL
   b. POWER SWITCH . . . . STBY

4. AFCS STAB AUG
   a. PITCH . . . . ON
   b. ROLL . . . . ON
   c. YAW . . . . ON

DEPARTING MARSHAL

1. DESCENT AT 4000 ft/min AND 250 KIAS
2. REPORT
   a. MODEX
   b. COMMENCING
   c. STATE

PLATFORM 5000 FEET

1. SLOW DESCENT TO 2000 ft/min
2. AIRSPEED 250 KIAS
3. REPORT:
   a. MODEX
   b. PLATFORM

10-MILE DME FIX 1200 FEET

1. SPN-154 . . . . POWER
2. REPORT:
   a. MODEX
   b. 10-MILE GATE
3. TRANSITION TO LANDING CONFIG

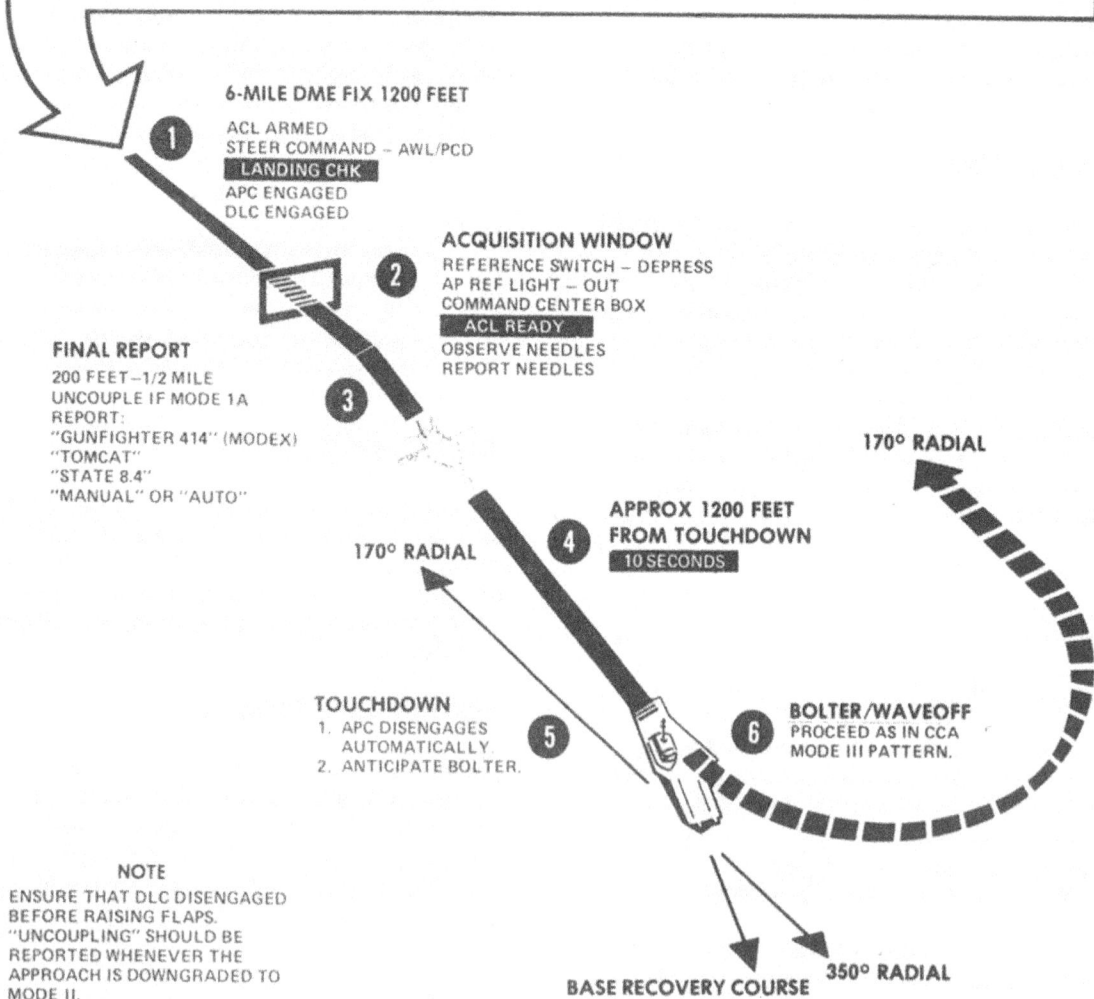

6-MILE DME FIX 1200 FEET
① ACL ARMED
STEER COMMAND – AWL/PCD
LANDING CHK
APC ENGAGED
DLC ENGAGED

ACQUISITION WINDOW
② REFERENCE SWITCH – DEPRESS
AP REF LIGHT – OUT
COMMAND CENTER BOX
ACL READY
OBSERVE NEEDLES
REPORT NEEDLES

FINAL REPORT
③ 200 FEET–1/2 MILE
UNCOUPLE IF MODE 1A
REPORT:
"GUNFIGHTER 414" (MODEX)
"TOMCAT"
"STATE 8.4"
"MANUAL" OR "AUTO"

170° RADIAL

APPROX 1200 FEET
④ **FROM TOUCHDOWN**
10 SECONDS

170° RADIAL

TOUCHDOWN
⑤ 1. APC DISENGAGES
AUTOMATICALLY
2. ANTICIPATE BOLTER.

BOLTER/WAVEOFF
⑥ PROCEED AS IN CCA
MODE III PATTERN.

BASE RECOVERY COURSE

350° RADIAL

NOTE
- ENSURE THAT DLC DISENGAGED BEFORE RAISING FLAPS.
- "UNCOUPLING" SHOULD BE REPORTED WHENEVER THE APPROACH IS DOWNGRADED TO MODE II.

5-F50-159-0

Figure 6-6. ACLS Mode I and II Approaches

data (VDI selection switch to ACL position) to be displayed as soon as it becomes available; that is, from the radar acquisition window on. Note that this configuration is desirable but not mandatory.

Landing Phase

As the aircraft continues its approach and passes through the 4-nmi ACLS radar acquisition window, a smooth transition, not requiring pilot action, occurs. If TACAN or ILS display information has previously been selected (for the approach phase), the pilot could use this information to land. Assume however, that AWL/PCD has been selected, ILS information is being displayed on the HUD and the VDI selection switch is in the ACL position. Until AN/SPN-42 data becomes available, the VDI displays BLS.

At the radar acquisition window, the AN/SPN-42 radar acquires the aircraft with the aid of the airborne radar beacon augmentor, and the system automatically sends a discrete indicating radar lock-on, which illuminates the ACL READY light. Transmission of vertical and lateral glide slope error, derived by the AN/SPN-42 radar, commences. These glide-slope error signals drive the precision course detector steering of the ACL display on the VDI while the HUD continues to display ILS (AN/SPN-41) error information.

The HUD and VDI symbology has thus been determined for the landing phase and no further pilot selection is required (unless a system malfunction occurs). The mode of operation for this phase of the landing is a function of the type of equipment used. In particular, there are three modes of landing applicable: Mode I, Mode II, and Mode III.

MODE I LANDING SEQUENCE

Note

Pending completion of system development, ACLS Mode I/IA approaches are not authorized. ACLS Mode II and III, and AN/SPN-41 or AN/ARA-63A ILS approaches are authorized.

The landing system (CATCC) (figure 6-5) generates a coupler available discrete which illuminates the A/P CPLR light and indicates that the pilot has the option of coupling the AFCS to data link commands of pitch and bank. At this time the aircraft should be in a landing configuration with APC and DLC engaged. The AWG-9 should be in STBY or PULSE search to avoid beacon interference problems. The AFCS should be armed in the ACL relief mode with the A/P REF

light illuminated; indicating that a pilot relief mode (in this case, ACL) has been selected but not engaged. The pilot can couple the AFCS to the data link by means of the autopilot engage button on his control stick, at which time, if the AFCS is functioning properly and the ACL interlock is true, the A/P REF light will go out and CATCC is automatically advised of engagement by the data link reply messages. The controller then sends a discrete message command control, which illuminated the CMD CONTROL light. The NTDS begins transmitting data link and pitch and bank commands to the aircraft. The autopilot (AFCS) actuates the appropriate control surface to execute the desired command, while the autothrottle (APC) maintains approach angle of attack by controlling the throttle setting.

Whenever the aircraft exceeds the MODE I flight-path control envelope the system automatically sends a signal to uncouple the AFCS (A/P CPLR light goes out). The approach may be continued in MODE II or MODE III. If the flightpath error increases to the point where a large maneuver is required to bring the aircraft back on course, the controller will send a waveoff message which is displayed on the VDIG and illuminates the WAVE-OFF light. This discrete also disconnects the autopilot (if engaged) and the AFCS reverts to stability augmentation. The controller then transfers the guidance of the aircraft to the bolter/waveoff controller, who directs the pilot back into the landing sequence.

If the information stored in the data link is not updated within any 2-second period during the descent, the TILT light illuminates (missed message) and the AFCS automatically disconnects and reverts to STAB AUG. The pilot can continue the descent in Mode II (ILS), or in Mode III.

At 12.5 seconds from touchdown (approximately 2,200 feet from the touchdown point) the 10 SECOND light illuminates indicating deck motion data are being added to the glidepath commands. This information is in the form of a slight increase (or decrease) in aircraft altitude to adjust for the movement of the touchdown point caused by the ship's motion (roll, pitch, and heave). Between 12.5 and 1.5 seconds from touchdown, the CATCC sends an automatic waveoff if any part of the carrier-based equipment fails and up to 5 seconds from touchdown if the aircraft exceeds the AN/SPN-42 flightpath control envelope. Waveoff signals also may be issued by the final controller (between lockon and touchdown) and the Landing Signal Officer (LSO) between 1-mile and touchdown. Approaches must be waved off at weather minimums (200-feet altitude and 1/2-mile visibility) if the pilot cannot see the meatball.

At 1.5 seconds from touchdown, the landing system freezes the pitch and bank commands and the AFCS holds the aircraft's attitude to touchdown, unless the pilot elects to override the AFCS either by maneuvering the control stick or by manually disengaging the AFCS and assuming control. If the aircraft bolters or if the pilot decides to go around, the AFCS is disengaged automatically by means of overriding the control stick, and the pilot enters the bolter/waveoff pattern.

Note

Use of paddle switch to disengage AFCS for Mode IA landing is not recommended as DLC is automatically stowed and must be reselected.

MODE II LANDING SEQUENCE

The early phases of a Mode II descent (figure 6-7) are identical to a Mode I descent sequence. The aircraft to be recovered is directed through the marshalling area, received LANDING CHK, and arrives at the ACLS radar acquisition gate. When the lock-on discrete (ACL READY) message is received, the pilot continues to fly the aircraft manually (using APC as desired) in response to VDIG and/or HUD displays.

If there is an equipment failure, the system (CATCC) will send a voice discrete signal, which illuminates the VOICE light, and the AN/SPN-42 error information displayed on the VDIG will be invalid. The pilot then expects to receive standard voice commands and will probably use the redundant ILS information or switch to TACAN display.

As long as the aircraft is located within the AN/SPN-42 flightpath control envelope for Mode II, the descent is continued until visual contact is made with the Fresnel Lens Optical Landing System (FLOLS) meatball. All waveoffs in Mode II are given by the final controller or the LSO. Approaches are terminated at weather minimums (200-feet altitude and 1/2-mile visibility) if the pilot cannot see the meatball.

At any time before 12.5 seconds from touchdown, the pilot can switch from a Mode II manual to a Mode I automatic flightpath control, providing the coupler available discrete is being received and the ACL interlock is true.

MODE III LANDING SEQUENCE

Mode III descents follow the same general sequence as that of Mode I and II, but Mode III approaches are talkdown landings; that is, all flightpath corrections are provided by voice and no computerized discrete signals are sent. The use of APC is optional. Approaches are terminated at the weather minimums if the FLOLS (meatball) is not visible to the pilot for continuing the landing.

SPN-41 ILS APPROACH

MARSHAL
ARA-63 – SET
PDCP – SET

1

170°

350°

250 KIAS
4000 ft/min

2

DEPARTING MARSHAL
"GUNFIGHTER 401" (MODEX)
"COMMENCING"
"STATE 10.5"

250 KIAS
2000 ft/min

3

PLATFORM 5000 FEET
"GUNFIGHTER 401"
"PLATFORM"

10-MILE DME FIX 1200 FEET

"GUNFIGHTER 401"
"10-MILES"
TRANSITION TO LANDING
CONFIGURATION
STEER COMMAND – AWL/PCD

4

5

ACQUISITION
WHEN REQUIRED BY CATCC,
REPORT NEEDLES (POSITION
OF GLIDE-SLOPE AND COURSE
DEVIATION SYMBOLS ON
HUD/VDI). FOR EXAMPLE,
"GUNFIGHTER 401 FLY UP,
FLY RIGHT".

6-MILE DME FIX 1200 FEET

APC ENGAGED (IF DESIRED)
DLC ENGAGED

6

APPROXIMATELY 3 MILES 1200 FEET

7

NEEDLES SHOULD BE
CENTERED, BEGIN
DESCENT

170° RADIAL

MEATBALL

8

"GUNFIGHTER 401"
"TOMCAT"
"BALL/CLARA"
"STATE 8.4"
"MANUAL" OR "AUTO"

170°

BOLTER/WAVEOFF
PATTERN

350° RADIAL

NOTE
ENSURE THAT DLC DISENGAGED
BEFORE RAISING FLAPS.

BASE RECOVERY COURSE

5-F50-158-0

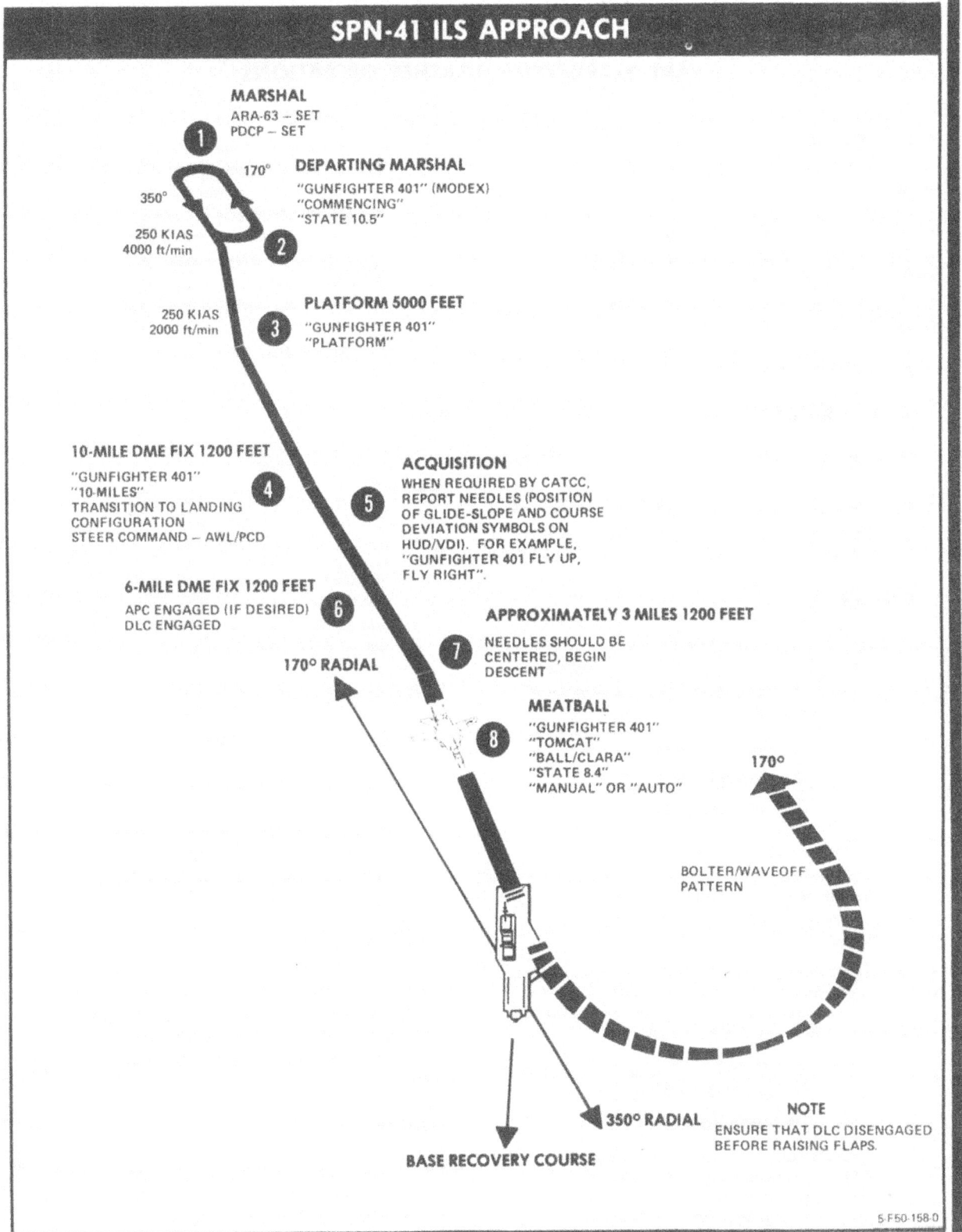

Figure 6-7. SPN-41 ILS Approach

PART 2 — EXTREME WEATHER OPERATIONS

ICE AND RAIN

Icing

Icing conditions should be avoided whenever possible. Before flight, check freezing levels and areas of probable icing from weather service. If ice starts to form on the windshield or wing leading edge, proceed as follows:

1. ENG PROBE ANTI-ICE switch . . . . ORIDE

2. CABIN AIR DEFOG lever . . . . FWD DEFOG

3. Windshield DEFOG switch . . . . MAX

4. Altitude . . . . .CHANGE AS REQUIRED
Climb or descend to an altitude where icing does not exist.

5. Engine instruments. . . . . . .MONITOR FREQUENTLY
Carefully monitor tachometer and turbine inlet temperature indicator. A reduction of RPM or an increase in TIT accompanied by a loss of thrust is an indication of engine icing.

WARNING

If turbine inlet temperature increases with loss of thrust, the throttle should be retarded. Low airspeed and high engine speed are conducive to engine icing.

A low approach into an area of moderate to severe icing should be considered an emergency approach. If there is any ice accumulation on the aircraft, an effort should be made to eliminate the ice before descending.

WARNING

Flight through light to moderate icing or visible frozen precipitation may cause heavy ice build-up above the no. 3 ramp, aft of the bleed door actuator. Field or carrier arrestment may dislodge this

accumulation and cause extensive engine FOD or failure. Flight through areas with these conditions should be avoided. If operational necessity requires flight in these conditions, or it is unavoidable and moderate to severe ice accumulation occurs, a straight-in field landing is preferred to an arrestment. Minimum power setting in flight and after landing is recommended.

Rain

Whenever rain is encountered, turn engine probe anti-ice switch to AUTO.

Note

In heavy rainfall, maintain a minimum engine power setting of 70% RPM. This will assure adequate acceleration margin and prevent possible engine speed hangup.

WINDSHIELD RAIN REMOVAL

Note

The chemical rain repellent is not a windshield washer and should not be used on a dry surface or during light rain. When applied to a dry windshield or during a light rain, a milky haze will form on the windshield surface.

TAKEOFF IN RAIN

Takeoffs performed with standing water on the runway may result in unstable engine operation due to water ingestion. Engine stall margin is reduced during engine transients, particularly afterburner staging.

LANDING IN RAIN

The AIR position of the WINDSHIELD switch controls a blast of air that blows rain off the windshield. Be aware of the possibility of flameout in a heavy rain and of reduced braking action due to a wet runway.

HYDROPLANING

Operations on wet or flooded runways may produce four conditions under which tire traction may be reduced to an insignificant value.

- Dynamic hydroplaning

- Viscous hydroplaning

- Reverted rubber skids

- Combined viscous and dynamic hydroplaning

Note

Hydroplaning has been experienced in the F-14A at speeds down to 40 knots.

Dynamic Hydroplaning

Dynamic hydroplaning is a condition in which a fluid separates the tires from the runway surface. When standing water on a wet runway is not displaced by the tire fast enough to allow contact over the complete footprint area of the tire, the tire rides on a wedge (or film) of water over all or part of the footprint area. Total dynamic hydroplaning occurs when the pressure between the tires and the runway lifts the tires off the runway surface to the extent that a nonrotating tire will not spin up (landing) or a rolling, unbraked tire will slow in rotation and may actually stop (takeoff). Total dynamic hydroplaning speed is represented by the following mathematical formulas: 9 times the square root of the tire inflation pressure for a rotating tire (as in takeoff); 7.7 times the square root of the tire inflation pressure for a nonrotating tire (as in landing).

Dynamic hydroplaning is insensitive to vertical load changes (weight), but is greatly affected by tire inflation pressure and tire wear. Since the fluid cushion is incapable of developing any appreciable shear forces, braking and sideforce coefficients become almost nonexistent.

Viscous Hydroplaning

Viscous hydroplaning occurs when the tires are separated from the runway surface by a thin film. Viscous fluid pressures in the tire-ground contact zone of rolling tires build up with speed to the danger levels required for hydroplaning only when water covered pavements are smooth or smooth acting, as when contaminants considerably more viscous than water coat the pavements. Since a tire operating on a surface with rubber deposits, paint, fuel, or oil can only partially displace the trapped water film, considerably higher hydroplaning pressures will be developed in the tire footprint area with these more viscous fluids. Even slight amounts of precipitation, e.g., a heavy dew that coats the pavement with a thin film of fluid, can produce this effect. Because the tire footprint separates from the runway with less fluid depth and at a lower relative ground speed than dynamic hydroplaning speed, viscous hydroplaning is potentially more dangerous than dynamic hydroplaning and is not greatly affected by changes in vertical tire load or tire inflation pressure. Grooved tires offer a greater advantage than smooth tires in reducing the effects of viscous hydroplaning. The runway pavement surface texture is also an important factor in combating viscous hydroplaning effects.

Combined Dynamic And Viscous Hydroplaning

Loss of tire friction with increasing or decreasing speed on wet or flooded runway pavements can be caused by the combined effects of viscous and dynamic hydroplaning. Figure 6-8 shows a pneumatic tire rolling at medium speed across a flooded pavement in a partial hydroplaning condition. The first zone shows the fraction of the tire footprint that is supported by bulk water (dynamic); the second zone, the fraction supported by a thin film of water (viscous); and the third zone, the fraction essentially in dry contact with the peaks of the pavement surface texture. The length of the first zone represents the time required for a rolling tire in this speed condition to expel bulk water from under the footprint; correspondingly, the length of the second zone represents the time required for the tire to squeeze out the residual thin water film remaining under the footprint after the bulk water has been removed. Since fluids cannot develop shear forces of appreciable magnitude, it is only in the third zone (essentially dry region) that fraction can be developed between the tire and the pavement for steering, decelerating, and accelerating a vehicle. The ratio of the dry contact area (third zone) to the total tire footprint area (zones 1, 2, and 3) multiplied by the coefficient the tire develops on a dry pavement, yields the friction coefficient the tire develops for this flooded pavement and speed condition. As speed is increased, a point is reached where the third zone disappears and the entire footprint is supported by either bulk water or a thin film. This speed condition is called combined viscous and dynamic hydroplaning. As speed is further increased, a point is reached where bulk water penetrates the entire footprint; this condition is called dynamic hydroplaning. If the runway is not flooded (no bulk water), such as on a runway covered

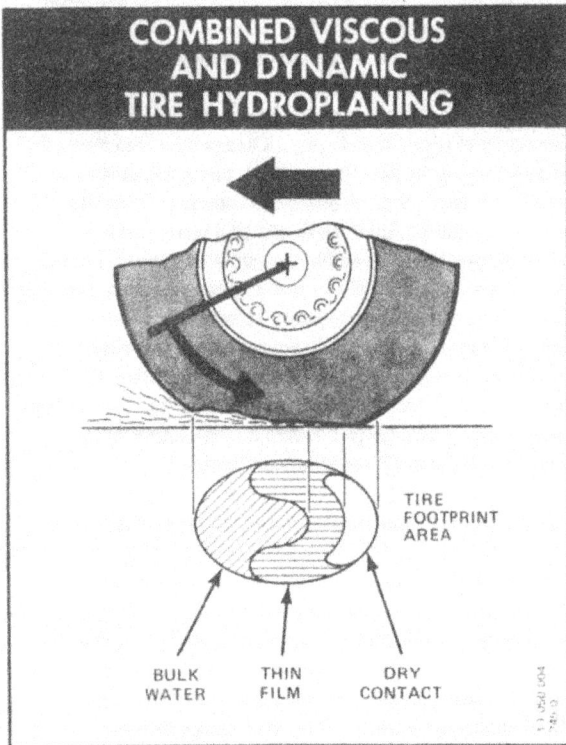

COMBINED VISCOUS AND DYNAMIC TIRE HYDROPLANING

TIRE FOOTPRINT AREA

BULK WATER   THIN FILM   DRY CONTACT

Figure 6-8. Combined Viscous and Dynamic Hydroplaning

with heavy dew, it is possible for the second zone to cover the entire footprint as speed is increased or decreased. The pavement would have to be smooth or smooth acting, as in the case where contaminants are present, for this to take place; this is called viscous hydroplaning.

Reverted Rubber Skids

A reverted rubber hydroplaning condition (also called reverted rubber skid) takes place when a wheel skid has started on a wet runway and enough heat is produced to turn the entrapped water to steam. The steam in turn melts the rubber in the tire footprint. The molten rubber forms a seal preventing the escape of water and steam. Thus, the tire rides on a cushion of steam which greatly reduces the coefficient of friction. On inspection of the portion of the tire involved, a patch of rubber would show signs of reverting to its uncured state and hence the name, reverted rubber. Once established, this condition may persist to very low ground speeds. The characteristic marks on a pavement for the reverted rubber skid are white, as opposed to the black marks left on the pavement during a dry skid. These white marks are associated with the cleaning process of super heated steam and high pressures that are present in the skid. The reverted rubber condition tends to make all runway surfaces smooth acting. Pavement surface texture, which has a large effect on traction losses from dynamic and viscous hydroplaning, has but little effect for the reverted rubber case with the possible exception of grooved surfaces. NASA research confirms the theory that the reverted rubber skid is the most catastrophic for aircraft operational safety because of the low braking friction and the additional fact that tire cornering capability drops to zero when the wheels rotation is stopped.

Landing On Wet Runway

Refer to Section III Landing Discussion.

TURBULENCE AND THUNDERSTORMS

Unless the urgency of the mission precludes a deviation from course, intentional flight through thunderstorms should be avoided to preclude the high probability of damage to the airframe and components by impact of ice, hail, and lightning. Flame-outs due to water ingestion or compressor stalls caused by rapid changes in flight attitudes could also occur. Radar provides a means of navigating between or around storm cells. If circumnavigating the storm is impossible, penetrate the thunderstorm in the lower third of the storm cell, away from the landing edge of the storm cloud, if possible. It is recommended that the auto pilot functions of the AFCS be disengaged. Structural damage could result with the automatic functions operating.

In The Storm

Maintain a normal instrument scan with added emphasis on the attitude gyro (VDI). Attempt to maintain a constant pitch attitude and, if necessary, accept moderate altitude and airspeed fluctuations. In heavy precipitation, a reduction in engine speed may be necessary due to the increased thrust resulting from water ingestion. If compressor stalls or engine stagnation develops, attempt to regain normal engine operation by momentarily retarding the throttle to IDLE then advance to the operating range. If the stall persists, shut down the engine and attempt to relight. If the engine remains stagnated at reduced power and the TIT is within limits, maintain reduced power until clear of the thunderstorm. While in the storm, the longitudinal feel trim, angle of attack, total temperature, windshield overheat, static pressure correction, and cabin pressurization systems may experience some abnormalities due to rain, ice, or hail damage. No difficulty should be encountered in maintaining control of the aircraft; however, the rapid illumination of numerous warning lights may be somewhat distracting to the pilot if he is not prepared.

IF NECESSARY TO PENETRATE A THUNDERSTORM:

1. Slow to between 275 to 300 KIAS.

2. ENG PROBE ANTI-ICE switch . . . . AUTO

3. AUTO PILOT switch . . . . OFF

4. Loose equipment . . . . SECURED

5. Tighten lamp belt and lock shoulder harness.

6. Cockpit lights . . . . ON BRIGHT

7. Fly attitude and heading indicators primarily while in extreme turbulence, because altimeter and airspeed will fluctuate.

Note

During severe icing conditions, the pilot can expect to lose airspeed indications even with the pitot heat on. GCI stations, if available, can aid the pilot with tracking assistance through thunderstorm areas.

Severe turbulent air at high altitudes may cause the inlet airflow distribution to exceed acceptable limits of the engine, thereby inducing compressor stalls. To avoid compressor stalls during flight due to turbulent air, maintain 275 to 300 KIAS at all altitudes.

COLD-WEATHER OPERATIONS

A careful preflight will eliminate many potential hazards found in cold weather operations. Inspect engine intakes for accumulation of ice and snow. If

possible, preheat the engine for easier engine starts. When removing ice and snow from the aircraft surfaces, be careful not to damage the aircraft. Also, use precautions not to step on any no-step surfaces which could be covered with ice or snow. Check the pitot-static tube for ice as well as the fuel pressurization ram/air intakes, and yaw, pitch, and angle-of-attack transducers.

Moisture in the fuel system greatly increases operational problems in cold weather. At lower temperatures, the water-dissolving capacity of fuel is greatly reduced and will result in considerable more water accumulation (as much as several gallons of water to 1,000 gallons of fuel). If the water separation occurs at below freezing temperatures, the water will crystallize on fuel drains and internal valves. Any water accumulation will settle to the bottom of the tanks and freeze up the fuel drains.

Normal operating procedures as outlined in Section III, Normal Procedures, should be adhered to with the following additions and exceptions:

Preflight

1. Check entire aircraft to ensure that all snow, ice, or frost is removed.

WARNING

Snow, ice, and frost on the aircraft surface are a major flight hazard. The result of this condition is a loss of lift and increased stall speeds.

2. Shock struts and actuating cylinders . . . . FREE OF ICE AND DIRT

3. Fuel drain cocks . . . . FREE OF ICE AND DRAIN CONDENSATION

4. Pitot tubes . . . . ICE AND DIRT REMOVED

5. Exterior protective covers . . . . REMOVED

Engine Start

Be sure that the aircraft is adequately checked before engine start.

Note

When operating in sub-freezing temperatures, moisture in the air entering the aircraft from the starting unit may freeze, causing ECS malfunctions. Starting the aircraft with the AIR SOURCE in the OFF position will prevent the problem. The AIR SOURCE in the BOTH ENG position should be selected after both engines have been started and the starter air disconnected. ECS malfunctions after engine start may still occur due to moisture internally present in the aircraft. If this occurs, select:

- TEMP mode selector switch — MAN

- TEMP control thumbwheel — Full hot (14)

- WSHLD DEFOG switch — MAX (if possible)

- Both engines at 75%. The ECS should thaw in about 20 minutes. During this warmup period leave all avionics and AWG-9 off.

In severely cold weather, allow a short time for warmup before increasing RPM out of the idle range. If oil pressure is low or fails to come up in a reasonable length of time, shut down. Attempt another start after heating the engines.

WARNING

If abnormal sounds or noises are present during starting, discontinue starting and apply intake duct preheating for 10 to 15 minutes.

Taxiing

Avoid taxiing in deep or rutted snow since frozen brakes will likely result.

To ensure safe stopping distance, and prevent icing of aircraft surfaces by melted snow and ice blown by jet blast of a preceding aircraft, increase spacing between aircraft while taxiing at sub-freezing temperatures.

Takeoff

When operating from runways which are covered with excessive water, snow, or slush, high-speed aborts may result in engine flame-out due to precipitation ingestion. The probability of flame-out is highest when throttles are chopped from afterburner to IDLE at speeds above 100 knots. With a double flame-out, normal braking, anti-skid and nosegear steering will be lost. Check applicable takeoff distance charts, Part 2 of Section XI.

Thrust available will be noticeably greater in cold temperatures during the take-off run.

CAUTION

Prior to initial take-off roll, ensure that all instruments are sufficiently warmed up. After takeoff, cycle landing gear a few times to prevent the possibility of the gear freezing in the wheel wells.

Landing

Use anti-skid during the landing roll.

Note

Hard braking on ice or wet runway, even with ANTI- SKID on, could result in dangerous skidding.

After Landing

During operations where the temperature is below freezing with heavy rain, or expected to drop below freezing with heavy rain, the aircraft may be parked with wings spread forward (20°), and flaps in the full down position.

Before Leaving Aircraft

Weather permitting, leave the canopy partially open to allow for air circulation. This will help prevent canopy cracking from differential cooling and decrease the possibility of windshield and canopy frosting.

HOT-WEATHER AND DESERT OPERATIONS

Check for accumulation of sand or dust in the intakes. Normal starting procedures will be employed.

Normal operating procedures as outlined in Section III, Normal Procedures, should be adhered to with the following additions and exceptions:

1. Expect higher temperatures than normally obtained in operating ranges.

2. Engine ground operation should be minimized as much as possible.

Taxiing

While taxiing in hot weather, the canopies may be opened, if necessary, to augment crew comfort. Do not operate the engines in a sand or dust storm, if avoidable. Park the airplane crosswind and shut down the engines to minimize damage from sand or dust.

Takeoff

The required takeoff distances are increased by a temperatuer increase. Check the applicable Takeoff Distance charts, Part 2 of Section XI.

CAUTION

Do not attempt takeoff or engine operation in a sand storm or dust storm, if avoidable. Park aircraft crosswind to prevent sand or dirt from blowing into the intake or exhaust ducts, and subsequently causing engine damage.

Landing

Anticipate a slightly longer landing distance and the possibility of turbulence due to thermal action of the air close to the ground. Use the defogging system if necessary, in warm, humid weather.

F-14A TOMCAT

THIS PAGE INTENTIONALY LEFT BLANK.

SECTION VII -- COMMUNICATIONS-NAVIGATION EQUIPMENT AND PROCEDURES

TABLE OF CONTENTS

PART 1 -- COMMUNICATIONS

COMMUNICATIONS AND ASSOCIATED EQUIPMENT

Figure 7-1 lists the communication-navigation-identification (CNI) equipment associated with the aircraft weapons system. For information on the AN/AWG-9 Weapons Control System, defensive electronic countermeasures (DECM) equipment, and data link system, refer to NAVAIR 01-F14AAA-1A.

```
CAUTION
```

Operation of electronic equipment for more than 5 minutes without adequate cooling will permanently damage the equipment.

COMMUNICATIONS ANTENNAS

Four UHF/L-band dual-blade antennas provide omnidirectional coverage for UHF voice. UHF data link, TACAN, and IFF/SIF transponder (APX-72) operation. TACAN and UHF voice communications use one set of antennas; the data link and IFF transponder, another set of antennas. Refer to GENERAL ARRANGEMENT

illustration (pages FO-1 and FO-2) for antenna locations. The IFF interrogator antenna (APX-76) is an integral part of the WCS antenna.

Each individual system is connected to the appropriate portion of an upper or lower antenna through a coaxial switch and duplexer. In aircraft BUNO 159468 and earlier without AFC 338, the COMM ANT switch on the integrated control panel may be used to select the upper or lower antenna manually or to select automatic actuation and lock-on of the first antenna to receive a usable signal for UHF voice communications. In aircraft BUNO 159588 and subsequent and those aircraft incorporating AFC 338, the UHF ANT switch must be used to select the upper or lower antenna manually; there is no automatic actuation function in these aircraft.

Note

To minimize mutual UHF communication/data link interference in aircraft BUNO 159468 and earlier without AFC 338, the COMM ANT switch on the integrated control panel should be used to manually select an antenna opposite to the data link antenna selected when data link is in use.

COMMUNICATIONS AND ASSOCIATED EQUIPMENT

TYPE AND DESIGNATION	FUNCTION	RANGE	OPERATOR	LOCATION OF CONTROLS
INTERCOM (LS-460)	PROVIDES VOICE COMMUNICATIONS BETWEEN CREW-MEMBERS AND BETWEEN COCKPIT AND GROUNDCREW, ALSO VARIOUS WARNING AND WEAPON TONES	WITHIN THE AIRCRAFT AND GROUND-CREW PERSONNEL	PILOT, RIO, AND GROUNDCREW PERSONNEL	PILOT AND RIO LEFT CONSOLE AND IN THE NOSEWHEEL WELL
TACAN AN/ARN-84(V)	DIGITAL NAVIGATION ALSO PROVIDES BEARING AND DISTANCE INFORMATION	LINE OF SIGHT UP TO 300 MILES, DEPENDING ON ALTITUDE	BOTH	PILOT RIGHT CONSOLE, RIO LEFT CONSOLE <hr> PILOT LEFT CONSOLE △2
UHF DATA LINK AN/ASW-27B	PROVIDES TWO-WAY DIGITAL MESSAGE COMMUNICATION	LINE OF SIGHT UP TO 180 NAUTICAL MILES	BOTH	RIO RIGHT CONSOLE
UHF COMMUNICATIONS SET AN/ARC-51	PROVIDES TWO-WAY VOICE COMMUNICA-TIONS AND ADF RECEIVER	LINE OF SIGHT UP TO 180 NAUTICAL MILES	BOTH	RIO LEFT CONSOLE <hr> RIO ONLY △1△2
UHF COMMUNICATIONS SET AN/ARC-159 △1△2△3	PROVIDES TWO-WAY VOICE COMMUNICATIONS	LINE OF SIGHT UP TO 180 NAUTICAL MILES	BOTH	PILOT LEFT CONSOLE △1△2△3 <hr> PILOT AND RIO LEFT CONSOLE △3
UHF AUXILIARY RECEIVER AN/ARR-69 △4	ADF AND AUXILIARY VOICE RECEIVER	LINE OF SIGHT UP TO 180 NAUTICAL MILES	BOTH	PILOT AND RIO LEFT CONSOLE
UHF DIRECTION FINDER AN/ARA-50	PROVIDES BEARING INFORMATION TO SELECTED UHF STATIONS	LINE OF SIGHT UP TO 180 NAUTICAL MILES	BOTH <hr> RIO ONLY △1△2	PILOT AND RIO LEFT CONSOLE <hr> RIO ONLY △1△2
VOICE SECURITY EQUIPMENT KY-28	CRYPTOGRAPHIC EN-CODING AND DECODING OF VOICE COMMUNI-CATIONS. USED WITH MAIN UHF COMMUNI-CATIONS RADIO(S)	LINE OF SIGHT UP TO 180 NAUTICAL MILES DEPENDING ON ALTITUDE	RIO	RIGHT CONSOLE <hr> LEFT CONSOLE △2
IFF TRANSPONDER AN/APX-72	RESPONDS TO INTER-ROGATIONS BY OTHER AIRCRAFT OR GROUND STATIONS	LINE OF SIGHT	RIO	RIGHT CONSOLE
IFF INTERROGATOR AN/APX-76A	REQUESTS IDENTIFI-CATION FROM OTHER AIRCRAFT	LINE OF SIGHT	RIO	DDD AND RIGHT CONSOLE
RECEIVER DECODER AN/ARA-63A	PROVIDES GLIDE-SLOPE SIGNALS FOR CARRIER LANDING SYSTEM	LINE OF SIGHT UP TO 20 NAUTICAL MILES	PILOT	RIGHT CONSOLE
RADAR ALTIMETER AN/APN-194	DISPLAYS HEIGHT ABOVE EARTH'S SURFACE	0 TO 5000 FEET	PILOT	RADAR ALTIMETER INDICATOR ON PILOT'S INSTRUMENT PANEL
RADAR BEACON AN/APN-154	AIDS IN TRACKING BY SHIP AND GROUND-BASED X-BAND RADARS. PROVIDES DOWN LINK FOR AUTOMATIC CARRIER LANDING SYSTEM	LINE OF SIGHT	RIO	RIGHT CONSOLE

△1 AIRCRAFT INCORPORATING AFC 338   △2 AIRCRAFT BUNO 159588 AND SUBSEQUENT   △3 AIRCRAFT BUNO 159637 AND SUBSEQUENT   △4 DELETED ON AIRCRAFT BUNO 159588 AND SUBSEQUENT

6-F50-125-o

Figure 7-1. Communications and Associated Equipment

The data link antennas are similarly selected manually: upper or lower antenna is selected by means of D/L ANT switches on either configuration of the data link control panels. In aircraft BUNO 159588 and subsequent and aircraft incorporating AFC 338, the ARC-159 antenna is shared with the data link antenna system and is always on the opposite antenna from the one selected by the D/L ANT switch.

The upper TACAN antenna is aft of the canopy on the turtle back, and the lower antenna is imbedded in the bottom left vertical fin. Only one antenna is used at a time. Automatic switching between antennas prevents loss of TACAN information. If a signal is lost or is too weak to hold receiver lock-up, the TACAN automatically cycles between the two antennas every 6 seconds seeking a stronger signal. During this cycling and search period, memory circuits retain range tracking for 8 to 12 seconds and bearing tracking for 8 seconds.

The IFF antenna lobing switch is controlled by the IFF ANT switch on the RIO's right outboard console. In the AUTO position, the lobing switch cycles the receiver-transmitter between upper and lower antenna pattern coverage. In the LWR (lower) position, only the lower antenna pattern is used to receive and transmit interrogation signals. The upper antenna pattern has a slight forward tilt; the lower pattern a slight aft tile.

Note

It is often necessary to select the LWR position to improve ground station reception.

INTERCOMMUNICATIONS (ICS)

The ICS provides normal, backup, or emergency communications between crewmembers. It also combines and amplifies audio signals received from other electronic receiving equipment (ECM, Sidewinder tone, IFF/SIF, radar altimeter, and voice radios, etc.

Identical ICS control panels (figure 7-2) are on the pilot and RIO left-side consoles. The ICS consists of four amplifiers, two at each cockpit station, which permit duplex operation during normal operation. If one amplifier fails, it may be bypassed by selecting either the B/U (backup) or EMERG (emergency) position on the ICS control panel. This permits continued ICS operation.

Note

If two amplifiers fail at the same station, intercommunication is impossible.

The external interphone connection is on the right side of the forward bulkhead in the nosewheel well. When either ICS switch is set to HOT MIC, ground personnel can communicate with the cockpit stations.

Audio Warning Signals

Audio warning signals from the weapon system are available to either or both crewmen through the ICS. Each signal has a distinct tone (figure 7-3). A visual display accompanies every audio signal so that the flight-crew can expect the tone and interpret its meaning. Most audio signals may be attenuated or turned off if not required, allowing the flightcrew to concentrate on more critical tones. Critical warning tones cannot be attenuated by any mode of ICS operation.

In aircraft BUNO 158637 and earlier and aircraft without AFC 561, the TONE VOLUME control panel (figure 7-4) has three volume controls for regulating audio signals from the APR-25, APR-27, and a Sidewinder missile lock on (SW). In aircraft BUNO 158978 and subsequent and earlier aircraft incorporating AFC 561, the TONE VOLUME control panel has three volume controls for regulating audio signals from the ALR-45, ALR-50, and the Sidewinder missile lockon (SW).

Note

- The EMERG position allows the pilot to use the RIO's amplifiers.

- The RIO can obtain a Sidewinder and engine overtemp tone by selecting the EMERG position on his ICS panel. This allows the RIO to use the pilot's amplifier.

Figure 7-3 provides a glossary of audio warning signals available within the aircraft weapon systems. Two 28 volts dc circuit breakers, ICS NFO (6F3) and ICS PILOT (6F2), control power to and provide circuit protection for the ICS. Power to both circuit breakers is from dc essential bus No. 1. Approximately 1 minute of warm-up is required in order to achieve normal operating temperature.

ICS System Checkout

When the CSDC, under OBC control, applies a BIT command to the DECM systems, they generate a tone in the crew headsets through the ICS. When the ICS performs this function, it transmits data on system performance back to the CSDC to be formatted. If a

NAVAIR 01-F14AAA-1

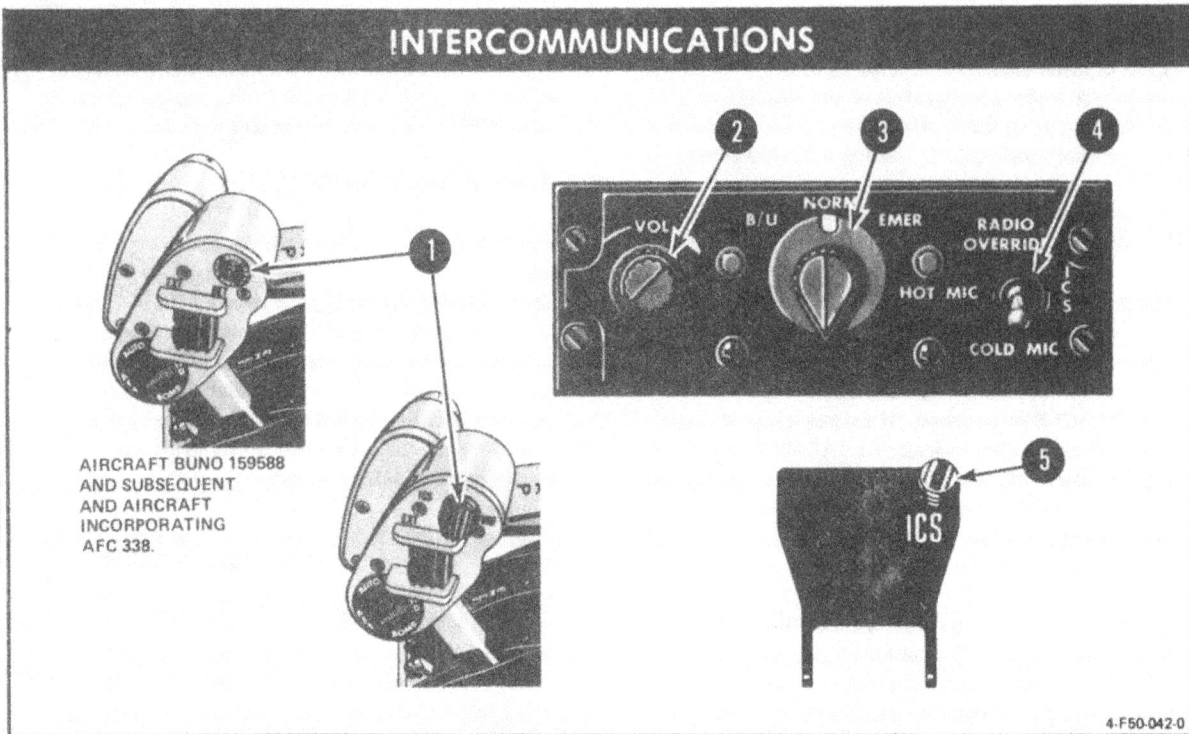

INTERCOMMUNICATIONS

AIRCRAFT BUNO 159588
AND SUBSEQUENT
AND AIRCRAFT
INCORPORATING
AFC 338.

4-F50-042-0

NOMENCLATURE	FUNCTION
① Pilot's UHF/ICS button	Momentary switch. Aft position permits intercommunication when COLD MIC is selected on function selector control. Additional positions permit transmissions on UHF No. 1, UHF No. 2, or both.
② VOL control	Controls intercommunication audio level at that cockpit station. Audio level at other station not affected.
③ Amplifier selector	B/U — (Backup) used to bypass a faulty amplifier and uses a backup output amplifier at own station.
	NORM — (Normal) used when all amplifiers are functioning properly.
	EMERG — (Emergency) used to bypass faulty amplifier, and makes use of input amplifier of other station. No HOT MIC.
	Note
	This feature allows reception of an ICS audio tone not normally routed to own station.
④ Function selector	RADIO OVERIDE — Attenuates noncritical radio audio to emphasize intercommunication when urgent.
	HOT MIC — Intercommunication without keying.
	COLD MIC — Intercommunication only when pilot actuates ICS keying switch on inboard throttle or flight officer actuates keying switch on left foot rest.
⑤ RIO's ICS button	Permits intercommunication if COLD MIC is selected on the function selector control.

Figure 7-2. Intercommunications

GLOSSARY OF TONES

TONE	CREW POSITION	CONTROLS	FUNCTION	CHARACTERISTICS
SIDEWINDER	PILOT	TONE VOLUME PANEL	MISSILE ACQUISITION	HIGH FREQUENCY, INCREASES IN INTENSITY WITH POSITIVE LOCKON
ALR-45	PILOT AND RIO	TONE VOLUME PANEL (PILOT) ECM PANEL (RIO)	THREAT RADAR WARNING	LOW TO HIGH FREQUENCY, DETERMINED BY SCAN RATE AND PRF OF THREAT RADAR
ALR-50 APR-27	PILOT AND RIO	TONE VOLUME PANEL (PILOT) ECM PANEL (RIO)	MISSILE LAUNCH WARNING	LOW- TO HIGH-FREQUENCY WARBLE WHEN TONE IS PRESENT
AN/ALQ-100 AN/ALQ-126 ⚠	RIO	DECM CONTROL PANEL	THREAT RADAR WARNING	RAW PRF SOUND
RADAR ALTIMETER	PILOT AND RIO	RADAR ALTIMETER INDICATOR (PILOT)	LOW-ALTITUDE WARNING	1000 Hz TONE, MODULATED AT 2 PULSES PER SECOND, LASTING FOR 3 SECONDS
APX-72	RIO	IFF CONTROL PANEL	VALID MODE 4 INTERROGATION	PRF OF INTERROGATION PULSE 2000 AND 6000 Hz
TACAN	PILOT AND RIO	TACAN CONTROL PANEL	STATION IDENTIFICATION	INTERNATIONAL MORSE CODE WITH THREE-LETTER DESIGNATION
AN/ARC-159	PILOT	UHF CONTROL PANEL (PILOT)	AIRCRAFT DF LOCATOR	1020 Hz
	PILOT AND RIO	BOTH UHF CONTROL PANELS ⚠2		
ENGINE OVERTEMPERATURE	PILOT	NONE	ENGINE TIT	MODULATED 320 Hz FOR 10 SECONDS MAXIMUM OR UNTIL FAULT IS REMOVED, WHICHEVER COMES FIRST.

⚠1 AIRCRAFT BUNO 161168 AND SUBSEQUENT   ⚠2 AIRCRAFT BUNO 159637 AND SUBSEQUENT

Figure 7-3. Glossary of Tones

TONE VOLUME CONTROL PANEL

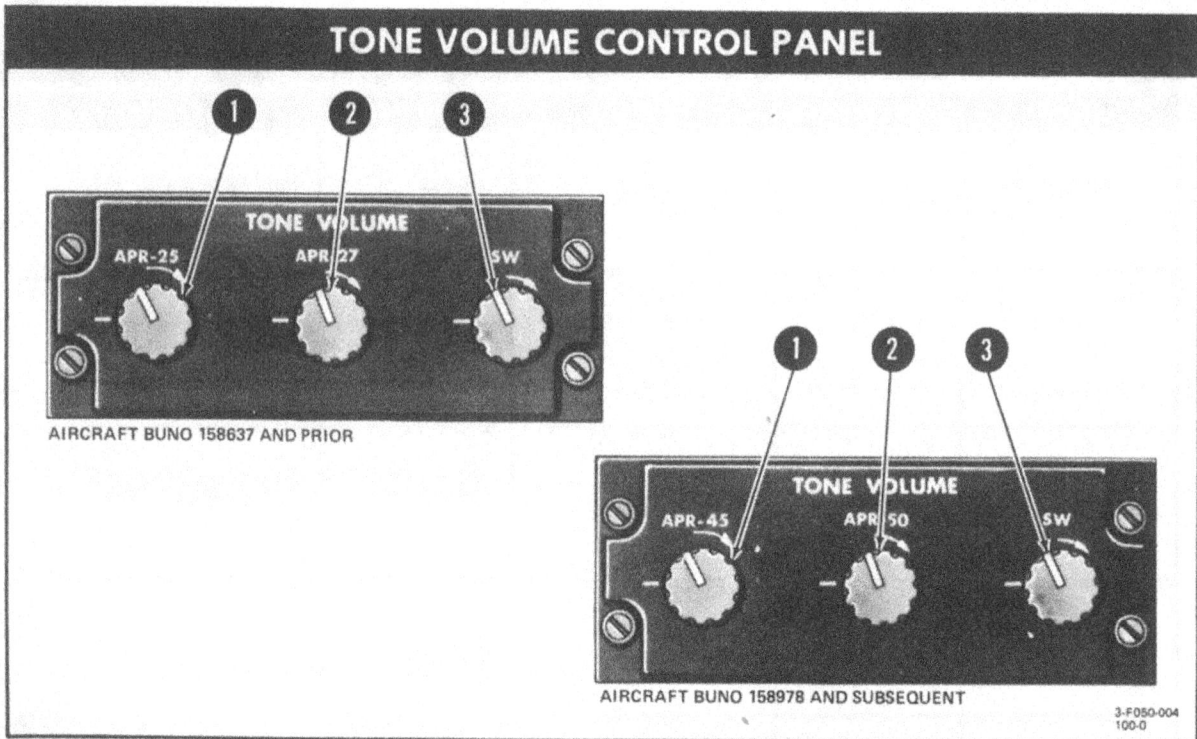

AIRCRAFT BUNO 158637 AND PRIOR

AIRCRAFT BUNO 158978 AND SUBSEQUENT

3-F050-004
100-0

NOMENCLATURE	FUNCTION
① ALR-45 volume control	Clockwise rotation increases tone in pilot's headset. Full counterclockwise rotation turns tone low.
② ALR-50 volume control	Clockwise rotation increases tone in pilot's headset. Full counterclockwise rotation turns tone low.
③ SW volume control (Sidewinder)	Clockwise rotation increases missile tone in pilot's headset. Full counter-clockwise rotation turns tone low.

Note

RIO can hear Sidewinder tone by
selecting EMERG ICS.

Figure 7-4. Tone Volume Control Panel

failure exists, the CSDC routes ICS failure data codes to the WCS, which displays the acronym of the failed unit on the TID and on the HSD.

DUAL UHF COMMUNICATIONS

In aircraft BUNO 159637 and subsequent, AN/ARC-159(V) and AN/ARC-159A(V)5 UHF radios are installed in the pilot's and RIO's cockpits, respectively; the pilot's UHF is referred to as UHF 1 and the RIO's UHF as UHF 2.

In aircraft BUNO 158612 through 159636 incorporating AFC 338 UHF communication consists of an AN/ARC-51A located on the RIO's left side console and an AN/ARC-159 located on the pilot's left side console. However, either crewman can listen to, or key, the other radio through the use of his UHF COMM Select panel.

In aircraft BUNO 158612 through 159468 without AFC 338, the AN/ARC-51 is the only type UHF radio installed in both cockpits.

Note

Transmissions on both UHF radios, while operating on the same frequency, may result in a squeal. This is a normal condition caused by r-f interaction between two radios operating on the same frequency in close proximity to each other.

UHF Radio (AN/ARC-51A)

The ARC-51A radio (figure 7-5) provides air-to-air and air-to-surface voice communications. Radio frequency range extends from 225.00 to 399.95 MHz. The equipment allows transmission and reception on any one of 20 preset channels and a guard channel (243.0 MHz). Guard frequency may be monitored simultaneously with any other frequency selected. Manual selection of 3,500 channels in 50-kHz increments is also provided. Dual antenna installations provide reliable line-of-sight communications up to 180 nautical miles (depending on altitude), with an average rf output of 20 watts. The ARC-51A radio may also be used with the automatic direction finder (AN/ARA-50) for navigation and with the KY-28 for secure voice communication. Plain or secure voice capability is provided by selecting either plain or cipher mode on the KY-28 panel. A jumper is provided for enabling the ARC-51A radio when the KY-28 is not installed.

Note

When the KY-28 is installed, the ARC-51A radio will not operate if the KY-28 power switch is OFF.

In aircraft BUNO 158612 to 159468, not incorporating AFC 338, identical ARC-51 control panels (figure 7-6) are installed at both the pilot's and RIO's left side consoles. A remote indicator on each instrument panel indicates the preset channel selected by the crewman who has commanded. The pilot's ARC-51A keying switch is on the inboard throttle grip; the RIO's keying switch is on the right foot well. The ARC-51A control panel is the integrated control panel which controls the ARC-51A receiver-transmitter, automatic direction finder, UHF auxiliary receiver, UHF data link transceiver, and antenna system of each of these units.

Note

- In aircraft incorporating AFC 338, an ARC-51 control panel is installed only at the RIO's left side console.

- A period of 5 minutes is required for normal warmup. During the warmup time failure indications should be disregarded.

In aircraft BUNO 158612 to 159468, not incorporating AFC 338, the UHF command pushbutton on the COMM/NAV CMD panel transfers control of the UHF function from one cockpit station to the other. Each crewmember may either give or take control; the active frequency is the frequency selected by the crewman who has control. The remote indicator at each cockpit station displays the present channel being used, G for guard channel, or OFF for manual frequency. The crewmember who has ARC-51A control is indicated by the green UHF pushbutton light on his COMM/NAV CMD panel. Actuation of controls on the integrated control panel is effective only if that cockpit has control; the volume control, however is individually controlled for each cockpit station and either ON/OFF switch will turn the set on. UHF and ICS volume are independent of each other and must be adjusted to permit reception of both simultaneously.

UHF Radio (AN/ARC-159)

The ARC-159 radio provides air-to-air and air-to-surface voice communications. Radio frequency range extends from 225.000 to 399.975 MHz. The equipment allows transmission and reception on any of the 20 preset channels and a guard channel (243.000 MHz). Guard frequency may be monitored

AN/ARC-51A UHF 2 CONTROL PANEL

AIRCRAFT BUNO 158612 THROUGH 159636
INCORPORATING AFC 338.

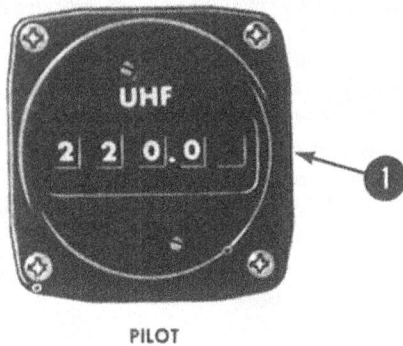

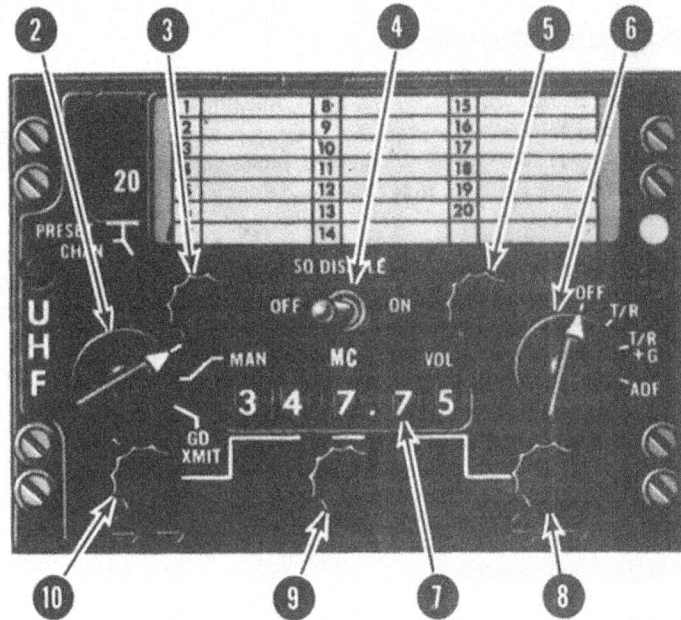

4-F50-041-2

NOMENCLATURE	FUNCTION
(1) Pilot's remote channel/ frequency indicator	Displays the channel selected by the RIO in PRESET CHAN mode, displays frequency selected in the MAN mode or G if GD XMIT is selected.
(2) Mode selector	PRESET — Permits selection of one of 20 preset UHF channels on preset channel selector.
	MAN — Permits selection of one of 3,500 possible frequencies (225.00 thru 399.95 MHz) selected by the manual frequency selector.
	GD XMIT — Permits rapid selection of guard channel.
(3) Preset channel selector	Permits selection of one of 20 preset channels. Mode selector must be in PRESET CHAN.
(4) SQ DISABLE switch	ON — Receiver noise is heard. Maximum sensitivity.
	OFF — Receiver is quiet.
(5) UHF VOL control	Rotary control, clockwise rotation increases main UHF audio level.
(6) Function selector	OFF — Secures radio
	T/R — Transmit and receive

Figure 7-5. AN/ARC-51A UHF 2 Control Panel (Sheet 1 of 2)

NOMENCLATURE	FUNCTION
	T/R + G — Transmit and receive; guard receive
	ADF — ADF receive
(7) Frequency selector indicator	Shows manual frequency selected from 225.00 to 399.95 MHz.
(8)(9)(10) Frequency selectors	The three rotary switches provide manual frequency selection when mode selector switch is set at MAN. The left switch is the tens and hundreds control and has 18 switch positions at 10 MHz steps from 220 to 390 MHz. The center switch is the unit digit control and has ten positions from 0 through 9. The right switch is the hundreds and tenths switch and provides 20 positions, 00 through 95. The difference between consecutive switch settings is 50 MHz.

Figure 7-5. AN/ARC-51A UHF 2 Control Panel (Sheet 2 of 2)

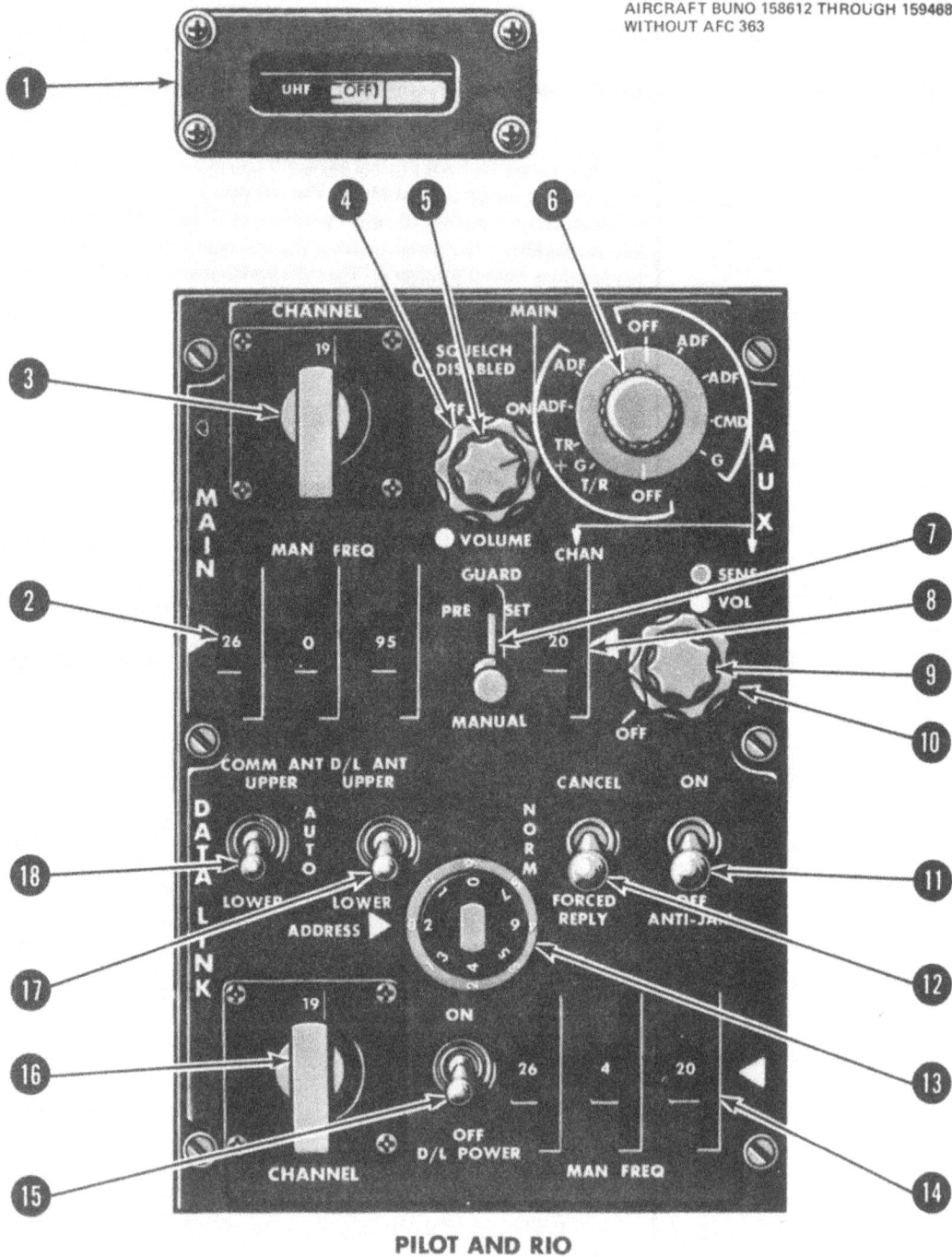

Figure 7-6. AN/ARC-51 UHF Control Panel (Sheet 1 of 2)

NOMENCLATURE	FUNCTION
(1) Remote channel indicator	Displays channel selected by crewmember who has UHF command if preset mode is used, or OFF in manual mode, or G if guard channel is used. Readout is identical at both cockpit stations.
(2) MAN FREQ selector	Manual UHF frequency set by rotation of three thumbwheels, provides 3,500 possible frequencies. GUARD-PRESET-MANUAL switch must be in MANUAL.
(3) Preset channel selector	Permits selection of one of 20 preset UHF channels. GUARD-PRESET-MANUAL switch must be in PRESET.
(4) UHF VOLUME control	Rotary control, inner knob; clockwise rotation increases main UHF audio level.
(5) SQUELCH DISABLED	Rotary control, outer knob; clockwise rotation increases UHF receiver sensitivity.

(6) Function selector (MAIN and AUX)

POSITION		CAPABILITY	
MAIN	AUX	MAIN	AUX
OFF	OFF	None	None
T/R	ADF	Transmit and receive	Receive ADF
T/R &G	ADF	Transmit and receive; guard receive	Receive ADF
ADF	CMD	ADF receive and UHF transmit with ADF interrupt.	Receive voice within frequency range
ADF	G	ADF receive; UHF transmit with ADF interruption.	Receive guard voice

(7) GUARD-PRESET-MANUAL frequency selector

GUARD	—	Permits rapid selection of guard channel.
PRESET	—	Permits selection of one of 20 preset UHF channels on the preset channel selector.
MANUAL	—	Permits selection of one of 3,500 possible UHF frequencies (225.0 thru 399.95) selected by the manual frequency selector, or manual selection of D/L frequencies.

(8) AUX CHANNEL selector — Thumbwheel selects one of 20 preset UHF AUX channels in 265 to 285 MHz range.

(9) AUX VOLUME control — Clockwise rotation of inner knob increases UHF AUX receiver audio level.

(10) AUX SENS control — Clockwise rotation of outer knob increases UHF AUX receiver sensitivity.

(11) ANTI-JAM switch

ON	—	All incoming messages are received including those with incorrect parity.
OFF	—	Messages with incorrect parity are rejected.

Figure 7-6. AN/ARC-51 UHF Control Panel (Sheet 2 of 2)

NOMENCLATURE	FUNCTION
(12) FORCED REPLY switch	CANCEL — Inhibits reply message transmission. NORM — Allows control message transmission.
(13) ADDRESS selector	Allows selection of last two digits of five-digit address. First three digits are preset by maintenance.
(14) MAN FREQ selector	Permits manual selection of data link operating frequency. Manual must be selected on channel selector.
(15) D/L POWER switch	ON — Applies power to data link communications set and DDI. OFF — Removes power from system.
(16) CHANNEL selector	Permits selection of one of 20 preset data link channels. Allows selection of manual frequencies.
(17) D/L ANT	UPPER — Selects upper data link antenna. LOWER — Selects lower data link antenna. AUTO — Automatically selects usable rf signal receiver of either antenna.
(18) COMM ANT switch	UPPER — Selects upper UHF antenna and lower D/L antenna. LOWER — Selects lower UHF antenna and upper D/L antenna. AUTO — Automatically selects usable rf signal receiver by either antenna.

Figure 7-6. AN/ARC-51 UHF Control Panel (Sheet 3 of 3)

simultaneously with any other frequency selected. The ARC-159 has a possible 7,000 frequencies available by manually tuning in 25-kHz steps.

The ARC-159 and the ARC-159(V)1 radios are solid-state, self-contained units with an average rf output of 10 watts. All controls for operation of the radio are on the front panel of the radio. The radio is located on the pilot's left console (figures 7-7 through 7-9). The ARC-159(V)1 has the additional capability of driving the remote indicator.

Note

- The ARC-159 and ARC-159(V)1 ADF positions are not functional; use the ADF mode of ARC-51A.

- The ARC-159 UHF 1 antenna is shared with the data link antenna system and always on the opposite antenna from the antenna position selected on the D/L ANT switch. To minimize mutual interference between UHF 1 and UHF 2, when using dual UHF communications capability, opposite antenna selection is recommended or frequency separation greater than 55 MHz. To minimize mutual interference between UHF communications and data link operation, when using data link capability, opposite antenna selection for UHF 2 and data link and a frequency separation greater than 55 MHz is recommended, along with turning UHF 1 or UHF 2 radio OFF. UHF communications interference with the data link may cause the TILT light to illuminate and the autopilot ACL or VEC/PCD mode to disengage. Data link interference with the UHF radios may cause audible chirping on the data link message reply rate.

The ARC-159A(V)5 UHF 2 radio is a 30-watt, solid-state receiver-transmitter with its control panel located on the RIO's left console (figure 7-10). Any frequency or channel selected by the RIO is displayed to the pilot on the UHF 2 remote indicator located on the pilot's instrument panel. All operating controls are on the front panel of the radio.

UHF Auxiliary Receiver (AN/ARR-69)
(Aircraft BUNO 158612 to 159468)

The UHF auxiliary receiver is primarily for ADF reception; its secondary use is for emergency voice reception. The AUX receiver provides ADF and voice on any one of 20 preset channels in the 265.0- to 284.9-MHz frequency range plus guard channel (243.0 MHz). It has line-of-sight range, varying with altitude. Controls are on the integrated control panels (figure 7-6). The auxiliary receiver audio is available only to the RIO; however, the volume controls are adjustable by each crewman. Approximately 1 minute warmup is required for normal operation.

Note

Cipher audio cannot be received with the UHF auxiliary receiver; it is incompatible with the KY-28.

UHF Automatic Direction Finder (AN/ARA-50)

The UHF automatic direction finder is used with the ARC-51A and ARC-159A(V)5 radio or with the AUX receiver. ADF provides relative bearings to transmitting ground stations or other aircraft. It can receive signals on any one of 20 preset channels or on any manually set frequencies in the 225 to 400 MHz range, depending on which receiver is used. The system is integrated with the appropriate receiver.

Note

In aircraft BUNO 159588 and sub-sequent and earlier aircraft incorporating AFC 338, ADF audio is available to the pilot and the RIO.

The system has a line-of-sight range, varying with altitude. Operating power is 115 volts ac from the essential no. 2 bus, 28 volts dc from the essential no. 2 bus, and 26 volts ac through the RIO circuit-breaker panels. The system requires a 5-minute warmup period. During the warmup time, failure indications should be disregarded. The system uses the ARA-48 ADF antenna. Bearing to transmitting stations is displayed on the pilot/RIO BDHI (No. 1 needle), pilot HSD, and RIO ECMD. The ADF signal is interrupted during UHF transmissions.

AN/ARC-159 UHF CONTROL PANEL

AIRCRAFT BUNO 158612 THROUGH 159636
INCORPORATING AFC 338.

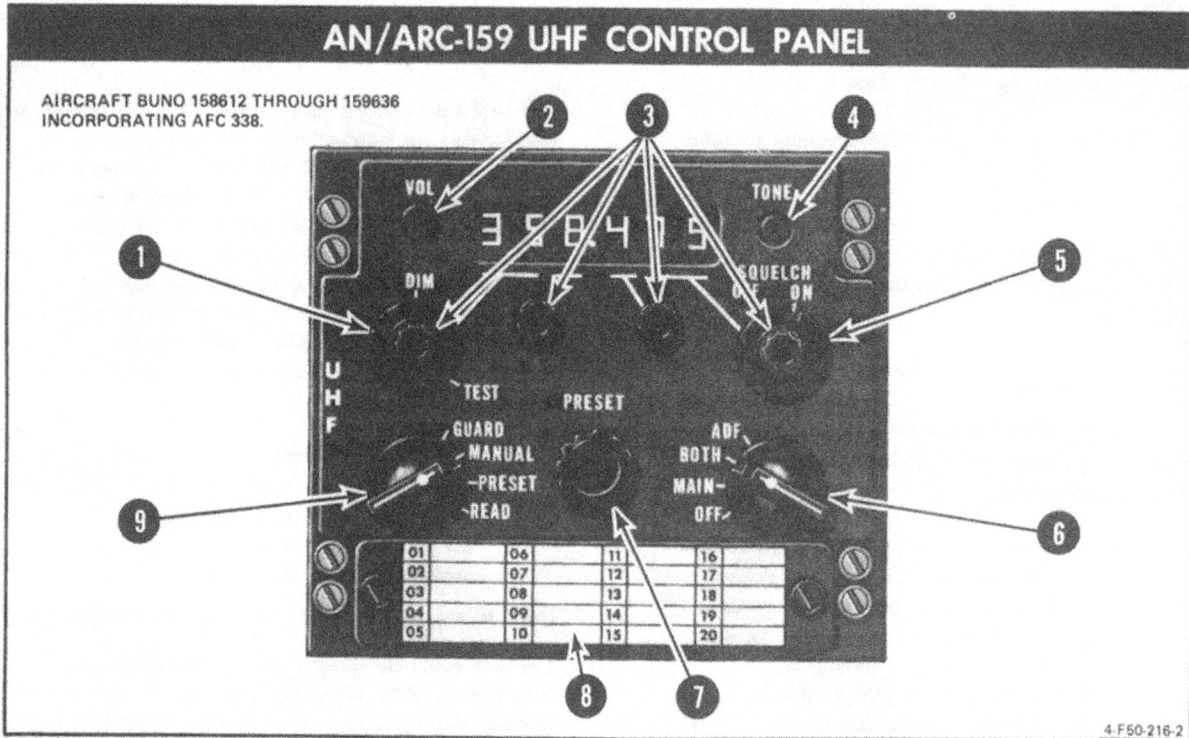

4-F50-216-2

NOMENCLATURE	FUNCTION
① DIM-TEST switch	DIM — Outer ring of the left frequency knob (3). Controls brightness of the frequency or channel readout. Full counterclockwise rotation provides the dimmest readout.
	TEST — Permits the frequency readout to be tested. The frequency 888.888 will appear.
② VOL control	Adjusts level of audio signal. Full counterclockwise rotation provides minimum volume.
③ Frequency tuning controls	Four frequency tuning controls are used to tune the transceiver when the mode selector switch is set to MANUAL. The left knob (inner control of the DIM switch) controls the hundreds and tens digits, the second knob controls units, the third controls tenths, and the right knob (inner control of the SQUELCH switch) controls hundredths and thousandths.
④ TONE pushbutton (spring return)	Depressing pushbutton causes a steady tone (1020 Hz) to be transmitted on the frequency or channel selected.
⑤ SQUELCH switch	Outer ring of the right frequency knob.
	ON — Squelch circuit is operational and background noise is removed by automatically reducing receiver gain.
	OFF — Disables the squelch circuit, restoring the receiver to full gain.

Figure 7-7. AN/ARC-159 UHF 1 Control Panel (Sheet 1 of 2)

NOMENCLATURE	FUNCTION
⑥ Function selector	OFF — Secures radio. MAIN — Main transceiver is energized, permitting normal transmission and reception. Receive or transmit function is selected by microphone push-to-talk switch. BOTH — Energizes both the main transceiver and GUARD receiver. ADF — The ARC-159 ADF function is not operational; use the ADF mode of the ARC-51A.
⑦ PRESET channel selector	Selects any one of 20 preset frequency channels when the tuning selector switch is set to PRESET.
⑧ Frequency chart	Used to record preset channel frequencies.
⑨ Mode selector switch	GUARD — Main transceiver is energized and shifted to guard frequency of 243.0 MHz permitting transmission and reception on the guard frequency. In this position both preset and manual frequency selections are not available. MANUAL — Frequency tuning controls are used to tune the main transceiver to any frequency (7,000 available) within the range of the set. The frequency selected is displayed in the readout window. In this position, PRESET selections are not available. PRESET — Used to tune the transceiver to any of 20 preset channels, using the preset channel selector. The selected channel is displayed on the readout window. READ — Permits the operator to read the frequency of the selected preset channel in the readout window.

Figure 7-7. AN/ARC-159 UHF 1 Control Panel (Sheet 2 of 2)

AN/ARC-159(V)1 UHF 1 CONTROL PANEL

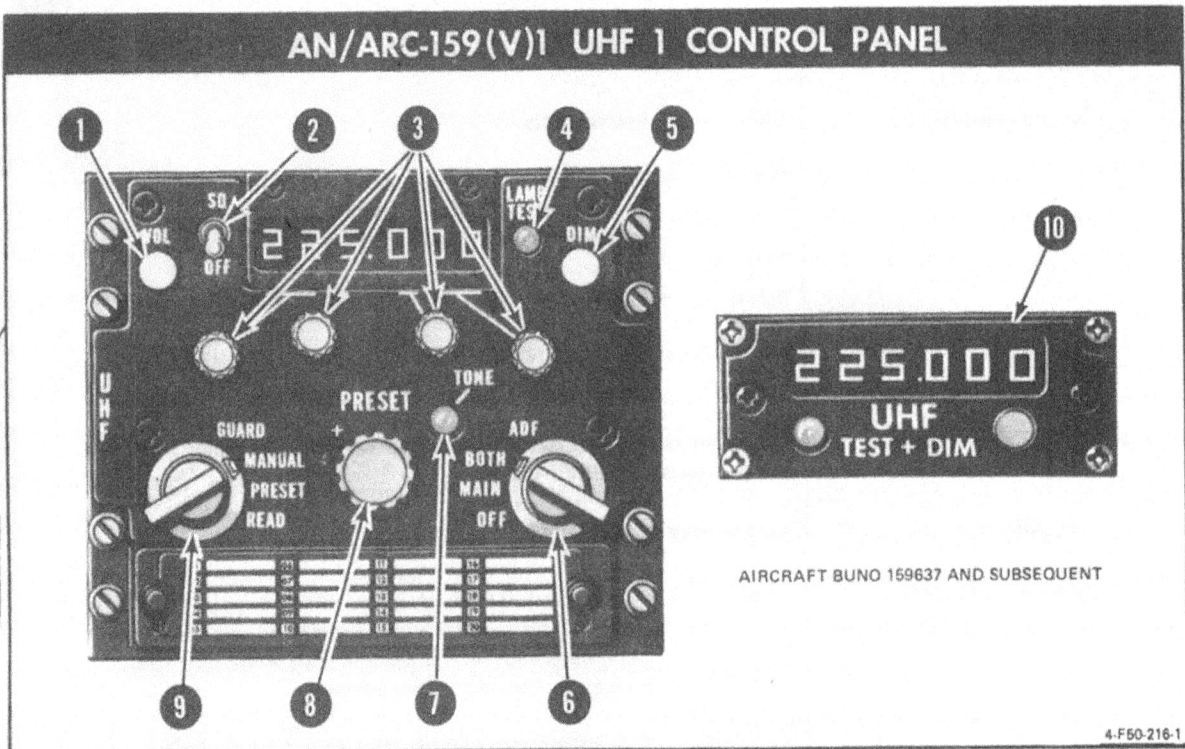

AIRCRAFT BUNO 159637 AND SUBSEQUENT

4-F50-216-1

NOMENCLATURE	FUNCTION
(1) VOL control	Adjusts level of audio signal. Full counterclockwise rotation provides minimum volume.
(2) SQ switch	SQ — Squelch circuit is operational and background noise is reduced by automatically reducing receiver gain.
	OFF — Disables the squelch circuit, restoring the receiver to full gain.
(3) Frequency tuning controls	Four frequency tuning controls are used to tune the transceiver when the tuning selector switch is set to MANUAL. The left knob controls the hundreds and tens digits, the second knob controls units, the third controls tenths, and the right knob controls hundredths and thousandths.
(4) LAMP TEST pushbutton (spring return)	Depressing pushbutton permits frequency readout to be tested. The frequency 888.888 will appear.
(5) DIM switch	Controls brightness of frequency or channel readout. Full counterclockwise rotation provides dimmest readout.

Figure 7-8. AN/ARC-159(V)1 UHF 1 Control Panel (Sheet 1 of 2)

NOMENCLATURE	FUNCTION
⑥ Function selector	OFF — Secures UHF 1 radio. MAIN — Main transceiver is energized, permitting normal transmission and reception. Receive or transmit function is selected by the microphone push-to-talk switch. BOTH — Energizes both the main transceiver and guard receiver. ADF — The ARC-159 ADF function is not functional; use the ADF mode of the ARC-51A or the ARC-159A(V)5.
⑦ TONE pushbutton (spring return)	Depressing pushbutton causes a steady tone (1020 Hz) to be transmitted on the frequency or channel selected.
⑧ PRESET channel selector	Selects any one of 20 preset frequency channels when the tuning selector switch is set to PRESET.
⑨ Mode selector switch	GUARD — Main transceiver is energized and shifted to guard frequency of 243.0 MHz permitting transmission and reception. In this position, both preset and manual frequency selections are not available. MANUAL — Frequency tuning controls are used to tune the main transceiver to any frequency (7,000 available) within the range of the set. The frequency selected is displayed in the readout window. In this position, PRESET selections are not available. PRESET — Used to tune the transceiver to any of 20 preset channels using the PRESET channel selector. The selected channel is displayed on the readout window. READ — Permits the operator to read the frequency of the selected preset channel in the readout window.
⑩ Remove channel/frequency indicator (2 each)	Displays pilot-selected frequency or channel to both the RIO and the pilot.

Figure 7-8. AN/ARC-159(V)1 UHF 1 Control Panel (Sheet 2 of 2)

AN/ARC-159 UHF CONTROL PANEL

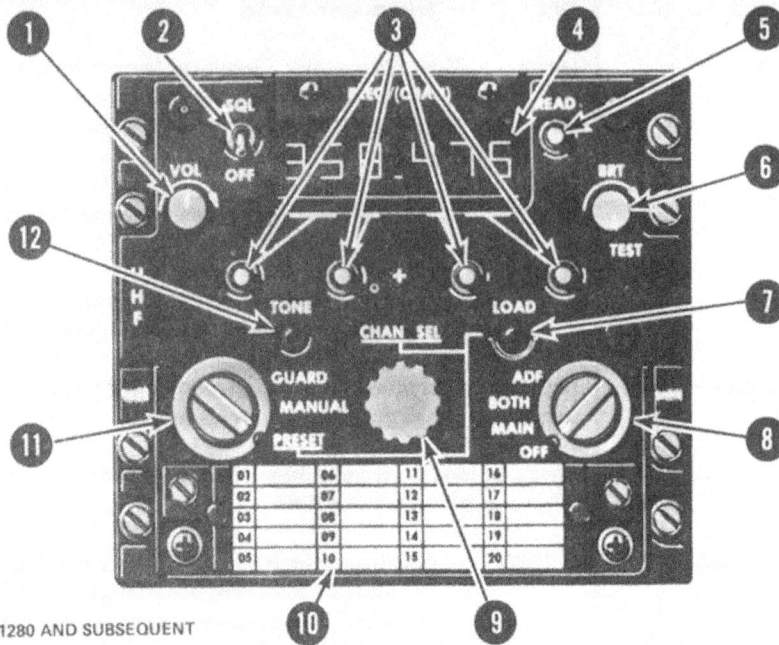

AIRCRAFT BUNO 161280 AND SUBSEQUENT

2-F50-216-4

NOMENCLATURE	FUNCTION
(1) VOL control	Adjust level of audio signal. Full counterclockwise rotation provides minimum volume.
(2) Squelch switch	SQL — Squelch circuit is operational, and background noise is removed automatically by reducing receiver gain. OFF — Disables the squelch circuit, restoring the receiver to full gain.
(3) Frequency tuning switches (spring return)	Four frequency tuning switches are used to tune transceiver when the mode selector switch is set to MANUAL. The left switch controls the hundreds and tens digits, the second switch controls units, the third switch controls tenths, and the right switch controls hundredths and thousandths. Forward deflection of the switch increases the numeric reading, and aft deflecting decreases the numeric reading.
(4) FREQ/CHAN	Displays frequency when mode selector switch is in MANUAL, or displays UHF channel when mode switch is in PRESET.
(5) READ switch	Deflection of the switch causes the frequency display to read the preset channel frequency.

Figure 7-9. AN/ARC-159 UHF   Control Panel (Sheet 1 of 2)

NOMENCLATURE	FUNCTION
⑥ BRT/TEST control	Clockwise rotation increases the brightness of FREQ/(CHAN) readout: counterclockwise rotation provides dimmest readout. Turn past full bright to read 888.888.
⑦ LOAD pushbutton (spring return)	Depressing pushbutton loads the display frequency of the preset channel.
⑧ Function selector	ADF — The ARC-159 ADF function is not functional; use the ADF mode of the ARC-51A or the ARC-159A(V)5. BOTH — Energizes both the main transceiver and the guard receiver. MAIN — Main transceiver is energized permitting normal transmission and reception. Receive or transmit function is selected by the microphone push-to-talk switch. OFF — Secures UHF 1 radio.
⑨ CHAN SEL control	Selects any one of 20 preset frequency channels when the tuning selector switch is set to PRESET.
⑩ Frequency chart	Used to record preset channel frequencies.
⑪ Mode selector switch	GUARD — Main transceiver is energized and shifted to guard frequency of 243.0 MHz permitting transmission and reception. In this position, both preset and manual frequency selections are not available. MANUAL — Frequency tuning controls are used to tune the main transceiver to any frequency (7,000 available) within the range of the set. The frequency selected is displayed in the readout window. In this position, PRESET selections are not available. PRESET — Used to tune the transceiver to any of 20 preset channels using the PRESET channel selector. The selected channel is displayed on the readout window. READ — Permits the operator to read the frequency of the selected preset channel in the readout window.
⑫ TONE pushbutton (spring return)	Depressing pushbutton causes a steady tone (1,020 Hz) to be transmitted on the frequency or channel selected.

Figure 7-9. AN/ARC-159 UHF 1 Control Panel (Sheet 2 of 2)

AN/ARC 159A(V)5 UHF 2 CONTROL PANEL

AIRCRAFT BUNO 159637 AND SUBSEQUENT

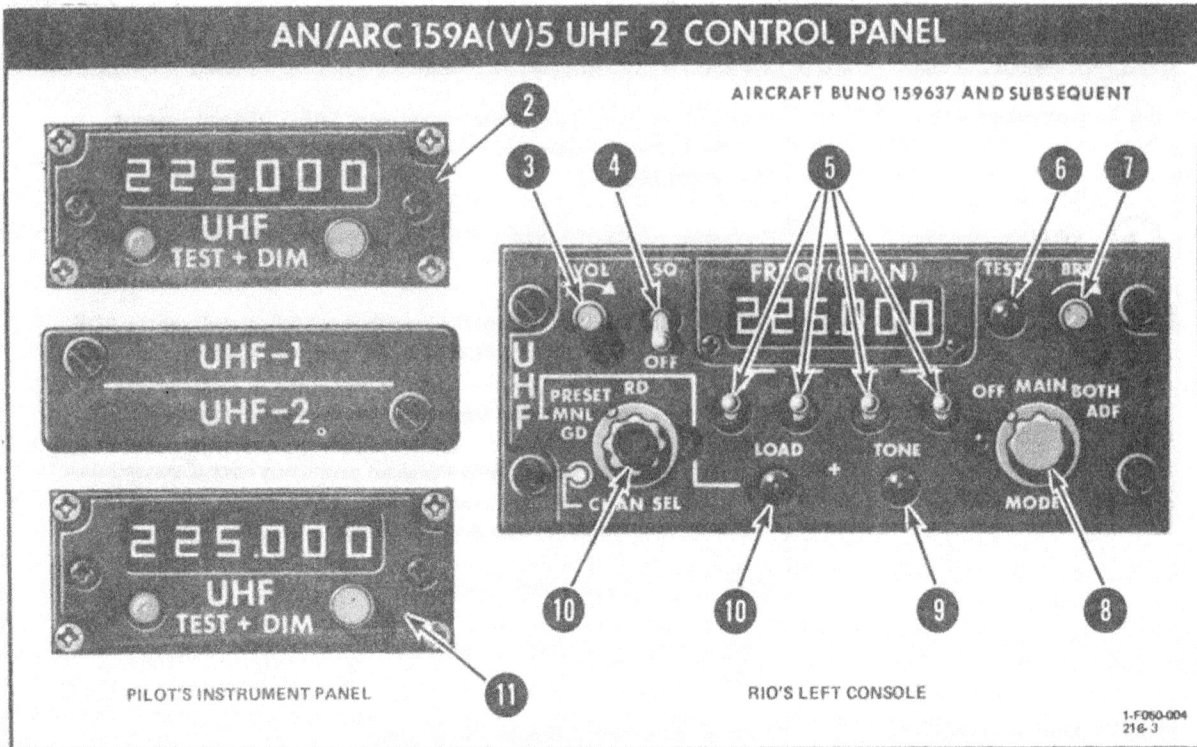

PILOT'S INSTRUMENT PANEL

RIO'S LEFT CONSOLE

1-F050-004
216-3

NOMENCLATURE	FUNCTION
① UHF 2 remote channel/ frequency indicator	Displays frequency or channel selected by RIO to the pilot.
② UHF 1 remote channel/ frequency indicator	Displays frequency or channel selected by pilot to the pilot.
③ VOL control	Adjusts level of audio signal. Full clockwise rotation increases UHF audio level.
④ SQ switch	SQ — Squelch circuit is operational and background noise is reduced by automatically reducing receiver gain. OFF — Disables the squelch circuit, restoring the receiver to full gain.
⑤ Frequency tuning switches (spring return)	Four frequency tuning switches are used to tune transceiver when tuning selector switch is set to MNL (manual). The left switch controls the hundreds and tens digits, the second switch controls units, the third switch controls tenths, and the right switch controls hundredths and thousandths.
⑥ TEST switch (spring return)	Depressing pushbutton causes the frequency display to read 888.888.

Figure 7-10. AN/ARC-159A(V)5 UHF 2 Control Panel (Sheet 1 of 2)

NOMENCLATURE	FUNCTION
(7) BRT switch	Controls brightness of frequency or channel readout. Full counterclockwise rotation provides the dimmest readout.
(8) Function selector	OFF — Secures UHF 2 radio.
	MAIN — Main transceiver is energized, permitting normal transmission and reception. Receive or transmit function is selected by microphone push-to-talk switch.
	BOTH — Energizes both main transceiver and guard receiver.
	ADF — Automatic direction finding equipment associated with UHF 2. Both main and guard receiver are enabled.
(9) TONE pushbutton (spring return)	Depressing pushbutton causes a steady tone to be transmitted on the frequency selected.
(10) LOAD pushbutton (spring return)	Depressing pushbutton loads the display frequency on the selected preset channel.
(11) Mode selector switch	GD — Main transceiver is energized and shifted to guard frequency (243.0 MHz) permitting transmission and reception. Preset and manual frequency or channel selections are not available in this position.
	MNL — Frequency tuning switches are used to tune main transceiver to any frequency within range of the set. Frequency selected is displayed in readout window.
	PRESET — Used to tune the transceiver to any one of 20 preset channels. Channel selected is displayed in readout window. To preset a channel: select desired channel to be tuned, select MNL, slew in frequency desired, and depress LOAD pushbutton.
	RD — Permits the operator to read the frequency of the selected preset channel in the readout window.

Figure 7-10. AN/ARC-159A(V)5 UHF 2 Control Panel (Sheet 2 of 2)

UHF Communication Selection Panel

Aircraft BUNO 159468 and earlier incorporating
AFC 338 use the UHF COMM SELECT panel (figure
7-11) to select either UHF 1 (ARC-51A) or UHF 2
(ARC-159) as the operating communications system.
Individual COMM SELECT panels are on the left console
in both the pilot and RIO cockpits. The XMT SEL
switch points to the desired UHF radio selected for audio
and transmission. The VOL knob controls the listener's
audio level of the UHF radio installed in the other
cockpit. A two position selector button on the inboard
throttle grip (figure 7-10) allows the pilot to select UHF
or ICS.

In aircraft BUNO 159588 and subsequent and earlier
aircraft incorporating AFC 338, the RIO has the
UHF 1 volume control, XMTR SEL toggle switch,
and the UHF 2 ANTENNA selection on his COMM/
TACAN panel on his left console (figure 7-11). The
pilot has a volume control knob for UHF 2 on his
COMM/TACAN panel (figure 7-12). A four-position
selection button on the inboard throttle grip (figure
7-11) allows the pilot to select UHF 1, UHF 2, both
UHF radios, of ICS.

COMM/NAV Command Control Panel

In aircraft BUNO 158612 to 158637 without AFC 338,
individual communication/navigation command
(COMM/NAV CMD) control panels (figure 7-12), on
the left side console at each cockpit station, transfer
TACAN, DATA LINK, and UHF control between the
pilot and RIO. Control is indicated by a green light in
the center of the pushbutton selected. The frequency
or radio channel selected by crewmen in control of the
particular radio is always available to both aircrew
positions. Thus, either crewman can switch between
the two radio channels by changing the command
position without reselecting radio frequencies.

In aircraft BUNO 158612 to 159468 with AFC 338
incorporated, the data link and UHF functions are
deleted.

In aircraft BUNO 159588 and subsequent and earlier
aircraft incorporating AFC 338, the pilot's TACAN
control panel (figure 7-12) provides for TACAN
transfer control between pilot and RIO and a
volume control knob for UHF 2. The RIO's TACAN
control panel (figure 7-12) provides selection of either
UHF radio, both UHF radios, UHF 2 antenna selection,
UHF 1 volume control, and TACAN transfer control
between pilot and RIO.

Voice Security Equipment (TSEC/KY-28)

The KY-28 is integrated, and operates, with the UHF
communications sets to permit secure voice in a hostile
environment. It shall be operated as directed by
appropriate authority. Theory of operation and practical
application are covered in the operation manual,
KAO-124B/TSEC.

The KY-28 control panel (figure 7-13, sheets 1 and 2)
on the RIO's left side console has the only cockpit
controls for operating the UHF receiver-transmitter in
either cipher or plain language modes. Electrical
power is from the dc essential bus no. 2 with circuit
protection on the RIO's dc essential no. 2 circuit breaker
panel (7C7).

The KY-28 has two basic modes of operation: plain (P)
and cipher (C). The plain mode is used during normal
UHF communications. The cipher mode is used when
secure voice communications are desired. The UHF set
must be ON to attain KY-28 operation. The receiving
station must be properly equipped to receive transmissions
in the cipher mode.

KY-28 OPERATION

PRELAUNCH. (Aircraft BUNO 159468 and prior
aircraft not incorporating AFC 338)

1. Determine that the proper code has been set in
   the KY-28 equipment by qualified
   personnel.

2. UHF radio .... ON

3. Power switch .... ON

4. Cipher/plain switch .... P

5. If a ground test of the equipment is desired,
   after a 2-minute warmup period, establish
   two-way plain text UHF radio communications
   with a suitable ground station and request an
   equipment check.

6. Cipher/plain switch C

   a. Listen for a steady, unbroken tone in the
      headset, followed by:

   b. A double-pitched broken tone.

UHF COMMUNICATION SELECTION PANEL

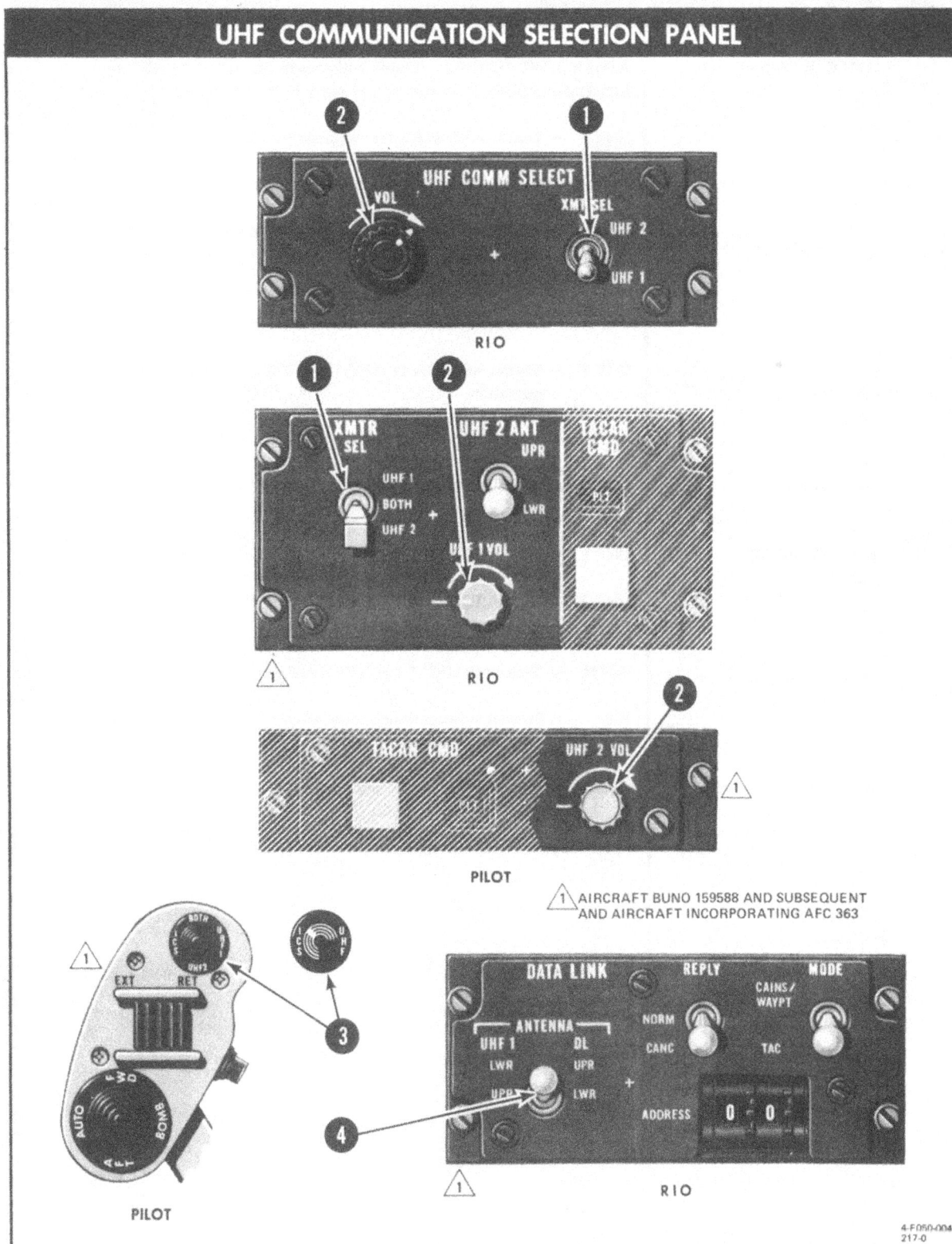

⚠️ AIRCRAFT BUNO 159588 AND SUBSEQUENT
AND AIRCRAFT INCORPORATING AFC 363

Figure 7-11. UHF Communication Selection Panel (Sheet 1 of 2)

NOMENCLATURE	FUNCTION
① XMTR SEL switch	Aircraft BUNO 158613 to 159468 and aricraft incorporating AFC 338. Selects desired UHF communications radio for transmission. UHF 1 — Selects ARC-51A UHF radio set for operation. UHF 2 — Selects ARC-159 UHF radio set for operation. Aircraft BUNO 159588 through 159636 and those aircraft incorporating AFC 363. Aircraft BUNO 159637 and subsequent, in which the ARC-159A(V)5 replaces the ARC-51A. UHF 1 — Selects ARC-159(V)1 UHF radio set for operation. UHF 2 — Selects ARC-51A or ARC-159A(V)5 UHF radio set for operation. BOTH — Selects both UHF 1 and UHF 2 radio sets for operation.
② VOL control	Controls listener's audio level of other crewman's radio.
③ Pilot's COMM/ICS button	UHF 1 — Keys ARC-159(V)1 UHF radio set for operation. UHF 2 — Keys ARC-51A 6r ARC-159A(V)5 UHF radio set for operation. BOTH — Keys both UHF 1 and UHF 2 radio sets. ICS — Permits intercommunication when COLD MIC is selected on function selector control.
④ UHF 1 ANTENNA select	LWR — Selects lower UHF 1 and upper D/L ANTENNA. UPR — Selects upper UHF 1 and lower D/L ANTENNA.

Figure 7-11. UHF Communication Selection Panel (Sheet 2 of 2)

COMM/NAV CMD CONTROL PANEL

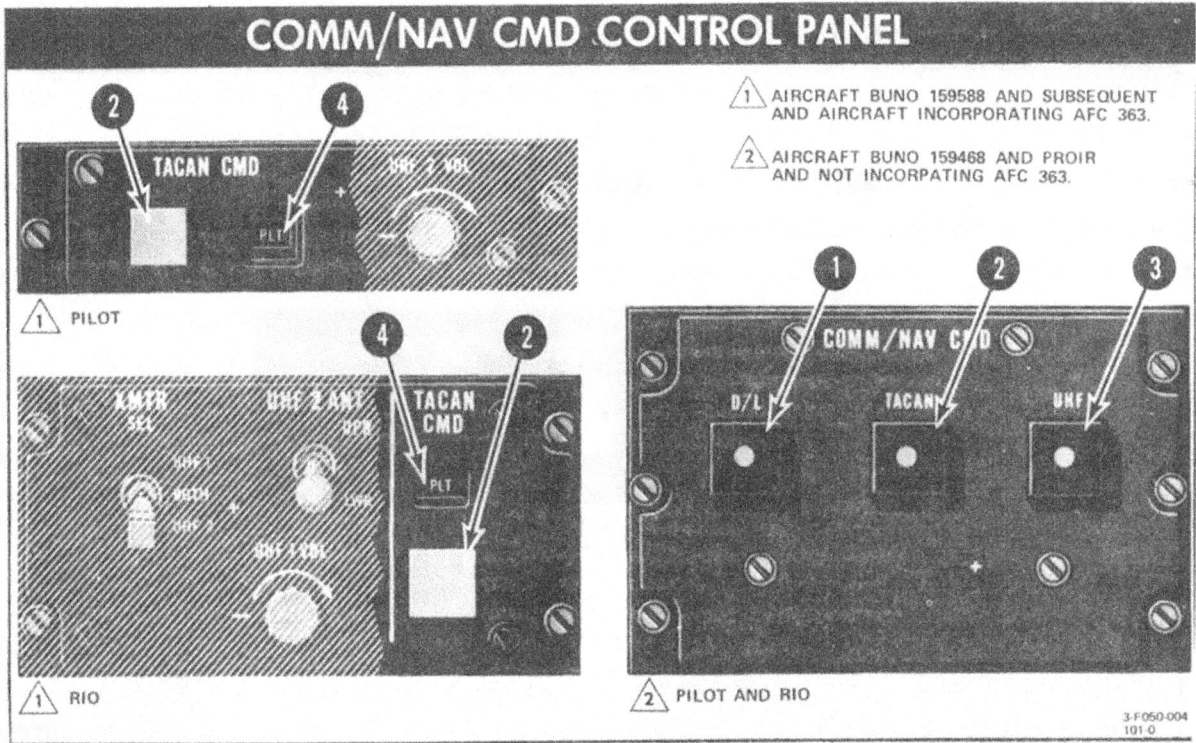

AIRCRAFT BUNO 159588 AND SUBSEQUENT AND AIRCRAFT INCORPORATING AFC 363.

AIRCRAFT BUNO 159468 AND PROIR AND NOT INCORPATING AFC 363.

3-F050-004
101-0

NOMENCLATURE	FUNCTION
1 D/L (data link) button	Alternating pushbutton to give and take control of data link; green light in button indicates cockpit station having control. Not in aircraft incorporating AFC 338.
2 TACAN button	Alternating pushbutton to give or take control of TACAN; green light in button indicates cockpit station having control. 2 No green light in button. 1
3 UHF button	Alternating pushbutton to give or take control of main UHF radio; green light in button indicates cockpit station having control. Not in aircraft incorporating AFC 338.
4 TACAN CMD	Flip-flop indicators show RIO or PILOT which crew position is in command in TACAN. 1

Figure 7-12. COMM/NAV Command Control Panel

KY-28 CONTROL PANEL

AIRCRAFT BUNO 159468 AND PRIOR AND
AIRCRAFT NOT INCORPORATING AFC 363.

3-F050-004
077-1

NOMENCLATURE	FUNCTION
① ZEROIZE switch	ZEROIZE — Guard lifted. The preset codes are erased and must be reset on the ground by qualified personnel before the cipher (C) mode can be used.
② Cipher-plain switch	C — Used to transmit and receive secure voice communications over the UHF radio in accordance with the preset codes.
	P — UHF radio is used as a plain language receiver/ transmitter.
③ Power switch	OFF — Removes power from the system.
	ON — Applies operating power to KY-28 system.
	RLY — Relay position: retransmits information between other facilities. (Relay position is not operational.)

Figure 7-13. KY-28 Control Panel (Sheet 1 of 2)

KY-28 CONTROL PANEL

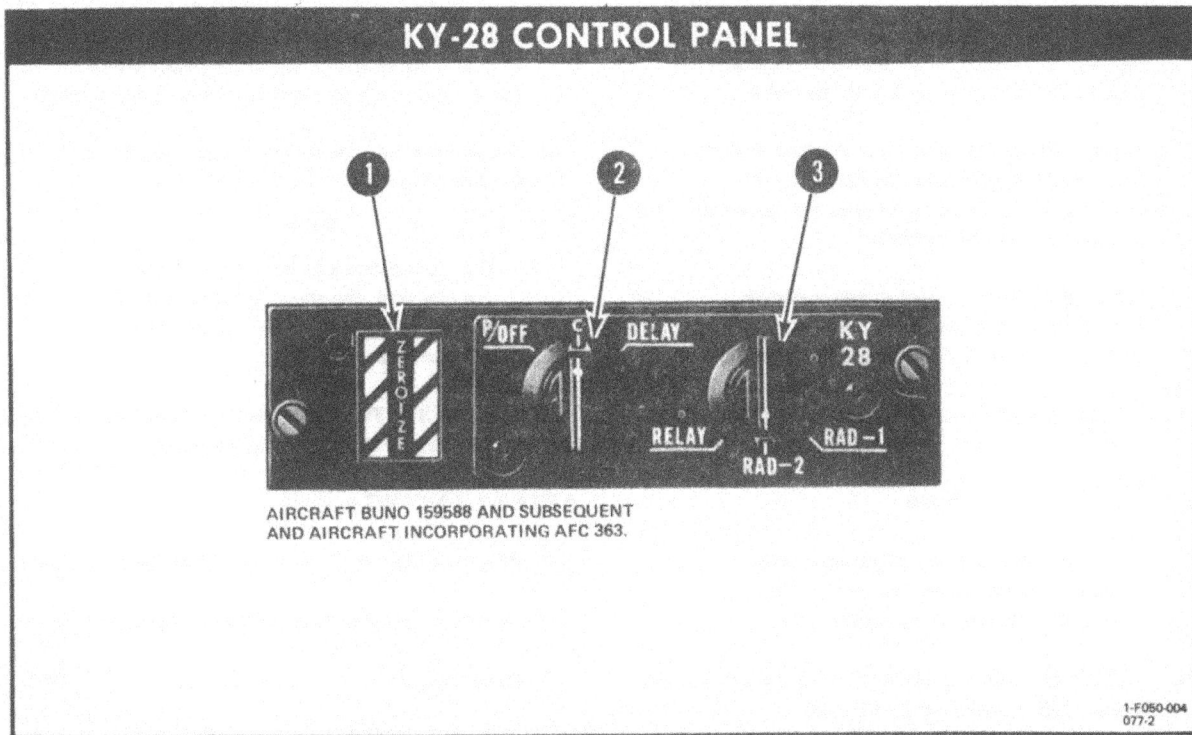

AIRCRAFT BUNO 159588 AND SUBSEQUENT
AND AIRCRAFT INCORPORATING AFC 363.

1-F050-004
077-2

NOMENCLATURE	FUNCTION	
① ZEROIZE switch	ZEROIZE	— Guard lifted. The preset codes are erased and must be reset on the ground by qualified personnel before the cipher (C) mode can be used.
② Power-Mode switch	P/OFF	— Removes power from the KY-28. UHF is used as a plain language receiver/transmitter.
	C	— Used to transmit and receive secure voice communications over the UHF radio in accordance with preset codes.
	DELAY	— Provides a time delay between push to talk and actual transmit.
③ Radio selector	RAD-1	— Selects UHF 1 for cipher operation.
	RAD-2	— Selects UHF 2 for cipher operation.
	RELAY	— Not operational.

Figure 7-13. KY-28 Control Panel (Sheet 2 of 2)

7. Depress MIC button, hold for approximately 2 seconds, and release. The double-pitched broken tone will cease and no sound will be heard.

8. Depress MIC button and hold. A single beep tone will be heard in approximately 1-1/2 seconds. When this tone is heard, the equipment is ready to cipher transmission.

9. After beep is heard, establish two-way cipher UHF communications with a cooperating ground station and check for readability and signal strength.

10. Set cipher/plain switch in accord with tactical situation.

Note

If a ground check is not practical, the above procedures may be used to perform an inflight check of the equipment.

PRELAUNCH. (Aircraft BUNO 159588 and subsequent and those aircraft incorporating AFC 338)

1. Determine that proper code has been set by personnel qualified in KY-28 equipment.

2. UHF radios . . . . ON

3. Power-mode switch C

4. Radio selector . . . . RAD-1 or RAD-2

5. If a ground test of equipment is desired, establish two-way plain text UHF radio communications on the de-selected radio with a suitable ground station and request an equipment check.

6. After a 2-minute warmup period, on the selected radio listed for a steady, unbroken tone in the headset followed by a double-pitched broken tone.

7. Key the appropriate radio selected for transmission, hold for approximately 2 seconds and release. Double-pitched broken tone will cease and no sound will be heard.

8. Key radio and hold. A single beep tone will be heard in approximately 1-1/2 seconds. When this tone is heard, the equipment is ready to cipher transmission.

9. After beep tone is heard, establish two-way cipher UHF communications with a cooperating ground station and check for readability and signal strength.

10. Set power-mode and radio selector switches in accord with the tactical situation.

Note

If a ground check of the equipment is not practical, the above procedures may be used to perform an in-flight check of the equipment.

POSTLAUNCH. The speech security equipment will be operated as directed by appropriate authority.

AFTER LANDING

1. ZEROIZE switch . . . . ZEROIZE (as briefed)

   Zeroize the code as directed by appropriate authority.

2. Power switch . . . . OFF

(Aircraft BUNO 159588 and subsequent and those aircraft incorporating AFC 363)

1. ZEROIZE switch . . . . ZEROIZE (as briefed)

   Zeroize the code as directed by appropriate authority.

2. Power-mode switch . . . . P/OFF

IN-FLIGHT VISUAL COMMUNICATIONS

Communications between aircraft are visual whenever practicable. Flight leaders shall insure that all pilots in the formation receive and acknowledge signals when given. The visual communications section of NWP-41 should be reviewed and practiced by all pilots and RIO's. Common visual signals applicable to flight operation are listed in figure 7-14, sheets 1 and 2.

GROUND HANDLING SIGNALS

Communications between aircraft and ground personnel are visual whenever practicable, operations permitting. The visual communications section of NWP-41 should be reviewed and practiced by all flightcrew and groundcrew personnel. For ease of reference, visual signals applicable to deck/ground handling are listed on page FO-19. During night operations, flashlights or wands shall be substituted for hand and finger movements.

IN-FLIGHT VISUAL COMMUNICATIONS

GENERAL CONVERSATION

MEANING	SIGNAL	RESPONSE
Affirmative (I understand.)	Thumb up, or nod of head	
Negative (I do not know.)	Thumb down, or turn of head from side to side	
Question (repeat) ; used in conjunction with another signal, this gesture indicates that the signal is interrogatory.	Hand cupped behind ear as if listening	As appropriate
Wait	Hand held up with palm outward	
Ignore last signal	Hand waved in an erasing motion in front of face, with palm forward	
Perfect, well done	Hand held up, with thumb and forefinger forming an 0 and remaining three fingers extended	
Numerals, as indicated	With forearm in vertical position, employ fingers to indicate desired numerals 1 through 5. With forearm and fingers horizontal, indicate number which, added to 5, gives desired number from 6 through 9. A clenched fist indicates zero.	A nod of the head (I understand). To verify numerals, addressee repeats. If originator nods, interpretation is correct. If originator repeats numerals, addressee should continue to verify them until they are understood.
Take over communications	Tap earphones, point to plane, and hold up one finger.	Execute

CONFIGURATION CHANGES

MEANING	SIGNAL	RESPONSE
Lower or raise landing gear.	Rotary movement of hand (flashlight at night) in cockpit, as if crankings wheels, pause, drop below canopy rail.	Execute when hand/flashlight drops.
Speed brakes	Open and close four fingers and thumb Flashlight at night – a series of flashes followed by a steady light; light out for execution.	Execute.
Extend or retract flaps.	Open and close four fingers and thumb Flashlight at night – a series of flashes followed by a steady light; light out for execution.	Execute when hand/flashlight drops.
Sweep wings aft.	Hand held up, palm aft, and swept aft along canopy rail; at night, flashlight swept aft along canopy rail	Execute on head nod/light out.
Sweep wings forward.	Hand held up, palm forward, and swept forward along canopy rail; at night, flashlight swept forward along canopy rail	Execute on head nod/light out.

FUEL AND ARMAMENT

MEANING	SIGNAL	RESPONSE
How much fuel have you?	Raise fist with thumb extended in a drinking position.	Indicate fuel in tens of gallons or hundreds of pounds by finger numbers.
Arm or safety missiles and ordnance.	Pistol cocking motion with either hand	Execute and return signal.

3-F050-004
068-1

Figure 7-14. In-Flight Visual Communications (Sheet 1 of 2)

FORMATION

MEANING	SIGNAL	RESPONSE
1 - OK	1 - Section leader gives thumbs-up signal.	1 Stands by for reply from wingman, holding thumbs-up until answered
2 - Commence take off power turnup.	2 - Leader gives a two-finger turn-up signal.	2 - Wingman returns two-finger signal and executes.
3 - I have completed my takeoff checklist and am, in all respects, ready for (section) takeoff; your aircraft appears to be properly configured and ready for takeoff.	3 Leader raises arm vertically in combination with a thumbs-up signal.	3 - Wingman returns thumbs-up signal if his takeoff checklist is complete; he is ready for takeoff in all respects; and lead aircraft appears properly configured and ready for takeoff.
4 - Takeoff path is clear. I am commencing takeoff.	4 Leader lowers arm.	4 Executes section/flight leader separation takeoff
Leader shifting lead to wingman	Leader pats self on head and points to wingman. At night, leader A/C switches lights to bright, and turns anti-collision light on. In the event of an external light failure, leader shines flashlight on hard hat, then shines light on wingman.	Wingman pats head and assumes lead. At night, wingman puts lights on dim, and turns grimes light off when he accepts the lead. In the event of an external light failure, wingman shines flashlight at leader, then on his hard hat, turns external lights to dim, and assumes lead.
Leader shifting lead to division designated by numerals	Leader pats self on head, points to wingman, and holds up two or more fingers.	Wingman relays signal; designated division leader assumes lead.
Take cruising formation	Thumb waved backward over the shoulder	Execute.
I am leaving formation	Any pilot blows kiss.	Nod (I understand.)
Aircraft pointed out, leave formation	Leader blows kiss and points to aircraft.	Execute.
Directs plane to investigate object or vessel	Leader beckons wing plane, then points to eye, then to vessel or object.	Wingman indicated blows kiss and executes.
Refers to landing of aircraft, generally used in conjunction with another signal; 1—I am landing 2—directs indicated aircraft to land	Landing motion with open hand; 1—followed by patting head 2—followed by pointing to another aircraft	Execute
1—Join up or break up, as appropriate 2—On GCA final: Leader has runway in sight.	Flashing external lights	1—Comply. 2—Wingman repeats, indicating runway/ship in sight. When runway conditions preclude a safe section landing leader will wave-off.
Wingman cross under	Leader raises forearm vertically.	Execute.
Section cross under	Leader raises forearm vertically and moves arm in pumping motion.	Execute.

AIR REFUELING

MEANING	SIGNAL	RESPONSE
Extend Drogue	Form cone-shape with hand, and move hand aft.	Tanker execute
Retract Drogue	Form cone-shape with hand, and move hand foward.	Tanker execute
Secure Turbine	One finger turn-up signal followed by cut signal.	Tanker execute

FORMATION SIGNALS — MADE BY AIRCRAFT MANEUVER — COMBAT OR FREE CRUISE

MEANING	SIGNAL	RESPONSE
Single aircraft cross under in direction of wing dip.	Single wing dip	Execute
Section cross under.	Double wing dip	Execute
Close up.	Series of small zooms	Execute
Join up; join up on me.	Series of pronounced zooms	Expedite join-up

2-F050-004
068-2

Figure 7-14. In-Flight Visual Communications (Sheet 2 of 2)

PART 2 — NAVIGATION

NAVIGATION SYSTEM

The primary navigation system is the inertial navigation system (AN/ASN-92), which consists of the following components: inertial measurement unit (IMU), power supply unit, and pilot and RIO navigation controls and displays (figure 7-15). Additionally, the inertial navigation system (INS) operates with the AWG-9 computer (WCS computer) and the computer signal data converter (CSDC), (page FO-20). Other associated equipment includes the attitude and heading reference system (AHRS), central air data computer (CADC), radar altimeter, instrument landing system, and TACAN.

The WCS computer and CSDC perform the computations necessary for aligning the INS, using various alignment routines stored on magnetic tape. The alignment routines stored in the WCS computer are referred to as the SMAL (Single Mode Alignment) program. When alignment is commanded, the alignment routines are placed in the WCS computer's destructive readout memory from magnetic tape. This process is referred to as tape read-in and designated by an M or B on the TID. During the course of aligning the IMU platform, the WCS computer talks to the CSDC and addresses specific CSDC navigation routines.

In the alignment modes, the CSDC supplies velocity data to the WCS computer, which calculates wander angle and platform alignment correction terms. These terms are continually updated and returned to the CSDC throughout the alignment. The CSDC adds these corrections to the inertial equations and generates IMU gyro-torquing pulses to correct and maintain the leveling of the platform. Alignment is complete when the computer calculations of wander angle error are within certain tolerances.

Upon completion of alignment, the system is ready to enter INS mode. When INS is entered, the last used and computed values of calibration data and of wander angle are stored by the CSDC. The WCS will now accept wander angle, velocity data, and position data generated by the CSDC.

The WCS also performs the following general navigation computations.

- Own aircraft groundspeed and ground track.

- Range, bearing, command course, command heading, and time-to-go to selected destination positions.

- Heading (true and magnetic).

- Windspeed and direction (true and magnetic).

- Backup present position.

- Magnetic variation (MAG VAR).

The information is presented on the TID, HSD, ECMD, HUD, and VDI, depending on the mode selected by the pilot and RIO. If an IMU or navigation computer failure occurs, two backup modes are available: IMU airmass or AHRS airmass.

In the INS mode, the IMU is the prime sensor supplying velocity pulses to the CSDC, which computes all inertial outputs. In the IMU/AM mode, the system uses true airspeed (TAS) from the CADC, either stored or entered wind, and true heading to compute the navigation parameters in the WCS. In both INS and IMU/AM modes, roll and pitch output are from the IMU. In the AHRS/AM mode, the system also uses TAS from the CADC and either stored or entered wind. However, true heading is derived from magnetic heading and stored or entered magnetic variation. The navigation computations are also performed in the WCS, but roll and pitch outputs are from the AHRS in this mode.

INERTIAL NAVIGATION SYSTEM (AN/ASN-92(V))

An important feature of the INS is its rapid alignment capabilites over a wide temperature range. The INS is a dead-reckoning system; it derives speed as a function of aircraft accelerations. Two accelerometers are used to measure acceleration in the horizontal plane. These outputs result in X and Y velocity components after coriolis corrections and integration inputs. Further integration about the north and east axes provide increments of latitude and longitude. Navigation in this manner provides precise knowledge of aircraft position, direction, and velocity at all times.

An accelerometer is basically a restrained mass that is free to move along one axis. An acceleration along that

NAVIGATION CONTROLS

NOTE

TACAN AND ADF ARE NOT SHOWN.
CONTROLS FOR THOSE SYSTEMS
ACCOMPANY SYSTEM DESCRIPTION.

STORED
HEADING
ALIGN-
MENT
SELECT
IS NOT
PLACARD-
ED

PILOT — DISPLAYS CONTROL PANEL

NOTE
EITHER PUSHTILE MAY BE PRESENT
AVC 2102 INCORPORATED

RIO — COMPUTER ADDRESS
PANEL

RIO — NAVIGATION CONTROL AND
DATA READOUT PANEL

RIO — DATA LINK REPLY AND
ANTENNA CONTROL PANEL

RIO — ECM DISPLAY
CONTROL PANEL

7-F50-187-0

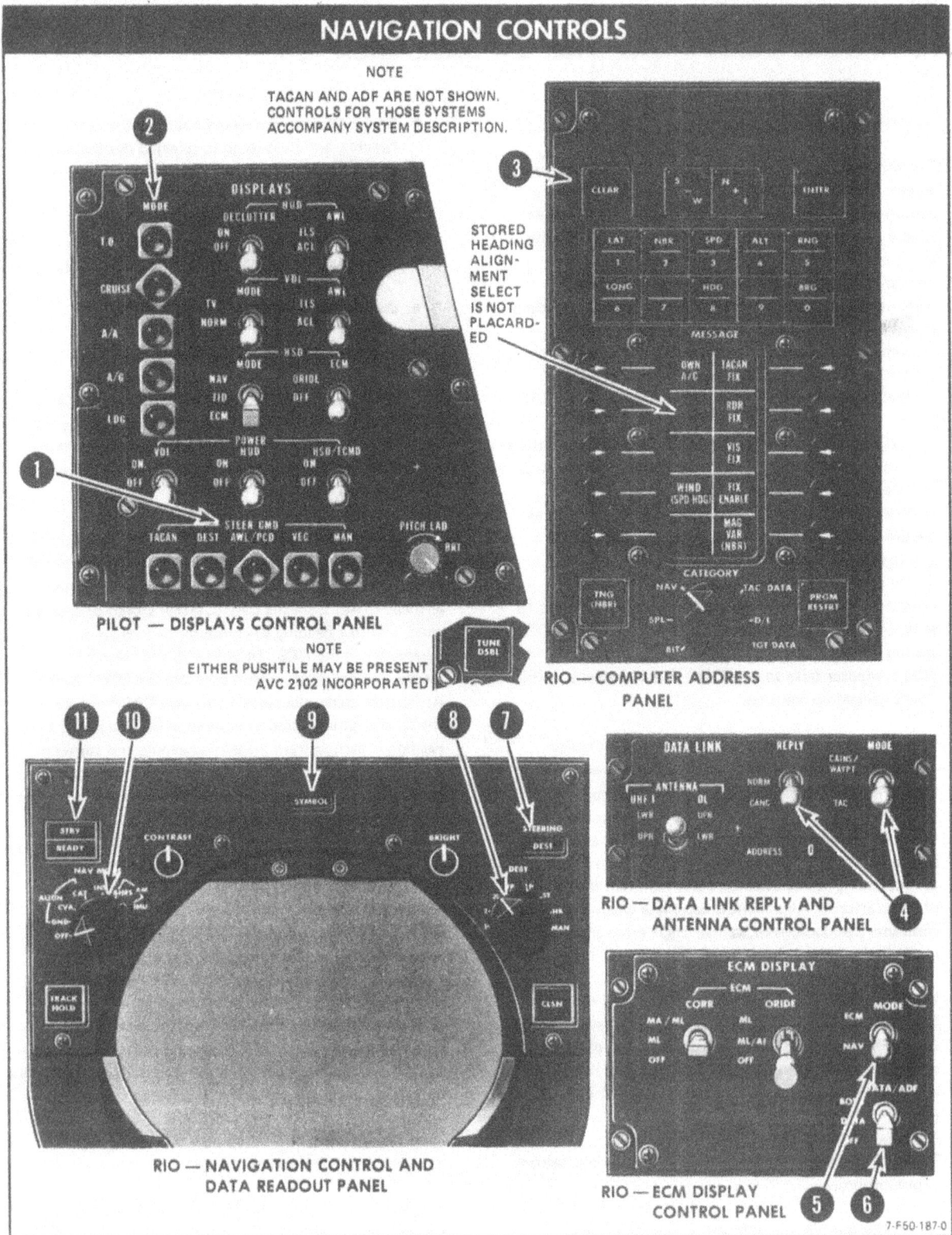

Figure 7-15. Navigation Controls (Sheet 1 of 5)

NOMENCLATURE	FUNCTION
① STEER CMD pushbutton	Provides selection of the type of steering commands to be displayed. These pushbuttons are mutually exclusive.
	TACAN — Provides TACAN steering and deviation from the selected TACAN radial.
	DEST — Provides course to selected preset destination point.
	AWL/PCD — Provides glideslope information during landing or precision course direction (vector) information during air-to-ground.
	VEC — Provides data link deviation steering.
	MAN — Displays manually selected course and heading.
② MODE pushbutton	Mutually exclusive and rotates to identify the mode selected.
	TO — Selects takeoff symbology for the HUD and VDI.
	CRUISE — Selects cruise symbology for the HUD and VDI.
	A/A — Selects air-to-air attack symbology for the HUD and VDI.
	A/G — Selects air-to-ground attack symbology for the HUD and VDI.
	LDG — Selects landing (ILS, ACL) symbology on the HUD and VDI.
③ Computer address panel	Enter, display, and update navigation data. CATEGORY switch is set to NAV for display of NAV matrix in MESSAGE window.
④ CAINS/WAYPT-TAC switch	CAINS/WAYPT — May be selected during data link alignment or when receiving waypoint data. If in INS or CVA ALIGN mode, only SINS data is processed until alignment is complete or may be inserted through data link before starting alignment.
	TAC — Allows manual selection of frequencies.
	Note
	• If switch is left in CAINS/WAYPT position when aircraft takes off, switch will automatically change to TAC.
	• If during a CVA alignment the switch is unlatched to the TAC position by a power transient, the INS will revert to a hand set alignment.
	• If the switch is unlatched while in suspend align, the system will revert to hand-set alignment.
	• If switch is in CAINS/WAYPT position during OBC, DLS test results are invalid.
⑤ DATA/ADF switch	BOTH — Enables display of ADF bug and navigation data block.
	DATA — Enables display of navigation data block. ADF bug is not displayed.
	OFF — ADF bug and navigation data are not displayed.
⑥ ECM DISPLAY-MODE switch	Selects either ECM or navigation presentations on ECMD. Must be set to NAV in order to display navigation data on ECMD.

Figure 7-15. Navigation Controls (Sheet 2 of 5)

°NOMENCLATURE	FUNCTION
(7) STEERING indicator	An 11-position mechanical drum that indicates the steering mode selected. The pilot and RIO may jointly or independently configure the WCS, the navigation system, or both to provide steering signals to the pilot. The feedback provided by this indicator facilitates cooperative efforts by the flightcrew. MAN — Indicates that the pilot has selected CRUISE mode and MAN steering on display control panel. Manual steering is presented on HUD, VDI, and HSD as a command heading symbol read against compass rose. Command course and heading symbols are set by CRS control on HSD. DEST — Indicates that pilot has selected CRUISE mode and DEST steering on the display control panel. Steering signals for the RIO-selected destination (DEST switch setting) along great circle course. This includes range, bearing, and command heading. TACAN — Indicates that pilot has selected CRUISE and TACAN. TACAN steering presented on VDIG and HSD as the deviation from the command course set based on a TACAN station. **Note** The following STEERING indications pertain to weapons system steering. For a full description of weapons system steering, refer to Section VIII of NAVAIR 01-F14AAA-1A. D/L — Indicates that pilot has selected CRUISE or A/A mode and VEC steering. WCS is not in TWS, PD STT, or PSTT. Data link command heading on VDIG and HSD. The aircraft is following TDS command steering. Command course and heading are disabled on the HSD and ECMD. LD CLSN — Indicates that the AWG-9, pilot displays, and weapons selection are configured to provide the displayed steering. PURST — Indicates that pilot has selected A/A mode, has not selected a weapon on the stick grip, and WCS is in PD STT or PSTT. Horizontal and vertical pursuit steering attempts to keep the nose of the aircraft pointed at the target. LD PURST — Indicates that pilot has selected A/A mode and has selected lead-pursuit steering (for an AIM-9G, AIM-7E/7F, or AIM-54A weapon), and the WCS is in PD STT or PSTT mode. Horizontal and vertical lead-pursuit steering to the target is supplied to the VDIG. CLSN — Indicates that RIO has selected the CLSN pushbutton.

Figure 7-15. Navigation Controls (Sheet 3 of 5)

NOMENCLATURE	FUNCTION
⑧ DEST selector	Selects one of the seven destinations prestored by RIO either during preflight or in flight. These are the only points to which destination steering is provided. When selected, steering information is provided to the selected point and displayed on the MDIG and VDI (when pilot has selected DEST steering). Steering information is range, command heading, command course, wind, TAS, and GS on the MDIG. The VDI displays only command heading.
	1, 2, and 3 — Detects stored waypoint
	FP — Selects fixed point.
	IP — Selects initial point.
	ST — Selects surface target.
	HB — Selects home base.
	MAN — Used in TARPS mode only.
⑨ DATA READOUT source indicator	Indicates source of data displayed on TID readouts. WAY PT, ST, FIX PT, IP, home base (HB), OWN A/C, and TGT 1 refer to their corresponding data sources. SYMBOL indicates that a radar, data link, defended point, or hostile area has been designated for readout with specific data source indicated by brightening of the symbol. Blank indication indicates a message hook exists for which there is no symbol (for example, WIND).
⑩ NAV MODE selector	Initiates alignment and/or manually selected operating mode of the navigation system. Any position other than OFF energizes the IMU/PSU and CSDC.
	Note
	Alignment can be selected to be concurrent with OBC (SAT — Simultaneous Align and Test).
	OFF — Turns off the power supply to the IMU.
	ALIGN — Selection of any one of the three ALIGN positions (GND, CVA, or CAT) initiates the alignment program. However, each position has a specific function and requires specific inputs in order to perform a valid alignment.
	GND — Initiates alignment while at shore-based facility. Only own-aircraft latitude and longitude are required for initialization.
	CVA — Initiates alignment for aircraft carrier operations with or without SINS data. Without SINS data, this position requires manual entry of the ship's latitude, longitude, true heading, and speed.
	CAT — Allows the alignment to continue with the aircraft on the catapult and with the handbrake released.

Figure 7-15. Navigation Controls (Sheet 4 of 5)

NOMENCLATURE	FUNCTION
	INS — Selects INS navigation.
	AHRS/AM — Selects attitude and heading reference system roll and pitch, and backup navigation using magnetic heading (AHRS), stored or entered magnetic variation, true airspeed from CADC, stored or entered wind, and angle of attack from the CADC.
	IMU/AM — Selects IMU roll, pitch, and backup navigation using stored or entered true heading, wind, and true airspeed from CADC.
(11) INS status indicators	Green STBY light illuminated — power applied to INS, but alignment is not complete. When the STBY light goes off and READY light illuminates, the alignment has reached minimum Phoenix launch criteria. READY light goes off when the INS mode is selected on NAV MODE switch.

Note

> INS accuracy is improved by prolonged alignment in the READY state.

Combinations of the STBY and READY lights are shown in figure 7-16 along with appropriate interpretations and corrective action. However, if both lights are on simultaneously, it indicates an initialization, IMU, CSDC or AHRS failure. The RIO can identify whether the IMU or NAV COMP failed by noting the CAUTION ADVISORY panel. If the lights go out in the AHRS/AM mode, the IMU failed; if in the IMU/AM mode, the CSDC NAV COMP function has failed.

Note

> The RIO can also verify whether IMU, (CSDC) or NPS failure occurred by CM/OBC monitoring. Failure of the NAV COMP, IMU or AHRS is also displayed on the pilot and/or RIO advisory light panels.

Figure 7-15. Navigation Controls (Sheet 5 of 5)

LIGHT STATUS	INTERPRETATION	ACTION REQUIRED
STANDBY ON READY ON	• SYSTEM CANNOT PERFORM SATISFAC- TORILY IN THE SELECTED MODE DUE TO IMU, NAV COMP, NPS OR AHRS FAILURE. • NORMAL DISPLAY FOR FIRST 45 SECONDS OF ALIGNMENT INITIALIZATION.	• SELECT ANOTHER MODE. • NONE
STANDBY ON READY OFF	• NORMAL DURING ALIGNMENT AFTER INITIALIZATION. • NORMAL DURING ERECTION OF IMU WHEN IMU/AM IS SELECTED PRIOR TO COMPLE- TION OF COARSE ALIGN.	• LEAVE NAV MODE SWITCH IN SELECTED POSITION TO OBTAIN NAV SYSTEM ALIGNMENT. • LEAVE NAV MODE SWITCH IN SELECTED POSITION. AHRS/AM MODE WILL BE PROVIDED AUTOMATICALLY UNTIL IMU IS ERECTED.
STANDBY AND READY FLASHING	ALIGNMENT NOT INITIATED DUE TO SUS- PENDED ALIGNMENT.	SET PARKING BRAKE OR CHECK BRAKE PRESSURE.
STANDBY FLASHING READY OFF	ALIGNMENT SUSPENDED.	SET PARKING BRAKE TO CONTINUE ALIGNMENT.
STANDBY OFF READY ON	• SATISFACTORY ALIGNMENT OBTAINED TO MEET MINIMUM WEAPONS LAUNCH REQUIREMENTS. • WITH AHRS/AM SELECTED, IMU/AM OR INS MAY BE AVAILABLE.	• LEAVE NAV MODE SWITCH IN SELECTED POSITION FOR IMPROVED ALIGNMENT OR SWITCH TO INS. • SELECT DESIRED NAV MODE.
STANDBY OFF READY OFF	SYSTEM OPERATING SATISFACTORILY IN THE SELECTED MODE OR NAV SYSTEM OFF.	NONE
STANDBY OFF READY FLASHING (AFTER 5 SECONDS BOTH OFF)	OCCURS ONLY WHEN IMU/AM IS SELECT- ED AND PLATFORM IS ALIGNED. IF ANOTHER MODE IS NOT SELECTED IN 5 SECONDS, ALIGN- MENT WILL BE LOST AND INS WILL NOT BE AVAILABLE.	SELECT ANOTHER MODE TO AVOID ALIGNMENT LOSS
READY FLASHING STANDBY OFF	ALIGNMENT SUSPENDED PAST MISSION ALERT CRITERIA (SECOND TIC) WITH PARKING BRAKE OFF.	SELECT INS, SET PARKING BRAKE, OR SELECT DESIRED MODE OF OPERATION.

Figure 7-16. Standby and Ready Light Logic

axis causes the mass to be displaced. This displacement is sensed by pickoff coils that develop a signal that is amplified, then applied to a torquer that restores the mass to its null position. The magnitude of torquing current required is proportional to the acceleration.

Because an accelerometer cannot distinguish between aircraft accelerations and the effect of gravity, the sensitive axis of the accelerometer must be kept perpendicular to the gravity vector at all times. To accomplish this, the accelerometers are mounted on a platform whose three axes (X, Y, and Z) are stabilized by gyros that are mounted on the same platform with their sensitive axes mutually perpendicular.

The gyros sense angular rotation about their sensitive axes. Gyro pickoffs provide electrical outputs that are proportional to platform deviation from the desired orientation. The outputs are amplified and used to drive platform torquer motors to null the sensed rotation. This keeps the platform space-stabilized while the aircraft rotates around it. Accelerometers on this platform are oriented relative to the earth's gravity.

Inertial Measurement Unit

The IMU is the heart of the aircraft inertial navigation system. It is a three-axis, four-gimbal, all-attitude unit containing gyros, accelerometers, and associated electronics. The accelerometers provide the basic inertial measurements necessary for the primary navigation computations. The four-gimbal structure allows the accelerometers to be maintained in their proper orientation through all aircraft maneuvers.

The IMU is first leveled to the local vertical by action of the x-axis and y-axis accelerometers which sense local gravity and torque the gyros until the accelerometer outputs are zero. The determination of wander angle and fine leveling is performed in the WCS computer using x- and y-accelerometer outputs and gyro-torquing computations as processed by the CSDC.

IMU BIT

IMU validity signals are continuously monitored by the CSDC. Absence of validity may indicate a failure, which causes the CSDC to automatically select another mode. IMU BIT monitors temperature, internal error signals, and electrical characteristics of the IMU. When the CSDC detects a failure in the IMU, it informs the WCS computer of the failure. The computer then displays on the TID the IMU acronym indicating the component of the INS that failed. At the same time the IMU advisory light illuminates on the RIO caution/advisory panel.

NAV COMP Light

If the NAV COMP advisory light illuminates with the NAV MODE switch in INS, there is a failure in the INS or CSDC; the navigation system automatically shifts to a backup mode. The NAV COMP light remains illuminated and the RIO should set the NAV MODE switch to the IMU/AM position. The NAV COMP advisory light indicates that the inertial navigation system is operating in a degraded mode (IMU/AM) as a result of manual selection by the RIO using the NAV MODE switch, or automatic selection due to a failure of the CSDC computer, the IMU quantizer, the IMU clock, or inflight refueling static discharge.

IMU Light

If the IMU advisory light illuminates, there is a failure in the IMU; the navigation system automatically switches to AHRS/AM mode and navigation accuracy may be degraded. Attitude information for the VDIG and missile control system are now provided by the AHRS. The IMU light remains on and the RIO should select the AHRS/AM position. With an AHRS light, the flightcrew should verify computed magnetic veriation (vC) and update if necessary.

AHRS Light

If the AHRS advisory light illuminates, the AHRS self-test has detected a failure. The magnetic heading on the HUD and VDI is now commanded by the WCS computer, using back-up heading computations. These computations use the last known value of magnetic variation. Consequently, over long distances and time, heading will be degraded unless new values of magnetic variation are inserted as required. IFR flight should be avoided.

Navigation Power Supply

The navigation power supply (NPS) converts primary electrical power into the voltages required for the IMU and CSDC. A nickel-cadmium battery provides the power to the IMU and CSDC for up to 10 seconds if there is a power interruption or transient. The battery charging and switching circuits are controlled automatically.

INS ALIGNMENT MODES

Before the INS can be used for navigation, the inertial platform must be leveled relative to local vertical and its orientation relative to true north. These procedures constitute the alignment of the INS and are accomplished automatically in two phases: coarse alignment and fine alignment. Alignment can be initiated either concurrently with OBC (SAT mode) or consecutively with OBC (either alignment or OBC first). The coarse alignment phase is entered upon completion of the initialization sequence and performs initial coarse estimates of IMU platform wander angle. The successful completion of this phase requires a predetermined accuracy of local level error necessary to proceed to the fine alignment phase. The fine alignment phase continues until all alignment criteria are met.

During alignment, power is applied to the NPS and to the IMU. The IMU heaters supply heat to all IMU elements that require warmup. In addition, the IMU gimbals (roll, pitch, azimuth) are caged through their respective synchros to the IMU case (airframe reference). The IMU gyros are brought up to running speed, and coarse leveling is performed using the accelerometer outputs. Concurrent with application of power to the NPS and IMU, the alignment program (SMAL) from the bulk storage tape is read into the WCS computer DRO memory. The alignment program includes estimates of wander angle, velocity errors, and gyro-torquing correction signals and supplies these quantities to the CSDC to align the IMU and to initialize the CSDC NAV program.

The following assemblies are used during alignment: IMU, NPS, CSDC, WCS computer, CAP and navigation control and data readout panel. Additionally, the data link receiver-processor is used for carrier alignment.

There are four primary modes of alignment (SAT ground and carrier and NON SAT ground and carrier alignment) which are selectable with the NAV MODE switch on the TID panel. Any of the alignment modes can be executed in either of two modes: SAT and non SAT. SAT operation allows OBC testing concurrently with the alignment. The basic TID display format is shown in figure 7-17, inset A. Non SAT remains unchanged. The automatic sequence of events is the same for all modes; however, during CVA ALIGN, ship's motion is inserted by the data link. The CAT ALIGN position overrides the requirement for the parking brake to be on (suspend align). There are two other submodes of alignment (hand-set and stored heading) which may be selectable with the NAV MODE switch. The hand-set mode is used when CVA ALIGN is desired, but SINS data is not available. The stored heading mode uses a previous alignment (reference alignment) to align the system rapidly.

Note

The parking brake must be on during initialization of any alignment. When the parking brake is released during coarse alignment the STBY and READY lights flash, and the align program will reinitialize; if the parking brake is released during fine alignment, a suspend align discrete is sent to the CSDC, the STBY or READY light blinks, and the time into alignment clock on the TID stops.

Non SAT Alignments

GROUND ALIGNMENT

For shore-based operations, the IMU is aligned using the ground alignment procedure. Own-aircraft or homebase latitude, longitude, and altitude are entered into the WCS computer via the CAP. This may be accomplished before or after selecting GND align. Selecting GND ALIGN on the NAV MODE switch initiates the align operation.

When GND ALIGN is selected, the CSDC computer interrogates the IMU for the stored instrument calibration parameters. After approximately 20 seconds, the STBY/READY lights illuminate. This is the time required to load the DRO SMAL program from the bulk storage tape unit. If the calibration data is valid, the alignment program is initialized. If the calibration data is not valid, a C acronym will appear on the TID and the CSDC will use the last stored calibration data. In this case, alignment would continue and should be reliable, unless the CSDC has been changed or its memory altered in the interim so that the stored calibrated data does not correspond to that of the IMU. If there is no stored calibrated data in the CSDC, alignment will not continue with invalid calibrated data in the IMU. This situation should be reported to maintenance.

Note

If fine align complete has not been achieved, an own-aircraft latitude entry will reinitialize the alignment.

On completion of the alignment program read-in, the alignment display appears on the TID (figure 7-17). During initialization, the TID display (figure 7-17-B) will have an alignment time of 07 displayed.

°TID ALIGNMENT DISPLAYS

A TID ALIGN/OBC

DRUM READOUTS
AND
FLYCATCHER

ALIGN
DATA

OBC
DISPLAY
(IF
SELECTED)

B DRUM READOUTS AND FLYCATCHER
DURING ALIGNMENT INITIALIZATION

FLYCATCHER
ADDRESS

GS 002 TC 210
00517 04024020

01

TIME IN
ALIGNMENT
(MINUTES
AND
TENTHS)

C NON SAT DISPLAY — ALIGN DATA

COARSE ALIGN
COMPLETE
MARKER

ALERT
MARKER

DIAMOND

TIME IN
ALIGNMENT
(MINUTES
AND TENTHS)

FINE
ALIGN
COMPLETE
MARKER

GS 000 TC 214
00517 10424 60

69

TELLTALE
STATUS
INDICATOR

C = CAL DATA
     FAIL
T = TEMP
     (COLD IMU)
S = SINS DATA
     INVALID
O = OBSERVABLE
     (STALLED ALIGN)

D SAT DISPLAY — COARSE ALIGN

CARET

GS 000 TC 063

20
      S2   RAMP

ECM   AFC   RAW   TCN
IFX   AIC   RDA   DDI
CIR   APC   GCS
BAG   CAD   DLS
                  SAM
                  IFB

AIC

NOTE

CARET BECOMES DIAMOND (◇)
UPON PASSING THRU COARSE
ALIGN COMPLETE MARKER.

E SAT DISPLAY BEFORE READY SIX

STBY-ON
READY-OFF

GS 000 TC 210
00517 40024020

39
      S2

ECM   AFC   RAW   TCN
IFX   AIC   RDA   DDI
CIR   AFC   GCS
BAG   CFD   DLS
                  SAM
                  IFB

DLS

F ASH SAT DISPLAY FINE ALIGN (AFTER READY SIX)

CHANGES TO 7 WHEN
DIAMOND PASSES THRU
FINE ALIGN COMPLETE
MARKER AND DOT
APPEARS IN
DIAMOND (◈)

STBY-OFF
READY-ON

LN 2325 LW11052
00517 60024060

29
      S2   1 ASH

ECM   AFC   RAW   TCN
IFX   AIC   RDA   DDI
CIR   APC   GCS
BAG   CAD   DLS
                  SAM
                  IFB

RDA

AUTO
STORED
HEADING
ALIGN

NOTE
OBC PORTION OF THIS
DISPLAY CAN BE EITHER
AN IN-PROCESS DISPLAY
OR A FAULT DISPLAY.

2-F050-004
249-0

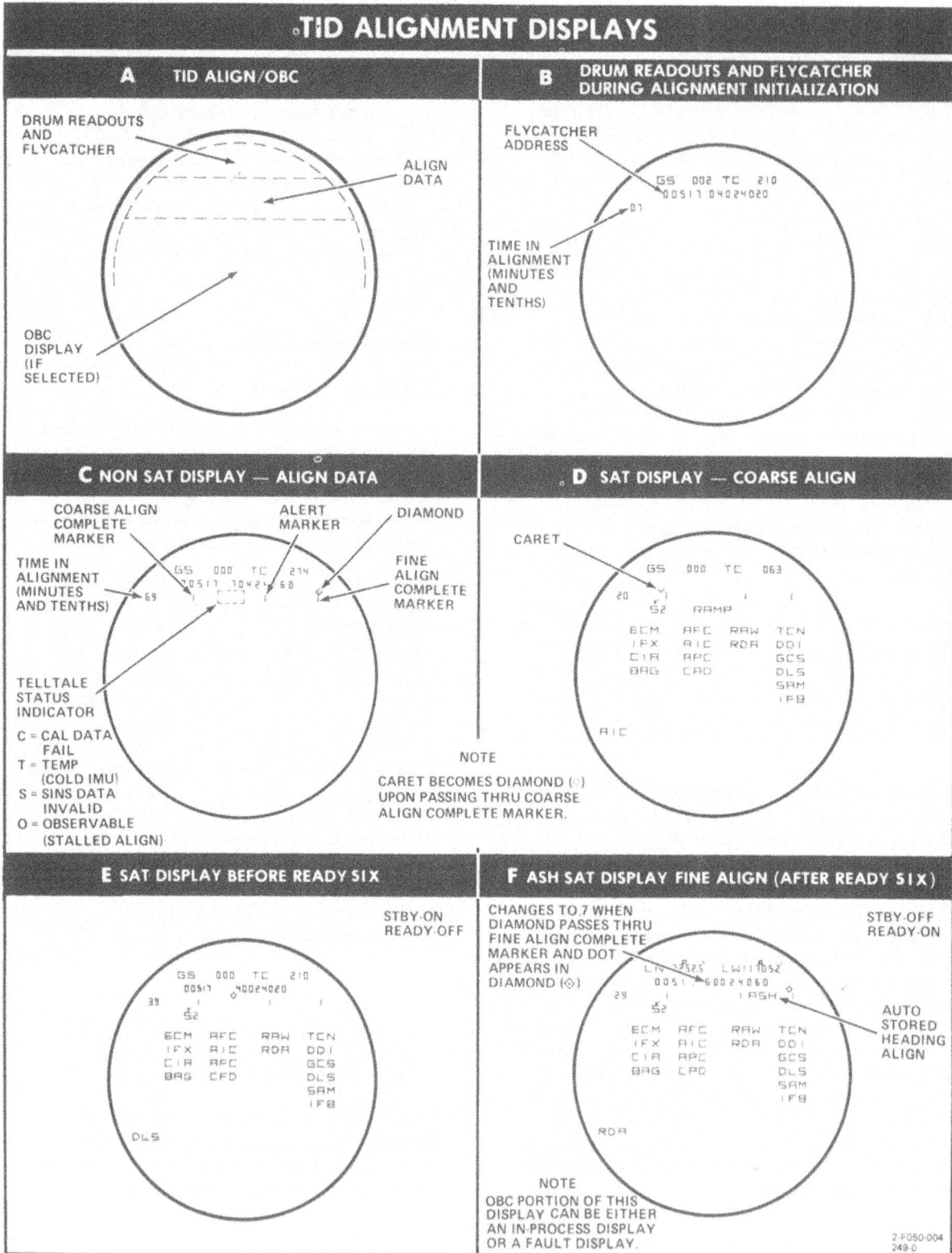

Figure 7-17. TID Alignment Displays

After approximately 42 to 45 seconds, the NAV COMP light on the caution/advisory panel will go out, indicating that the IMU has entered the ready state; the READY light will also go out. The alignment program will begin computation of the alignment parameters. At this time, the SMAL program issues an initialize code to cause the CSDC to initialize its own parameters.

An alignment status indicator, called a caret ($\vee$), moves from left to right. The status of the alignment is indicated by where the caret appears in relationship to three alignment-tick indicators (figure 7-17-C). The first tick indicator is called the coarse-align complete marker, the second is the alert marker and the third indicator is the fine-align complete marker. An elapsed time indicator provides a minutes-and-tenths display of alignment time. The clock indicator will begin with 0.7 displayed and continue after a 42-second delay. After 9.9 minutes of elapsed time, the clock display will pass through zero and begin again. If the alignment is suspended, the clock will stop counting until alignment is resumed. Between the first and second ticks are the telltale status indicators which incidate a failure of one of four systems: C = cal data fail, T = Temp (cold imu), S = SINS data invalid, and O = observeable (alignment data bad, that is, LAT, LONG, SPEED, etc.). If a letter appears, there is a failure in that system. If a C appears, there is a failure in the transfer of calibration data between the IMU and the CSDC; any resulting alignment with this condition should be suspect. The T appears normally at the start of alignment and disappears when the IMU has reached operating temperature. If the T does not disappear, there is a failure in the system, and alignment will not progress. The S may appear at the start of any alignment and will normally disappear shortly after. If the S does not disappear, there is a failure and a bad alignment will result. The S is also present if incoming SINS data is invalid, and the alignment should be suspect.

Note

The S will appear during a suspect align but will disappear upon re-entry to the align mode. The O appears if a problem has been detected in the alignment data; the alignment will be bad and may stall. Recycle alignment if O is present.

If a problem has been detected in the alignment data, the alignment will be bad and may stall.

During coarse alignment (figure 7-17-D) the alignment caret moves in discrete jumps because it is based on the wander angle error. At completion of the coarse alignment, the $\vee$ is directly above the first tick. As the program transitions to fine alignment, the $\vee$ changes to a diamond ($\diamondsuit$), indicating to the pilot that he may release the parking brake and taxi (if OBC is complete), which will suspend alignment (figure 7-17-E). The diamond will then continue to step to the right as alignment improves.

At the second tick, which indicates that alignment meets the minimum weapons launch criteria, the STBY light will go off and the READY light will illuminate. The INS mode may be selected at this time. If INS is not selected, the diamond continues to move to the right. When it reaches the third tick, it indicates that fine alignment is complete (flycatcher 517 first number changes from 6 to 7), and a dot will appear in the diamond ($\diamondsuit$). If the system is left in the alignment mode, a progressively more accurate alignment will result.

Note

During suspend align, if the aircraft is taxied for a distance of more than 4000 feet, the quality of alignment will be unknown and the INS performance could be unreliable.

If the alignment caret ($\vee$ or diamond) stops moving while in an alignment mode, the program has stopped aligning. If the symbol stops somewhere between the first and third ticks (coarse and fine alignment), it indicates that alignment has been suspended. The clock stops counting as soon as the alignment has been suspended. The clock resumes counting as soon as the alignment program starts running again and continues until switched out of alignment by the NAV MODE switch or the parking brake is released again.

Note

The alignment display will not progress past coarse align complete tick until the IMU temperature has reached 165°. When this temperature is reached, the telltale T symbol will be removed. During a stored heading alignment, the temperature interlock is bypassed. Since the stored heading alignment is usually completed before 2 minutes, a degraded alignment could result if the IMU is not preheated.

Selecting INS will turn off the READY light; alignment is terminated, the tactical tape appears, and the normal navigation display is available.

Note

● When the NAV MODE switch is set to INS, the CSDC is in navigation mode and the READY light goes out.

● The aircraft may take off before the 18-seconds have elapsed or before INS is selected.

The RIO then observes an IN acronym on the attitude status readout on the TID.

If it becomes necessary to reinitialize the program during alignment, the RIO must deselect GND ALIGN, enter LAT and LONG again, and reselect GND ALIGN.

ALIGNMENT QUALITY

The most accurate alignment can be achieved by allowing the alignment to continue to fine align complete (◇). Shore-based, closeout velocity errors should not exceed 6 knots ground speed, as long as a fine alignment was attained before flight. Closeout ground speeds greater than 6 knots indicate that the IMU should be reported to maintenance if the AWG-9 and CSDC performed satisfactorily during alignment and in flight. Air combat maneuvering (ACM) should have little effect on the alignment quality. Aboard ship, ground speed closeout velocity may vary, depending on the ship speed and quality of SINS data during alignment.

STORED HEADING ALIGNMENT

The stored-heading alignment is an additional feature of the aircraft's inertial navigation system (INS) that allows for quick-reaction response. To do this, the aircraft is parked and tied down in the alert position. A reference alignment is required so that aircraft heading can be stored prior to powering-down the aircraft. When the aircraft is powered-up for a mission, the system uses the stored heading to align the INS in less than 2 minutes. When align is first selected and a reference align is available, the TID will display an automatic stored heading, solid (ASH) (Figure 7-17E), acronym in the alignment display. The ASH indicates to the RIO that a stored heading has been initiated automatically. The ASH align will continue without RIO action and the ASH will remain as an advisory. One depression of the NAV category function button No. 2 (FB2) will terminate the ASH align, initiate a normal alignment, and extinguish the ASH acronym. A second depression of FB2 will reinstate a stored heading align without ASH displayed on the TID. Consequently, using this technique requires adherence to several constraints in order to achieve a valid reference alignment.

- Reference alignment must be made to the fine-align complete stage (third tick mark on the TID), dot in the center of the alignment diamond.

- The aircraft must not be moved after the reference alignment heading is stored.

Note

Monitor STBY/READY lights for simultaneous illumination. If simultaneous illumination occurs after initial 42 to 45 seconds, a failure has caused the alignment to reinitiate and an erroneous alignment may result. RIO must turn NAV MODE switch to OFF for 1 second, then restart alignment, using normal navigation alignment procedures.

AUTO STORED HEADING ALIGNMENT. With platform warm up period and AWG-9 initialization:

1. WCS switch . . . . STBY

2. NAV MODE switch . . . . CBA/GND
   (Two minutes maximum without cooling air)

3. DATA LINK switch. . . . . ON (CVA ops only)

4. D/L MODE switch. . . . . CAINS/WAYPT

Note

Power transients will cause CAINS to be deselected as indicated by a flashing HS acronym.

5. NAV MODE switch . . . . INS at the
   READY light or a ◇
   on the TID.

Note

Ensure T (gyro float temperature indicator) is not displayed to the right of the first tick mark before selection of INS.

REFERENCE ALIGNMENT

The reference alignment can be made, using internal or external power. The reference alignment is made by entering latitude and longitude via the CAP into the own-aircraft file. This is accomplished by either an auto transfer from homebase, entry into own-aircraft before GND ALIGN is selected, or entry into own-aircraft after GND ALIGN is selected. The alignment is carried out to a complete fine alignment. CVA align may be used to establish a reference alignment if SINS data is available. A reference alignment is present in both GND and CVA when a dot appears in the diamond.

It is not necessary to select INS after the reference alignment is established.

Note

A handset alignment will not produce a valid reference alignment even if continued to fine alignment completion.

QUICK-REACTION ALIGNMENT

A quick-reaction alignment using stored heading may be initiated as soon as the aircraft is on internal or external power. Prior to WCS power-up, the NAV MODE switch should be set to GND/CVA. The WCS power switch is set to STBY. At the completion of AUTO BIT 2, the align status symbol will appear as in coarse align (∨). However, after program read-in, the symbol will appear to the right of the first tick mark (coarse-align complete). Approximately 2 minutes after stored heading is initiated the diamond will appear to the left of the third tick mark. At this time, the RIO may select INS and taxi may be initiated if OBC is complete.

Carrier Alignment Non SAT

The carrier alignment procedure is used when the INS is to be aligned on the deck of a carrier where ship's latitude, longitude, speed, and heading are changing. There are three methods of aligning the INS on a carrier deck: rf data link alignment, deck-edge cable alignment, and manual (hand-set) alignment. All three methods use the same TID alignment display as described under GROUND ALIGNMENT. Refer to section III, part 3, Ship-Based Procedures for step-by-step procedures.

CARRIER DATA LINK ALIGNMENT

The rf data link alignment (CAINS) is the primary carrier alignment mode. This mode uses the Ship's Inertial Navigation System to align the IMU. The inertial inputs are received and transmitted by the rf data link to the WCS computer. These inputs include ship's longitude, latitude, north and east velocity, incremental north and east position changes, roll, pitch, heading, heading rate, and octant of geographic position.

The inertial data are in a data link message and are transmitted by the shipboard data link equipment. To align the INS by the CVA alignment method, turn on the power to the data link system, turn the WCS power to the STBY position, set the D/L mode on the DATA LINK control panel to CAINS/WAYPT, and select CVA ALIGN on the NAV MODE switch. This procedure must be used for rf or SINS cable data transmission. The received message is processed by the data link equipment in the aircraft and fed to the WCS computer. The computer compares the IMU data with the received CAINS data and after processing sends correction signals to the CSDC for fine alignment.

CVA ALIGN is accomplished in a similar manner as in GND ALIGN, with the exception that reference parameters are received through the data link set.

If the aircraft is moved during alignment, a signal from the parking brake is sent to CSDC to suspend alignment. Suspend alignment is indicated by flashing STBY and/or READY lights and the stopping of the time into alignment clock. When the suspend alignment discrete is removed by resetting the parking brake, the alignment will either reinitialize and continue (if the suspend occurred during coarse align (∨), or the alignment will continue (if the suspend occurred during fine align (◇)).

Note

If SINS data link is lost during taxi, a flashing HS will appear on the TID. This will disappear when data link is reacquired; however, due to align timing requirements it may remain flashing up to 8 seconds after data link is reacquired. If the HS flashing does not stop 8 seconds after resetting the parking brake, SINS data is lost but the alignment can continue by entering carrier speed and true heading into the own aircraft file and completing the align in hand-set mode. If data link is reacquired during this period, the HS will disappear from the TID and a normal data link CVA align will continue.

At the completion of alignment, the INS position is selected on the NAV MODE switch. A properly functioning and aligned INS is indicated by both the STBY and READY lights off and the appearance of the IN acronym in the attitude status readout on the TID.

Note

- System performance will degrade after a CVA alignment if the INS mode is selected while the ship is in a turn, even though the system indicated fine-align complete. This degradation is caused by a lag in wander azimuth angle between the aircraft inertial platform and the SINS platform during the ship's turn. Since there is no lag or resultant degradation when the ship is steady on course, the navigation system will operate normally if the selection of INS is delayed until the ship's turn is completed. Stored heading alignments are not affected under the above operating conditions, and the system will operate normally if the INS mode is entered after alignment while the ship is in a turn. Once the transition from ALIGN to INS mode is completed, there are no further comparisons made of aircraft IMU and SINS/D/L parameters.

- If during a CVA alignment the CAINS/WAY PT — TAC switch is unlatched to the TAC position by a power transient, or data link signal is lost, the INS will revert to a hand-set alignment (HS).

CARRIER CABLE ALIGNMENT

The deck-edge cable alignment (SINS) is an alternate for the rf data link alignment. When in CVA ALIGN, SINS inputs can also be sent over a secure cable to the data link from the deck-edge outlet box on the carrier. Switching from rf data link to cable inputs is achieved automatically upon cable hookup.

HAND-SET ALIGNMENT

The hand-set (HS) alignment mode is a manual shipboard alignment option available should SINS data from rf data link or cable be unavailable, inaccurate, or interrupted (indicated by the illumination of the TILT light on the DDI). The HS mode is similar to ground alignment, but the RIO must insert more data and the computer requires more processing to account for ship movement.

If SINS data is not available and CVA ALIGN is selected with the NAV MODE switch, a flashing HS acronym appears between the second and third tick marks of the alignment display on the TID. For all cases where the flashing HS

acronym appears before the alignment has started and the RIO elects to align the system with the HS option, he must insert the following appropriate ship's data:

- Latitude
- Longitude
- Speed
- Ship's true heading

Note

If the HS acronym is not flashing, proper SINS data has been entered into the system.

For aircraft data, use the OWN A/C pushbutton on the the CAP NAV DATA matrix, and the LAT and LONG prefix pushbuttons; for carrier heading and speed, use own-aircraft HDG and SPD pushbuttons. After this data is entered, the HS acronym stops flashing. At this point, the alignment progresses as with ground alignment, but the alignment may take three times as long.

Note

The success of this particular alignment will depend upon the carrier maintaining a constant heading and speed during alignment.

REINITIALIZATION

During CVA ALIGN, if it is decided that the alignment should be restarted for any reason, the capability exists to reinitialize the alignment without going through the power-up and tape read cycles. This is accomplished by cycling the NAV MODE switch to GND and back to CVA. This will reinitialize only the computer's alignment processing, starting with the initial values.

CATAPULT ALIGNMENT

With the incorporation of the SMAL program, the primary function of the CAT ALIGN mode is to inhibit suspend align while positioned on the catapult, when the parking brake has been released. Therefore, the function of the catapult align mode is to prolong normal CVA ALIGN as much as possible.

With CAT ALIGN selected, large ship roll, pitch, heading, or velocity inputs can cause the program to automatically switch to INS.

SAT Alignments

The computer program makes possible simultaneous running of INS alignment (SMAL) and OBC with weight on wheels only. The only differences between a SAT and a NON SAT alignment are the entering and initializing of the alignment and the reinitializing and exiting of the alignment mode. The SAT mode is entered automatically by selecting an align mode on the TID NAV MODE switch prior to the successful completion of power-up auto BIT sequence 2. If auto BIT sequence 2 fails, SAT will not be entered. SAT may also be manually initiated at the conclusion of auto BIT sequence 2. Either SMAL or OBC may be initiated first, but the second mode initiation causes the SAT entry. If auto BIT sequence 2 is completed before selection of an alignment mode, independent execution of SMAL and OBC is still possible. There are two ways in which to reinitialize the SAT alignment: (1) by depressing the PRGM RESTRT tilebutton with the NAV MODE switch in its present position (this will exit SAT) and (2) cycling the NAV MODE switch to any position (other than OFF or CAT ALIGN) and then back to the original position. This is the preferred method, because SAT will not be exited.

Note

Cycling the NAV MODE switch to OFF and then back to its original position during SAT alignment may result in a degraded alignment.

To have the system accept the alignment, select INS, or have weight-off-wheels condition with a diamond (◇). Figure 7-18 outlines the methods and condition for SAT and NON SAT alignments. After INS selection, depress PRGM RESTART to exit SAT display and to obtain tactical display.

NAVIGATION CONTROLS AND DISPLAYS

Navigation Displays

Tactical navigational information is displayed on the VDIG, MDIG, and BDHI. The type of information displayed is predicted on the flight mode and steering command selected. System navigation information is displayed on the TID and HSD. Figure 7-19 is a summary of system outputs available to the displays. Specific presentations for each navigation mode are presented in the navigation modes and steering section. All displays provide navigation information with respect to magnetic north, unless alignment is in progress.

Navigation Controls

The INS is controlled with the navigation control and data panel and the computer address panel (figure 7-15). The selection made at the navigation control and data readout panel determine the operation mode, alignment mode, and the desired destination point. The computer address panel permits the insertion of navigation parameters and the selection of information to be displayed on the TID. The navigation control and data readout panel also has platform STBY and READY advisory lights, which report the status of the alignment program and of the navigation system. Failure indicators for the major assemblies of the navigation system are on the caution/advisory light panels in both cockpits.

Method	Enter	Exit	Reinitialize
SAT-Automatic	Set NAV MODE switch to ALIGN (GND, CVA, CAT) during auto BIT sequence 2. Alignment and OBC run concurrently upon completion of Auto BIT sequence 2.	Set NAV MODE switch anywhere but ALIGN and depress the PRGM RESTRT pushtile manual selection of BIT sequence 2.	Depress PRGM RESTRT pushbutton.
SAT-Manual	Set NAV MODE switch to ALIGN (GND, CVA, CAT) after auto BIT sequence 2 completed or OBC initiated; second mode entry causes SAT.	Weight-off-wheels completes alignment, if diamond (◇) present.	Set NAV MODE switch from any position back to original ALIGN mode position.
Non SAT	Set NAV MODE switch to ALIGN (GND, CVA, CAT).	Set NAV MODE switch to INS, IMU, AHRS or OFF.	Set NAV MODE switch from any position back to original ALIGN mode position.
Note:	Reinitializing by depressing PRGM RESTRT pushbutton will exit SAT. Preferred method is to cycle NAV MODE switch.		

Figure 7-18. Alignment Operation

NAVIGATION DISPLAYS SUMMARY

NAVIGATION SYSTEM OUTPUT	DISPLAY	
	PILOT	RIO
ADF BEARING	HSD, BDHI	ECMD, BDHI
ALTITUDE (A/C)	HUD, ALTIMETER	TID
BEARING TO DESTINATION	HSD	ECMD, TID
COMMAND COURSE TO DESTINATION	HSD	ECMD
COMMAND HEADING TO DESTINATION	HSD, VDI	ECMD
COMMAND ALTITUDE AND AIRSPEED	VDI	—
GROUNDSPEED (A/C)	HSD	ECMD, TID
GROUND TRACK (A/C)	HSD	TID
LATITUDE AND LONGITUDE (A/C)	HSD (TID REPEAT)	TID
MAGNETIC HEADING (A/C)	HUD, VDI, HSD, BDHI	ECMD, BDHI, TID
MAGNETIC VARIATION (COMP)	HSD (TID REPEAT)	TID
RANGE TO DESTINATION	HSD	ECMD, TID
ROLL AND PITCH (A/C)	HUD, VDI, SAI, HSD	SAI, TID, DDD
STEERING ERROR TO DESTINATION	VDI, HSD	ECMD
TACAN DEVIATION	HUD, VDI, HSD	ECMD
TACAN RANGE AND BEARING	HSD, BDHI	ECMD, BDHI
TIME TO GO	HSD, (TID REPEAT)	TID
TRUE AIRSPEED (A/C)	HSD, AIRSPEED MACH INDICATOR	ECMD, TID
TRUE HEADING (A/C)	HSD (TID REPEAT)	TID
VERTICAL SPEED	HUD, VSI	—
WIND SPEED AND DIRECTION	HSD	ECMD, TID

5-F050-004
129-0

Figure 7-19. Navigation Displays Summary

Note

Except when attempting to clear CSI or CSD acronyms in accordance with procedures in the Troubleshooting Guide in NAVAIR 01-F14AAA-1B, do not pull the CSDC circuit breaker or secure aircraft electrical power unless the NAV MODE switch is OFF. Failure to follow this procedure results in unnecessary battery discharge.

Control of the pilot's displays (HUD, VDI, and HSD) and the flight officer's ECMD for navigation is achieved through the pilot's display control panel and the ECMD control panel, respectively.

Note

For detailed information on CAP operation, refer to controls and displays in NAVAIR 01-F14AAA-1A.

For navigation, the computer address panel is used to request computer data for readout and to insert data for computation or display. Data inserted or read out from the computer is displayed on the TID. The CATEGORY switch on the lower end of the panel affects the function of the MESSAGE pushbutton. For navigation only, two categories are considered: NAV and TAC DATA.

NAVIGATION CATEGORY

When the CATEGORY switch is set to NAV, the following matrix appears in the MESSAGE windows:

OWN AC	TACAN FIX
— —	RDR FIX
— —	VIS FIX
— —	FIX ENABLE
WIND (SPD/HDG)	MAG VAR (HDG)

For each window, there is a pushbutton. Depressing a pushbutton tells the WCS computer which function of the matrix is being initiated. When OWN A/C, WIND, or MAG VAR is depressed, data can be entered or displayed concerning each.

Normally, own-aircraft airspeed and magnetic heading are displayed on the TID. If own-aircraft data file is hooked using the TID cursor, heading will be magnetic. If OWN A/C pushbutton was selected (hooked) via the CAP, own-aircraft true heading, speed (ground speed), altitude, or course can be displayed on the TID by depressing the appropriate prefix pushbutton (figures 7-19 and 7-20). Either the LAT or LONG pushbutton will display own-aircraft latitude and longitude. Depressing the SPD pushbutton displays groundspeed and magnetic course. However, true airspeed and true heading are displayed when the HDG prefix pushbutton is

7-46

TID READOUT PAIRS

LEFT READOUT	RIGHT READOUT	HOOK REQUIRED
LN (LATITUDE NORTH)/ LS (LATITUDE SOUTH)	LE (LONGITUDE EAST)/ LW (LONGITUDE WEST)	
* GS (GROUNDSPEED)	MC (MAGNETIC COURSE)	YES
AS (TRUE AIRSPEED)	* MH (MAGNETIC HEADING)	OWN A/C VIA TID CURSOR
AS (TRUE AIRSPEED)	* TH (TRUE HEADING)	OWN A/C VIA CAP
GS (GROUNDSPEED)	** MH (MAGNETIC HEADING)	YES
VM (MAGNETIC VARIATION)	MH (MAGNETIC HEADING)	MAG VAR (MANUAL)
VC (MAGNETIC VARIATION)	MH (MAGNETIC HEADING)	MAG VAR (COMPUTED)
RA (RANGE)	* MB (MAGNETIC BEARING)	YES
* RA (RANGE)	TG (TIME TO GO)	NAV POINT
* RA (RANGE)	BLANK (IF NO LAR)	SENSOR TARGET
--	OR RM (MAXIMUM LAUNCH RANGE)	--
--	OR RO (OPTIMUM LAUNCH RANGE)	--
--	OR RI (MINIMUM LAUNCH RANGE)	--
GS (WINDSPEED)	MD (MAGNETIC DIRECTION)	WIND
--	AN (ANGLE)	CAT 1, 2, 3, OR 4
* AL (ALTITUDE)	--	YES
* AD (ALTITUDE DIFFERENCE)	--	IP TO TARGET
* NB (NUMBER)	--	SPECIAL TEST

* INDICATES PREFIX SELECTED   ** READS TRUE HEADING IF SELECTED DURING ALIGNMENT

Figure 7-20. TID Readout Pairs

depressed. Altitude is displayed on the left TID readout (right is blank) when the ALT pushbutton is used. To change or enter own-aircraft latitude, longitude, true heading, or altitude, the procedure is to depress the respective prefix pushbutton followed by the desired quantity. While the data is being entered, it is displayed on the upper middle readout on the TID. At the same time, the present data is being displayed on the two lower readouts. If the new data is correct, the RIO depresses the ENTER pushbutton, and the new value appears on the readout.

If the WIND pushbutton is depressed, the TID will display present windspread (left readout) and magnetic direction (right readout). If an entry is desired, the WIND pushbutton is depressed, then either the SPD or HDG prefix pushbutton, and then the appropriate numbers; knots (0 to 512) for speed or degrees (000 to 359) for magnetic direction are entered. The ECMD data readout of WIND direction is always displayed as true.

The MAG VAR pushbutton is used for displaying and entering magnetic variation (MAG VAR). Depressing it displays alternating values of computed MAG VAR (vC) and manual MAG VAR (vM) on the left readout and displays magnetic heading (MH) on the right readout (figure 7-21). The two values alternate every 2 seconds. As indicated on the CAP sign/direction buttons, plus corresponds to east variation and minus to west variation.

To enter manual MAG VAR, depress the MAG VAR pushbutton. Then depress HDG, E, or W, the angle in degrees and tenths, and ENTER. Tenths of a degree must be entered even if zero. Computed MAG VAR is continuously calculated in the AWG-9 by comparing true heading from the IMU with magnetic heading from AHRS. The difference is stored as computed MAG VAR.

The MAG VAR source used by the computer for displays and CAP entries is summarized in figure 7-22.

Computed MAG VAR and manual MAG VAR are continuously compared by the AWG-9 computer. If they differ by $5°$ or more, the acronym MV appears alternately with the IN or IM navigation mode acronym on the TID and HSD. The acronym is cleared when the difference becomes less than $5°$.

Note

When operating in SLAVED or COMP mode near a magnetic disturbance, such as aboard a carrier, the MV acronym should be expected to appear.

If the MV acronym appears in flight with no AHRS, IMU, or NAV COMP failure lights, the RIO should call up the MAG VAR display and determine which is in error, vC or vM. If vM is in error, entering the correct MAG VAR should clear the acronym.

MAG VAR READOUT

MANUALLY ENTERED

vM + 11.5 MH 015

CALCULATED

vC + 11.5 MH 015

1-F060-004
255-0

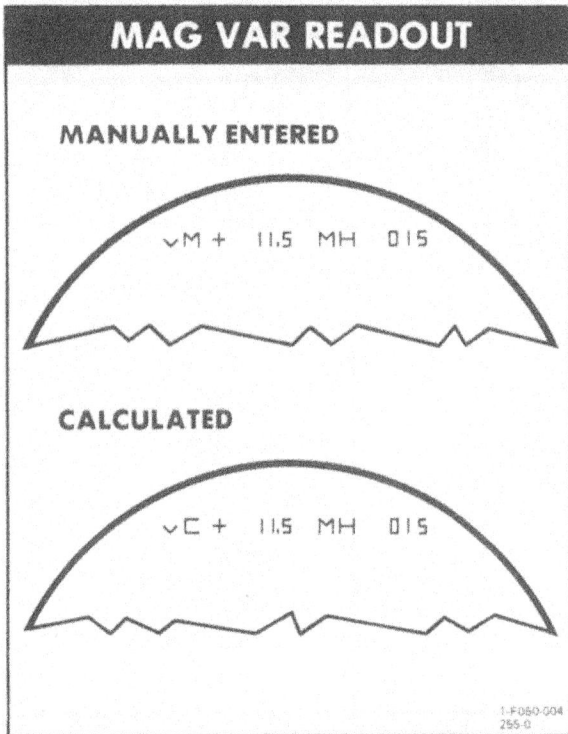

Figure 7-21. MAG VAR Readout

If vC is in error, it is caused by incorrect heading information from either the AHRS or the IMU. While in the SLAVED compass mode, INS or IMU/AM navigation mode, and level unaccelerated flight, the pilot should compare the heading on VDI or HUD with the standby compass. If they agree, the problem is probably in the IMU. To confirm this, switch to COMP mode on the AHRS compass controller panel. VDI and HUD headlines are now IMU true heading with manual MAG VAR applied. If they now disagree with the standby compass, the IMU heading is erroneous. The RIO should now select the AHRS/AM navigation mode, then enter the correct MAG VAR.

If VDI/HUD heading and standby compass heading did not initially agree, the problem is in the AHRS. Synchronize the AHRS by depressing the HDG pushbutton. If it synchronizes correctly, the MV acronym will clear. If it will not synchronize properly or stay synchronized, switch to the COMP mode and remain there. All computer and

CRT display functions will now use IMU true heading with manual MAG VAR applied. The MV acronym may or may not clear. The BDHI will be receiving unfiltered heading information from the flux valve and may or may not be correct, depending on what has malfunctioned in the AHRS.

TACTICAL DATA CATEGORY

When the CATEGORY switch is set to TAC DATA, the following matrix appears in the MESSAGE windows:

WAY PT 1	HOME BASE
WAY PT 2	DEF PT
WAY PT 3	HOST AREA
FIX PT	SURF TGT
IP	IP TO TGT

Each function in this category has an accompanying TID symbol, except IP to TGT. When any one of these MESSAGE pushbuttons is depressed, the TID symbol brightens and the activated MESSAGE pushbutton illuminates, indicating completion of a hook. The RIO can then perform the functions for which hooking was required. Additionally, data concerning the hooked point may be displayed on the TID readouts by depressing the appropriate prefix pushbutton. Also, the latitude, longitude, and altitude of the hooked point may be entered by depressing either the LAT, LONG, or ALT pushbutton, followed by the desired numerals. As before, the desired numerals should appear on the TID, and if correct, the RIO then enters the data into the WCS computer by depressing the ENTER pushbutton.

Navigation Lights

The caution advisory panel on the RIO's right knee panel has three advisory lights that indicate failures within the navigation system (IMU, NAV COMP, AHRS). The panel also has two other advisory lights, C&D HOT and AWG-9 COND, that are indirectly related to navigation system operation. Illumination of either or both of these lights could mean degraded navigation operation and would require further investigation of the WCS. Refer to NAVAIR 01-F14AAA-1A.

CONDITION	MAG VAR SOURCE
COMP mode selected by pilot.	Manual MAG VAR (vM)
RIO enters manual MAG VAR after selecting AHRS navigation mode.	Manual MAG VAR (vM)
RIO updates MAG VAR after IMU or AHRS failure.	Manual MAG VAR (vM)
All other conditions	Current or last value of computed MAG VAR (vC)

Figure 7-22. MAG VAR Source Logic

NAVIGATION UPDATING

Updates are used to correct an error of latitude or longitude in the computer position of the aircraft. Updating is important in the backup modes (AHRS AN and IMU AM) because of the estimated winds and magnetic variation changes. The navigation updating techniques use a ground reference point (latitude and longitude) position. The range and bearing of this position to present aircraft position is used for correction. The navigation system may be updated by either a radar fix, TACAN fix, or visual fix.

Prior to updating, the latitude and longitude of the desired update point (radar, TACAN, or visual) must be stored in one of eight navigation point locations (three WP's, FIX PT, HOME BASE, HOST AREA, DEF PT, and IP). This data may be stored prior to flight by data link or by manual insertion. The point selected for the update must be hooked. The prestored latitude and longitude will be displayed on the TID. The CATEGORY select switch is rotated to NAV and the desired type of update selected.

Radar Update

For a radar update, the WCS computes own-aircraft position by measuring radar range and bearing from the reference point coordinates in the track file.

Once the update point is called up, its latitude and longitude are verified on the TID readouts. The same point is then located on the DDD, using the hand control with the radar operating in the pulse search mode (1 BAR and STAB IN, ground mapping). DDD CURSOR is selected on the hand control and half-action is selected so that the DDD cursors are presented on the DDD. Once the cursors overlay the selected point, FULL ACTION is selected. This tells the computer the point selected. When the RADAR FIX pushbutton is depressed, the computer will compute the present position of the aircraft by measuring the range and bearing from the selected point. The difference between the computer position and the position determined by the INS is then displayed on the TID. If it is desired to enter this delta into the navigation computations, the FIX ENABLE pushbutton is depressed. However, if the observed delta does not appear to be correct, the computer and the readout can be cleared by depressing the RADAR FIX pushbutton. The fix may then be attempted again.

Note

OBC continuous monitor does not test own aircraft altitude system. System altitude errors over 10,000 feet could exist without any indication to the flightcrew. The RIO should perform periodic checks of own-aircraft system altitude and update the altitude, as required.

Radar updating is performed as follows:

1. TID CURSOR/CAP . . . . HOOK DESIRED
                    NAVIGATION POINT FOR UPDATE

2. PULSE SRCH pushbutton . . . . DEPRESS

3. On sensor control panel:

    a. STAB switch . . . . IN

    b. EL BARS switch 1

    c. AZ SCAN switch . . . . AS DESIRED

4. RDR FIX button . . . . DEPRESS

5. DDD CURSOR pushbutton . . . . DEPRESS

6. Action switch . . . . HALF ACTION
                                            (first detent)

7. Cursor is displayed on DDD.

8. Manipulate hand control DDD cursor over desired ground map point.

9. Action switch . . . . FULL ACTION
                                            and RELEASE

Note

This causes the DDD cursor to remain at the selected position.

10. Observe delta of LAT and LONG on TID.

11. If readouts are unsatisfactory, deselect RDR FIX and repeat steps 4 through 12.

12. FIX ENABLE pushbutton . . . . DEPRESS

13. Update is accomplished.

Note

To clear previous hooked DDD cursor
position, go to half action and then
release prior to initiating FULL ACTION
for new position hook.

TACAN Update

Updating the navigation system by TACAN requires that
the prestored point in a track file be the same latitude and
longitude as the TACAN station to be used for the update.
The TACAN channel that corresponds to the station
selected must be selected and verified by listening to the
identifier (coded tone) in the headset, and lockon using
the BDHI or HSD. To update the aircraft position with
respect to the station, the TACAN FIX pushbutton is
depressed. The WCS computer then computes own-aircraft
position error based on the range and bearing from the
TACAN station. The delta is then observed and entered
into the computer in the same manner as for radar updating.

TACAN updating is performed as follows:

1. Select a TACAN chnnel whose latitude and
   longitude correspond to an update point.

2. Hook desired update point
   (WAY PT 1, FIX PT, HOME BASE, etc.)

3. CATEGORY switch . . . . NAV

4. TACAN FIX pushbutton . . . . .DEPRESS

5. Observe present position delta readout.

6. Cross-check present position latitude and
   longitude with BDHI.

7. If delta is unsatisfactory, deselect TACAN FIX
   and repeat steps 2 through 8.

8. FIX ENABLE pushbutton. . . . . .DEPRESS

Note

● The delta update is entered into the computer
  by depressing FIX ENABLE pushbutton.

● When performing a TACAN update, MAG
  VAR must be correct; otherwise, the update
  will be in error.

Visual Update

The visual update is performed by flying over the prestored
point and depressing the VIS FIX pushbutton. A timing
estimate must be made since the aircraft nose and fuselage
may obscure the fix point for some time during the
overflight. Also, it is difficult to estimate when directly
overhead a ground reference point when altitude is
greater than 10,000 feet. The delta then appears on the
TID. Again this delta may be entered into the computer
by depressing FIX ENABLE.

Visual updating is accomplished as follows:

1. Hook desired update point
   (WAY PT, HB, IP, etc.)

2. Select NAV cateogry on CAP.

3. Overfly the selected prestored point and
   when over the point, depress the VIS FIX
   button on the cap.

4. If a satisfactory delta is displayed, depress
   the FIX ENABLE button; this causes the delta
   correction of own-aircraft position to be inserted
   into the computer.

Note

● If the delta is not satisfactory, reselecting
  the VIS FIX button will clear the delta
  readouts from the TID and the procedure
  may be repeated.

● Navigation updating while in the INS,
  and to a lesser degree in IMU, will most
  often enter a greater navigational position
  error than already exists. Navigational
  updating has its greater application, and
  is required for more often, in the AHRS
  mode.

Position Marking

The SURF TGT position of the TAC DATA category
may be used to mark the position of a pulse radar
target, a visual target, or a TACAN station for display
on the TID. When displayed on the TID, latitude,
longitude, range, bearing, and steering data are
displayed, using the CAP or the navigation destination
control, or both.

Note

The SURF TGT position is not to be used
for updating the navigation computer.
The surface target position symbol is
repositioned with respect to own-
aircraft vice own-aircraft being updated
in reference to the surface target.

When a pulse radar target is to be marked and displayed
on the TID, the method is to first select the SURF
TGT pushbutton. Then complete a normal RADAR
FIX. For visual targets, the method is the same, but
a visual fix is required. A TACAN station that is
providing good bearing and range may also be marked
using the same method and following the TACAN
FIX procedures. When any of the above procedures
are completed, the SURF TGT symbol is displayed
on the TID at the computed latitude and longitude
coordinates.

The surface target symbol may also be used as a
destination point. If its position has been previously
entered, the symbol will appear on the TID. One
method for special position marking is to hook any
point on the TID and select SURF TGT. The surface target symbol now appears over the hooked point
and its position is stored in the WCS computer.
For example (a radar fix), select the DDD cursor,
using the pulse system for radar mapping, designate
the point of interest by placing the cursor over the
point and select FULL ACTION. Then select
RADAR FIX, which will present a delta from the
hooked point to the surface target. Ignore the delta
and select FIX ENABLE, which will position the
surface target symbol over the previously hooked
radar position. Now, a very accurate readout of
latitude, longitude, and steering information is
available to the point. This procedure may also be
used for visual or TACAN fixing.

ATTITUDE AND HEADING REFERENCE SET (A/A24G-39)

In the absence of attitude information from the
inertial navigation system, the attitude and heading
reference set (AHRS) provides backup pitch and roll
information to the CSDC and to the WCS computer.
At all times, the AHRS provides prime magnetic
heading to the BDHI for direct analog display and
to the CSDC where it is converted to digital information
for the VDIG, MDIG, and the WCS. The AHRS also
provides heading reference for the autopilot.

Note

The BDHI is the only analog cockpit display of magnetic heading. The other
cockpit displays (HUD, VDI, TID, HSD,
and ECMD) are digital and receive their
inputs from AHRS through the CSDC.
Therefore, should there be a CSDC
failure, the only magnetic heading is
displayed on the BDHI.

Basic components of the AHRS include a two-gyro
platform (vertical and directional displacement gyros),
an electronic control amplifier, and a compass controller
(figure 7-23). Also associated with the AHRS is a
magnetic azimuth detector (MAD) and an electronic
compensator. The platform consists of gyros, level
sensors, gimbals, and associated electronics. The
platform is unlimited in roll, but limited to 82° in pitch.
If the IMU fails, the CSDC automatically selects AHRS
attitude information for display and autopilot control.
The directional gyro is also used to filter the flux valve
heading signal in the SLAVED mode or to provide a
direct heading reference in the DG mode. The resulting
heading is subsequently transmitted to and used by the
BDHI, the CSDC, and the WCS.

Note

- IMU heading is available in the INS
  mode for display. The IMU heading
  must be converted to magnetic by adding
  or subtracting MAG VAR, which is a
  function of AHRS. It is more accurate
  to use AHRS heading for all magnetic
  displays.

- AHRS does not have an all-attitude capability and will precess if pitch attitudes
  exceed ±82°. A gradual precession in
  roll, pitch and heading can also be expected
  in sustained turns at slow rates (less than 6°
  per minute). Large roll and pitch precession
  errors can be corrected by flying straight and
  level, without accelerating, and pressing and
  holding the HDG set pushbutton on the compass controller panel. Pressing and holding this
  button corrects precession errors at a rate
  of 12° per minute minimum. The HDG set
  pushbutton should be held for at least 3
  minutes. Before repeating the 3-minute
  cycle, it should be released for at least
  1 minute.

COMPASS CONTROLLER

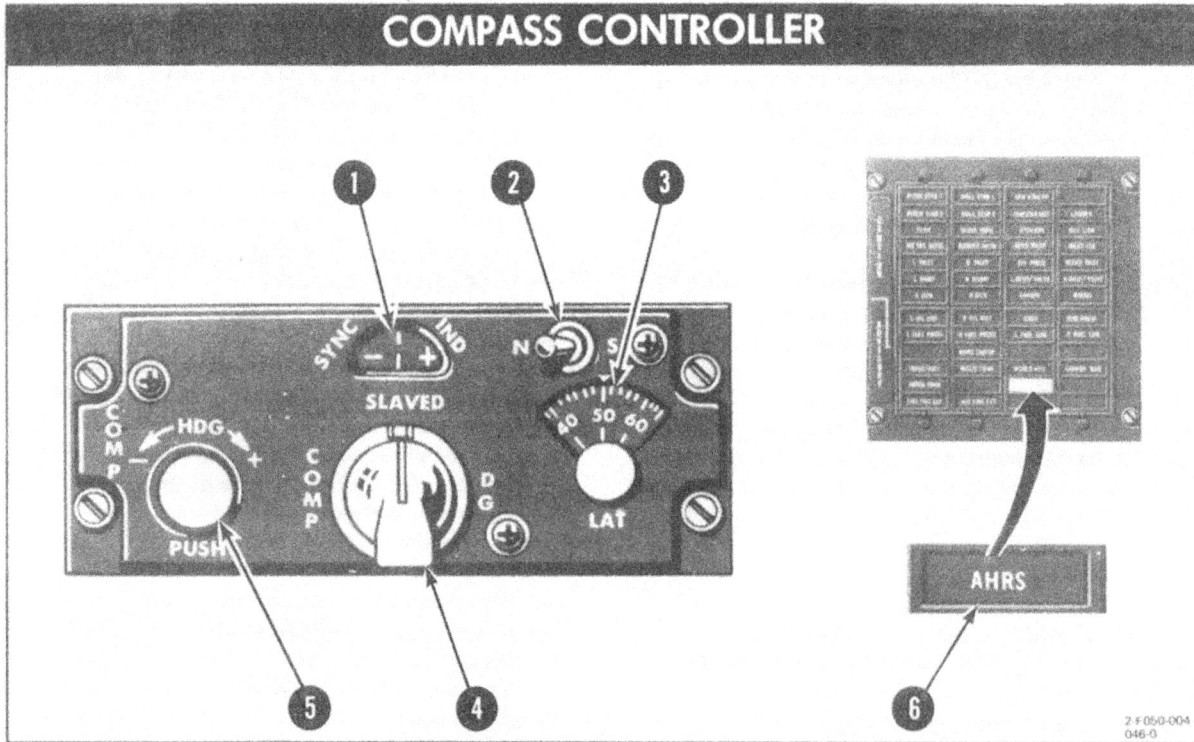

2-F-050-004
046-0

NOMENCLATURE	FUNCTION
(1) SYNC IND	Indicates synchronization between AHRS directional gyro and magnetic azimuth detector. Indicator is deactivated in DG and COMP modes.
(2) N-S hemisphere select switch	In DG and SLAVED modes, hemisphere (N or S) in which aircraft is operating must be preselected to provide proper earth-rate correction.
(3) LAT correction control	Used for DG and SLAVED modes; latitude from 0° to 90° must be preselected to allow for earth rate corrections.
(4) COMP-SLAVED-DG mode switch	COMP — Compass heading is obtained directly from magnetic detector without stabilization by the platform directional gyro. This position is used if the gyro malfunctions and for emergencies only. The HUD, VDI, HSD, and ECMD use manual magnetic variation (vM) automatically in this mode.

SLAVED — Normal position, gyro-stabilized magnetic heading is obtained from magnetic azimuth detector.

DG — Directional gyro mode uses selected gyro heading; magnetic azimuth detector is not used. |

Figure 7-23. Compass Controller (Sheet 1 of 2)

NOMENCLATURE	FUNCTION
⑤ HDG pushbutton	In SLAVED mode, the pushbutton synchronizes the directional gyro with the magnetic azimuth detector and sets magnetic heading on the BDHI. Pushbutton must be depressed and held until the synchronization indicator needle is bracketing the null mark. (Synchronization cannot be accomplished if aircraft is accelerating or decelerating by more than 75 knots per minute.)
	In DG mode, the pushbutton shall be depressed and rotated CW or CCW to set the selected heading on the BDHI.
	The pushbutton has no function in COMP mode.
	Fast erection in pitch and roll (12° per minute minimum) is obtained by depressing and holding pushbutton for up to 3 minutes (button must be released for at least 1 minute before repeating the 3-minute cycle).
⑥ AHRS advisory light	Green light illuminates when attitude or heading information from AHRS is unreliable.

Figure 7-23. Compass Controller Panel (Sheet 2 of 2)

Displacement Gyro Assembly

The displacement gyro assembly includes a directional and a vertical gyroscope. The directional gyroscope establishes a reference from which heading deviation is measured. The vertical gyro provides a horizontal reference from which roll and pitch outputs are determined. These outputs are transmitted to the CSDC and the AFCS. The AFCS receives steering error signals from the CSDC, based on changes in AHRS heading.

Electronic Control Amplifier

The amplifier supplies power to the displacement gyro and is the interface between the displacement gyro and aircraft subsystems. The amplifier also handles AHRS malfunction detection, gyro erection, and compass mode logic. It incorporates the electronics for slaving the heading gyro, for compensation of magnetic heading errors, for the compass mode affected, and for erection.

Magnetic Azimuth Detector (MAD)

The MAD provides the magnetic heading for the aircraft. It interfaces with the electronic control amplifier and gyro, which acts to stabilize and amplify the magnetic heading signals. The MAD is in the left vertical tail section.

Compass Controller Panel

The compass controller panel has the controls for selecting one of three compass modes when the AHRS is used as the heading reference (figure 7-23).

Where magnetic heading references are unreliable, the system should be operated in DG mode. In areas where magnetic references are reliable, the system shall be operated in the SLAVED mode. When DG or SLAVED modes are inoperable, the COMP mode shall be employed for emergencies.

The AHRS operates from the ac essential no. 2 bus through the AHRS PH A, PH B, PH C circuit breakers (3C6, 4C1, 4C5) on the RIO's left side circuit breaker panel.

Note

If both the IMU and the AHRS gyros fail, pitch and roll attitude indications are removed from the HUD, TID, and DDD, and the IMU and AHRS advisory lights illuminate. Selecting COMP mode on the compass controller panel may restore valid magnetic heading information to the HUD, VDI, and HSD, and the AHRS advisory lights will go off. However, invalid pitch and roll attitude information will be restored to the HUD and VDI and should be disregarded.

AHRS Operation

As a compass, the AHRS operates in three modes: the directional gyro (DG) mode provides a free-gyro heading reference with earth-rate correction; the SLAVED mode provides a gyro-stabilized magnetic heading; and the compass (COMP) mode provides an emergency magnetic heading from the compass transmitter only. The random drift (precession rate) of the gyro in the DG mode does not exceed 1.5° per hour. This mode may be used at all latitudes, but is more useful when operating in regions where the magnetic field is weak or distorted. When the COMP mode is selected, the AFCS is automatically disengaged to prevent erratic steering commands. The COMP mode does not provide a sufficiently stable heading signal for AFCS operation and shall be used only for emergency operation. To erect the AHRS, press and hold the HDG set pushbutton on the compass controller (3 minutes on, 1 minute off cycle) until the needle of the synchronous indicator is bracketing the null mark.

AHRS does not have an all-attitude capability, and will precess in pitch and/or roll if pitch attitudes exceeding $\pm 82^\circ$ are attained. If the navigation system is operating in the INS or IMU/AM modes, attitude displays will continue to indicate properly when the AHRS pitch limit is exceeded, but all displays of magnetic heading will be in error and the advisory lights may be on or off. If this condition is encountered, accurate and stable magnetic heading displays on the HUD, VDI, HSD, TID and ECMD can be regained immediately by the following procedure:

1. Pull the AHRS PH A, B, and C circuit breakers (3C6, 4C1, 4C5).

2. Insert proper MAG VAR via the computer address panel.

This procedure will render the AHRS and BDHI magnetic heading indication inoperative and provide IMU derived magnetic heading for the HUD, VDI, HSD, TID and ECMD displays.

AHRS RESET PROCEDURE

To regain the AHRS for both magnetic heading information and backup navigation/attitude capability, the following procedure is recommended:

1.   While flying straight and level, reset the AHRS PH A, B, and C circuit breakers (3C6, 4C1, 4C5). (These breakers must be pulled for a period of at least 30 seconds before resetting).

2.   Allow three minutes of AHRS erection time, and then sync magnetic heading via the HDG pushbutton on the compass control panel.

A gradual AHRS precession in roll can be expected when in sustained turns (exceeding $3°$ per minute for extended periods of time). AHRS pitch and roll precession errors can be corrected by flying straight and level and pressing and holding the HDG pushbutton on the compass controller panel. Holding this button in erects the AHRS at a rate of $12°$ per minute.

```
CAUTION
```

If an undetected AHRS failure occurs, an erroneous value of MAG VAR will be computed, and the Mv acronym will appear on the TID. The flightcrew should isolate the navigation failure and correct it.

AHRS BIT

A validity signal from the AHRS is continuously monitored by the CSDC. Absence of this signal indicates an AHRS failure. The BIT monitors internal error signals and electrical characteristics of the AHRS; when the BIT detects an AHRS failure, the AHRS light on the caution advisory panel illuminates. When the CSDC detects this failure, the WCS computer is informed and the failure is displayed by the TID status lights. The WCS computer also informs the CM routines of the failure, providing an AHR acronym display on the TID OBC CM readout.

COMPUTER SIGNAL DATA CONVERTER (CSDC)

The computer signal data converter (CSDC) performs the inertial navigation computations after the INS is complete and is also the interface between the various navigation subsystems and the auxiliary equipment (FO-19). Specifically, the CSDC consists of an analog-to-digital and digital-to-analog converter and a miniature general-purpose computer.

The CSDC receives accelerometer inputs, roll, pitch, and azimuth information, calibrated data, as well as numerous discretes from the IMU. The CSDC also receives magnetic heading from the AHRS; radar altitude from the radar altimeter; TACAN range and bearing from the TACAN set; ADF bearing from the UHF/ADF system; true airspeed, Mach number, airstream air temperature, and pressure altitude from the CADC. Groundspeed, range-to-destination, and relative command course and heading are received from the WCS. The CSDC converts or rescales all these inputs to the digital format required by the CSDC computer to perform the following computations for outputs and displays:

● Pitch and yaw rate

● Platform velocities (Vx, Vy, Vz)

● Present position (latitude and longitude)

● System altitude

● True or magnetic heading

● Wander angle

● TACAN deviation and relative bearing

● Command airspeed error

● Command altitude error

The CSDC calculates the wander angle from a matrix based on the direction cosines from two coordinate frames. One is the earth's coordinate frame; the other is established by the IMU. The earth's reference frame is established by inserting a known latitude and longitude (for example, home base). The IMU reference frame is established after warmup.

The system rotates the two coordinate frames so that they coincide, thus generating the necessary direction cosines. From this the CSDC calculates present position and wander angle through simple computational techniques. The matrix is continuously updated by the CSDC when it calculates own-aircraft velocities.

The CSDC furnishes the following data to WCS
computer for weapons system steering.

- Aircraft present latitude and longitude

- System altitude

- Inertial velocities (Vx, Vy, Vz)

- Vertical acceleration

- Wander angle and magnetic heading

- Pilot's selected manual command heading,
  course, or both

The CSDC also operates with the flight crew displays to
display various navigational parameters (figure 7-19).

Note

A CSD acronym on the TID may
indicate failure of the primary attitude
reference. When an acronym appears,
the VDI and the standby attitude
indicator should be cross-checked to
determine if the primary reference
system is degrading. If the VDI is
erroneous, the standby attitude indicator should be used as the primary
reference instrument and instrument
flight should be avoided.

CSDC BIT

The CSDC continuously monitors its own program and
electrical characteristics in addition to those of other
systems. If a failure is detected, it is displayed on the
TID by the WCS computer and if the navigation functions
fail, the WCS computer lights the NAV COMP light on
the caution advisory panel.

Note

CSDC failures are not always evident
unless OBC continuous monitoring has
been initiated. The acronyms include:
CSD (hard CSDC failure), CSI
(standard serial interface failed), and
WOW (weight-on-wheels).

TACAN SYSTEM (AN/ARN-84)

The TACAN system provides continuous indications of
slant range to 0.1 nmi and bearing of 0.5° to any surface
station selected. Slant range is available to other aircraft
equipment with an air-to-air (A/A) mode. Operating
range is line-of-sight to approximately 300 nautical miles.

The system provides 126 operating channels in each of
two modes. Receiving frequencies for surface-to-air
operation are 962 to 1024 MHz and 1151 to 1213 MHz;
for air-to-air operations the frequencies are from 1025
to 1150 MHz. All transmitting frequencies are within
1025 to 1150 MHz. TACAN uses two aircraft antennas,
automatically switching between the two at 6-second
intervals until a threshold signal is received.

Note

The aircraft permits use of the NORMAL/
X mode only. The Y mode is for future
development of 126 additional TACAN
channels.

The system, when operating in the REC or T/R modes,
is capable of receiving valid signals from a ground station
simultaneously with 99 other aircraft. When in the A/A
mode, the system is capable of transponding with each of
five cooperating aircraft, providing slant range information
to each; however, the system will interrogate and lock on
to only one. In the A/A mode, the second aircraft must
be 63 channels apart. Only range is received because the
aircraft antenna complement is not configured to receive
or transmit bearing information. Identical TACAN control panels (figure 7-24) are on the pilot's and RIO's left
consoles. Individual TACAN CMD buttons (figure 7-24)
on both the pilot's and RIO's left consoles provide for
transfer of TACAN control. Control of the TACAN is
indicated by a flip-flop indicator in each cockpit showing
PLT (pilot) or RIO. Either crewman may adjust the audio
level of the identification signal.

Bearing and distance information is displayed on the
BDHI (bearing, distance, heading indicator), the HSD,
and the ECMD. Deviation, which is defined as the
difference between actual TACAN bearing and
manually-selected course bearing, is computed by the
CSDC and displayed on the HUD and VDI (VDIG) and
the HSD and ECMD (MDIG). The MDIG presentation
of TACAN information includes TACAN bearing marker,

TACAN CONTROLS AND DISPLAYS

NOTE
- HSD AND ECMD HAVE SAME PRESENTATION.
- HUD DISPLAY OF TACAN DEVIATION BAR NOT SHOWN.

RNG
203.0

CRS
030

IN

RIO

PILOT

DISPLAYS

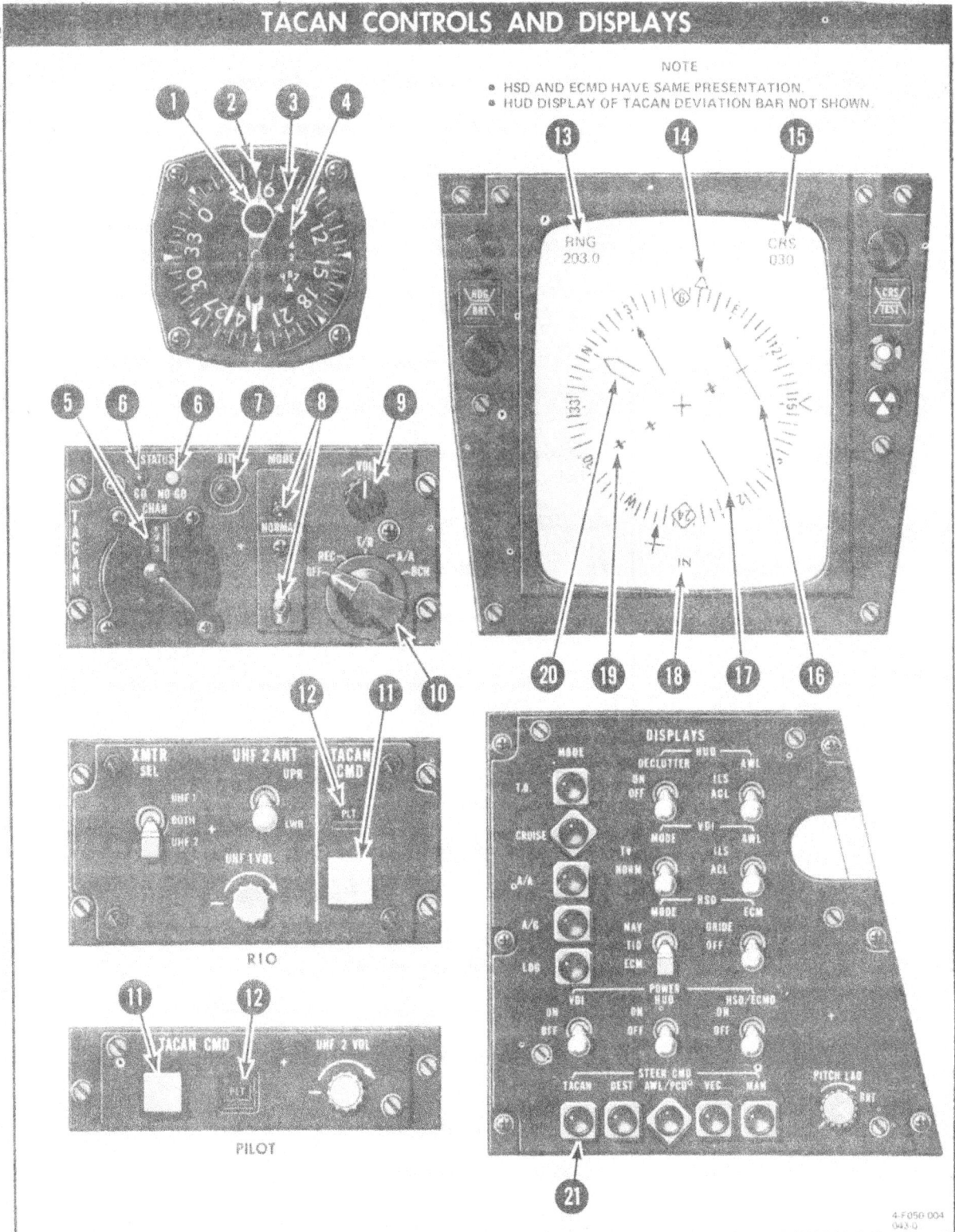

Figure 7-24. TACAN Controls and Displays (Sheet 1 of 3)

NOMENCLATURE	FUNCTION
(1) No. 2 bearing pointer	Indicates magnetic course to selected TACAN station.
(2) Compass rose	Top of instrument indicates aircraft magnetic heading.
(3) No. 1 bearing pointer	Indicates bearing to UHF/ADF station selected.
(4) Distance counter	Indicates slant range (nautical miles) to TACAN station selected or to cooperating aircraft.
(5) TACAN CHAN switch	Outer control selects first two numbers and inner control selects last digit of desired TACAN channel (126 channels available).
(6) STATUS lights	Indicates GO or NO-GO status of TACAN system during TACAN BIT.
(7) BIT switch	Initiates interruptive self-test portion of built-in-test feature.
(8) MODE switch	NORMAL and X are fixed (other modes, INVERSE and Y, are not selectable).
(9) VOL control	Regulates audio level of station identification signal through ICS.
(10) Function selector	OFF — Sets power off.
	REC — Provides only bearing information from surface station selected.
	T/R — Permits reception of bearing and slant range from surface station selected.
	A/A — Slant range from another air-to-air equipped aircraft is available, if such aircraft have selected TACAN frequencies 63 channels apart.
	BCN — Not operational.
	Note
	Items 11 through 16 appear on both the HSD and ECMD, when selected.
(11) TACAN button	Alternating pushbutton to give or take control of TACAN.
(12) TACAN CMD	Flip-flop indicators show NFO or PLT, indicating which crewman is in command of TACAN.
(13) RNG readout	Indicates range to TACAN station in tenths of a mile.
(14) TACAN bearing pointer	Indicates direction to TACAN station relative to own-aircraft.
(15) CRS readout	Indicates TACAN source selected, and is set by pilot with CRS control.

Figure 7-24. TACAN Controls and Displays (Sheet 2 of 3)

NOMENCLATURE	FUNCTION
(16) Deviation bars	Indicates difference between course selected and TACAN bearing.
(17) Course bar	Indicates TACAN course selected, set by pilot with CRS control.
(18) Source of attitude reference acronym	Indicates the aircraft attitude reference source selected by the computer or RIO. Can be IN for INS, IM for IMU/AM mode, or AH for the AHRS/AM mode.
(19) Deviation ticks	Symbols used as scale for deviation bar. Each tick represents 6° of deviation.
(20) ADF bearing symbol	Indicates bearing to ADF station selected.
(21) TACAN-STEER CMD pushbutton	Enables TACAN steering. When selected during T.O., CRUISE, or LDG, enables presentation of TACAN data on the VDIG and MDIG.

Figure 7-24. TACAN Controls and Displays (Sheet 3 of 3)

deviation ticks, range-to-TACAN station, and course.
The HUD and VDI display provide a TACAN deviation
bar, which is coded (solid line-TO station, dashed line-
FROM station on the HUD; bright bar-TO station, dark
bar-FROM station for the VDI).

TACAN information is displayed on identical BDHI on
the pilot's and RIO's right upper instrument panels. The
bearing and distance functions of the BDHI are activated
when the TACAN mode select switch is set to T/R. In
the REC and T/R modes, magnetic bearings are displayed
by the no. 2 (large) needle, which unlocks and enters a
search mode whenever bearing information is unreliable.
Range information which is received in the T/R mode or,
when operating with another aircraft in the A/A mode,
is displayed in nautical miles on the distance counter.
An OFF flag covers the counter window when range
information is unreliable. TACAN information is also
displayed in conjunction with other system modes on the
pilot HSD, HUD, and VDI and on the RIO ECMD.

The TACAN system takes approximately 2 minutes to
warm up. If, after the warmup period, the range and
bearing indications continue to search when a reliable
station is selected, circuit breakers should be checked, or
another station checked.

The system has a memory feature so that tracking will not
be interrupted by momentary disruption of received
signals. A range signal that is lost after at least 10 seconds
of tracking will be sustained by memory for 9 to 12 seconds. A bearing signal that has been tracked for at least
15 seconds will be retained for 3 to 8 seconds after signal
loss. This allows automatic antenna switching without a
loss of TACAN displays.

Power requirements for the TACAN system are provided
by the ac essential bus no. 2 (phase A) through the
TACAN ARN- 84 circuit breaker (3D6) on the RIO's left
panel, the dc essential bus no. 2 through the TACAN/
BDHI circuit breaker (7E8) on the RIO's right aft panel,
and the ac essential bus no. 2 (phase C) through the
TACAN/BDHI INST PWR circuit breaker (4F6). During
the minimum warmup time, failure indications and
erroneous readouts should be disregarded and self-test
results may be inconclusive.

TACAN BIT

The TACAN system contains a built-in test that provides
continuous automatic monitoring and interruptive self-
test. The TACAN control unit has a momentary push
button (BIT switch) for starting a 22-second interruptive
self-test sequence and two status lights labeled GO (green)

and NO-GO (amber). The NO-GO light illuminates
whenever any of the continuous monitor functions are
NO-GO or if the results of the self-test cycle are NO-GO.
The GO light illuminates momentarily (6 to 18 seconds)
only at the completion of a satisfactory interruptive self-
test cycle. Light circuitry is tested at the start of the
self-test cycle. When the BIT switch is depressed, both
GO and NO-GO lights illuminate; they go out when the
BIT switch is released.

Note

BIT on TACAN stations within 2 nmi
may give an invalid indication. If a TCN
acronym or NO-GO response is observed,
while tuned to a local station, along with
normal TACAN azimuth and range, the
acronym and/or the NO-GO should be
disregarded.

The normal BIT sequence is as follows:

1. Set MODE switch to T/R; allow 2 minutes for
   warmup.

2. Press and hold BIT button.

   a. Both GO and NO-GO lights illuminate (light
      test).

   b. BDHI range OFF flag appears.

   c. BDHI bearing needle rotates counterclockwise.

3. Release button; both lights go out (self-test starts).

   a. After 5 to 6 seconds, BDHI and HSD range
      reads 2 nmi; BDHI and HSD bearing reads 4°.
      (Identify TACAN station.)

   b. After 22 seconds; if good, green GO light
      illuminates, if bad, amber NO-GO light
      illuminates.

Signal Data Converter

The TACAN system signal data converter furnishes
analog signals for display on the pilot's BDHI, HUD, VDI,
and HSD and on the RIO's BDHI and ECMD. In addition,
digital range and bearing signals are provided for use by
the WCS computer. The converter also interfaces with
the data link system, providing that system with range
and bearing in analog form, station number in binary
form, and discretes.

BEARING DISTANCE AND HEADING INDICATOR (BDHI)

A bearing distance and heading indicator (BDHI) is on the right side of the pilot and RIO instrument panels (figure 7-24). It is a remote-type heading indicator, which displays aircraft magnetic heading with navigation bearing data and range information. The compass card receives heading reference from the attitude heading reference set (AHRS). Controls on the compass panel permit the BDHI compass card to operate in a slaved or non-slaved (FREE DG) compass mode. A fixed index marker at the 12-o'clock position indicates magnetic heading.

Two servo-driven bearing needles indicate magnetic bearings to selected UHF (ADF) and TACAN stations. The No. 1 (single bar) needle receives signals from the UHF (ADF) system; the No. 2 (double bar) needle receives signals from the TACAN coupler. If the compass card is misaligned or a malfunction exists in the compass system, the No. 1 needle will continue to point toward the signal source; however, the bearing to the station is displayed on the indicator as a relative bearing, the top of the indicator bezel being 000°. Under the same circumstances, the No. 2 needle will continue to indicate magnetic bearing to the selected station or will revert to the search mode.

RADAR ALTIMETER SYSTEM (AN/APN-194)

The radar altimeter is a low-altitude (0 to 5,000 feet), pulsed, range-tracking radar that measures the surface or terrain clearance below the aircraft. Altitude information is developed by radiating a short-duration rf pulse from the transmit antenna to the earth's surface and measuring elapsed time until rf energy returns through the receiver antenna. The altitude information is continuously presented to the pilot, in feet of altitude, on an indicator dial. The system also outputs a digital signal to the VDIG for display of radar altitude on the HUD from 0 to 1,400 feet during takeoff and landings.

The radar altimeter has two modes of operation. In the search mode, the system successively examines increments of range until the complete altitude range is searched for a return signal. When a return signal is detected, the system switches to the track mode and tracks the return signal to provide continuous altitude information.

When the radar altimeter drops out of the track mode, an OFF flag appears and the pointer is hidden by a mask. The altimeter remains inoperative until a return signal is received, at which time the altimeter will again indicate actual altitude above terrain.

Reliable system operation in the altitude range of 0 to 5,000 feet permits close altitude control at minimum altitudes. The system will operate normally in bank angles up to 45° and in climbs or dives except when the reflected signal is too weak.

The system includes a height indicator (altimeter), a test light on the indicator, a low altitude warning tone, a radar receiver-transmitter under the forward cockpit, and two antennas (transmit and receive) one on each side of the IR fairing, in the aircraft skin. During descent, the warning tone is heard momentarily when the aircraft passes through the altitude set on the limit index. When the aircraft is below this altitude, the red low-altitude warning light on the indicator stays on.

Note

If radar altitude is unreliable, only the OFF flag is present.

The radar altimeter receives power from the ac essential bus no. 1 through the RADAR ALTM circuit breaker (5D6) and from dc essential bus no. 1 through the ALT LOW WARN circuit breaker (6B6). Both circuit breakers are on the RIO's right side circuit breaker panels. The radar altimeter has a minimum warmup time of 3 minutes. During this time, failure indications and erroneous readouts should be disregarded.

Radar Altimeter

The radar altimeter (figure 7-25) on the pilot's instrument panel has the only controls for the system. The indicator displays radar altitude above the earth's surface on a single-turn dial that is calibrated from 0 to 5,000 feet in decreasing scale to provide greater definition at lower altitudes. The control knob in the lower left corner of the indicator is a combination power switch, self-test switch, and positioning control for the low-altitude limit bug.

ALTIMETER BIT

Depressing and holding the control knob energizes the self-test circuitry; the green test light illuminates, the indicator reads 100±10 feet, and the HUD altitude scale reads approximately 100 feet. If the altitude limit bug is set above 100 feet, the aural and visual warnings are energized. Normal operation is resumed by releasing the control knob.

RADAR ALTIMETER

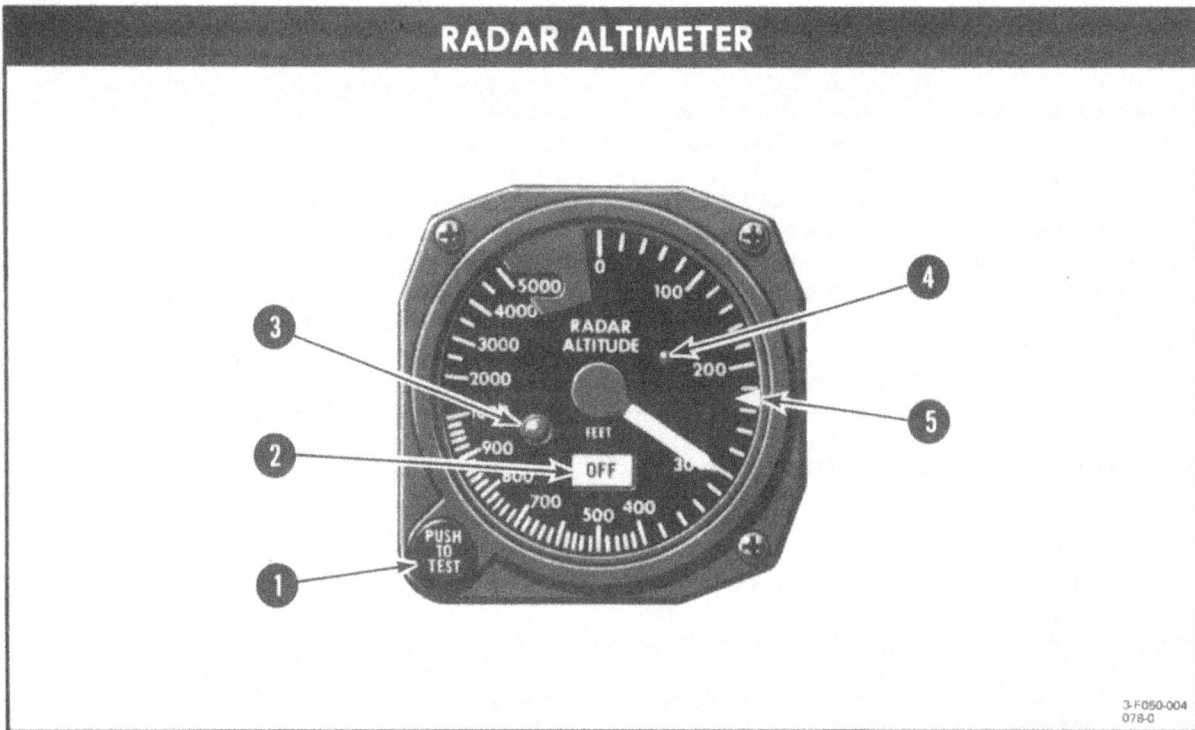

3-F050-004
078-0

NOMENCLATURE	FUNCTION
(1) Radar altimeter control knob (combination switch)	Initial clockwise rotation turns system power on; continued rotation increases altitude limit index setting. When depressed and held, it tests the system on the ground and in flight. Counterclockwise rotation decreases altitude limit index setting; fully counterclockwise turns system power off.
(2) OFF flag	OFF appears if altimeter is turned off, power is lost, or radar signal is unreliable.
(3) Low altitude warning light (red)	Illuminates as a warning whenever aircraft is below altitude set by the limit bug.
(4) Self-test light (green)	When control knob is depressed and held, light should illuminate and pointer should read 100±10 feet.
(5) Low-altitude limit index (limit bug)	Can be preset to low-altitude limit desired by turning control knob.

Note

Radio override has no effect on radar altimeter low-altitude warning tone.

Figure 7-25. Radar Altimeter

LOW ALTITUDE AURAL WARNING

A low-altitude aural warning alarm provides a 1000 Hz tone, modulated at 2 pulses per second, lasting for 3 seconds. The tone is available to both crewmembers when the aircraft descends below the altitude set on the low-altitude limit bug.

NAVIGATION SYSTEM INTEGRATION

Navigation Modes

The primary navigation mode is inertial and is achieved by the INS, employing the IMU (and PSU) and the CSDC. They provide the flightcrew with own-aircraft position, velocity, attitude, and heading information. Initial alignment of the IMU is computed by the WCS computer, which initializes the navigation computer section of the CSDC. The CSDC then maintains the alignment and performs all computations necessary for determining present position.

Position updates can be made during a flight, but may not improve a fully aligned platform position output for flights of less than several hours duration.

The inertial measurement unit/air mass (IMU/AM) is the backup navigation mode. Entry into this mode permanently degrades INS platform heading alignment. The third mode of navigation is the AHRS/air mass (AHRS/AM) mode which utilizes the AHRS attitude and heading information in place of the IMU.

INERTIAL NAVIGATION MODE

Following alignment, the INS mode should be entered. The READY light illuminates in GND and CVA alignment positions and stays on after launch in CAT alignment, indicating completion of alignment. If the INS mode is selected, both the STBY and READY lights will go out. However, if the INS mode is selected before fine alignment (before the second tick mark), both the STBY and READY lights will illuminate and the system will revert to a backup mode (IMU/AM).

When in the INS mode, the IMU and CSDC will provide the following outputs:

- Aircraft latitude and longitude.

- Aircraft magnetic or true heading (depending on CAP prefix pushbutton selected).

- System altitude (barometric damped inertial altitude).

- Platform wander angle.

- Velocity components (x, y, z).

- Vertical acceleration.

Aircraft magnetic heading is derived from the AHRS. If AHRS fails, magnetic heading is then derived by subtracting MAG VAR from true heading. The quantities available for display on the TID readouts are latitude, longitude, groundspeed, ground track, true airspeed, wind (speed and direction), MAG VAR, altitude, and aircraft true or magnetic heading. Other parameters available from the CSDC are sent to the WCS computer for general navigation computations.

The WCS computer makes calculations in true north coordinates for steering, etc., and uses the magnetic heading input from AHRS to update the computed value. Wind is computed from the difference between inertial velocities and air mass velocities.

Included in the general navigation functions performed in the WCS and in certain of the CSDC computations are the steering and cueing functions required for display to the flightcrew. The information can be presented on the TID, HSD, ECMD, HUD, and VDI, depending on the navigation and steering modes selected by the flightcrew (figure 7-15).

The destination or navigation points are way points 1, 2, or 3, fix point, home base, surface target, and identification point and may be designated by the DEST switch on the TID. Navigational points (latitude and longitude) may also be inserted by the RIO using the CAP or by the data link message (when on the deck) using either cable or the rf link. The course to set (heading to a selected navigational point), range, bearing, and time-to-go to a point are based on great circle calculations. The time-to-go assumes the aircraft is flown at its present groundspeed along the great circle heading to the selected point.

Note

If INS fails, the RIO should verify MAG VAR and WIND data before relying on backup navigation and mode data.

IMU/AM NAVIGATION MODE

The IMU/AM mode is automatically entered upon failure of the navigation computer section of the CSDC or certain failures in the IMU. These failures are indicated by both the STBY and READY lights illuminating and the NAV COMP light illuminating on the RIO's CAUTION/ ADVISORY panel. Mode entry is indicated by the IN acronym on the TID and HSD changing to IM. The RIO should select IMU/AM on the NAV MODE switch to extinguish the STBY and READY lights.

The IMU/AM mode can be entered manually by selecting it on the NAV MODE switch. If the switch is turned off before selecting IMU/AM, the computer will not enter the IMU/AM mode for approximately 3 minutes. This is necessary to allow the IMU to level itself after being turned off. During these 3 minutes, the aircraft must remain stationary on the ground or in level unaccelerated flight. The computer will use the AHRS/AM navigation mode until the IMU is leveled. If the IMU is already level (that is, alignment past course align complete), the entry into IMU/AM will occur immediately upon selection.

Note

If an aligned platform exists with no NAV COMP failure and the RIO switches to IMU/AM, the READY light will flash, indicating that if the switch was not returned to INS within 5 seconds, the INS mode cannot be reentered without completing a new alignment.

During the IMU/AM mode, the WCS computer performs dead-reckoning navigation, using heading information from the IMU and true airspeed from the CADC. The same general navigation functions are performed in the backup mode as for the INS mode. The accuracies of the computer outputs are degraded because of inferior speed and heading information. Wind is applied, using either the wind last computed in the INS mode or a wind manually entered through the CAP. In a similar manner, the IMU heading is referenced to the last computed INS heading or to a manually entered true heading. Since corrections are not made to the IMU from the CSDC, the azimuth gyro is operated in a free DG mode. IMU attitude information is displayed to the pilot.

Note

After entering the IMU/AM mode, check wind and MAG VAR values. If MV is in error, enter own-aircraft true heading. If winds are in error, update.

AHRS/AIR MASS MODE

The AHRS/AM mode of navigation is another backup mode. It uses the last known aircraft position, either the last navigation computer value or a flight officer inserted value, and extrapolates the present position of the aircraft. It is automatically selected when the IMU fails

or by switching to the AHRS/AM position on the NAV MODE switch. If an IMU failure occurs, it is indicated by the STBY and READY status lights and the IMU advisory light illuminating. Additionally, the attitude status readout on the TID changes to AH.

Note

Although the navigation mode automatically switches to AHRS when the IMU fails, the STBY and READY lights remain on until the AHRS/AM position on the NAV MODE switch is selected.

In this mode, the AHRS provides heading information required for DR navigation in place of the IMU platform. The CSDC provides barometric altitude, altitude rate, and true airspeed as in the IMU/AM mode. Updated wind speed and direction and magnetic variation may be entered using the CAP.

The AHRS itself may be operated in any of three subheading modes selected on the compass controller panel.

SLAVED — Magnetic north referenced (flux value), directional gyro is slaved to flux value, used where reliable magnetic heading reference is available.

DG — Free azimuth gyro, compensated for drift due to earth's (polar operations), used where magnetic reference is unreliable.

COMP — Magnetic north referenced direct (flux valve), no gyro damping. The HUD, VDI, HSD, and ECMD use manual magnetic variation (vM) automatically in this mode.

Note

- If both the IMU and the AHRS gyros fail, pitch, and roll attitude indications are removed from the HUD, TID, VDI, and DDD. Magnetic heading indications become erroneous, and the IMU and AHRS advisory lights illuminate. Selecting the COMP position on the compass control panel may restore valid magnetic heading information to the HUD, VDI, and HSD and turn off the AHRS advisory light. However, invalid pitch and roll attitude information will be restored to the HUD, VDI, TID, and DDD.

General navigation computations are performed in the WCS computer as in the INS and IMU/AM modes. In flight, the

RIO may switch from either the INS mode to AHRS/AM mode or from IMU/AM mode to AHRS/AM mode for comparison, without degradation, since the AHRS is a separate system. This cannot be done between the INS and IMU/AM modes since the IMU is used in both cases and permanent degradation to the IMU alignment will result. The navigation system will automatically operate in the AHRS/AM mode with the navigation control and data readout panel in the INS position in the event of an IMU failure, as long as the WCS receives AHRS heading and CADC airspeed.

Note

If takeoff is initiated in AHRS mode, MAG VAR and WIND must be manually inserted for proper navigational computations.

For the situation where the platform is aligned and the AHRS/AM backup navigation mode is selected, the STBY light is off but the READY light is on, indicating that the inertial navigation mode may be selected if desired.

The same functions and outputs for display are computed as in INS; however, different inputs are used for some calculations, and degraded navigation performance can be expected.

STEERING

Two basic types of steering are provided, navigation and attack. Navigation steering is computed on a great circle course to a fixed point on the earth's surface or as a deviation from a selected course or heading, which is equivalent to a rhumb line.

Note

For attack steering, refer to Section VIII in NAVAIR 01-F14AAA-1A.

NAVIGATION STEERING

Navigation steering is computed on either a great circle course or rhumb line to a fixed point on the earth's surface or as a deviation from a selected course or heading. In general, great circle computations are used for long ranges and rhumb line for short distances (where it is close to a great circle course). The point being steered may be the RIO's selected destination (three way points, fixed point, identification point, surface target, or home base), a TACAN station,

ADF information, ACLS information, or data-link selected points.

FLIGHT MODES AND STEERING SUBMODES

The pilot has the option of selecting any one of five VDIG display formats, depending on the flight phase, to provide him with the data necessary to accomplish that particular flight phase. These five flight modes are arranged as five vertical, mutually-exclusive pushbuttons on the pilot's display control panel. The five phases are takeoff (T.O.), CRUISE, air-to-air (A/A), air-to-ground (A/G), and landing (LDG).

Note

ACM selection overrides the CRUISE, A/A and A/G modes; however it does not override the T.O. or LDG modes.

In addition to controlling the VDIG formats, the flight mode selections also control AFCS, armament and WCS logic.

In addition to the essential data such as altitude, vertical speed indicator, etc., the VDIG format also provides steering cues. In each of the flight modes, the pilot has the capability of displaying several types of steering commands. Altogether there are five distinct steering command submodes: TACAN, destination (DEST), AWL/PCD, vector (VEC), and manual (MAN). The five selections are arranged horizontally along the bottom of the display control panel.

The five submodes determine the display format on the pilot's HSD and the RIO's ECMD. The HSD and ECMD present, in a horizontal plane, steering to the selected point. The HSD follows the five submodes when the pilot places the HSD-MODE switch to NAV. The RIO also performs the same function by setting the MODE switch on his ECMD control panel to NAV. Also, when LDG is selected, the pilot has the option of displaying ILS or ACL information via switches that can be used to individually and independently select the HUD and VDI for display. A typical choice would be to select ILS (SPN-41/ARA-63A) for the HUD and for the VDI (SPN-42/Data Link). Refer to Section VI for all-weather carrier landing system procedures.

Note

All steering commands (such as command course and command heading) are processed to some extent through the WCS computer.

The STEERING indicator on the navigation control and data readout panel provides a readout for the RIO to inform him of what steering submode the pilot has chosen (figure 7-15).

TAKEOFF STEERING

The takeoff flight mode is entered by depressing the T.O. pushbutton on the display control panel. The VDIG format displays a vertical speed indicator on the left side and an altitude scale on the right side (figure 7-26). Prior to takeoff, the pilot should check the magnetic heading on the top of the HUD and VDI against a known reference (that is, runway heading). The vertical speed indicator should be used to verify a positive climb after takeoff.

After takeoff, the navigation system normally computes wind and magnetic variation, which are needed for steering. For backup modes, the WCS uses the last computed or RIO-entered wind speed, direction, and magnetic variation.

TAKEOFF TACAN STEERING. The TACAN steering submode functions in the same way, whether used for takeoff, cruise, or landing, by providing the pilot with a TACAN deviation. This is the angular difference between the bearing to the TACAN station (TACAN radial) and the command course (TACAN course) selected by the pilot with the CRS control on the HSD. The TACAN displays are available on the HUD, VDI, HSD, and ECMD. The HSD and the CMD also display TACAN range as well as the relative bearing to a selected TACAN station.

To enter the submode, the pilot depresses the TACAN pushbutton on his display control panel. After selection of TACAN course, the HUD and VDI display the TACAN deviation symbol. In addition to TACAN deviation, symbols on the HSD, ECMD, HUD, and VDI indicate whether the TACAN course is toward or away from the TACAN station. If the TACAN deviation is less than 90°, TO symbology is shown; if greater than 90°, FROM symbology is shown. On the HSD and ECMD, an arrow on the deviation bar pointing in the same direction as the TACAN course indicates a course toward the station; an arrow pointing in the opposite direction, away from the station. On the HUD, a dashed line indicates FROM; a solid line, TO. On the VDI, a dark bar indicates FROM; a bright bar, TO.

On the HUD, the deviation symbol moves 3° (linear) in the field of view for a 6° deviation from the selected TACAN radial. These limits prevent the symbol from

leaving the field of view or interfering with the scales on the left and right side. On the VDI, the deviation symbol is scaled to move 1.5 inches (linear) for a 6° deviation.

TAKEOFF/MANUAL STEERING. The manual steering submode is similar to the basic takeoff mode. The mode is entered by depressing the MAN pushbutton and selecting a desired course with the CRS control on the HSD. The navigation system will then compute a command heading and display it on the VDI as a small diamond under the magnetic heading scale.

CRUISE STEERING

The cruise flight mode is entered by depressing the CRUISE pushbutton. There are four steering submodes available during cruise operations; destination, TACAN, manual, and vector. While it is physically possible to depress the AWL/PCD steering pushbutton on the display control panel, doing so has no operational value.

Note

Should the AWL/PCD submode be selected while in CRUISE, it will inhibit the display of other steering cues.

CRUISE/TACAN STEERING. This submode is similar to that for takeoff. The flightcrew displays are shown in figure 7-27 and are similar to the takeoff mode. Range to the selected TACAN station is given on the pilot's HSD and BDHI and on the RIO's ECMD and BDHI. Selected TACAN course is also numerically displayed on the HSD and ECMD. The VDIG format displays the TACAN deviation as described under TAKEOFF/TACAN STEERING. This deviation is also displayed on the HSD and ECMD over a scale for easy reference. TO and FROM indications are provided on all indicators, and station passage is readily discernible on all indicators.

CRUISE/DESTINATION STEERING. This submode provides steering to the pilot as a command heading symbol on the VDI and command heading pointer on the HSD to a point selected by the RIO (figure 7-28). To enter this submode, the pilot must depress the DEST pushbutton on the display control panel, and the RIO select the desired destination with the DEST switch on the navigation control and data readout panel. The destination can be any one of seven prestored points. The latitude/longitude of the destination may be changed by hooking the point on his TID and inserting the new data.

TAKEOFF STEERING DISPLAYS

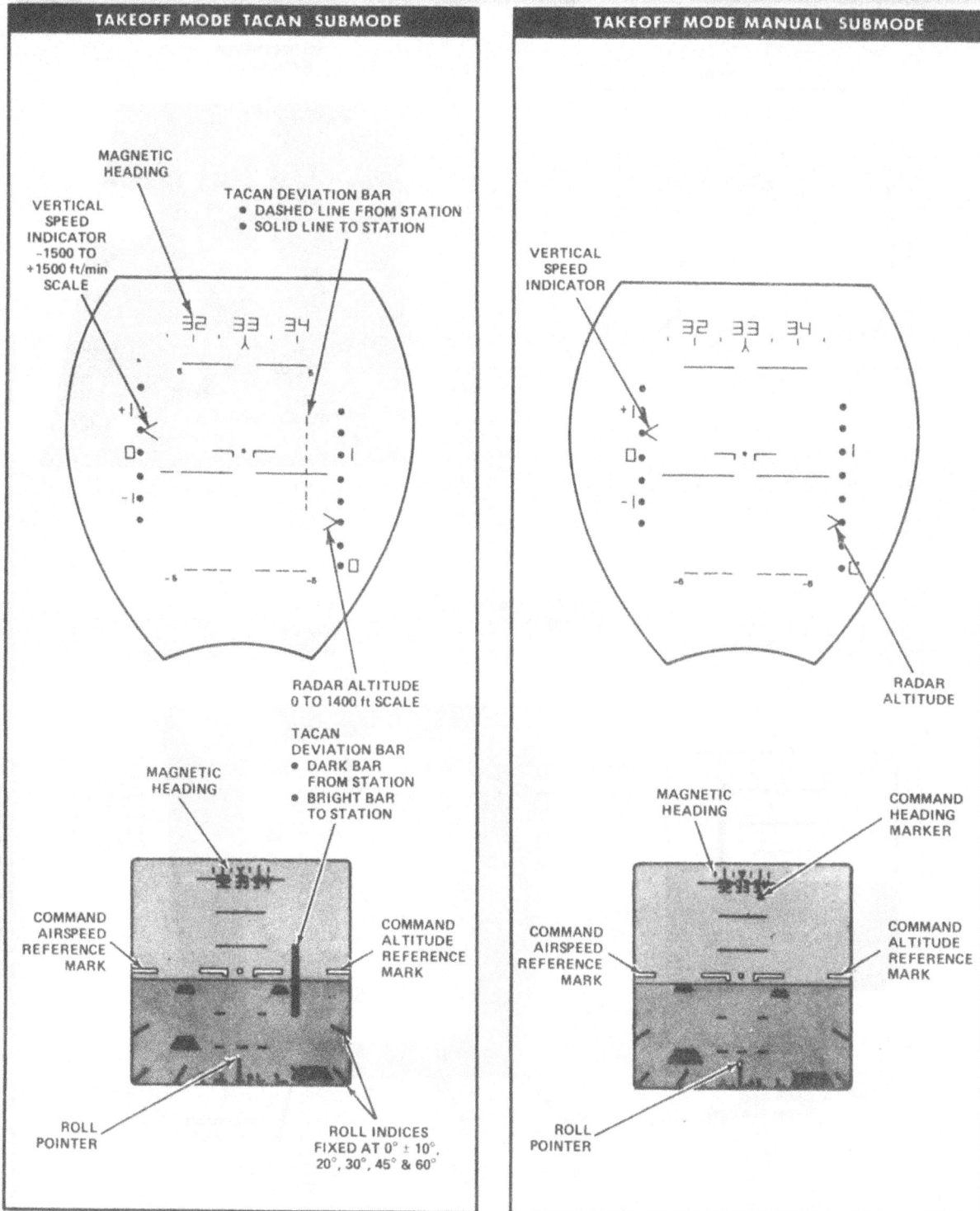

TAKEOFF MODE TACAN SUBMODE

MAGNETIC HEADING

VERTICAL SPEED INDICATOR
-1500 TO +1500 ft/min SCALE

TACAN DEVIATION BAR
- DASHED LINE FROM STATION
- SOLID LINE TO STATION

RADAR ALTITUDE
0 TO 1400 ft SCALE

TACAN DEVIATION BAR
- DARK BAR FROM STATION
- BRIGHT BAR TO STATION

MAGNETIC HEADING

COMMAND AIRSPEED REFERENCE MARK

COMMAND ALTITUDE REFERENCE MARK

ROLL POINTER

ROLL INDICES FIXED AT 0° ± 10°, 20°, 30°, 45° & 60°

TAKEOFF MODE MANUAL SUBMODE

VERTICAL SPEED INDICATOR

RADAR ALTITUDE

MAGNETIC HEADING

COMMAND HEADING MARKER

COMMAND AIRSPEED REFERENCE MARK

COMMAND ALTITUDE REFERENCE MARK

ROLL POINTER

Figure 7-26. Takeoff Steering Displays

2-F50-189-0

CRUISE TACAN STEERING DISPLAYS

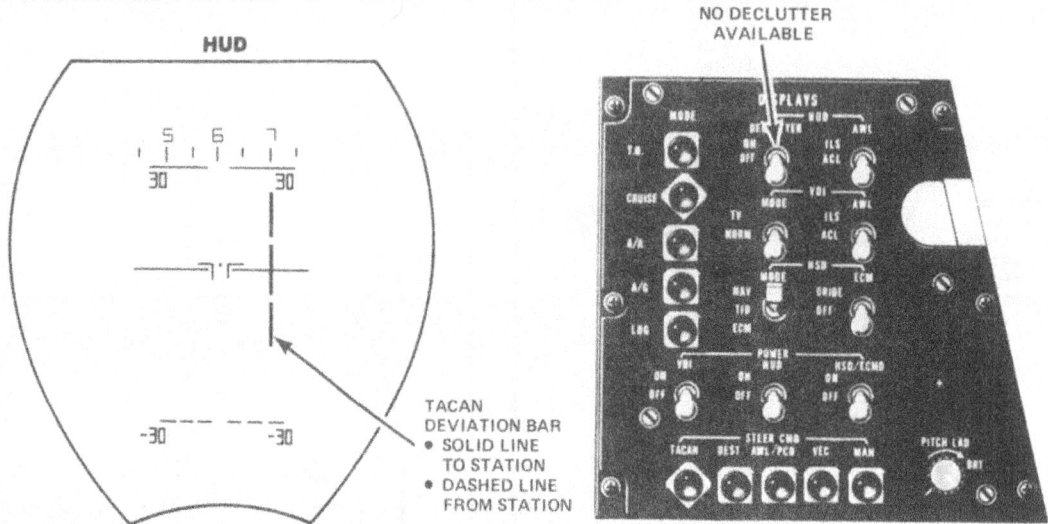

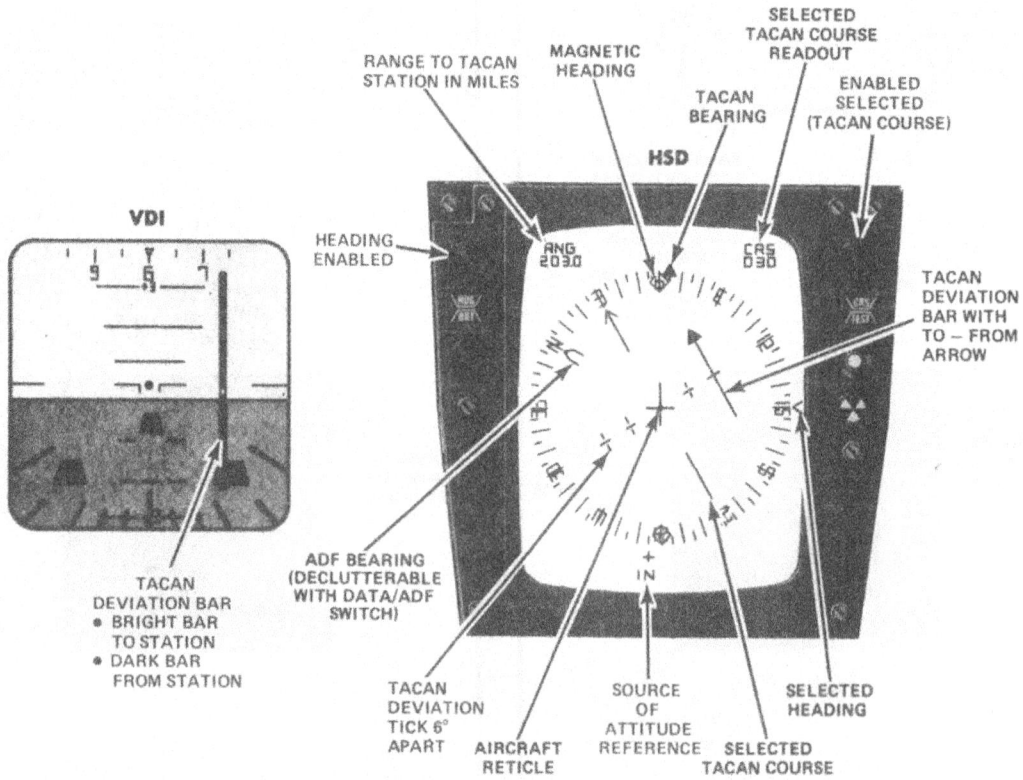

HUD

NO DECLUTTER
AVAILABLE

TACAN
DEVIATION BAR
● SOLID LINE
  TO STATION
● DASHED LINE
  FROM STATION

VDI

TACAN
DEVIATION BAR
● BRIGHT BAR
  TO STATION
● DARK BAR
  FROM STATION

RANGE TO TACAN
STATION IN MILES

MAGNETIC
HEADING

TACAN
BEARING

SELECTED
TACAN COURSE
READOUT

ENABLED
SELECTED
(TACAN COURSE)

HSD

HEADING
ENABLED

RNG
203.0

CRS
030

TACAN
DEVIATION
BAR WITH
TO – FROM
ARROW

ADF BEARING
(DECLUTTERABLE
WITH DATA/ADF
SWITCH)

TACAN
DEVIATION
TICK 6°
APART

AIRCRAFT
RETICLE

SOURCE
OF
ATTITUDE
REFERENCE

SELECTED
TACAN COURSE

SELECTED
HEADING

7-F50-137-0

Figure 7-27. Cruise TACAN Steering Displays

CRUISE DESTINATION STEERING DISPLAYS

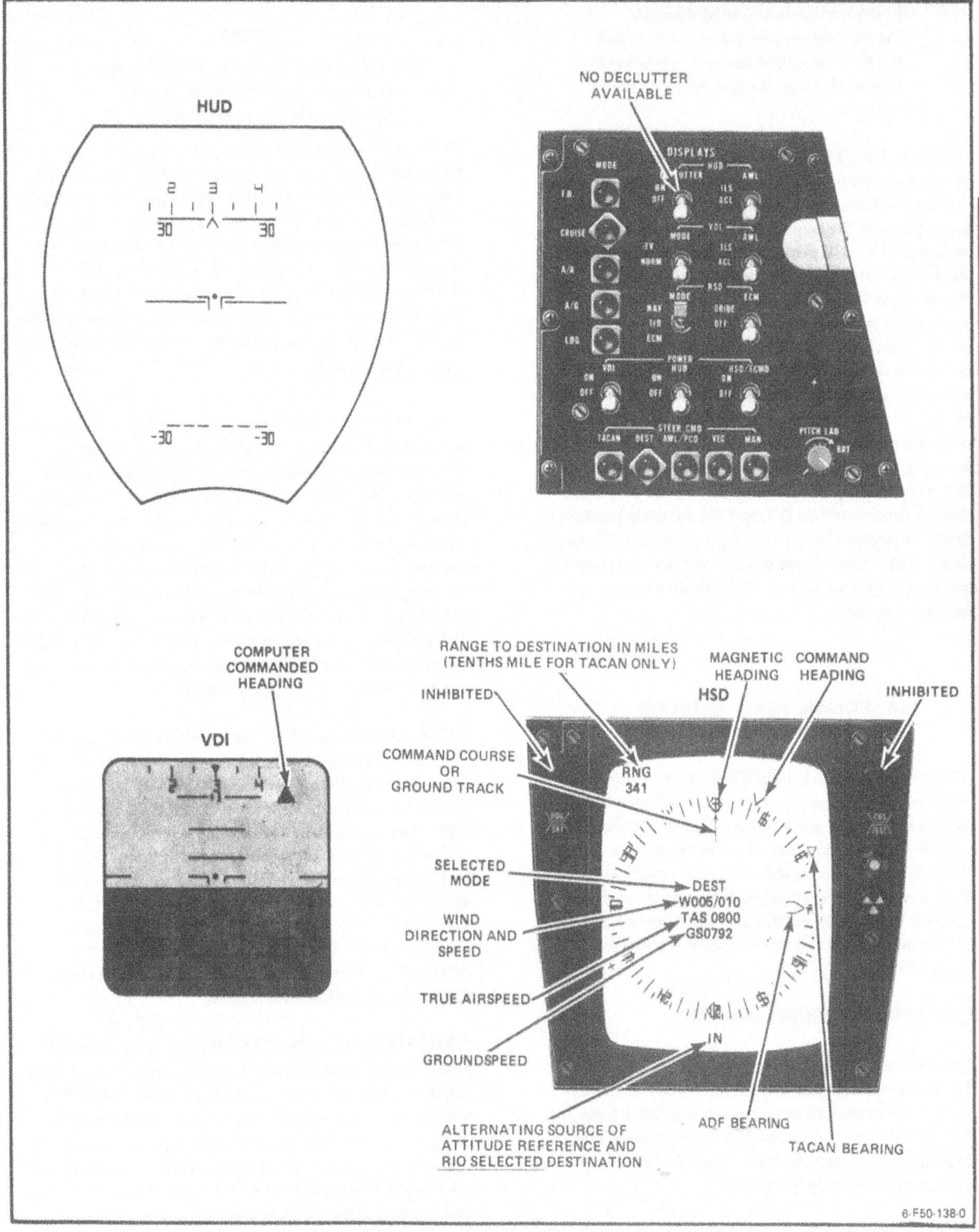
HUD
2 3 4
30 30
-30 -30
NO DECLUTTER AVAILABLE
DISPLAYS
COMPUTER COMMANDED HEADING
VDI
RANGE TO DESTINATION IN MILES (TENTHS MILE FOR TACAN ONLY)
MAGNETIC HEADING
COMMAND HEADING
HSD
INHIBITED
INHIBITED
COMMAND COURSE OR GROUND TRACK
RNG
341
SELECTED MODE
DEST
W005/010
TAS 0800
GS0792
WIND DIRECTION AND SPEED
TRUE AIRSPEED
GROUNDSPEED
IN
ALTERNATING SOURCE OF ATTITUDE REFERENCE AND RIO SELECTED DESTINATION
ADF BEARING
TACAN BEARING
6-F50-138-0

Figure 7-28. Cruise Destination Steering Displays

Note

In the destination steering submode,
the destination selected by the RIO and
the NAV MODE in use will be alternately
displayed on the bottom center of the
HSD.

CRUISE/VECTOR STEERING. This submode
utilizes the data link (D/L) system to provide command
heading, altitude, and airspeed (figure 7-29). In this
submode, both the pilot's HDG and CRS controls are
inhibited. The D/L provides the command steering to
the destination along with command altitude in the
thousands of feet and command airspeed (Mach) error.
If the D/L command airspeed is in knots, it is displayed
on the airspeed/Mach indicator on the pilot's center
console, and there is no HSD indication. Errors in either
altitude or airspeed are indicated in each case by a pair
of vertical lines. Zero error is indicated when the error
symbol is centered on the reference line. As the symbol
moves up or down, it indicates an altitude or airspeed
high or low error. The commanded course on ground
track as determined by D/L and the computed command
heading necessary for a reliable ground track is displayed
on the HSD. Wind direction, true airspeed, and groundspeed are displayed on the HSD and computed by the
navigation system.

Note

In VEC mode, no ADF or TACAN
information is provided on the MDIG.

CRUISE/MANUAL STEERING. This submode
functions in the same way as it does in takeoff. The
commanded course selected by the pilot with the CRS
control is displayed on the HSD and the ECMD. The
command heading required to make the selected command course good is displayed as command heading
markers on the HSD, ECMD, and VDI (figure 7-30). In
this mode, TACAN and ADF bearings are also provided,
but not range.

LANDING STEERING MODES

The landing steering mode is entered by the pilot
depressing the LSD pushbutton on the display control
panel. This mode can be engaged at any point from
Marshal point on. If a waveoff or bolter occurs, the
pilot has only to depress the T.O. pushbutton and the
takeoff steering mode is engaged.

The basic landing mode symbology is essentially the same
as the basic takeoff mode. Exceptions are the addition of

angle-of-attack error symbol on the HUD, and velocity
vector symbol and 5° pitch increments on the VDI.

Note

In all landing submodes, a VDIG breakaway symbol can be displayed upon
receipt of a D/L waveoff message.

There are three steering command submodes that are
applicable during landing. These are TACAN, VEC, and
AWL/PCD. For the TACAN or VEC submodes of LDG,
the HUD, VDI, and HSD displays are similar to the same
submodes, respectively, as when in CRUISE, except that
in LDG the HUD display includes the velocity vector
symbol as well as the radar altitude symbol and the vertical
speed indicator symbol. The HUD, VDI, and HSD displays
for the TACAN landing submode are shown in figure 7-31.

AWL STEERING

If ILS information from the SPN-41/ARA-63A is available
at the Marshalling area, the pilot may select the AWL/
PCD submode. To observe glideslope displays, the HUD
and VDI AWL switches on the pilot's display and control
panel should be placed in the ILS position. The HUD and
VDIG will then provide vertical and lateral precision
course vector symbols, forming crossed pointers, and
driven by the ILS (SPN-41/ARA-63A) (figure 7-32). On
the HUD, full-scale vector deflection is limited to 2°.
Full-scale vector deflection on the VDI is 1.5°. The HSD
display at this time (AWL/PCD submode of LDG) will
automatically show TACAN displays.

At the acquisition window, the pilot may continue with
the ILS display, or, if ACL information from the SPN-42
data link is available, he may select the ACL positions of
the AWL switches for either the VDI or HUD displays, or
both. The ACL display uses the same vertical and lateral
precision course vector symbols as the ILS, but these are
now driven by the SPN-42 data link. A typical display
combination during the final stages of landing is ILS on
the HUD and ACL on the VDI. With valid ACL data
available, the AFCS may be engaged by selecting the
ACL position on the VEC/PCD, OFF, and ACL switch
located on the AFCS control panel.

LANDING/VECTOR STEERING. When the VEC
pushbutton is depressed, the data link vector symbology is
added to the basic landing symbology (figure 7-33). The
command heading symbols now indicate the heading to
be flown to make good the data link command course. In
addition, altitude and airspeed error symbols are added to
the right and left sides of the VDI, respectively. When the
double bar is above the reference, the error means that
the aircraft is below the commanded altitude or slower
than the commanded airspeed.

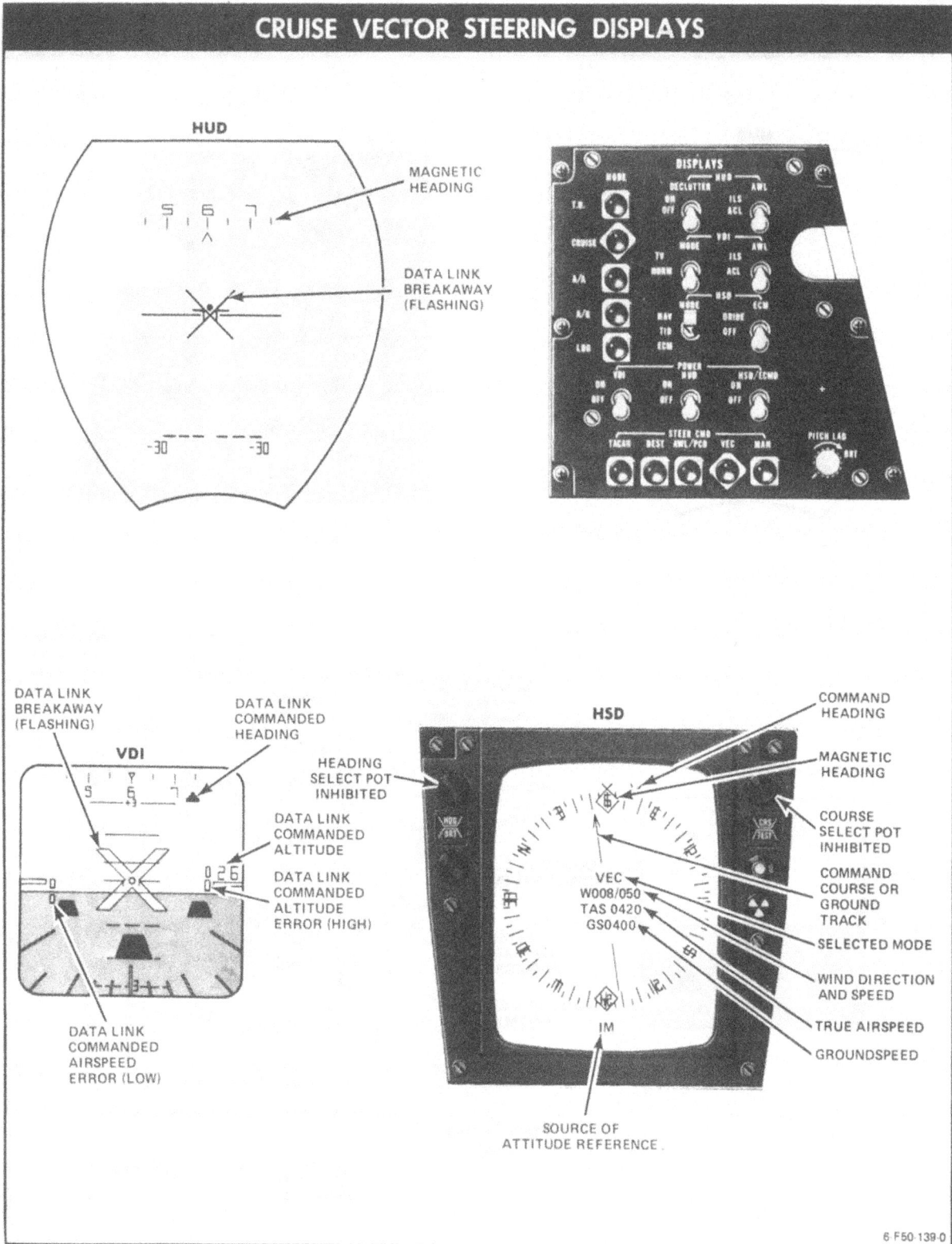

CRUISE VECTOR STEERING DISPLAYS

HUD

MAGNETIC
HEADING

DATA LINK
BREAKAWAY
(FLASHING)

DATA LINK
BREAKAWAY
(FLASHING)

DATA LINK
COMMANDED
HEADING

VDI

HEADING
SELECT POT
INHIBITED

DATA LINK
COMMANDED
ALTITUDE

DATA LINK
COMMANDED
ALTITUDE
ERROR (HIGH)

DATA LINK
COMMANDED
AIRSPEED
ERROR (LOW)

HSD

COMMAND
HEADING

MAGNETIC
HEADING

COURSE
SELECT POT
INHIBITED

COMMAND
COURSE OR
GROUND
TRACK

SELECTED MODE

WIND DIRECTION
AND SPEED

TRUE AIRSPEED

GROUNDSPEED

VEC
W008/050
TAS 0420
GS0400

IM

SOURCE OF
ATTITUDE REFERENCE

6-F50-139-0

Figure 7-29. Cruise Vector Steering Displays

CRUISE MANUAL STEERING DISPLAYS

HUD

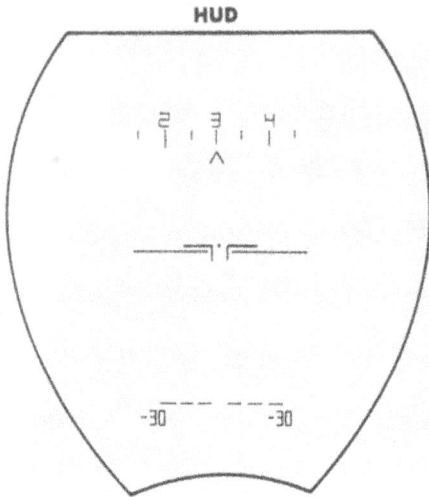

VDI

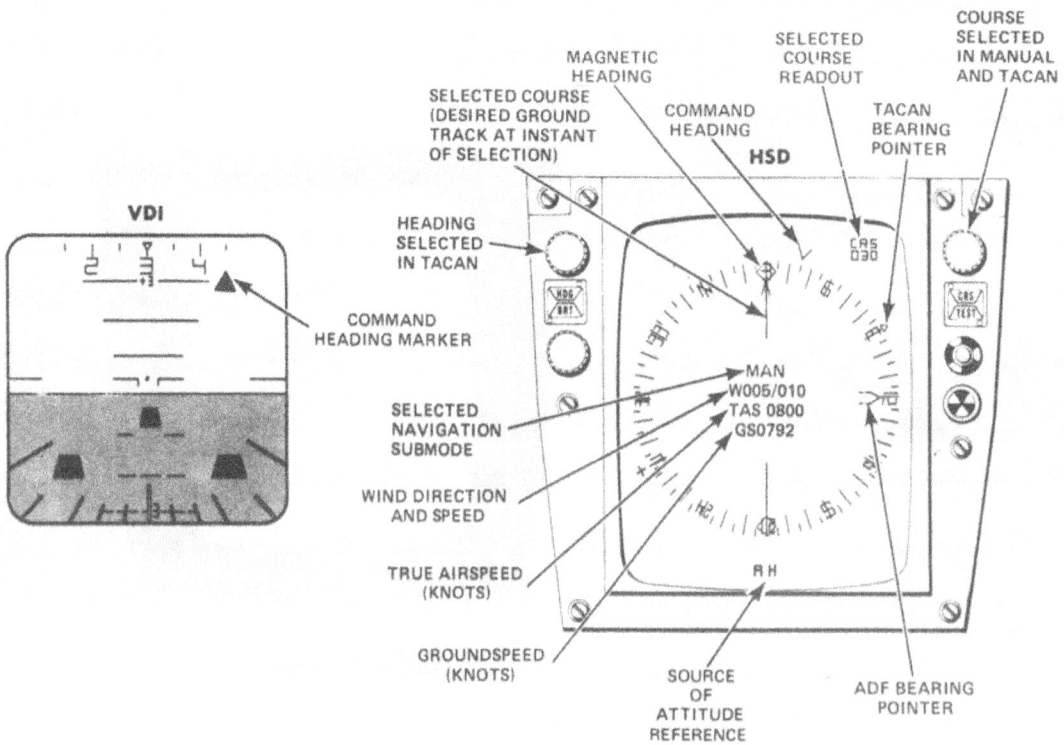

- COMMAND HEADING MARKER
- HEADING SELECTED IN TACAN
- SELECTED COURSE (DESIRED GROUND TRACK AT INSTANT OF SELECTION)
- MAGNETIC HEADING
- COMMAND HEADING
- SELECTED COURSE READOUT
- COURSE SELECTED IN MANUAL AND TACAN
- TACAN BEARING POINTER

HSD

- SELECTED NAVIGATION SUBMODE
- WIND DIRECTION AND SPEED
- TRUE AIRSPEED (KNOTS)
- GROUNDSPEED (KNOTS)
- SOURCE OF ATTITUDE REFERENCE
- ADF BEARING POINTER

MAN
W005/010
TAS 0800
GS0792

5-F50-140-0

Figure 7-30. Cruise Manual Steering Displays

LANDING TACAN STEERING DISPLAYS

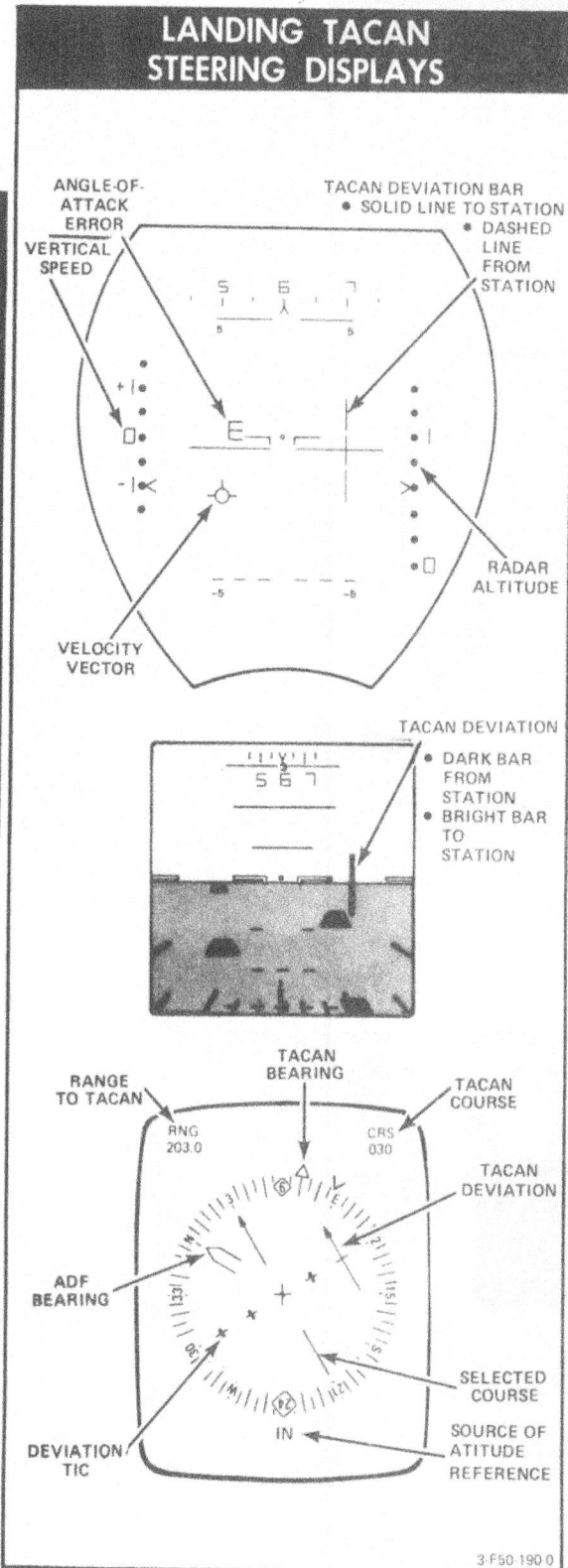

Figure 7-31. Landing TACAN Steering Displays

AWL STEERING DISPLAYS

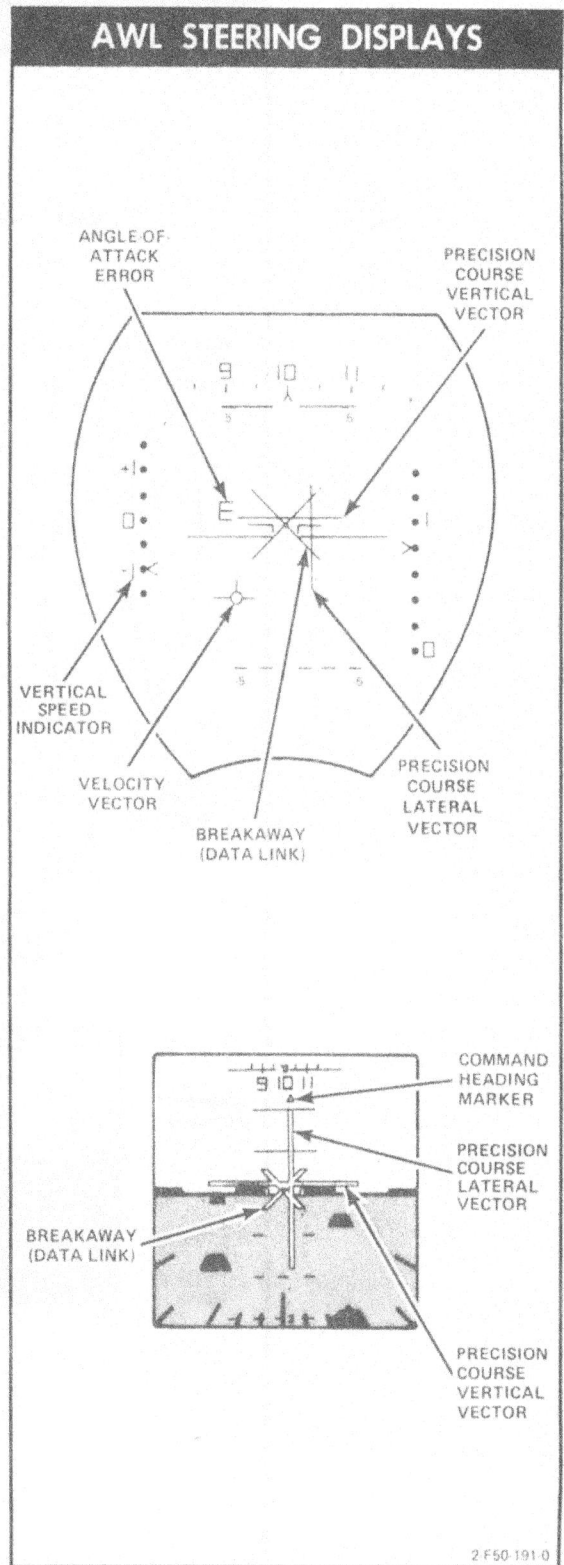

Figure 7-32. AWL Steering Displays

LANDING VECTOR
STEERING DISPLAYS

D/L
BREAKAWAY

ANGLE OF
ATTACK
ERROR

RADAR
ALTITUDE

VERTICAL
SPEED
INDICATOR

VELOCITY
VECTOR

COMMAND
HEADING
MARKER

COMMAND
AIRSPEED
ERROR

COMMAND
ALTITUDE
ERROR

COMMAND
ALTITUDE
REFERENCE
MARK

COMMAND
AIRSPEED
REFERENCE
MARK

D/L
BREAKAWAY

1-F050-004
192-0

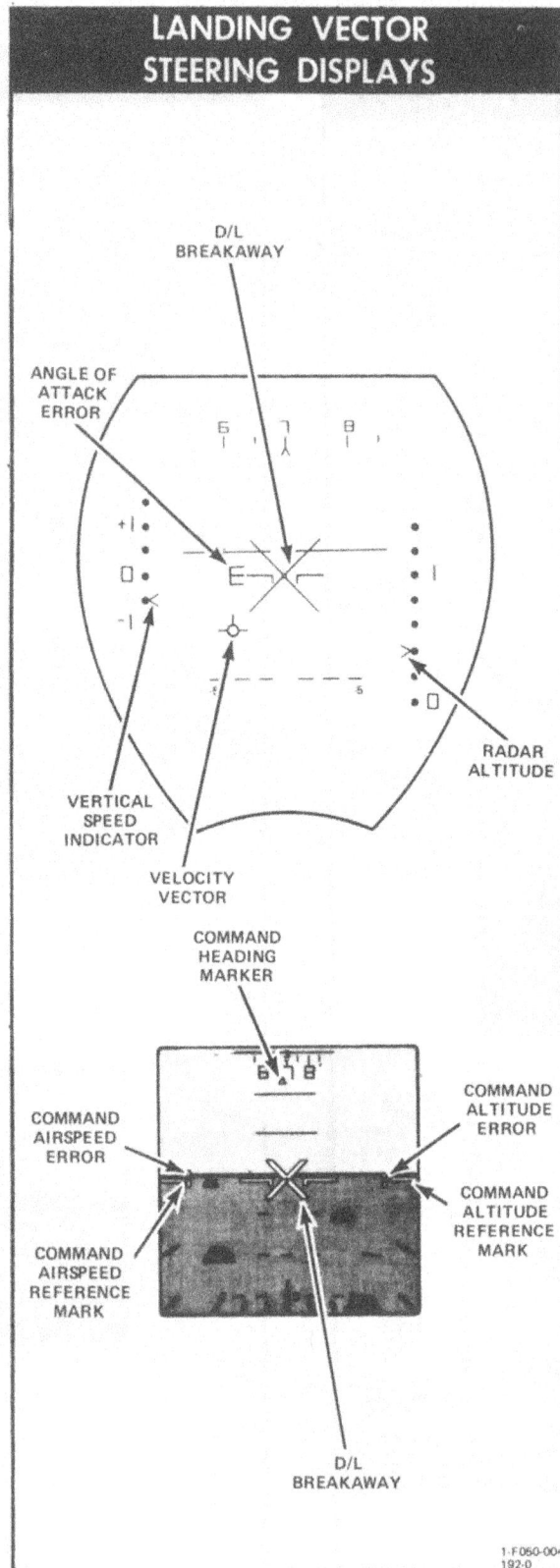

Figure 7-33. Landing Vector Steering Displays

PART 3 — IDENTIFICATION

IDENTIFICATION TRANSPONDER (IFF/SIF) (AN/APX-72)

IFF/SIF AIM Transponder

The AIMS transponder system is capable of automatically reporting coded identification and altitude signals in response to interrogations from surface (or airborne) stations so that the stations can establish aircraft identification, control air traffic, and maintain vertical separation. The system has five operating modes (1, 2, 3/A, C, and 4). Modes 1 and 2 are IFF modes, mode 3 (civil mode A) and mode C (automatic altitude reporting) are primarily air traffic control modes, and mode 4 is the secure (encrypted) IFF mode. The IFF control panel is in the rear cockpit (figure 7-34).

MASTER SWITCH

The MASTER switch applies power to all the AIMS transponder system components except the altimeter components. It is a five-position rotary switch placarded OFF, STBY, LOW, NORM, and EMER. The switch must be lifted over a detent to switch to EMER or to OFF. STBY should be selected for 2 minutes prior to switching to LOW or NORM to allow the transponder to warm up. In the NORM position, the transponder system is operational at normal receiver sensitivity. In the LOW position, the system is operational, but the transponder receiver sensitivity is reduced. In the EMER position, the transponder transmits emergency replies to mode 1, 2, or 3/A interrogations. The mode 3/A emergency reply includes code 7700. When EMER is selected, all modes are enabled regardless of the position of the selector switches. When the front seat ejects, a switch is tripped that automatically selects the emergency mode if the MASTER switch is in any position other than OFF.

IDENT-OUT-MIC SWITCH

The IDENT-OUT-MIC switch is a three-position toggle switch. The spring-loaded IDENT position adds an identification of position pulse to mode 1, 2, and 3/A replies for a period of 15 to 30 seconds. In the MIC position, the identification of position function is activated for 15 to 30 seconds each time the UHF microphone switch is pressed.

MODE 1, 2, AND 3/A CODE SELECTORS

The two-mode 1 thumbwheel selector switches allow selection of 32 mode-1 codes and the four mode-3/A thumbwheel selectors allow selection of 4096 mode-3/A codes. The mode-2 code cannot be changed during flight.

MODE SWITCHES

The four mode switches (M-1, M-2, M-3/A, and M-C) each have OUT, ON, and spring-loaded TEST positions. The center ON position of each switch enables that mode. To test the transponder, press the mode switch of each mode to the TEST position.

Illumination of the TEST light indicates proper operation of that mode. The MASTER switch must be set to NORM for the test function to operate. The modes not being tested should be OUT when testing on the ground to prevent unnecessary interference with nearby ground stations. If a malfunction exists during these self-tests, an IFX CM acronym will appear on the TID.

RAD TEST-OUT-MON SWITCHES

The MON position of the RAD TEST-OUT-MON switch is used to monitor the operation of modes 1, 2, 3/A and C. When MON is selected, the TEST light will illuminate for 3 seconds each time an acceptable response is made to an interrogation on a selected mode.

The spring loaded RAD TEST position is used for testing. It enables a mode-3/A code reply to a TEST mode interrogation from a ramp test set. It also enables a mode-4 reply to a VERIFY 1 interrogation from a surface station or a ramp test set. A VERIFY 1 interrogation is a modified mode-4 interrogation used for testing.

MODE 4 OPERATION

Mode-4 operation is selected by placing the MODE 4 toggle switch to ON, provided that the MASTER switch is NORM or LOW. Placing the MODE 4 switch to OUT disables mode 4.

IFF CONTROL PANELS

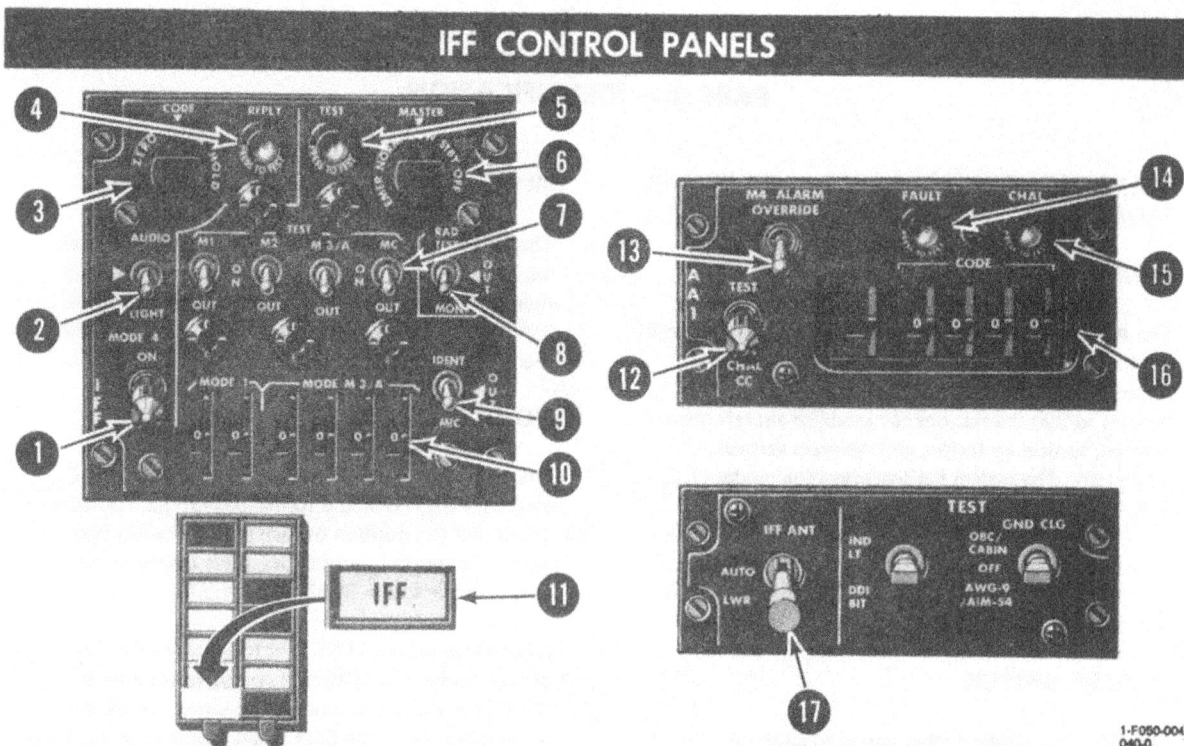

1-F050-004
040-0

NOMENCLATURE	FUNCTION
① MODE 4 switch	ON — Enables mode 4. OUT — Disables mode 4. See figure 7-35 for Mode 4 caution/reply light logic.
② MODE 4 AUDIO/ LIGHT switch	AUDIO — Selects aural monitoring when mode 4 is being interrogated and REPLY light monitoring of MODE 4 replies. ▶ — Disables light and aural monitoring of mode 4. LIGHT — Selects mode 4 reply light monitoring.
③ MODE 4 CODE switch	ZERO — Erases code 4 from KIR-1A and KIT-1A computers. B — Selects KIT-1A computer B code. A — Selects KIT-1A computer A code. HOLD — Retains code in KIR-1A and KIT-1A computers when landing gear is down or when system is turned off.
④ MODE 4 REPLY light	Illuminates to indicate mode-4 reply to interrogation.
⑤ TEST light	Illuminates when respective test switch is actuated; indicates proper operation of modes 1, 2, 3/A, and C. Master switch must be set to NORM.
⑥ MASTER switch	OFF — Deenergizes set. STBY — Energizes receiver-transmitter for immediate operation upon switching to an operating position. LOW — Provides reduced receiver sensitivity for response only to strong nearby interrogations. NORM — Allows full receiver sensitivity for response to maximum range interrogations.

Figure 7-34. IFF Control Panels (Sheet 1 of 2)

NOMENCLATURE	FUNCTION
⑦ MODE switches	EMER — Provides full receiver sensitivity and generates emergency replies to mode 1, 2, and 3/A (code 7777) and a normal reply to mode C, when interrogated, whether mode switches are on or off. TEST — TEST light illuminates if system is functioning properly. ON — Permits selection of interrogating modes to which the transponder will respond. OUT — Deenergized position.
⑧ RAD TEST-OUT-MON switch	RAD TEST — Not operational (for ground testing only). OUT — Deenergized position. MON — Allows RIO to monitor APX-72 reply (modes 1, 2, 3, and C). Test light will illuminate.
⑨ IDENT-OUT-MIC switch	IDENT — Momentary position provides IDENT reply for 15 to 30 seconds after releasing switch; replies to interrogation in modes 1, 2, 3/A. OUT — Deenergizes circuit. MIC — Transfers IDENT reply activation switch from Ident to radio microphone switch.
⑩ CODE selectors (MODE 1 and 3/A)	Code selectors are rotatable drums with imprinted numbers that appear in code selector windows, permitting selection of codes for mode 1 and 3/A.
⑪ IFF caution light	Indicates mode-4 interrogation was received, but system has not generated reply.
⑫ TEST-CHAL CC switch	Momentary two-position center-return switch. TEST — Onboard transponder is triggered by onboard interrogator. Both sets must have same code setting. IFF solid lines are displayed on DDD at 3 and 4 miles. CHAL CC — A SIF interrogation cycle starts the 5 to 10 second challenge period. Only correct modes and code replies are displayed (two brackets only on DDD).
⑬ M4 ALARM OVERRIDE switch	Disables the mode-4 tone alarm to the RIO's ICS.
⑭ FAULT light	Indicates a malfunction of APX-76 receiver-transmitter, caused by receiver, video, or transmitter signals.
⑮ CHAL light	Remains illuminated for the duration of a challenge period indicating correct operation.
⑯ CODE selectors	First thumbwheel selects mode, 1, 2, 3A, 4A, or 4B. Last four thumbwheel rotatable drums with imprinted numbers appearing in code selector windows, permit selection of desired interrogation code.
⑰ IFF ANT switch	AUTO — Antenna lobing switch cycles receiver-transmitter between upper and lower antenna pattern coverage. LWR — Only the lower antenna pattern below aircraft is used.

Figure 7-34. IFF Control Panels (Sheet 2 of 2)

The MODE 4 CODE switch is placarded ZERO, B, A, and HOLD. The switch must be lifted over a detent to switch to ZERO. It is spring-loaded to return from HOLD to the A position. Position A selects the mode-4 code for the present code period and position B selects the mode-4 code for the succeeding code period. Both codes are mechanically inserted into the transponder by maintenance personnel. The codes are mechanically held in the IFF, regardless of the position of the MASTER switch or the status of aircraft power, until the first time the landing gear is raised. Thereafter, the mode-4 codes will automatically zeroize anytime the MASTER switch or the aircraft electrical power is turned off. The code settings can be mechanically retained after the aircraft has landed (landing gear must be down and locked) by turning the CODE switch to HOLD and releasing it at least 15 seconds before the MASTER switch or aircraft electrical power is turned off. The codes again will be held, regardless of the status of aircraft power or the MASTER switch, until the next time the landing gear is raised.

The mode-4 codes can be zeroized any time the aircraft power is on and the MASTER switch not OFF by turning the CODE switch to ZERO.

An audio signal, the REPLY light, and the IFF caution light are used to monitor mode-4 operation. The AUDIO/▶-LIGHT switch controls the audio signal and the REPLY light, but not the IFF caution light. In the LIGHT position, the REPLY light illuminates as mode-4 replies are transmitted. In the AUDIO position, an audio tone in the RIO's headset indicates that valid mode-4 interrogations are being received and the REPLY light illuminates if mode-4 replies are transmitted. In the ON position, the audio indications and the REPLY light are inoperative and the REPLY light will not press-to-test. (CAUTION and REPLY light logic are shown in figure 7-35.)

IFF CAUTION LIGHT

The IFF caution light on the RIO's threat advisory lights panel illuminates to indicate that mode 4 is not operative. The light is operative whenever aircraft power is on and the MASTER switch is not OFF. However, the light will not operate if the mode-4 computer is not physically installed in the aircraft. Illumination of the IFF caution light indicates that: (1) the mode-4 codes have zeroized, (2) the self-test function of the KIT-1A/TSEC computer has detected

MODE 4 CAUTION AND REPLY LIGHT LOGIC

TRANSPONDER (APX-72)	INTERROGATOR (APX-76)	CAUTION	REPLY (APX-72)
4 OUT (A) STBY	A	ON	OFF
4 ON (A) STBY	A	ON	OFF
4 ON (A) NORM	A	OFF	ON
4 ON (A) NORM	B	ON	OFF
4 ON (B) NORM	A	ON	OFF
4 ON (B) NORM	B	OFF	ON
4 ON (B) STBY	B	ON	OFF
4 OUT (B) STBY	B	ON	OFF
4 ON (A) NORM RAD TEST	VERIFY BIT 1 (A)	OFF	ON
4 ON (A) NORM	VERIFY BIT 1 (A)	ON	OFF
4 ON (A) STBY	VERIFY BIT 1 (A)	ON	OFF
KIT ZERO	A OR B	ON	OFF

1-F050-004
136-0

Figure 7-35. Mode 4 Caution and Reply Light Logic

a faulty computer, or (3) the transponder is not replying to proper mode-4 interrogations.

If the IFF caution light illuminates, switch the MASTER switch to NORM (if in STBY) and insure that the MODE 4 toggle switch is ON. If illumination continues employ operationally-directed flight procedures for an inoperative mode-4 condition.

ANTENNA SWITCHING UNIT

The aircraft has two IFF antennas and an antenna switching unit, which electronically switches the transponder between the top and bottom IFF antennas at a nominal rate of 38 cycles per second. The aircraft also has an antenna switch placarded AUTO and LWR. The AUTO position automatically cycles the IFF receiver-transmitter between the upper and lower antenna. In the LWR position, only the lower antenna is used.

ALTITUDE COMPUTATIONS

Altitude computations are performed by the central air data computer (CADC).

The computer outputs are altitude-information corrected for static position error. The synchro output is supplied to the altimeter providing the crew with a corrected altitude indication. The digital output from the computer is applied to the transponder for transmission on mode C, is coded in increments of 100 feet, and is referenced to 29.92 inches of mercury.

The altimeter has a primary (servo) mode and a standby (pressure) mode of operation, controlled by a spring-loaded self-centering mode switch placarded RESET and STBY. In the primary (servo) mode, the altimeter displays altitude (corrected for position error) from the synchro output of the altitude computer. In the standby mode of operation, the altimeter displays altitude directly from the static system (uncorrected for position error) and operates as a standard pressure altimeter. A dc-power internal vibrator is automatically energized while in the standby mode to minimize friction in the display mechanism.

IFF INTERROGATOR (AN/APX-76)

The AN/APX-76 provides radar identification of airborne and surface Mark 10 IFF systems. It operates in conjunction with the AN/AWG-9 radar and is automatically turned on whenever the AWG-9 power switch is placed to any position except OFF. A minimum warmup time of 3 minutes is required before successful operation or BIT can be performed. The system requires 115 volts ac from the main AC bus through the IFF A/A AC circuit breaker (117) and 28 volts dc from the main DC bus through the IFF A/A circuit breaker (8F6). It is capable of interrogation and display of modes 1,2,3A and 4, and of displaying EMERG AND IDENT on the DDD.

The APX-76 system consists of five basic components: an antenna system, a receiver-transmitter, a switch amplifier, a synchronizer unit, and a control panel.

The IFF antenna is an integral part of the AWG-9 antenna. It consists of 12 dipole antennas mounted on the surface of the AWG-9 planar array antenna. The antenna azimuth and vertical coverage is the same as that of the AWG-9 antenna except that the beam width of the APX-76 is 13 degrees. The transmitter operates at a fixed frequency of 1030 MHz and the receiver operates at a fixed frequency of 1090 MHz.

The APX-76 operationally functions the same in low and high PRF modes of the AWG-9, however, trigger pulses to the APX-76 are generated from different sources, depending on the operational mode of the AWG-9. During pulse doppler modes of the AWG-9, including standby, the interface unit (WRA-461) provides the trigger pulses to the APX-76. These trigger pulses are delayed in the APX-76 and are sent back to the AWG-9. Replies from the interrogated transponder are decoded by the APX-76 and displayed in range and azimuth on the DDD. For pulse and pulse standby modes of the AWG-9, the trigger pulses to the APX-76 are generated in the radar synchronizer (WRA-010). The trigger pulses are delayed by the APX-76, then sent back and used to initiate a video display gate for normal IFF display on the DDD.

Note

The trigger pulses for low and high PRF modes of the AWG-9 are degenerated by different WRA's. If the interrogator fails to operate in one mode, APX-76 operation should be attempted in the other mode. Failure to work in either mode enhances the possibility that one of the components of the APX-76 is at fault.

A KIR-1A computer generates mode-4 interrogations and interpolates mode-4 replies. Display of mode 4 is the same as all other modes.

IFF Self-Test

Prior to APX-76 operation, self-test of the unit should be performed. The APX-76 contains a self-test function that provides closed loop testing in conjunction with the on-board APX-72 (IFF Transponder). To perform the self-test, the RIO must set the mode and code switches on the control panel to correspond with the mode and code switches of the APX-72. The APX-72 must be in an operating mode (low, normal, or emergency) before performing the test. The RIO may now initiate self-test by holding the TEST/CHAL CC switch in the TEST position for 5 to 10 seconds. Provided both the IFF and the APX-76 are functioning properly, two horizontal bars will be displayed across the DDD at approximately 3 and 4 miles. Illumination of the green CHAL light on the control panel, while the switch is being held in the test position, also indicates that the APX-76 made a valid interrogation. The bottom line on the DDD indicates that the APX-72 responded in mode and the top line indicates it responded in code. Both lines together indicate that the APX-76 is decoding properly. Biasing of the mode and code lines enables them to be spread across the entire DDD during test. If the first attempt to test the APX-76 fails because of lack of video on the DDD, or the amber fault light on the control panel illuminates, the RIO should initiate a valid challenge by momentarily holding the CHAL CC/TEST switch in the CHAL CC position in order to reset the BIT flags associated with the APX-76. The APX-76 normally powers up with the BIT flags in the fault position. The system will continuously fault until the flags are reset. The APX-76 antenna is checked during the test by receiving actual video from the APX-72 antenna. Failure of any part of the APX-76 closed loop test will cause IFI to be displayed in continuous monitor. A further breakdown as to what portion of the system has failed can be verified by calling up the maintenance file. Testing of all modes of the APX-76 should be performed independently. Failure of one mode does not necessarily mean that all modes are malfunctioning.

Portions of the APX-76 are tested during class III OBC. During OBC, CHALLENGE IFF is displayed on the TID in order to remind the RIO to reset the BIT flags by making a valid challenge. DDD video is not tested during OBC since that function requires the AWG-9 to be operating in a tactical mode.

APX-76 Operation

The operating mode of the APX-76 is determined by the setting of the six-position mode thumbwheel on the control panel. MODES 1, 2, and 3 are the normal operating modes. MODES 4A and 4B are crypto-secure modes requiring the installation of KIR equipment. The white position selects the APX-76 to standby and should be selected prior to AWG-9 power-down. Four-digital thumbwheels are located on the panel to provide code selections. Each thumbwheel has eight positions numbered from 0 to 7. The code selection is dependent on the mode selection. If mode 1 is selected, the code selection must be a multiple of 100 ranging from 0000 to 7300. If mode 2 or 3 is selected, the code selection may be any four-digit number from 0000 to 7777.

There are two ways to make a valid challenge of Mark 10 IFF systems. One is by depressing the IFF tile on the DDD. By depressing the tile, all targets replying in the mode selected on the control panel will cause a line to be displayed on the DDD just below the target's range and at the same azimuth as the target. Actual video return of the target may or may not be displayed. During the pulse search mode, target video may be displayed if the target falls within the radar beam width and produces enough skin return to be seen. If any of the targets are replying in the same code that the RIO has selected on the code thumbwheels, a second line will appear just above the target range and azimuth. IFF interrogation and the corresponding display continue for a maximum of 10 seconds or until the switch is released, whichever is shorter.

The second way to challenge a Mark 10 IFF transponder is to hold the CHAL CC/TEST switch in the CHAL CC (CHALLENGE CORRECT CODE) position. The mode and code of all targets replying to the mode and code that the RIO has set will be the only IFF video displayed. Mode return alone will never be displayed if the CHAL CC switch position is used. IFF interrogation and display continue for a maximum of 10 seconds or until the switch is released, whichever is shorter.

Note

Flickering CHAL and Fault lights may be observed when interrogating with the AWG-9 radar in the 5, 10, 20, and 50-pulse range scale. In addition, an IFI acronym may be observed in the OBC file. This condition does not interfere with live video IFF targets on the APX-76 loop test.

IFF Displays

Display formats for IFF operation are shown in figure 7-36. During IFF interrogation, the radar continues its normal operations. With the radar in a pulse mode (search, track, or acquisitions), the IFF returns are mixed with the radar video and appropriate pulse radar display formats are displayed on the DDD. The IFF range scale is defined by the radar range scale.

If the RIO commands an IFF interrogration during any single track mode (PDSTT or PSTT), or when a target is designated (hooked) on the TID, an expanded IFF range display is generated on the DDD. The nominal target range is displayed at the vertical center of the DDD. The range scale is indicated on the range scale indicator as ±10 and it cannot be changed. The azimuth of the current radar mode is retained in the expanded IFF format. By unhooking the STT target, normal range vs azimuth is displayed along with the IFF display.

With the radar operating in the pulse doppler mode, the DDD presentation is switched to a B-scan format. In this format, the pulse doppler antenna scan pattern is retained but IFF range sweeps are displayed. A computer-generated symbol representing actual target range is displayed along with the IFF video on the DDD. The IFF scale is indicated on the range indicator. It may be changed by pressing the range-select pushbutton. In the pulse-doppler, single-target track mode, the attack symbols are retained and the IFF sweep is displayed at the azimuth of the AWG-9 tracking antenna.

Note

During all STT modes, IFF video return may be difficult to detect because of DDD smearing.

When a target is detected, the RIO must tell the computer whether the target is friend or foe. The APX-76 cannot perform this function. It can only determine if Mark 10 IFF is being utilized by the target. If the sensor target has been data-link associated with a data link file, the IFF contact will be sent to the TDS as part of the R3A message.

IFF DISPLAY FORMATS

RIO HAS MODE 3 CODE 1200 SET

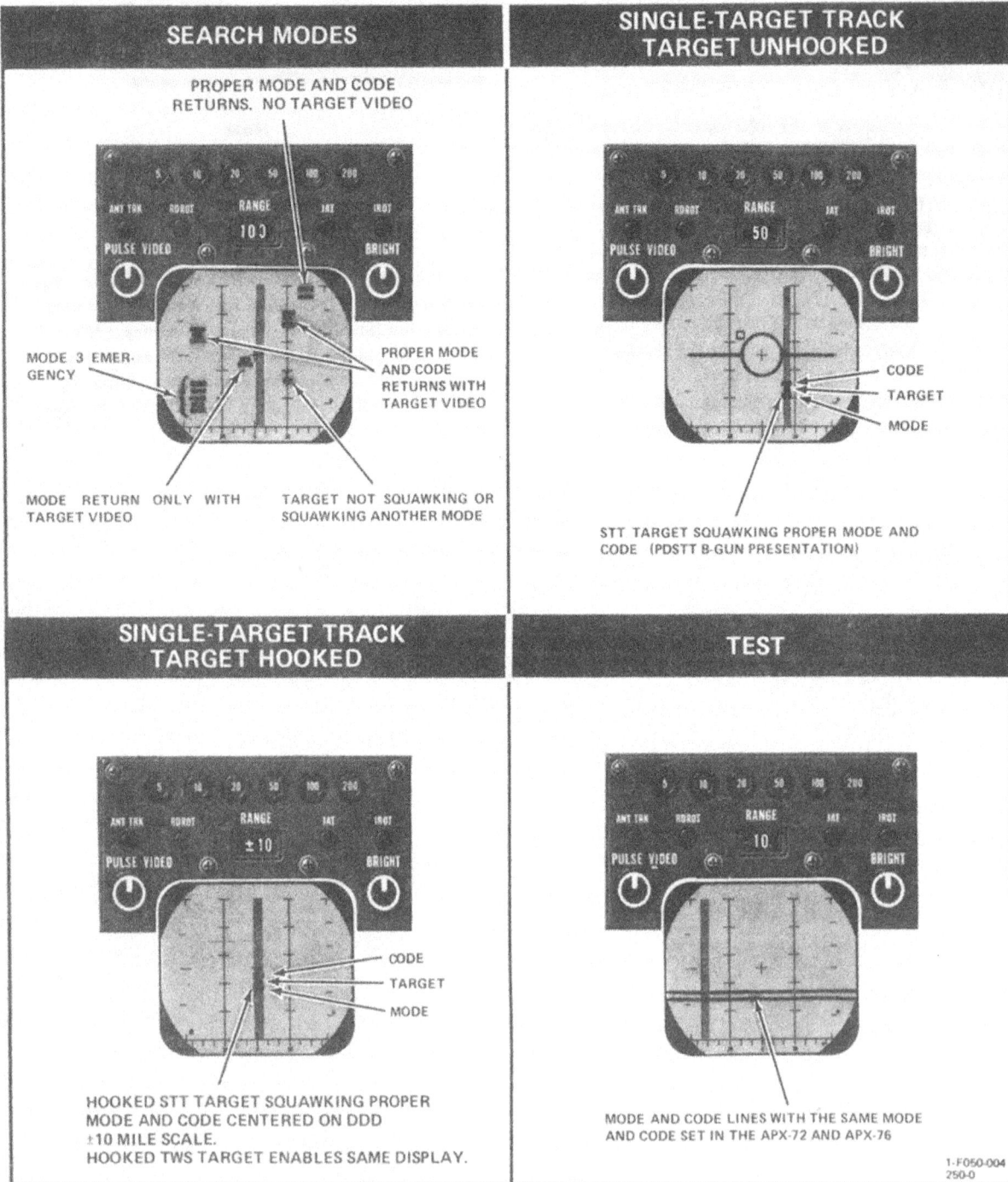

SEARCH MODES

PROPER MODE AND CODE
RETURNS. NO TARGET VIDEO

RANGE
100

MODE 3 EMER-
GENCY

PROPER MODE
AND CODE
RETURNS WITH
TARGET VIDEO

MODE RETURN ONLY WITH
TARGET VIDEO

TARGET NOT SQUAWKING OR
SQUAWKING ANOTHER MODE

SINGLE-TARGET TRACK
TARGET UNHOOKED

RANGE
50

CODE
TARGET
MODE

STT TARGET SQUAWKING PROPER MODE AND
CODE (PDSTT B-GUN PRESENTATION)

SINGLE-TARGET TRACK
TARGET HOOKED

RANGE
±10

CODE
TARGET
MODE

HOOKED STT TARGET SQUAWKING PROPER
MODE AND CODE CENTERED ON DDD
±10 MILE SCALE.
HOOKED TWS TARGET ENABLES SAME DISPLAY.

TEST

RANGE
10

MODE AND CODE LINES WITH THE SAME MODE
AND CODE SET IN THE APX-72 AND APX-76

1-F050-004
250-0

Figure 7-36. IFF Display Formats

SECTION VIII — WEAPONS SYSTEM

TABLE OF CONTENTS

PART 1 — VERTICAL DISPLAY INDICATOR GROUP (AN/AVA-12)

VERTICAL DISPLAY INDICATOR GROUP (VDIG)

The vertical display indicator group (VDIG) provides the pilot symbolic takeoff, cruise, air to air, air to ground, landing, and test information. Electronically generated symbology portrays aircraft attitude, command, and tactical information. The information displayed depends on mission phase and the mode of operation the pilot initiates for a given phase. Information for the mode selected is displayed simultaneously on a TV-like display called the vertical display indicator (VDI) and on a projected-image-type heads-up display (HUD), which is coincident with the pilot's forward vision through the windshield. The VDI is a head-down display, whereas the transition from the head-down display to visually acquire target information is provided on the HUD. Figure 8-1, sheets 1 through 11, show the various HUD and VDI presentations for each mode and submode.

Controls for the VDIG modes are on the pilot's display control panel. (See figure 8-2.) These modes provide symbolic display of information necessary to the particular mode. A VDIG converter controls symbol enabling, positioning, and status by mode logic and signals from the computer signal data converter (CSDC), AWG-9, or AWG-15, depending on the mode. Figure 8-3, sheets 1 through 3, shows the VDIG symbols and their functions, along with a pictorial comparison of the symbols as they appear on the HUD and VDI. During all the various modes except cruise, certain symbols may be removed from the HUD through the DECLUTTER switch on the display control panel. In addition, the VDIG has built-in test (BIT) capabilities for system checks on the ground and in flight.

Note

If information displayed on the HUD and VDI are different, determine which display is correct by use of the VDIG test modes (MASTER TEST switch set to INST) and by reference to other cockpit instruments. Eliminate the erroneous display with the corresponding power switch.

VDIG Data Freeze

Logic circuitry in the VDIG senses the receipt of pitch and roll information from the CSDC. If for any reason this information is not received by the VDIG converter for 200 milliseconds or more, the VDIG will turn off the steering symbol pitch lines on the HUD and the pitch ladder, steering symbol, horizon line, roll pointer, ground texture and moving elements, and sky plane on the VDI.

VERTICAL DISPLAY INDICATOR

The VDI provides an in-the-cockpit vertical display to the pilot during medium and long-range missile attacks, initiation of visual identification (VID) passes, data-link vectoring, automatic carrier landing, aircraft flight attitude, and navigation. (The VDI replaces the mechanical attitude director indicator of older aircraft systems as an aircraft attitude instrument pitch, roll, and heading. TV video can also be presented on the VDI from the television camera set (TCS), when installed.

Note

Selecting the TV position on the pilot's display control panel without a television sight unit installed will blank out the VDI presentation.

The direct-view VDI uses a combination of calligraphic, or stroke-writing, techniques with a television-like raster scan. The 525-line raster scan enables a great deal of

symbology to be generated; and the calligraphy adds sharp bright lines where needed for accuracy and high resolution. Steering information and pitch lines are generated calligraphically, for example, stroke writing is performed during the retract of the raster scan.

Attitude information is displayed on the VDI by an aircraft reticle, a horizon line, ground and sky texture, and a calligraphic pitch ladder. The aircraft reticle is fixed at the center of the display, and the horizon line and pitch ladder moves about it in accordance with the aircraft pitch and roll attitudes.

The flight parameters displayed include magnetic heading, data link-command airspeed (Mach number) and altitude, and vertical velocity. Ground texture elements superimposed on the ground plane simulate both motion and simple perspective. Command information includes steering, breakaway, and air-to-air attack steering. The circle used to indicate allowable attack steering error is programmed to diminish continually in size as steering error becomes progressively more critical during a maneuver. Tactical symbology provides such information as target position, azimuth position, target range and armament types, and quantities for the air-to-air attack mode.

HEADS-UP DISPLAY

The HUD is used for visual identification terminal approaches, short-range air-to-air weapon attacks (predominantly Sidewinder and gun), air-to-ground attacks, and automatic carrier landings. The instrument is used primarily when the pilot can see the target or landing zone.

The HUD provides a combination of real-world cues and flight direction symbology, projected directly on the windshield. The display is focused at infinity, thereby creating the illusion that the symbols are superimposed on the real world (and so that visual cues received from outside the aircraft are not obscured). The pilot usually steers based on his interpretation of the visually observed real world.

The HUD presentation is generated entirely by calligraphic means. The brightness of the symbols is set manually and automatically adjusted so that they are visible under any ambient conditions. Symbology for the HUD is basically attack oriented, and includes symbols for steering, target, target range, minimum range, release range, closure rate, and armament legends. A moving reticle, bomb impact line, and velocity vector are included for air-to-ground attacks. The moving reticle symbol is aircraft-stabilized in both manual and automatic weapon delivery modes. The HUD also displays attitude, flight situation, and command information, primarily for the landing mode.

The overall field of view of the HUD is 20°. The optics exit port is 8 inches in diameter. Parallax error has been minimized over the entire field of view in order to improve the accuracy of the gunnery and bombing modes.

HUD Circular Polarized Filter

A circular polarized filter is added to the windshield-projected HUD. The filter fits on top of the Fresnel wedge and is used to eliminate excessive sun reflections. However, installation of this filter does reduce HUD brightness somewhat and the HUD intensity must be set at a higher reading. Installation of the HUD filter will extend beyond the present outline of the indicator, but this will not affect over-the-nose visibility.

Pitch Ladder

The pitch ladder provides attitude reference for basic flight control. The symbol is driven by inputs from the inertial navigation system (INS) during normal operation and from the attitude heading and reference set (AHRS), if selected or when the inertial measurement unit (IMU) fails. The pitch lines maintain an orientation parallel to the horizon. Scaling depends on the flight mode selected. (See figure 8-4.) For takeoff, landing, and air-to-ground operations, the HUD has a 1:1 pitch line compression with a pitch line displayed every 5° up to ±30° and in 10° increments up to ±90°. The A/A and cruise HUD displays have a 1:4 pitch compression with 30° increments up to ±90°. Pitch lines above the HUD magnetic heading scale are blanked. The VDI, on the other hand, has its pitch lines determined by the total amount of pitch attitude displayed on the VDI. For A/A and A/G, 130° of pitch is displayed with the pitch lines every 30°. Takeoff, cruise, and landing display a total of 65°. However, takeoff and cruise have pitch lines graduated every 10° between ±60°; the landing displays are also graduated every 5° between ±30°.

TEST MODES

The test modes are initiated by the pilot when he selects INST on the MASTER TEST panel. During his instrument checks, the VDIG is tested by observing various displays on the HUD and VDI. These displays are static and are conditional on the mode selected. Figure 8-5 shows each of the four test displays. Selecting T.O. or CRUISE (test display 1) gives a presentation indicative of whether the deflection or brightness circuits of the indicator are operating correctly. Each of the other three modes (A/A, A/G, or LDG) generate presentations which depict all the symbols that are available in that mode and its submodes, and is a check of the converter unit.

If the VDI or HUD is blank, the pilot should cycle the POWER switches so that power is turned off and then back on. A blank screen may occur during a power interrupt and may last up to 20 seconds after a long interrupt.

SYSTEM CHECKOUT

When power is applied, the VDI and HUD converter and indicators continuously perform self-tests. If a failure is detected, the BIT indicator on the respective unit is turned on. In addition, results of the tests are applied to the CSDC. The CSDC formats the test data, and if a failure exists, routes the information to the weapons control system (WCS) computer under onboard checkout (OBC) program control for display of a vertical display indicator group (DIG) acronymn on the tactical information display (TID) and the HSD, if TID repeat is selected. The following table provides DIG acronym definitions.

DIG ACRONYM	DEFINITION
DIG C	VDI converter
DIG CH	HUD converter
DIG I	VDI indicator
DIG IH	HUD indicator

VDI AND HUD PRESENTATIONS

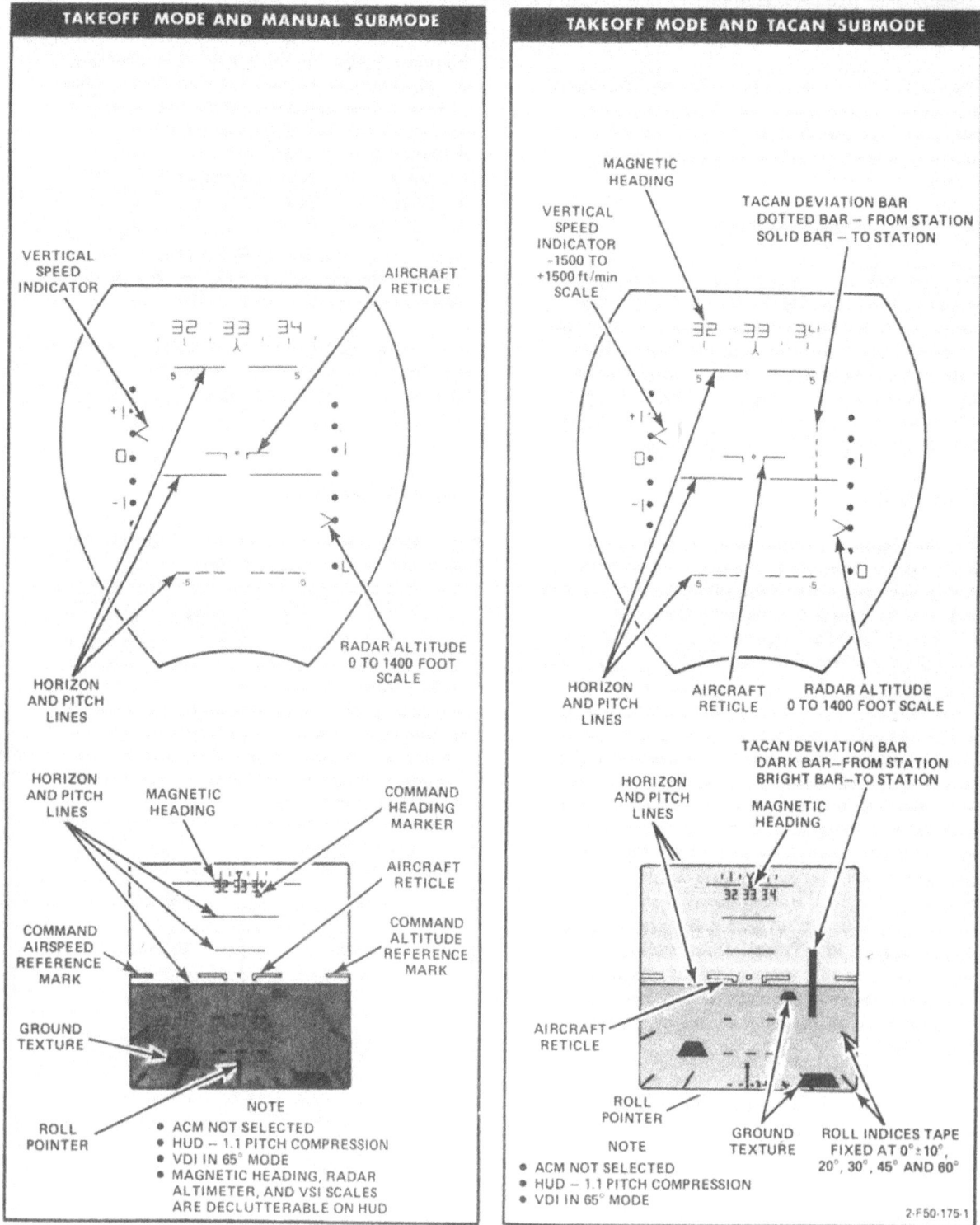

TAKEOFF MODE AND MANUAL SUBMODE

VERTICAL
SPEED
INDICATOR

AIRCRAFT
RETICLE

32 33 34

RADAR ALTITUDE
0 TO 1400 FOOT
SCALE

HORIZON
AND PITCH
LINES

HORIZON
AND PITCH
LINES

MAGNETIC
HEADING

COMMAND
HEADING
MARKER

AIRCRAFT
RETICLE

COMMAND
AIRSPEED
REFERENCE
MARK

COMMAND
ALTITUDE
REFERENCE
MARK

GROUND
TEXTURE

ROLL
POINTER

NOTE
- ACM NOT SELECTED
- HUD – 1.1 PITCH COMPRESSION
- VDI IN 65° MODE
- MAGNETIC HEADING, RADAR
  ALTIMETER, AND VSI SCALES
  ARE DECLUTTERABLE ON HUD

TAKEOFF MODE AND TACAN SUBMODE

MAGNETIC
HEADING

TACAN DEVIATION BAR
DOTTED BAR – FROM STATION
SOLID BAR – TO STATION

VERTICAL
SPEED
INDICATOR
-1500 TO
+1500 ft/min
SCALE

32 33 34

HORIZON
AND PITCH
LINES

AIRCRAFT
RETICLE

RADAR ALTITUDE
0 TO 1400 FOOT SCALE

TACAN DEVIATION BAR
DARK BAR—FROM STATION
BRIGHT BAR—TO STATION

HORIZON
AND PITCH
LINES

MAGNETIC
HEADING

32 33 34

AIRCRAFT
RETICLE

ROLL
POINTER

GROUND
TEXTURE

ROLL INDICES TAPE
FIXED AT 0°±10°,
20°, 30°, 45° AND 60°

NOTE
- ACM NOT SELECTED
- HUD – 1.1 PITCH COMPRESSION
- VDI IN 65° MODE

2-F50-175-1

Figure 8-1. VDI and HUD Presentations (Sheet 1 of 11)

VDI AND HUD PRESENTATIONS

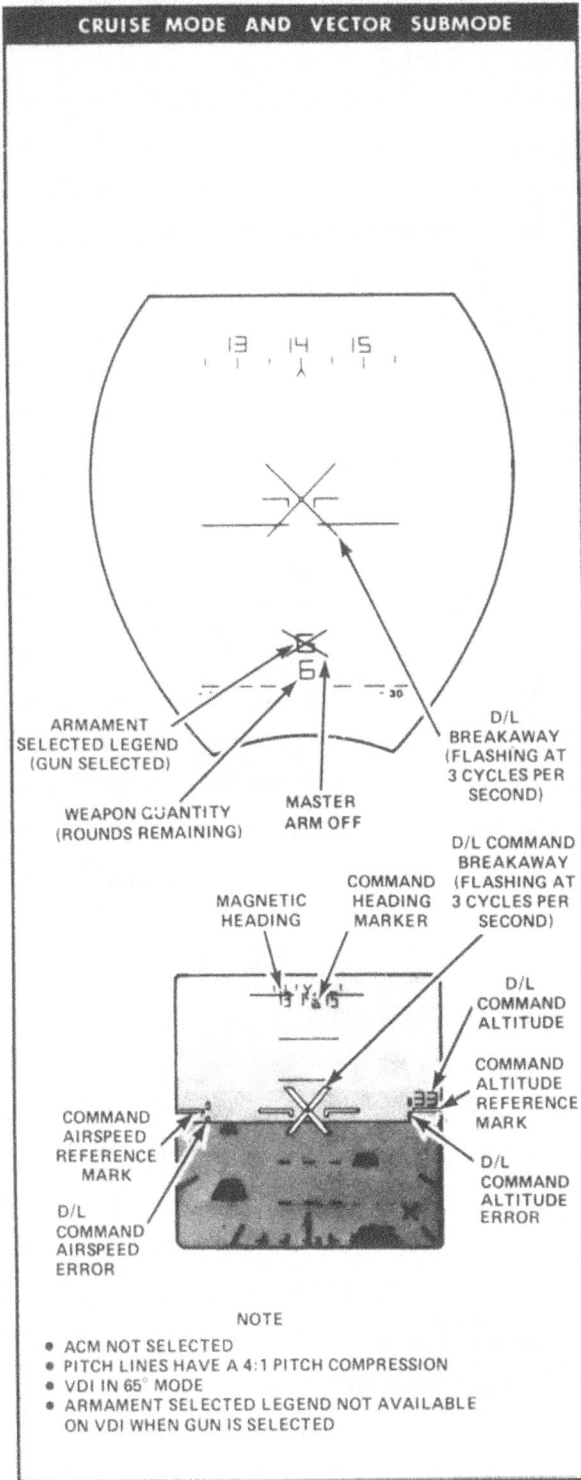

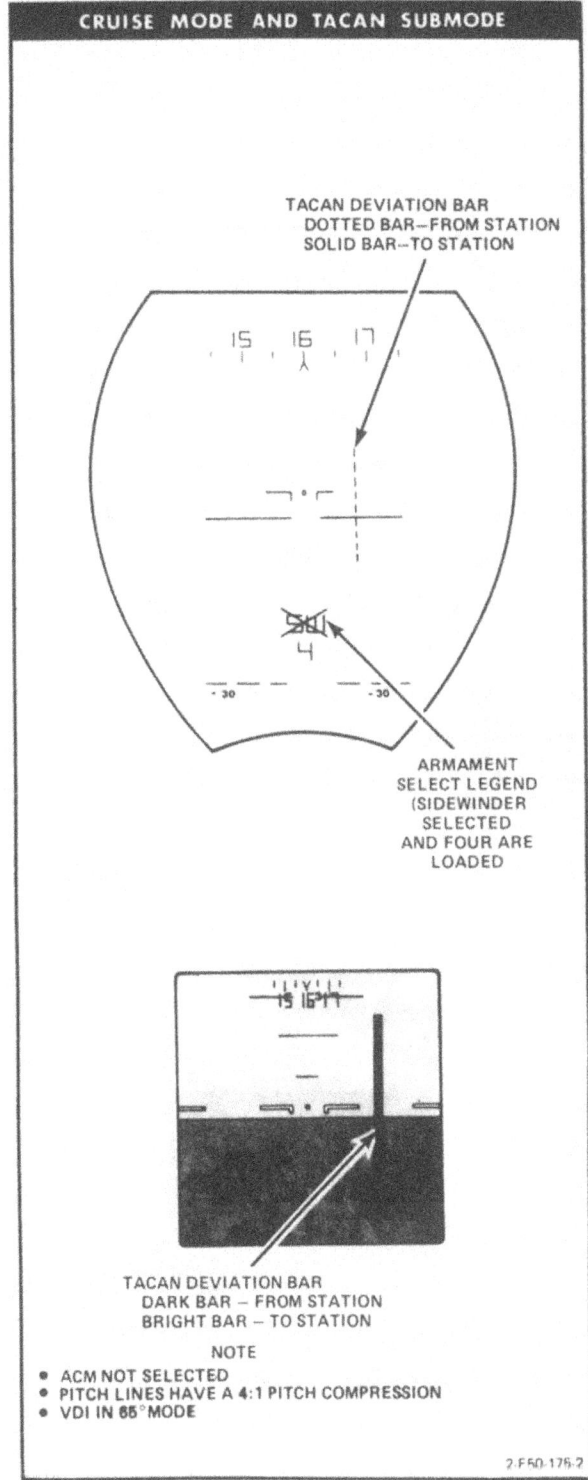

CRUISE MODE AND VECTOR SUBMODE

13   14   15

ARMAMENT
SELECTED LEGEND
(GUN SELECTED)

WEAPON QUANTITY
(ROUNDS REMAINING)

MASTER
ARM OFF

D/L
BREAKAWAY
(FLASHING AT
3 CYCLES PER
SECOND)

D/L COMMAND
BREAKAWAY
(FLASHING AT
3 CYCLES PER
SECOND)

MAGNETIC
HEADING

COMMAND
HEADING
MARKER

D/L
COMMAND
ALTITUDE

COMMAND
ALTITUDE
REFERENCE
MARK

COMMAND
AIRSPEED
REFERENCE
MARK

D/L
COMMAND
AIRSPEED
ERROR

D/L
COMMAND
ALTITUDE
ERROR

NOTE

- ACM NOT SELECTED
- PITCH LINES HAVE A 4:1 PITCH COMPRESSION
- VDI IN 65° MODE
- ARMAMENT SELECTED LEGEND NOT AVAILABLE
  ON VDI WHEN GUN IS SELECTED

CRUISE MODE AND TACAN SUBMODE

TACAN DEVIATION BAR
DOTTED BAR—FROM STATION
SOLID BAR—TO STATION

15   16   17

ARMAMENT
SELECT LEGEND
(SIDEWINDER
SELECTED
AND FOUR ARE
LOADED)

TACAN DEVIATION BAR
DARK BAR – FROM STATION
BRIGHT BAR – TO STATION

NOTE

- ACM NOT SELECTED
- PITCH LINES HAVE A 4:1 PITCH COMPRESSION
- VDI IN 65° MODE

2-F50-175-2

Figure 8-1. VDI and HUD Presentations (Sheet 2 of 11)

VDI AND HUD PRESENTATIONS

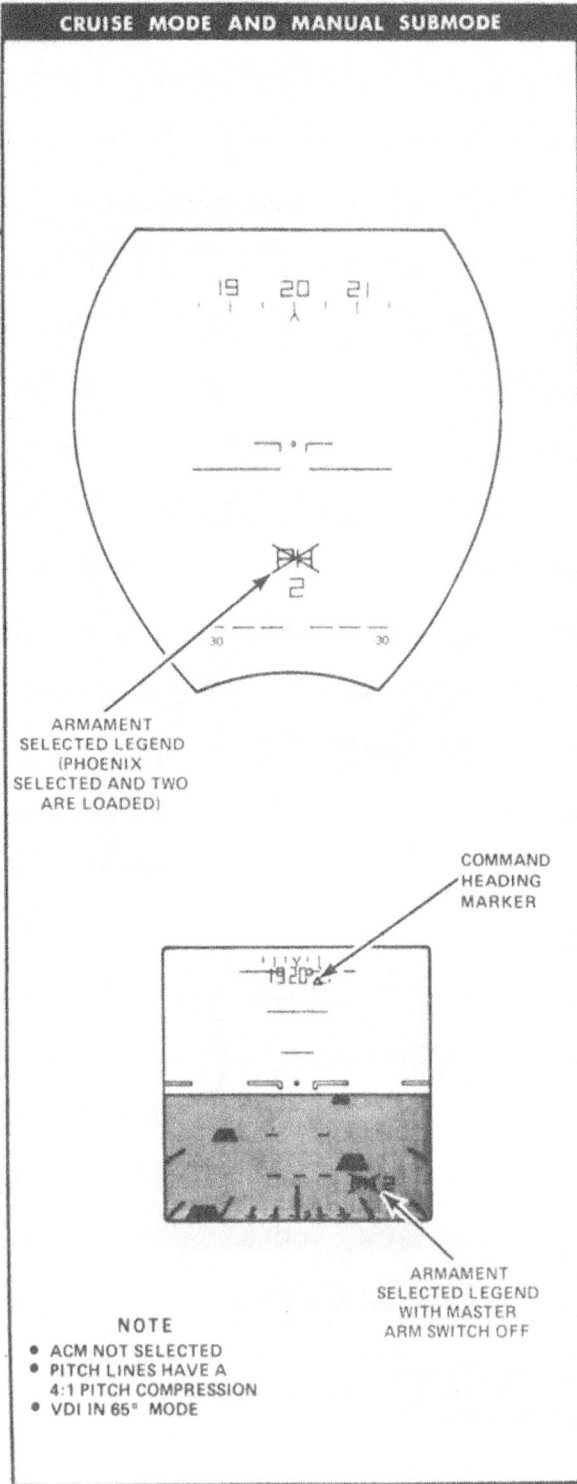

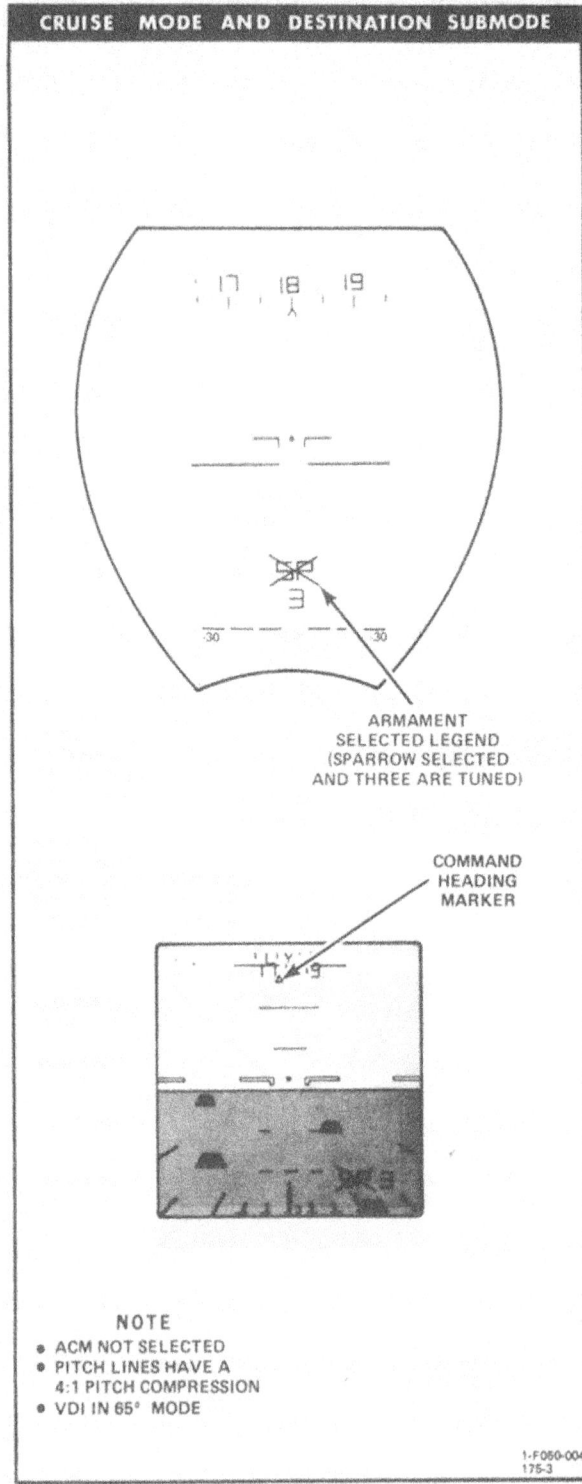

CRUISE MODE AND MANUAL SUBMODE

ARMAMENT
SELECTED LEGEND
(PHOENIX
SELECTED AND TWO
ARE LOADED)

COMMAND
HEADING
MARKER

ARMAMENT
SELECTED LEGEND
WITH MASTER
ARM SWITCH OFF

N O T E

• ACM NOT SELECTED
• PITCH LINES HAVE A
  4:1 PITCH COMPRESSION
• VDI IN 65° MODE

CRUISE MODE AND DESTINATION SUBMODE

ARMAMENT
SELECTED LEGEND
(SPARROW SELECTED
AND THREE ARE TUNED)

COMMAND
HEADING
MARKER

N O T E

• ACM NOT SELECTED
• PITCH LINES HAVE A
  4:1 PITCH COMPRESSION
• VDI IN 65° MODE

1-F050-004
175-3

Figure 8-1. VDI and HUD Presentations (Sheet 3 of 11)

VDI AND HUD PRESENTATIONS

A/A MANUAL GUN MODE

BORESIGHT
REFERENCE
(ADL)

MOVEABLE RETICLE
(SET BY ELEVATION
LEAD CONTROL)

30   30

o

o

30   30

WEAPON NUMERIC
(NUMERIC 0 TO 6 INDICATES
ROUNDS REMAINING IN 100'S)

COMMAND
HEADING
MARKER

0 1 2 3 4 5 6

+3

66

—3—

D/L COMMAND
AIRSPEED
ERROR

D/L COMMAND
ALTITUDE
ERROR

COMMAND
ALTITUDE
REFERENCE
MARK

NOTE
• THIS MODE CAN ALSO BE ENTERED BY SELECTED ACM
• GUN CAGE IS NOT SELECTED FOR THIS MODE
• PITCH LINES, HORIZON LINE, AND AIRCRAFT
  FIXED RETICLE ARE DECLUTTERABLE ON HUD

RTGS HUD DISPLAYS

BULLET PATH
AT 1000 FEET

30

+

30

10

-30

TARGET
RANGE
ONE BULLET
TIME OF
FLIGHT
AWAY (INSIDE
4000 FEET)

-30   G
      4

STT

BULLET PATH
AT 1000 FEET

-30

BULLET PATH
AT 2000 FEET

-30   G
      4

NO STT

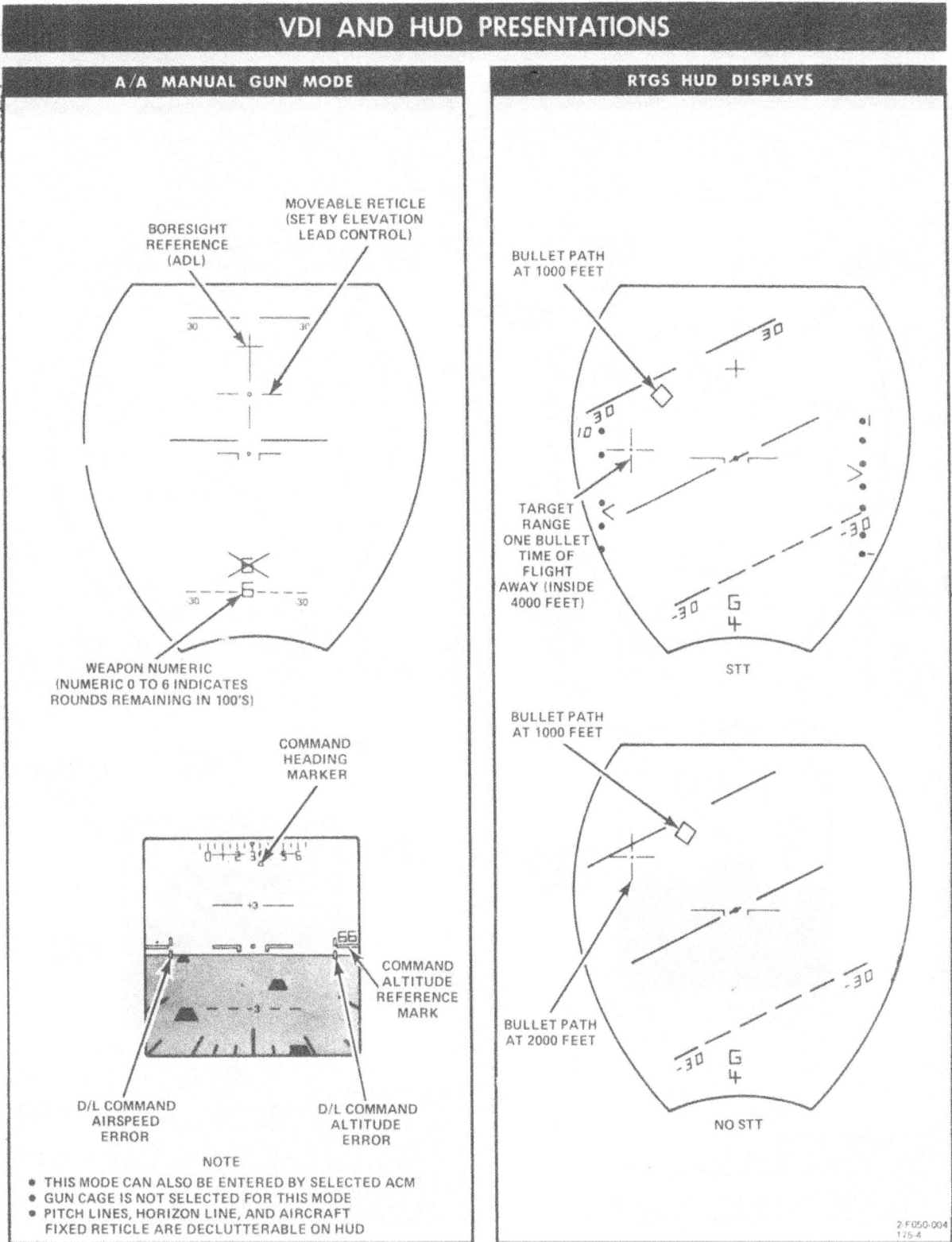

Figure 8-1. VDI and HUD Presentations (Sheet 4 of 11)

VDI AND HUD PRESENTATIONS

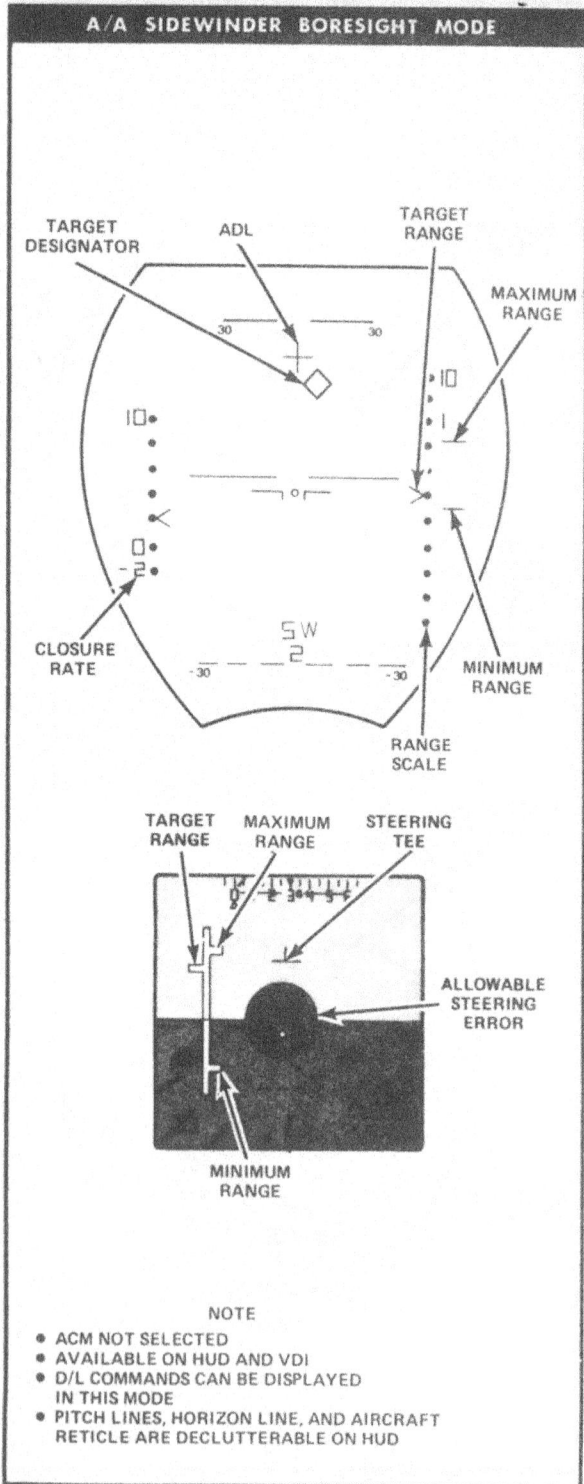

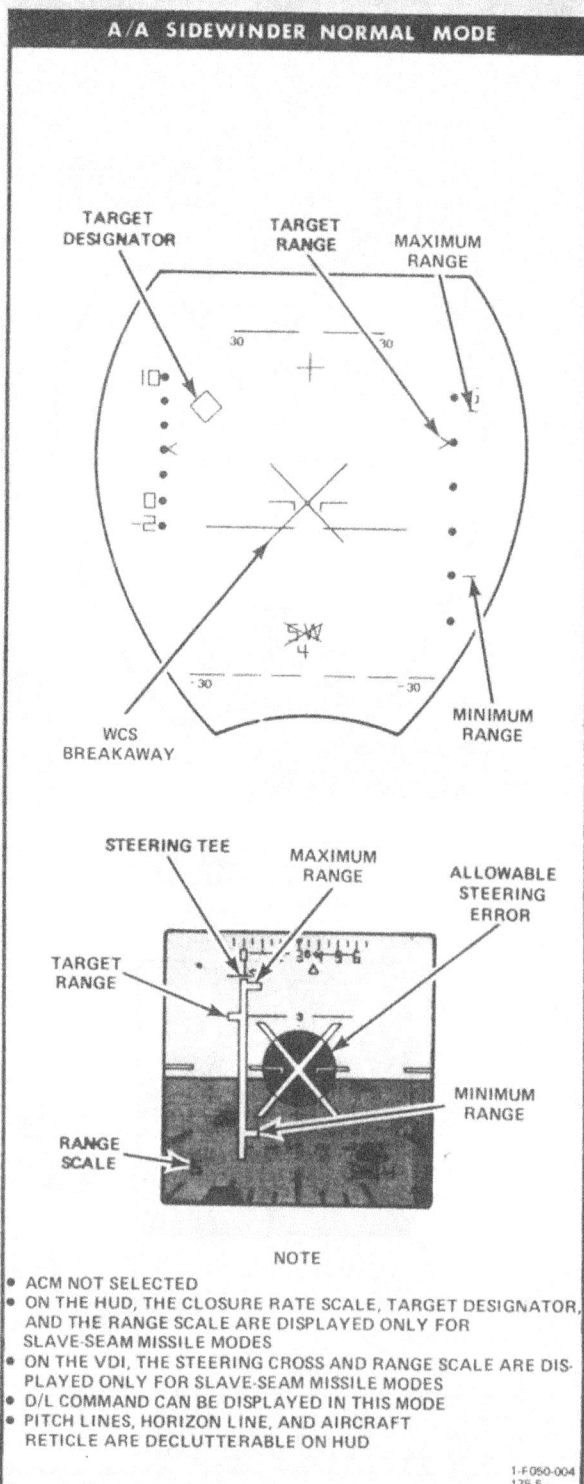

A/A SIDEWINDER BORESIGHT MODE

TARGET DESIGNATOR

ADL

TARGET RANGE

MAXIMUM RANGE

CLOSURE RATE

MINIMUM RANGE

RANGE SCALE

SW 2

TARGET RANGE   MAXIMUM RANGE   STEERING TEE

ALLOWABLE STEERING ERROR

MINIMUM RANGE

NOTE
* ACM NOT SELECTED
* AVAILABLE ON HUD AND VDI
* D/L COMMANDS CAN BE DISPLAYED IN THIS MODE
* PITCH LINES, HORIZON LINE, AND AIRCRAFT RETICLE ARE DECLUTTERABLE ON HUD

A/A SIDEWINDER NORMAL MODE

TARGET DESIGNATOR

TARGET RANGE

MAXIMUM RANGE

WCS BREAKAWAY

MINIMUM RANGE

SW 4

STEERING TEE

MAXIMUM RANGE

ALLOWABLE STEERING ERROR

TARGET RANGE

MINIMUM RANGE

RANGE SCALE

NOTE
* ACM NOT SELECTED
* ON THE HUD, THE CLOSURE RATE SCALE, TARGET DESIGNATOR, AND THE RANGE SCALE ARE DISPLAYED ONLY FOR SLAVE-SEAM MISSILE MODES
* ON THE VDI, THE STEERING CROSS AND RANGE SCALE ARE DISPLAYED ONLY FOR SLAVE-SEAM MISSILE MODES
* D/L COMMAND CAN BE DISPLAYED IN THIS MODE
* PITCH LINES, HORIZON LINE, AND AIRCRAFT RETICLE ARE DECLUTTERABLE ON HUD

1-F050-004
175-5

Figure 8-1. VDI and HUD Presentations (Sheet 5 of 11)

VDI AND HUD PRESENTATIONS

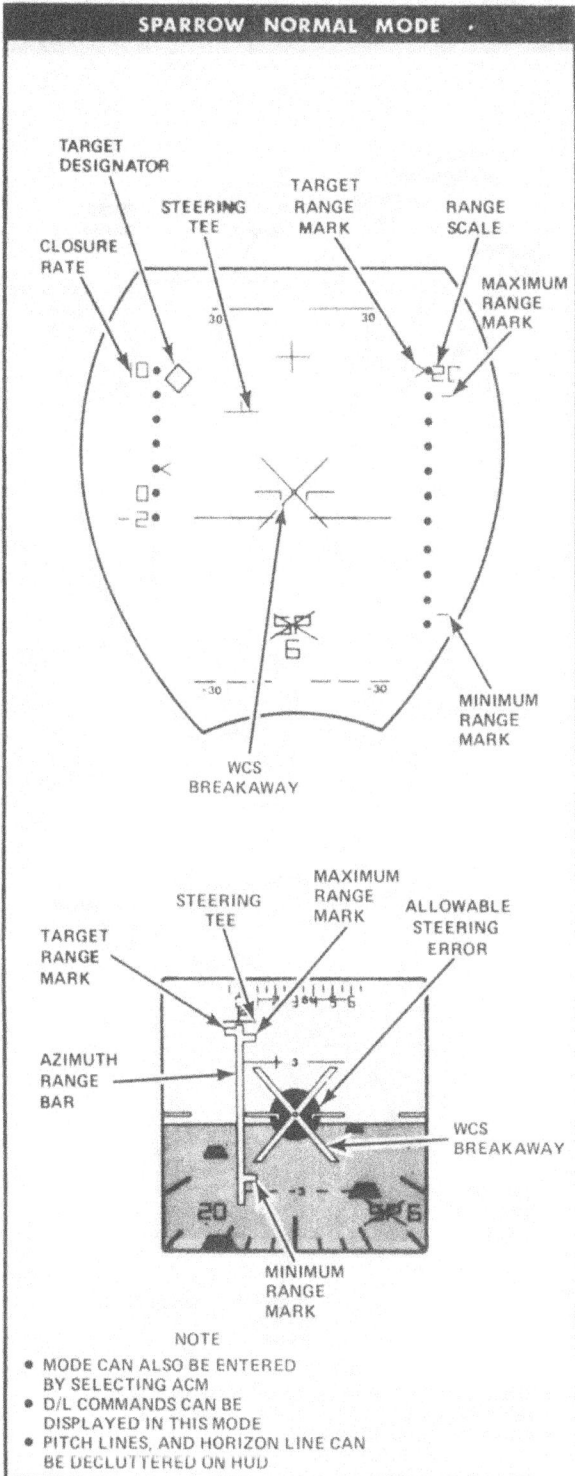

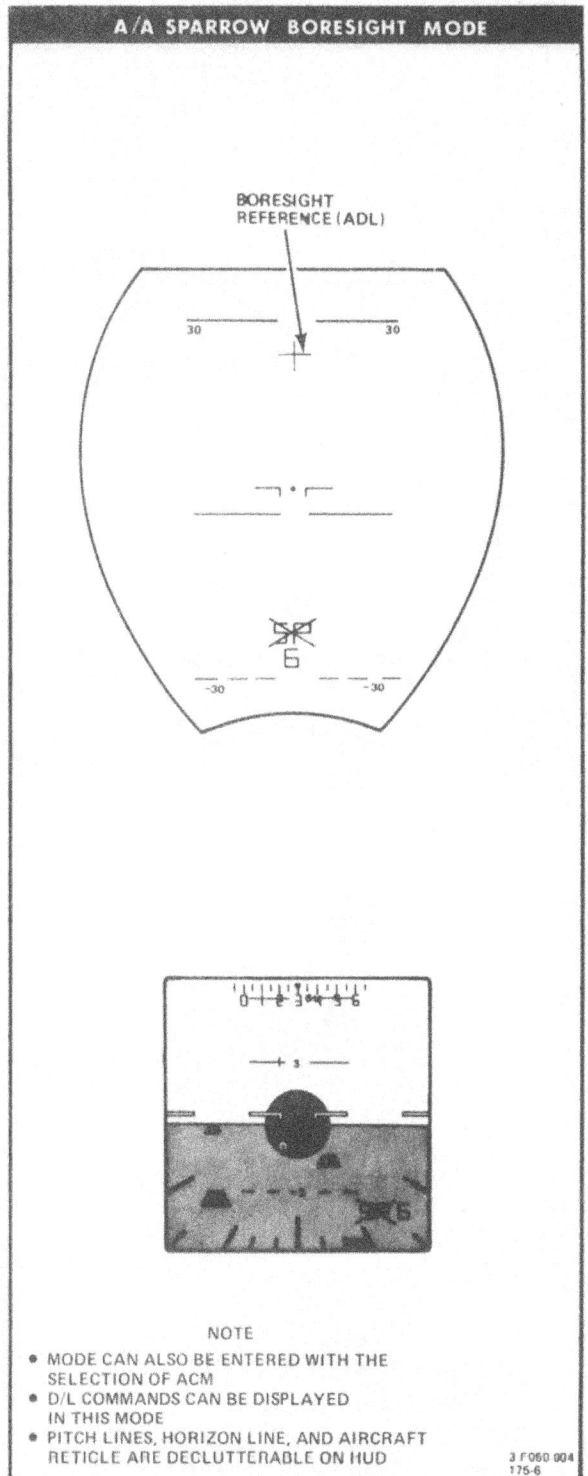

SPARROW NORMAL MODE

TARGET
DESIGNATOR

STEERING
TEE

TARGET
RANGE
MARK

RANGE
SCALE

CLOSURE
RATE

MAXIMUM
RANGE
MARK

WCS
BREAKAWAY

MINIMUM
RANGE
MARK

STEERING
TEE

MAXIMUM
RANGE
MARK

ALLOWABLE
STEERING
ERROR

TARGET
RANGE
MARK

AZIMUTH
RANGE
BAR

WCS
BREAKAWAY

MINIMUM
RANGE
MARK

NOTE
- MODE CAN ALSO BE ENTERED
  BY SELECTING ACM
- D/L COMMANDS CAN BE
  DISPLAYED IN THIS MODE
- PITCH LINES, AND HORIZON LINE CAN
  BE DECLUTTERED ON HUD

A/A SPARROW BORESIGHT MODE

BORESIGHT
REFERENCE (ADL)

NOTE
- MODE CAN ALSO BE ENTERED WITH THE
  SELECTION OF ACM
- D/L COMMANDS CAN BE DISPLAYED
  IN THIS MODE
- PITCH LINES, HORIZON LINE, AND AIRCRAFT
  RETICLE ARE DECLUTTERABLE ON HUD

3 F060 904
175-6

Figure 8-1. VDI and HUD Presentations (Sheet 6 of 11)

VDI AND HUD PRESENTATIONS

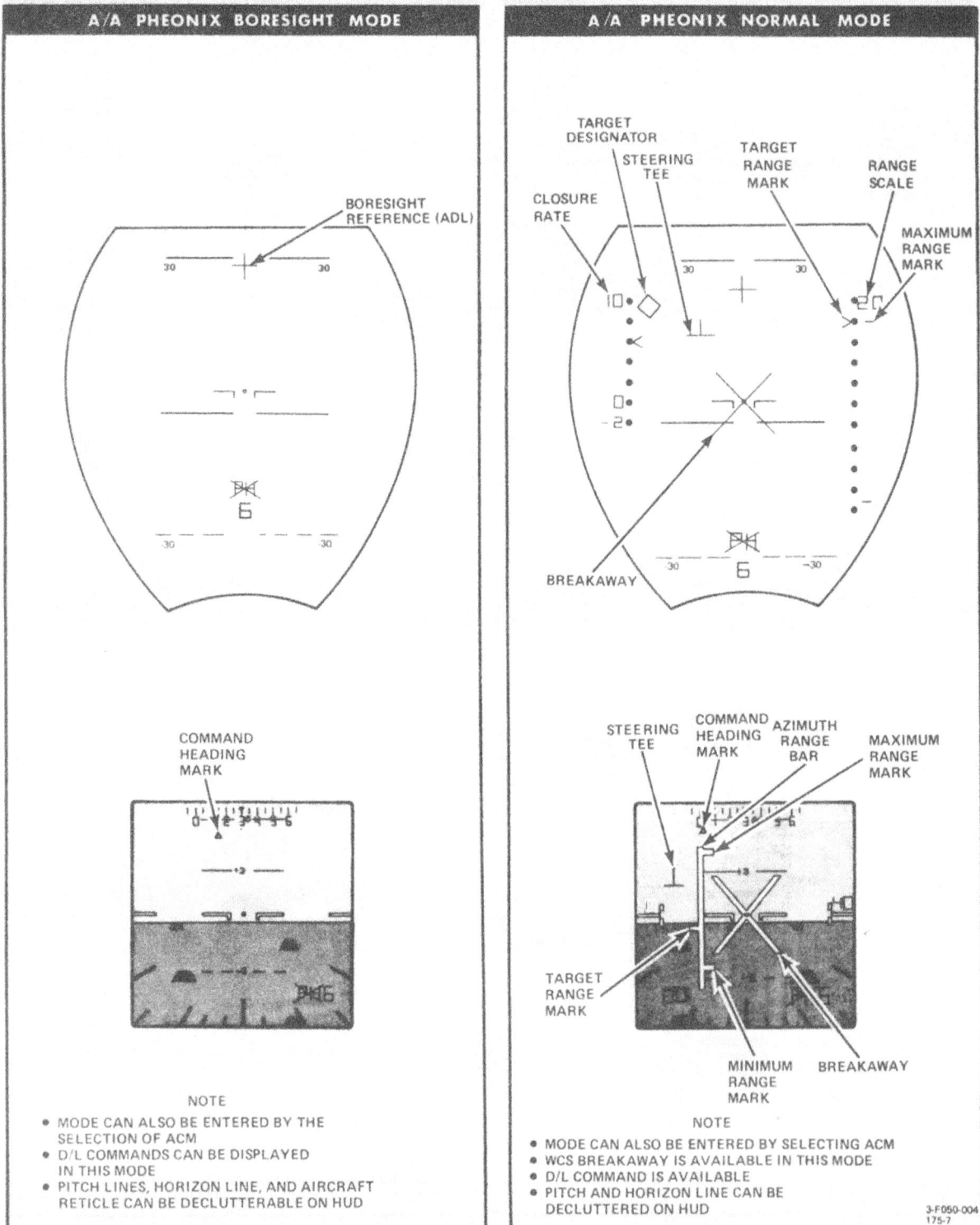

A/A PHEONIX BORESIGHT MODE

BORESIGHT
REFERENCE (ADL)

COMMAND
HEADING
MARK

NOTE

- MODE CAN ALSO BE ENTERED BY THE
  SELECTION OF ACM
- D/L COMMANDS CAN BE DISPLAYED
  IN THIS MODE
- PITCH LINES, HORIZON LINE, AND AIRCRAFT
  RETICLE CAN BE DECLUTTERABLE ON HUD

A/A PHEONIX NORMAL MODE

TARGET
DESIGNATOR

STEERING
TEE

TARGET
RANGE
MARK

RANGE
SCALE

CLOSURE
RATE

MAXIMUM
RANGE
MARK

BREAKAWAY

STEERING
TEE

COMMAND
HEADING
MARK

AZIMUTH
RANGE
BAR

MAXIMUM
RANGE
MARK

TARGET
RANGE
MARK

MINIMUM
RANGE
MARK

BREAKAWAY

NOTE

- MODE CAN ALSO BE ENTERED BY SELECTING ACM
- WCS BREAKAWAY IS AVAILABLE IN THIS MODE
- D/L COMMAND IS AVAILABLE
- PITCH AND HORIZON LINE CAN BE
  DECLUTTERED ON HUD

3-F050-004
175-7

Figure 8-1. VDI and HUD Presentations (Sheet 7 of 11)

VDI AND HUD PRESENTATIONS

A/G COMPUTER IP MODE

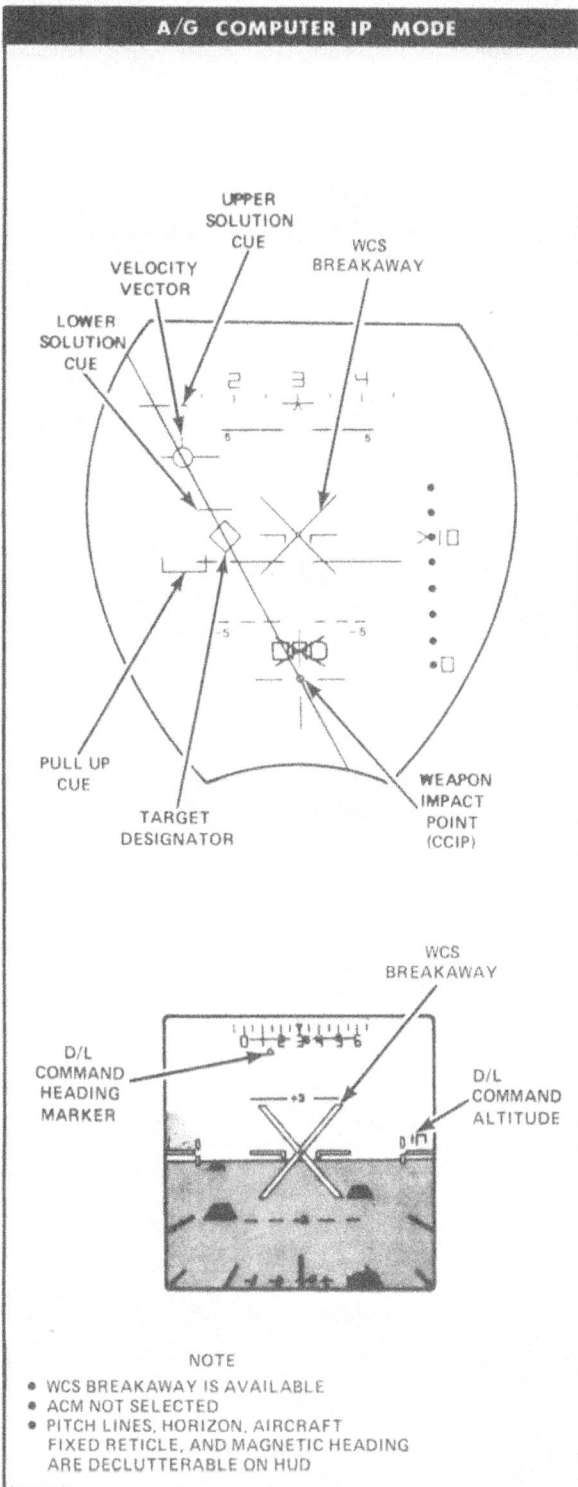

UPPER SOLUTION CUE
VELOCITY VECTOR
WCS BREAKAWAY
LOWER SOLUTION CUE
PULL UP CUE
TARGET DESIGNATOR
WEAPON IMPACT POINT (CCIP)

WCS BREAKAWAY
D/L COMMAND HEADING MARKER
D/L COMMAND ALTITUDE

NOTE
- WCS BREAKAWAY IS AVAILABLE
- ACM NOT SELECTED
- PITCH LINES, HORIZON, AIRCRAFT FIXED RETICLE, AND MAGNETIC HEADING ARE DECLUTTERABLE ON HUD

A/G COMPUTER TARGET MODE

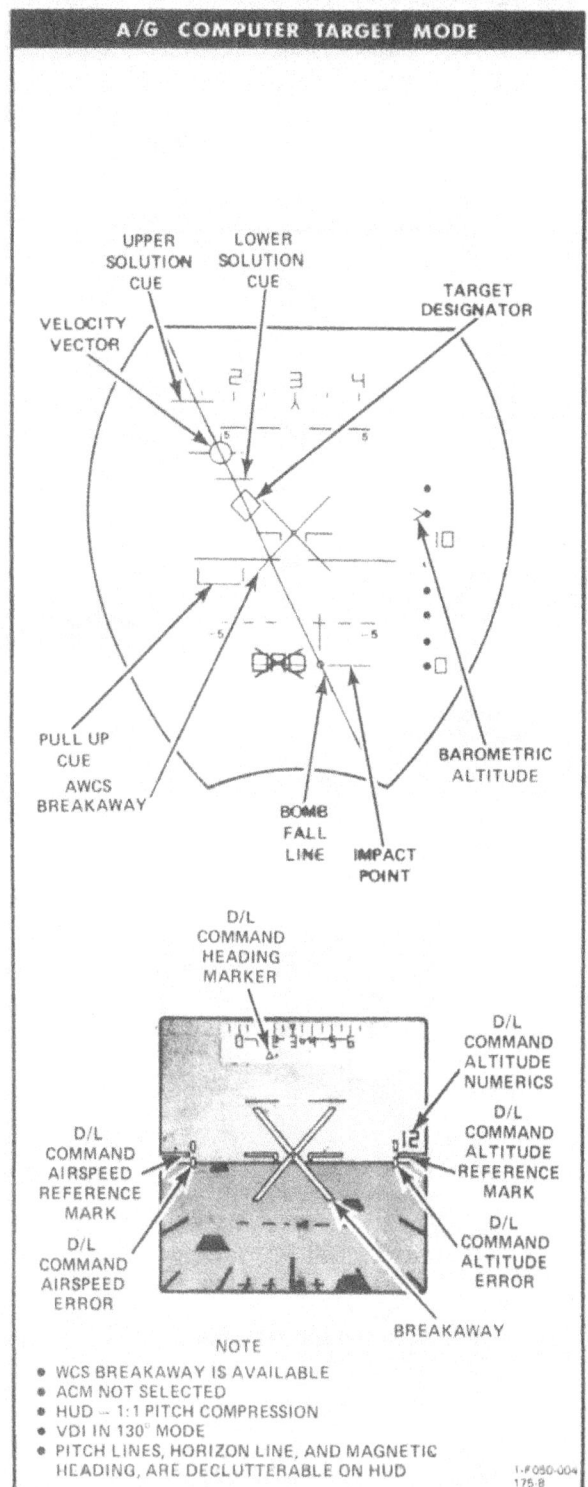

UPPER SOLUTION CUE
LOWER SOLUTION CUE
TARGET DESIGNATOR
VELOCITY VECTOR
PULL UP CUE
AWCS BREAKAWAY
BOMB FALL LINE
IMPACT POINT
BAROMETRIC ALTITUDE

D/L COMMAND HEADING MARKER
D/L COMMAND ALTITUDE NUMERICS
D/L COMMAND ALTITUDE REFERENCE MARK
D/L COMMAND AIRSPEED REFERENCE MARK
D/L COMMAND ALTITUDE ERROR
D/L COMMAND AIRSPEED ERROR
BREAKAWAY

NOTE
- WCS BREAKAWAY IS AVAILABLE
- ACM NOT SELECTED
- HUD — 1:1 PITCH COMPRESSION
- VDI IN 130° MODE
- PITCH LINES, HORIZON LINE, AND MAGNETIC HEADING, ARE DECLUTTERABLE ON HUD

1-F050-004
175-B

Figure 8-1. VDI and HUD Presentations (Sheet 8 of 11)

VDI AND HUD PRESENTATIONS

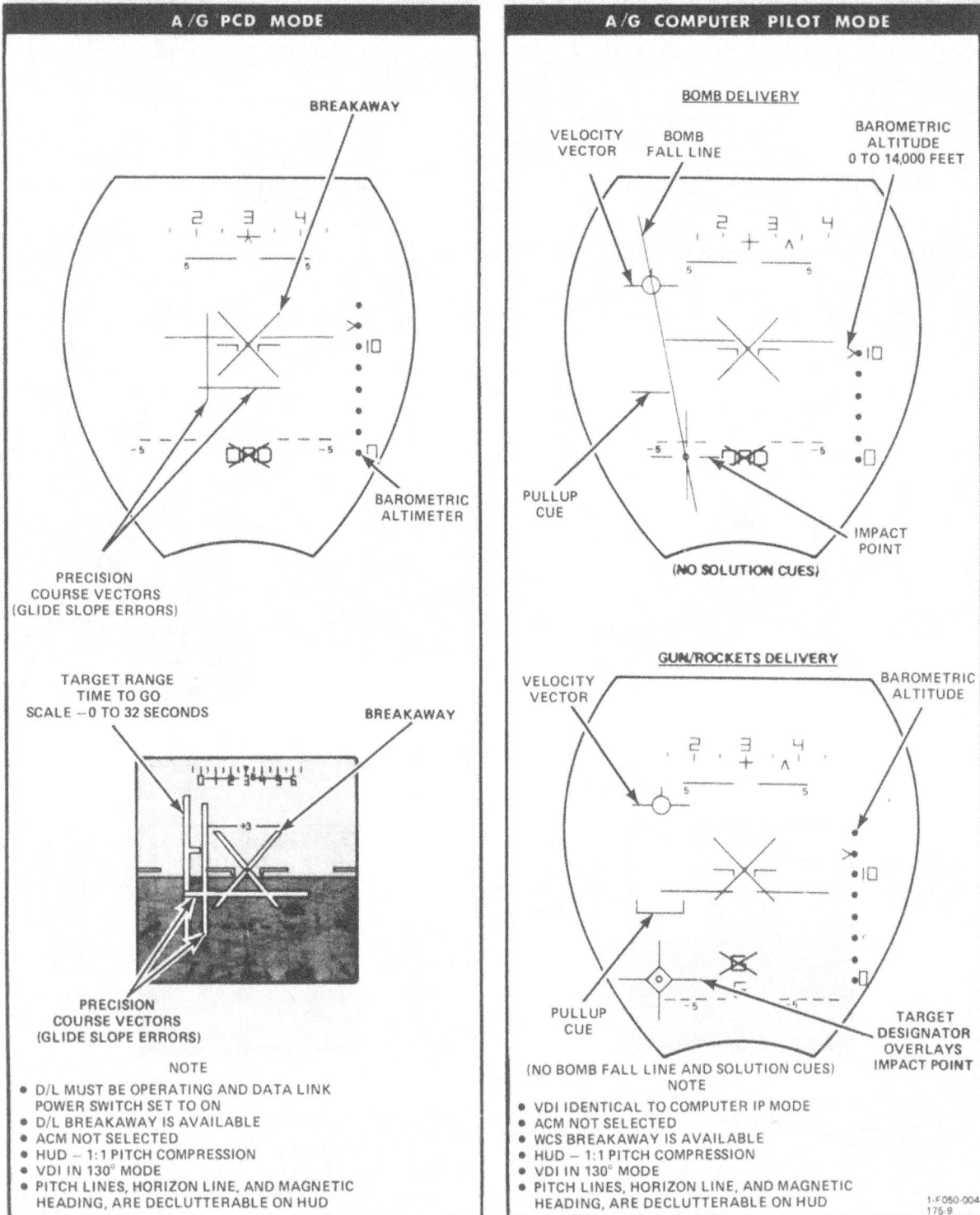

A/G PCD MODE

BREAKAWAY

PRECISION
COURSE VECTORS
(GLIDE SLOPE ERRORS)

BAROMETRIC
ALTIMETER

TARGET RANGE
TIME TO GO
SCALE – 0 TO 32 SECONDS

BREAKAWAY

PRECISION
COURSE VECTORS
(GLIDE SLOPE ERRORS)

NOTE
- D/L MUST BE OPERATING AND DATA LINK
  POWER SWITCH SET TO ON
- D/L BREAKAWAY IS AVAILABLE
- ACM NOT SELECTED
- HUD – 1:1 PITCH COMPRESSION
- VDI IN 130° MODE
- PITCH LINES, HORIZON LINE, AND MAGNETIC
  HEADING, ARE DECLUTTERABLE ON HUD

A/G COMPUTER PILOT MODE

BOMB DELIVERY

VELOCITY
VECTOR

BOMB
FALL LINE

BAROMETRIC
ALTITUDE
0 TO 14,000 FEET

PULLUP
CUE

IMPACT
POINT

(NO SOLUTION CUES)

GUN/ROCKETS DELIVERY

VELOCITY
VECTOR

BAROMETRIC
ALTITUDE

PULLUP
CUE

TARGET
DESIGNATOR
OVERLAYS
IMPACT POINT

(NO BOMB FALL LINE AND SOLUTION CUES)
NOTE
- VDI IDENTICAL TO COMPUTER IP MODE
- ACM NOT SELECTED
- WCS BREAKAWAY IS AVAILABLE
- HUD – 1:1 PITCH COMPRESSION
- VDI IN 130° MODE
- PITCH LINES, HORIZON LINE, AND MAGNETIC
  HEADING, ARE DECLUTTERABLE ON HUD

1-F050-004
175-9

Figure 8-1. VDI and HUD Presentations (Sheet 9 of 11)

VDI AND HUD PRESENTATIONS

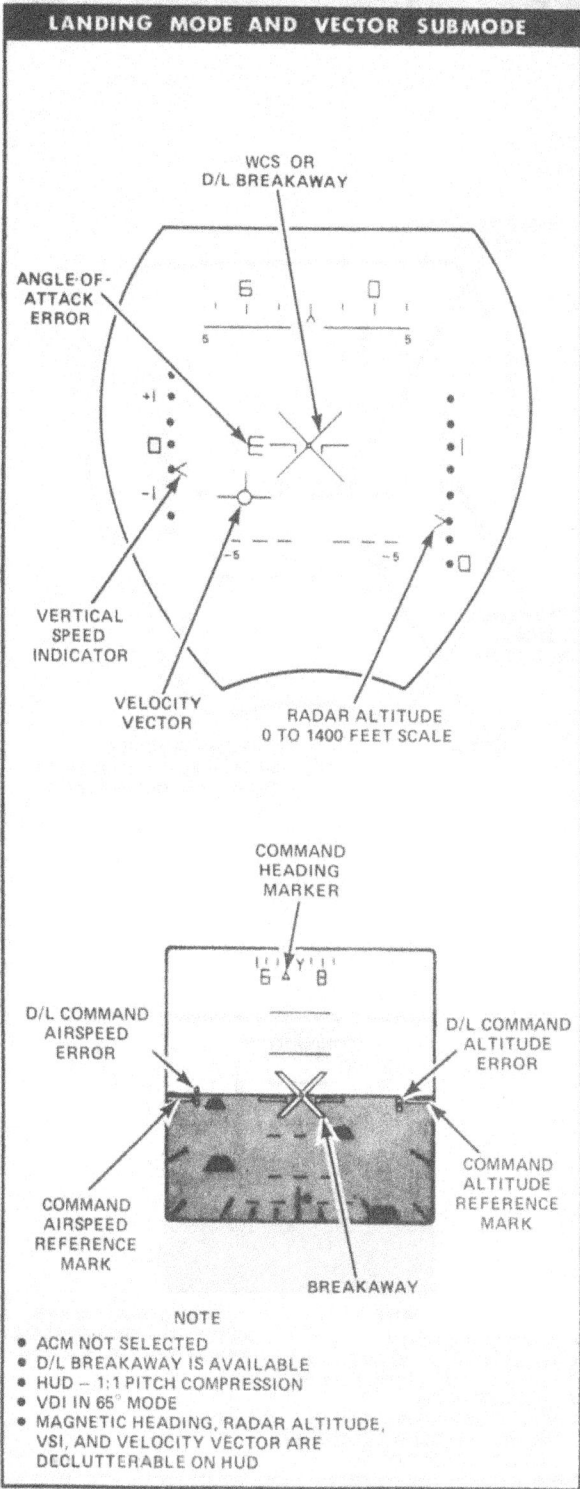

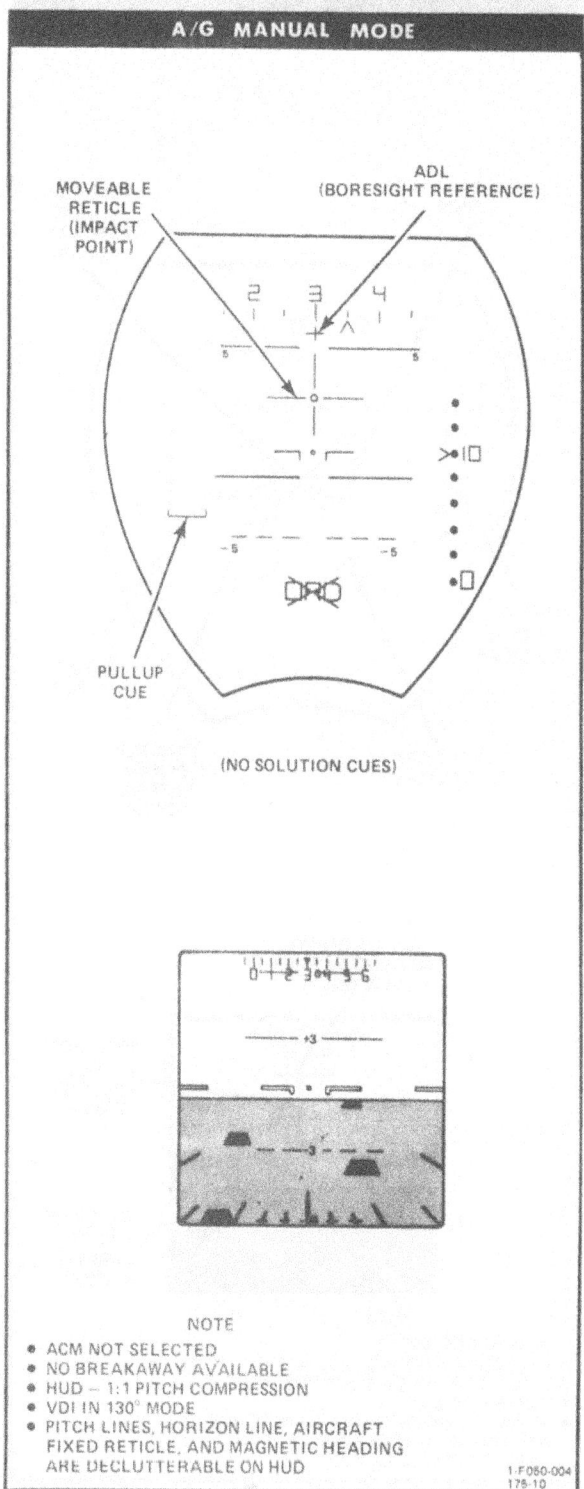

LANDING MODE AND VECTOR SUBMODE

WCS OR
D/L BREAKAWAY

ANGLE-OF-
ATTACK
ERROR

VERTICAL
SPEED
INDICATOR

VELOCITY
VECTOR

RADAR ALTITUDE
0 TO 1400 FEET SCALE

COMMAND
HEADING
MARKER

D/L COMMAND
AIRSPEED
ERROR

D/L COMMAND
ALTITUDE
ERROR

COMMAND
AIRSPEED
REFERENCE
MARK

COMMAND
ALTITUDE
REFERENCE
MARK

BREAKAWAY

NOTE
- ACM NOT SELECTED
- D/L BREAKAWAY IS AVAILABLE
- HUD – 1:1 PITCH COMPRESSION
- VDI IN 65° MODE
- MAGNETIC HEADING, RADAR ALTITUDE, VSI, AND VELOCITY VECTOR ARE DECLUTTERABLE ON HUD

A/G MANUAL MODE

MOVEABLE
RETICLE
(IMPACT
POINT)

ADL
(BORESIGHT REFERENCE)

PULLUP
CUE

(NO SOLUTION CUES)

NOTE
- ACM NOT SELECTED
- NO BREAKAWAY AVAILABLE
- HUD – 1:1 PITCH COMPRESSION
- VDI IN 130° MODE
- PITCH LINES, HORIZON LINE, AIRCRAFT FIXED RETICLE, AND MAGNETIC HEADING ARE DECLUTTERABLE ON HUD

1-F050-004
175-10

Figure 8-1. VDI and HUD Presentations (Sheet 10 of 11)

VDI AND HUD PRESENTATIONS

LANDING MODE AND AWL SUBMODE

ANGLE-OF-ATTACK
ERROR

PRECISION
COURSE
VERTICAL
VECTOR

VERTICAL
SPEED
INDICATOR

RADAR
ALTITUDE

VELOCITY
VECTOR

PRECISION
COURSE
LATERAL
VECTOR

D/L COMMAND
HEADING MARKER
(ACL ONLY)

PRECISION
COURSE
VERTICAL
VECTOR

PRECISION
COURSE
LATERAL
VECTOR

NOTE
- ACM NOT SELECTED
- D/L BREAKAWAY IS AVAILABLE
- HUD – 1:1 PITCH COMPRESSION
- VDI IN 65° MODE
- MAGNETIC HEADING, RADAR ALTITUDE,
  VSI, AND VELOCITY VECTOR ARE
  DECLUTTERABLE ON HUD

LANDING MODE AND TACAN SUBMODE

ANGLE-OF-ATTACK
ERROR

RADAR
ALTITUDE

VERTICAL
SPEED
INDICATOR

VELOCITY
VECTOR

TACAN DEVIATION BAR
DOTTED BAR—FROM STATION
SOLID BAR—TO STATION

TACAN DEVIATION BAR
DARK BAR – FROM STATION
BRIGHT BAR – TO STATION

NOTE
- ACM NOT SELECTED
- D/L BREAKAWAY IS AVAILABLE
- HUD – 1:1 PITCH COMPRESSION
- VDI IN 65° MODE
- MAGNETIC HEADING, RADAR ALTITUDE,
  VSI, AND VELOCITY VECTOR ARE
  DECLUTTERABLE ON HUD

3-F050-004
178-11

Figure 8-1. VDI and HUD Presentations (Sheet 11 of 11)

VDI AND HUD CONTROLS

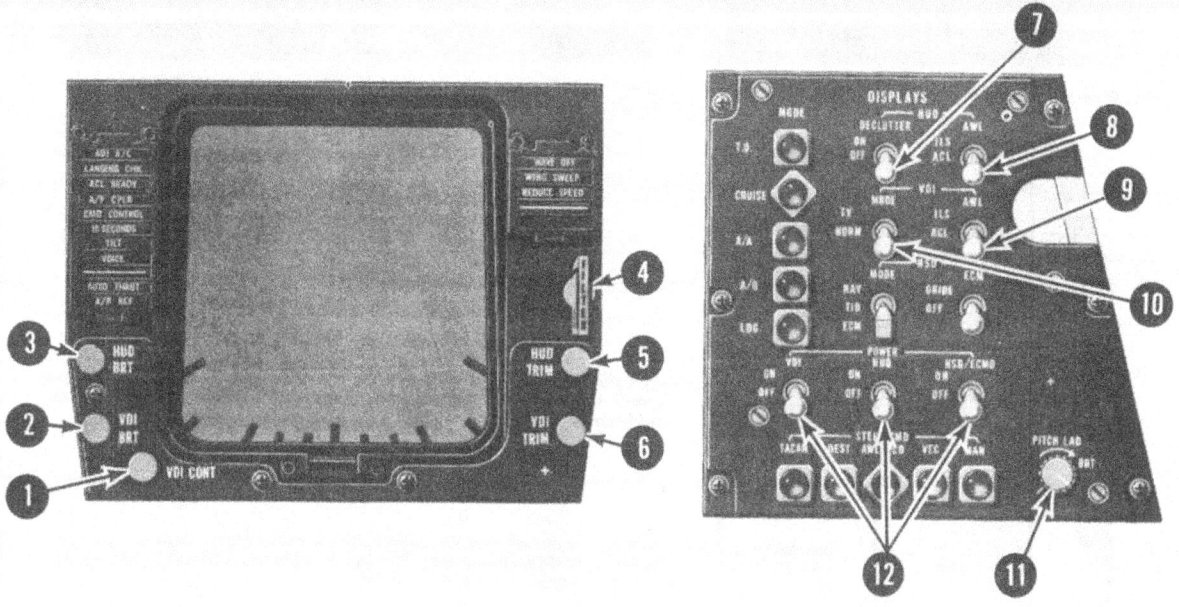

1-F050-004
176-0

NOMENCLATURE	FUNCTION
① VDI CONT control	Allows pilot to vary VDI contrast.
② VDI BRT control	Allows pilot to vary VDI brightness.
③ HUD BRT control	Allows pilot to vary HUD brightness.
④ FILTER pull slide	Places filter on VDI for night flying.
⑤ HUD TRIM control	Allows pilot to adjust pitch trim on HUD.
⑥ VDI TRIM control	Allows pilot to adjust pitch trim on VDI.
⑦ HUD DECLUTTER switch	Placing the switch to ON position reduces the amount of symbology on the HUD.
⑧ HUD AWL switch	ILS — Selects ILS (AN/SPN-41) display presentation on the HUD during landing phases. ACL — Selects ACL (AN/SPN-42) display presentation on the HUD during landing phases. Normally left in the ACL position.
⑨ VDI AWL switch	ILS — Selects ILS (AN/SPN-41) displays on the VDI. ACL — Selects ACL (AN/SPN-42) displays on the VDI. Normally left in the ACL position.
⑩ VDI MODE switch	TV — Inoperative. NORM — VDI displays format selected by display MODE switch.
⑪ PITCH LAD control	Controls the intensity of the pitch ladder on the HUD.
⑫ POWER switches	Three separate switches are provided for ON-OFF power control of the VDI, HUD, and HSD/ECMD. All three switches must be set to ON to satisfy OBC. In the event of display loss caused by electrical power transients, display may be restored by cycling the appropriate ON-OFF switch.

Figure 8-2. VDI and HUD Controls

VDIG SYMBOLOGY

HUD SYMBOL	VDI SYMBOL	NAME	FUNCTION
	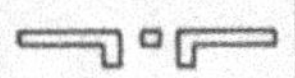	Aircraft reticle	Depicts own aircraft wings and when lined up with the horizon, the aircraft is in straight and level flight.
		Horizon	This is a demarcation point between ground and sky textures on the VDI. It represents the horizon with respect to the aircraft, and changes orientation with any change in aircraft pitch or roll.
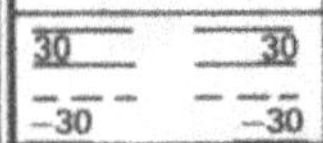 30 30 -30 -30	 3 -3	Pitch lines	Indicates with respect to aircraft reticle pitch attitude. In cruise and A/A mode, HUD pitch lines have a 4:1 compression ratio. Dotted lines indicate a negative pitch. Solid lines indicate positive pitch. Pitch lines above the magnetic heading scale are blanked.
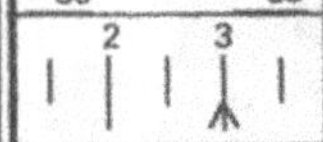 2 3	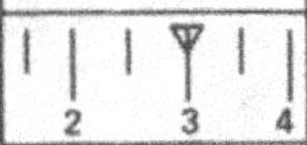 2 3 4	Magnetic heading (AHRS or WCS via CSDC)	Indicates magnetic heading with respect to index mark.
 1 0	N/A	Radar Altimeter Scale	Indicates altitude derived from radar altimeter. Scale is from 0 to 1400 feet in 200-foot increments and has a movable pointer. This symbol is only available in the takeoff and landing modes.
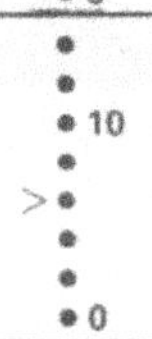 10 0	N/A	Barometric Altimeter Scale	Indicates altitude derived from the pressure altitude via the CADC. Scale is from 0 to 14,000 feet in 2000-foot increments. This scale is available in A/G and TARPS modes. The altitude is referenced to 29.92 inches Hg.
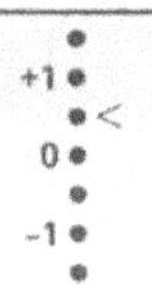 +1 0 -1	N/A	Vertical Speed Indicator	Indicates rate of altitude change. Scale is from -1500 to +1500 ft/min in 500-foot increments. This scale appears on the left side of the HUD in takeoff and landing modes.
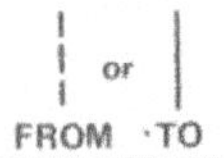 or FROM · TO	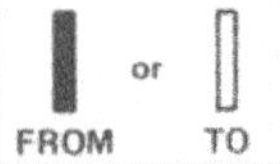 or FROM TO	TACAN Deviation Bar (TACAN via CSDC)	Indicates difference between bearing to TACAN and selected TACAN radial. Deviation is limited to ±5.625° TACAN deviation on the VDI, and ±3° TACAN deviation on the HUD. The symbol never leaves field of view. It will limit at edge nearest selected TACAN radial.
 ✕	 ✕	Breakaway	Appears as a flashing symbol at a 3-cycle per second rate in the center of field-of-view when range-to-go-to minimum or safe pullup point is zero. Symbol is commanded by the WCS computer or by D/L, depending on mode of operation.
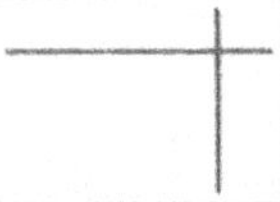	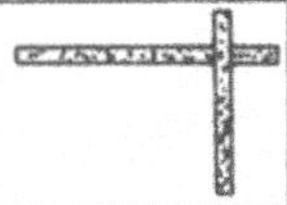	Precision course Vector	This symbol consists of two independent vectors (vertical and horizontal), which form a cross pointer. Elevation glide slope information positions the horizontal vectors, whereas the vertical vector is positioned by azimuth glide slope information. This symbol is also used in D/L bombing modes.
	N/A	Velocity Vector	Indicates direction of ground track velocity vector (where the aircraft is going, not where it's pointed).
E	N/A	Angle of attack error (CADC via CSDC)	Its position in relation to aircraft reticle indicates angle-of-attack error. Symbol is position by true AOA. Small center horizontal bar indicates zero error. When this symbol is in line with the aircraft reticle, the AOA is 15 units (10.31°). If symbol is below aircraft reticle, AOA is too high; above the aircraft reticle indicates AOA is too low. Symbol is displayed in landing mode only.
ORD, G, SW, SP, or PH 0,1,2,3,4,5,6	SW, SP, or PH only 0,1,2,3,4,5,6 (GUN and ORD are not displayed)	Armament Ready Legends	ORD – indicates bombs or rockets selected, or bombs and gun selected (A/G GUN switch on ACP set to MIXED) G – indicates gun is selected or gun and bombs are selected (A/G GUN switch on ACP set to OFF). Number under G indicates rounds remaining in hundreds (6, 5, 4, 3, 6 2, 1, 0). SW, SP or PH – indicates missile type selected (Sidewinder, Sparrow or Phoenix) and the numbers (0 to 6) indicates number of missiles ready for launch.

6-F50 177-1

Figure 8-3. VDIG Symbology (Sheet 1 of 3)

HUD SYMBOL	VDI SYMBOL	NAME	FUNCTION
		MASTER ARM switch off	An X symbol through armament ready legend indicates MASTER ARM switch on ACM panel is OFF. Disappears when MASTER ARM switch is set to ON.
		Steering Tee	Provides elevation and azimuth steering in the air-to-air modes when a single target track exists. Type of steering (pursuit, collision, etc.) is dependent on weapon selection and the mode selected on the TID by the RIO. May also provide azimuth steering only on the VDI in TWS. Aircraft steering is accomplished by aligning and maintaining the vertical and horizontal bar of the inverted T with the aircraft reticle center dot. Steering sensitivity on the HUD is 26.5° per inch; on the VDI 25° per inch.
	N/A	Boresight Reference	Symbol is a set of crosshairs which is fixed on the HUD and is used to represent the armament data line (ADL) of the aircraft. The reference is located 5.03° above aircraft reticle.
	N/A	Moveable Reticle (Impact Point)	Symbol serves as an optical sight for A/A gunnery and for A/G weapons delivery. In A/A gunnery, the symbol is used for lead angle determination. It can be positioned manually using the elevation lead control or automatically by the computer when the pilot selects the gun solution mode. The symbol is so designed that stadiametric ranging techniques can be employed during manual gun firing mode. This symbol in A/G modes indicates instantaneous weapon impact point. This symbol is positioned by the computer in all modes except manual. In the manual mode, it is positioned by the elevation lead control. The computer positions the symbol based on the ballistics of the bomb, wind conditions, and various aircraft parameters. This symbol must overlay the target at the moment of release.
	N/A	Target Designator	Indicates radar pointing angle. This symbol is used in all computer A/G modes. It is positioned in this case by the pilot using the TARGET DESIGNATE switch. Once the symbol is over the target, the switch is depressed and the computer knows the slant range to the target. In real-time gunsight mode of air-to-air gun, the target designator is positioned at the 1000-foot bullet solution. In all other weapon modes or with OFF selected, and with a valid sensor angle track on the target, the target designator symbol represents the approximate line of sight to the target. If the target is not within the HUD field of view (FOV) the symbol is positioned at the edge of the FOV in the direction of the target. Limits of designator are 10° in horizontal and vertical.
	N/A	Closure Rate	Indicates closing velocity from -200 to +1000 knots between aircraft and target.
		Target Range Scale (azimuth range bar) / Target Range Mark / Maximum Range Mark / Minimum Range Mark	Appears on right side of HUD during A/A modes. Scaling is determined by RIO selecting RANGE pushbuttons on the DDD panel. On the VDI, the range scaling is indicated in the lower left corner of the display. Limits are ±35° in horizontal. The symbol appears on the left side of the VDI during A/A modes, except in STT when it will be located at target azimuth. Indicates range to target. Indicates maximum range for weapon launch. Indicates minimum range for weapon launch.
	N/A	Upper Solution Cue	Cue is a measure of instantaneous weapon range. It is constrained to motion on the bomb fall line. Displayed with respect to velocity vector to indicate range-to-go to weapon release. When it crosses the velocity vector symbol the computer commands a weapon release.
	N/A	Lower Solution Cue	Cue is a measure of maximum range of weapon calculated from instantaneous aircraft position, constrained to motion on the bomb fall line and is used in conjunction with velocity vector to indicate range-to-go-to in range. In range when cue crosses velocity vector and indicates weapon can reach target if pilot executes a pullup. Cue appears after designate.
	N/A	Pullup Cue	Cue is a measure of range at which a 4 g pullup is required to clear weapon fragmentation pattern or ground. Positioned directly below velocity vector. Used in conjunction with velocity vector to indicate range-to-go to minimum safe pullup point, when one crosses the other.

Figure 8-3. VDIG Symbology (Sheet 2 of 3)

HUD SYMBOL	VDI SYMBOL	NAME	FUNCTION
(diagonal line)	N/A	Bomb Fall Line (BFL)	A line determined by velocity vector and weapon impact point used to acquire target in azimuth and present post designate steering in conjunction with velocity vector. Angle of line is an indication of wind direction and velocity.
N/A	(trapezoid)	Ground Texture	Simulated ground patterns to give better relationship between sky and ground. The symbol consists of dark green trapezoids on a lighter background. The sky texture is a uniform light green on the other hand. The size and spacing of the ground texture are arranged to give perspective to the display. Ground texture remains parallel to the horizon line and provides a basic A/C attitude reference compatible with heading change. The ground texture moves toward the pilot and emanates from the horizon to simulate motion.
N/A	(roll pointer symbol)	Roll Pointer and Indices	Indicates roll position. Indices are fixed at 0°, ±10°, ±20°, ±30°, ±45°, and ±60° and are permanently fixed to VDI face with opaque fluorescent red tape. The pointer is generated by the VDIG and moves across the indices.
N/A	(circle)	Allowable Steering Error (ASE)	Circle located around aircraft reticle which indicates steering error allowed for launching a Sidewinder, Sparrow or Phoenix in the normal mode. Size of the ASE circle is determined by the magnitude of the allowable error.
N/A	(square)	Target Symbol	IR tracked target. Symbol is a bright flashing square. Not used.
DATA LINKS SYMBOLS			
N/A	(bar)	Reference	Continuously displayed symbol which appears on each side of the display opposite the wings of the A/C reticle. The symbol is used as a reference in determining errors in D/L commanded altitude and airspeed.
N/A	(symbol)	Commanded Airspeed Error	Positioned relative to reference. Symbol (two vertically movable reference squares) can indicate either Mach or knots of error dependent on control station. Symbol is a fly-to-type, where if the symbol is above reference, it indicates airspeed is below commanded and therefore airspeed should be increased to bring symbol back down. When error symbol is bisected by referenced, the A/C is at the commanded airspeed.
N/A	(symbol .80)	Commanded Mach Numeric	Numeric printed on display above reference. It indicates commanded Mach to the nearest twentieth. Range is from 0.4 to 3.5 with 0.4 to 1.0 in increments of 0.05 and 1.0 to 3.5 in increments of 0.1. NOTE Will always differ from command Mach bug on the Mach/airspeed indicator by 0.2 or 0.3 because commanded Mach can only be transmitted by TDS in 0.05 increments starting at 0.38.
N/A	(symbol)	Commanded Altitude Error	Positioned relative to the reference symbol. Symbol is a fly-to type in that if the error symbol is below the reference, it indicates that A/C altitude is above commanded and the A/C must be pointed down in order to bring the symbol back up the reference.
N/A	(symbol 12)	Commanded Altitude Numeric	Numeric printed on display above reference. It indicates commanded altitude from 0 to 99,000 feet. However only 2 digits are displayed (0 to 99 numerics).
N/A	(triangle)	Command Heading	Positioned relative to magnetic heading scale. Symbol can be positioned by the WCS computer or D/L depending on steering submode selected. For destination steering, symbol indicates heading to steer to the destination selected by the RIO. Where commanded heading is beyond display scale limits, the symbol is pegged at the edge nearest to the commanded heading.
N/A	(symbol 1)	Time-To-Go	Positioned on the range bar during D/L bombing to indicate time-to-go in seconds before weapon release. However, target range is not displayed. Total length of the bar represents 32 seconds. This symbol is displayed at a fixed position 1.5 inches to the left of center.

Figure 8-3. VDIG Symbology (Sheet 3 of 3)

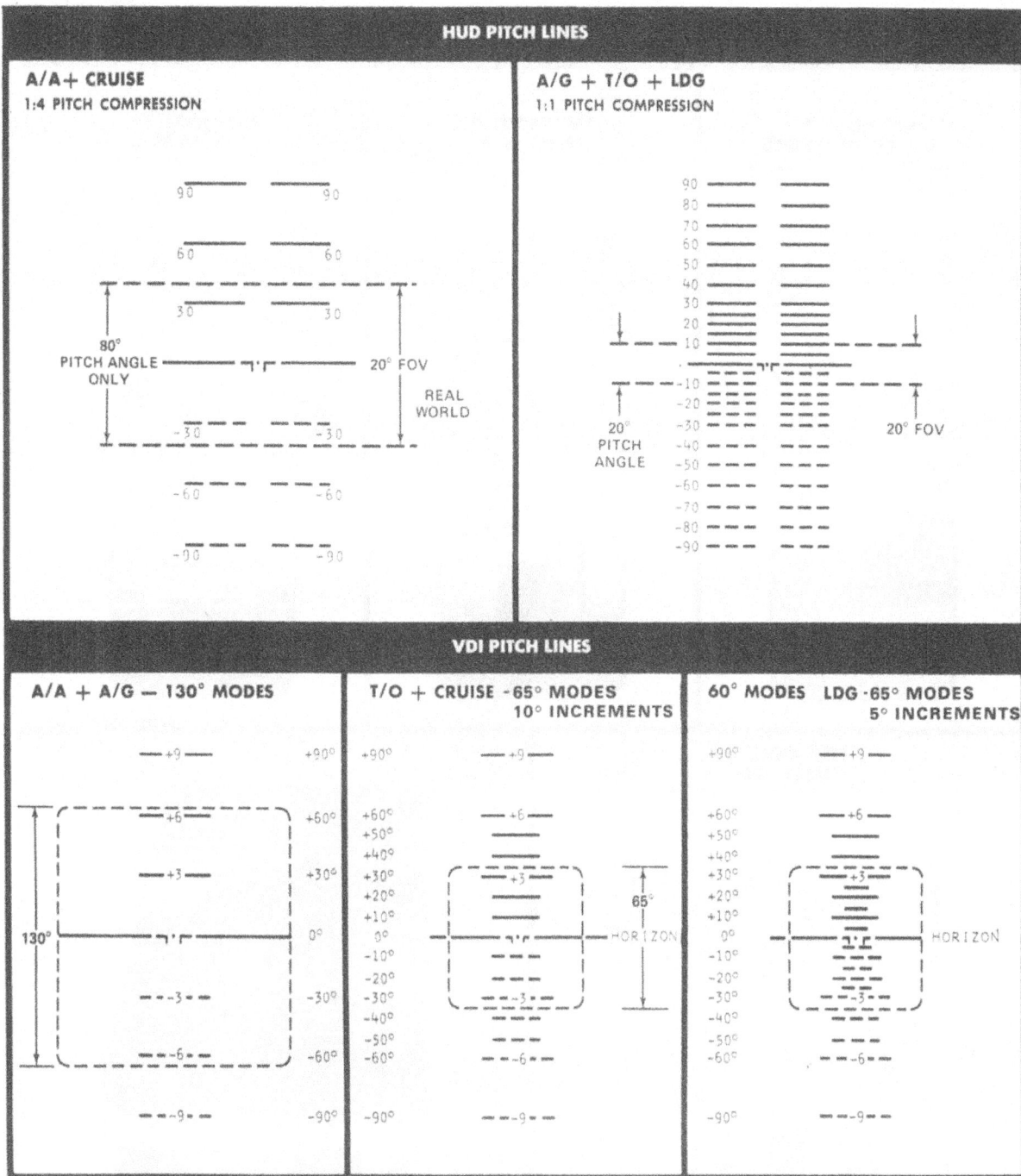

HUD PITCH LINES

A/A+ CRUISE
1:4 PITCH COMPRESSION

A/G + T/O + LDG
1:1 PITCH COMPRESSION

VDI PITCH LINES

A/A + A/G — 130° MODES

T/O + CRUISE -65° MODES
10° INCREMENTS

60° MODES   LDG -65° MODES
5° INCREMENTS

Figure 8-4. VDIG Pitch Modes

VDIG TEST DISPLAYS

TEST MODE 1
SELECT T.O. OR CRUISE

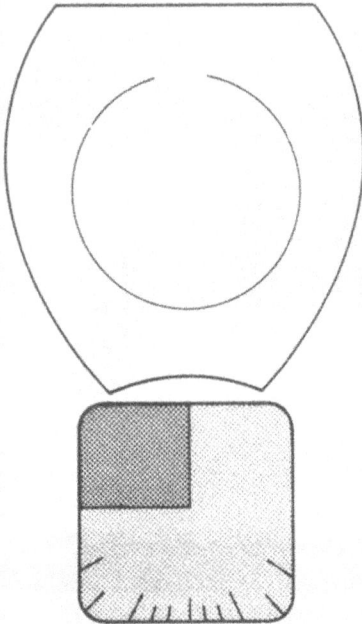

TEST MODE 2
SELECT A/A

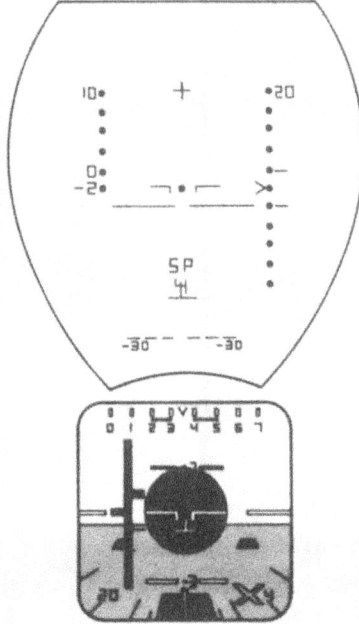

TEST MODE 3
SELECT A/G

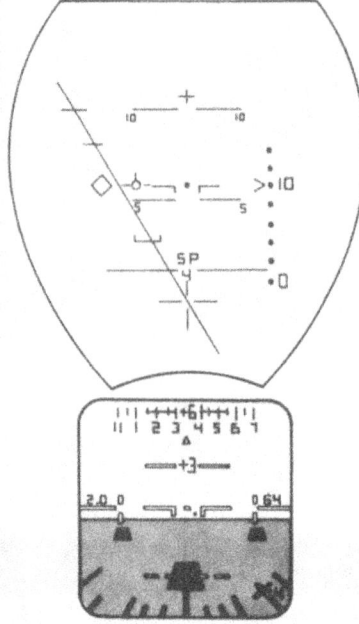

TEST MODE 4
SELECT LDG

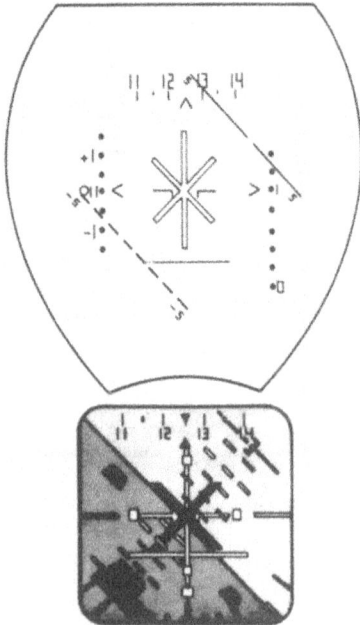

Figure 8-5. VDIG Test Displays

1-F050-004
179-0

PART 2 — MULTIPLE DISPLAY INDICATOR GROUP (AN/ASA-79)

MULTIPLE DISPLAY INDICATOR GROUP (MDIG)

The multiple display indicator group (MDIG) provides the pilot and RIO with navigation, tactical, or ECM data in alphanumeric and symbolic form. The MDIG is composed of the pilot's horizontal situation display (HSD), the RIO's electronic countermeasures display (ECMD), and a processor. The HSD displays all three types of data but the ECMD is capable of displaying only navigation anv ECM data. The HSD and ECMD can simultaneously display the same data or each can present different data. However, when both indicators operate in the navigation mode, and any one of four submodes, they display data for the same submode. The MDIG also has BIT capabilities for system checks on the ground and in flight.

DISPLAY PRIORITIES

Because of the type of information displayed, the ECM mode has the highest priority: it can override the navigation and tactical information display (TID) repeat modes. The following table provides a list of the priorities for the HSD and ECMD with all overrides enabled.

HSD	ECMD
	ECM ML/AI
ECM ML/AI	
ECM Mode	ECM Mode
Navigation Modes	
TID Repeat	Navigation Mode

HORIZONTAL SITUATION DISPLAY

The HSD is the pilot's primary navigation display. The HSD is also capable of displaying ECM information, and repeating the RIO's TID presentation.

The HSD indicator (and the ECMD) is formatted in horizontal PPI or in horizontal plane, depending on display mode. The HSD display consists of a cathode-ray tube (CRT) providing a 5-inch diameter (approximate) display format. The display format is dependent on the position of the HSD MODE switch (NAV, TID, or ECM). The HSD provides line-written symbols, which are internally generated for the navigation display, and which are received from associated systems as deflection and video signals during ECM and TID modes. The HSD presents the following navigation information: magnetic heading of the aircraft, command heading, command course, TACAN bearing, ADF bearing, data block, and range readout. The data block presents alphanumerics of the aircraft's true airspeed, windspeed and direction, and groundspeed. Range to TACAN station, an RIO-inserted destination, or an RIO manually set range is displayed on the HSD range readout. The aircraft heading (a compass rose read against a rubber line) is presented in all display modes.

ELECTRONIC COUNTERMEASURES DISPLAY

The ECMD displays navigation and ECM data to the RIO. The ECMD navigation and ECM display modes use the same format as the pilot HSD modes. Although the ECMD and HSD operate from a common display processor, the pilot and RIO can each select any desired mode.

Note

Refer to Section VIII, NAVAIR 01-F14AAA-1A, for detailed information on MDIG ECM functions.

PROCESSOR

The MDIG processor processes data inputs from other aircraft systems for display on the HSD and ECMD. Based on the mode selected and navigation submode selected, the processor sets the appropriate priority for each indicator.

Depending on the navigation submode selected, the unit processes data representative of TACAN deviation and bearing, groundspeed, windspeed and direction, range to destination, TACAN range, and true airspeed to enable display of alphanumerics, symbols, or both.

If correlate-select, strobe deflection, and strobe deflection enabling inputs are applied, and if the proper warning signals are present, the processor enables display of the correlated or ECM strobe, provided the indicators are operating in the ECM mode. The processor also generates the deflection signals necessary for display of the required data.

MDIG CONTROLS

Controls for the MDIG are provided for the pilot and RIO to give each control of his respective indicator. The pilot's controls for the HSD are on the pilot's display control panel. (See figure 8-6.) The display format depends on the position of the HSD MODE switch (NAV, TID, or ECM). Additional controls are on the panel surrounding the HSD tube face. The RIO's controls are on the ECM DISPLAY control panel. (See figure 8-7.) As with the HSD, the ECMD format depends on the position of the MODE switch (ECM or NAV). Additional controls are around the ECM display indicator.

MDIG MODES

The system operates in any of three modes; electronic, countermeasures (ECM), navigation, and TID repeat.

Electronic Countermeasures Mode

The ECM mode is selected by the pilot with the HSD MODE switch on the pilot's display control panel. The ECM mode can also be selected by the RIO with the MODE switch on the ECM DISPLAY control panel. This mode can also be initiated automatically for HSD when the ECM switch on the pilot's display control panel is set to ORIDE, and for the ECMD when the ORIDE switch on the ECM DISPLAY control panel is set to ML or ML/AI, and warning signals from the defensive electronic countermeasures systems are applied to the MDIG processor.

The ECM mode displays aircraft threats, which can consist of SAM activity, radar-controlled antiaircraft gun emplacements, or any enemy aircraft. The aircraft threats are displayed as three coded strobes, individually or in combinations thereof, and alphanumerics.

The correlate function, which displays only SAM threats, affects the ECM mode display on both indicators. This function is controlled by the RIO with the ECM CORR switch on the ECM DISPLAY control panel.

Note

For a typical ECM display, refer to
Section VIII, NAVAIR 01-F14AAA-1A.

Navigation Mode

The navigation mode is selected by the pilot with the HSD MODE switch on the pilot's display control panel. The mode can also be selected by the RIO with the MODE switch on the ECM DISPLAY control panel. When the navigation mode is initiated, any one of four navigation submodes (TACAN, destination, vector, or manual) can be selected. They are selected on the display control panel with the STEER CMD pushbuttons.

Note

Refer to section VII, part 2, for detailed description of navigation and steering modes.

The navigation mode provides symbolic and alphanumeric display of navigation information for the selected submode (figure 8-8). The DATA/ADF switch on the ECM DISPLAY control panel allows display of certain navigational information on both indicators.

TID Repeat Mode

The TID repeat mode provides display of the TID presentations on the HSD (figure 8-9), and is initiated with the MODE switch on the pilot's displays control panel.

Note

During TID repeat mode operation, the HSD may go blank in the course of any AWG-9 transients, recycles, etc. In this case, perform HSD reset by depressing the TEST pushbutton on the HSD front panel.

Test Mode

The HSD and ECM indicator can be tested for an indication of whether its deflection or brightness circuits are operating correctly. (See figure 8-10.) The test mode can be initiated by the pilot or RIO by depressing the TEST pushbutton on his indicator. For the pilot, the HSD will display a typical infrared (IR) display (for the rear-looking IR system, which may be installed at a later date). This is a test pattern and is not indicative of whether the IR system is functioning properly. For the RIO, initiation of the test will present a typical ECM display on his indicator. The display is static and does not represent whether the DECM systems are functioning correctly.

SYSTEM CHECKOUT

When power is applied to the MDIG, the HSD, ECMD, and the processor perform individual self-tests. If a failure is detected, the self-contained BIT indicator of the failed assembly goes on. In addition, self-test results are applied to the CSDC. The CSDC formats the data, and if a failure exists, routes the data to the WCS computer for display on the TID.

MDIG SYMBOLOGY

Figure 8-11, sheets 1 through 4, show a pictorial representation of the MDIG-generated symbology for ECM and navigation modes as they appear on the pilot's HSD and the RIO's ECMD. Symbology generated during TID repeat mode is discussed in NAVAIR 01-F14AAA-1A.

HORIZONTAL SITUATION DISPLAY CONTROLS

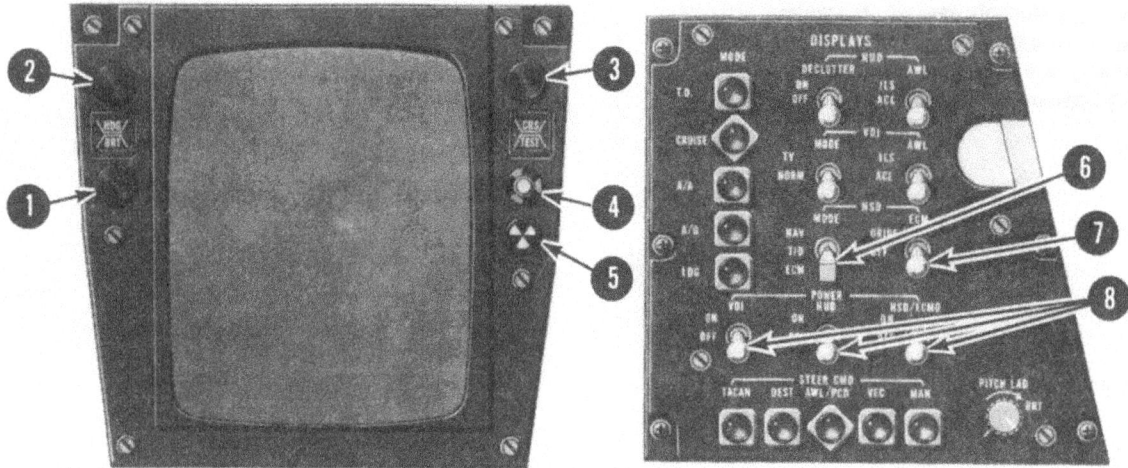

1-F050-004
180-0

NOMENCLATURE	FUNCTION
① BRT control	Allows pilot to vary brightness. Clockwise rotation increases brightness.
② HDG control	Enables pilot to rotate heading reference bug in TACAN mode.
③ CRS control	Enables pilot to set desired course in manual (MAN) and TACAN mode.
④ TEST pushbutton	Allows pilot to reset HSD power monitor when indicator goes blank due to overload.
⑤ BIT indicator	Indicates component status. Solid black indicates system is normal. Failure is indicated by white flags. Indicator is reset by rotating it clockwise.
⑥ HSD MODE switch	NAV — Presents navigation steering information associated with STEER CMD mode selected.
	TID — Repeats information from TID.
	ECM — Repeats ECM information from ECM display.
⑦ HSD ECM switch	ORIDE — ECM display will override navigation display if ML correlation or AI signal is present.
	OFF — Navigation display is not overriden.
⑧ HSD/ECMD POWER switch	Provides ON-OFF power control for both HSD and ECMD.

Figure 8-6. Horizontal Situation Display Controls

ECMD CONTROLS

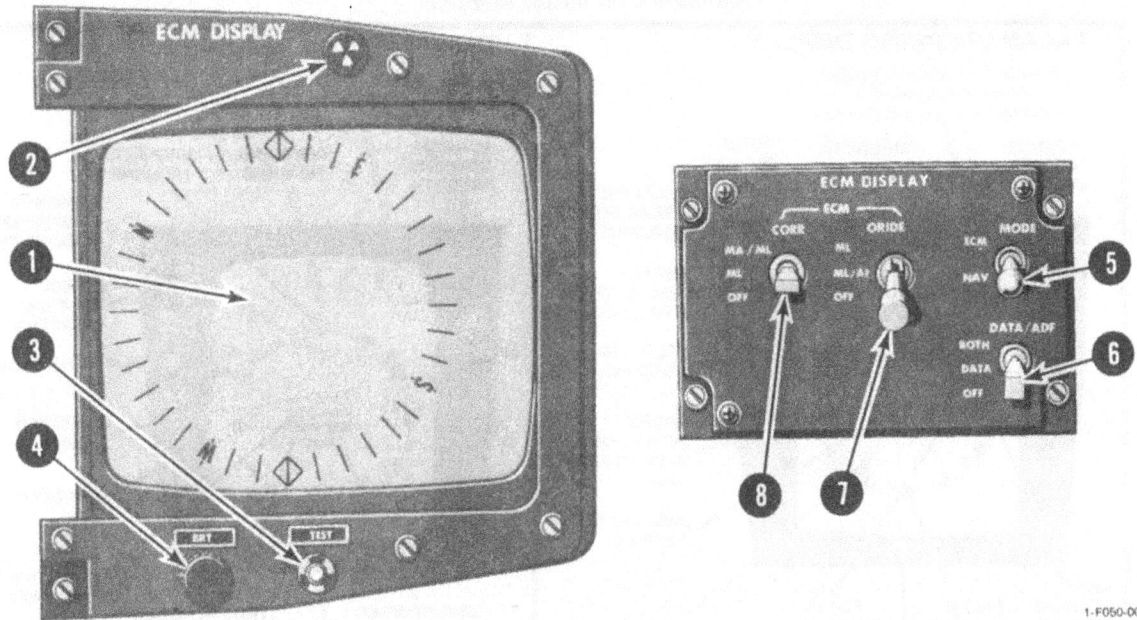

1-F050-004
181-0

NOMENCLATURE		FUNCTION	
①	Direct-view CRT	Electronic display of navigation (horizontal situation) or ECM (AN/APR-25) with acronyms.	
②	BIT indicator	Indicates component status. Solid black indicates system is normal. Failure is indicated by white flags. Indicator is reset by rotating clockwise.	
③	TEST pushbutton	Displays ECMD test pattern (ECM display).	
④	BRT control	Adjusts the brightness of the display. Clockwise increases brightness.	
⑤	MODE switch	ECM —	Selects ECM display (AN/APR-25, AN/APR-27, AN/ALQ-45, or AN/AOQ-50).
		NAV —	Selects NAV display.
⑥	DATA/ADF switch	BOTH —	Enables display of ADF bug and navigation data acronyms.
		DATA —	Enables display of navigation data acronyms. ADF bug is not displayed.
		OFF —	ADF bug and navigation data are not displayed.
⑦	ECM ORIDE switch	ML —	Overrides navigation displays with ECM displays.
		ML/AI —	Overrides navigation displays with ECM displays if ML correlation or AI signal is present.
		OFF —	No ECM display.
⑧	ECM CORR switch	MA/ML —	Allows correlation of MA or ML signals.
		ML —	Allows correlation of ML signals.
		OFF —	Disables the correlation function.

Figure 8-7. ECMD Controls

MDIG NAVIGATION FORMATS

NOTE
PILOT HAS NO VISUAL INDICATION OF WHICH
DESTINATION THE RIO HAS SELECTED.

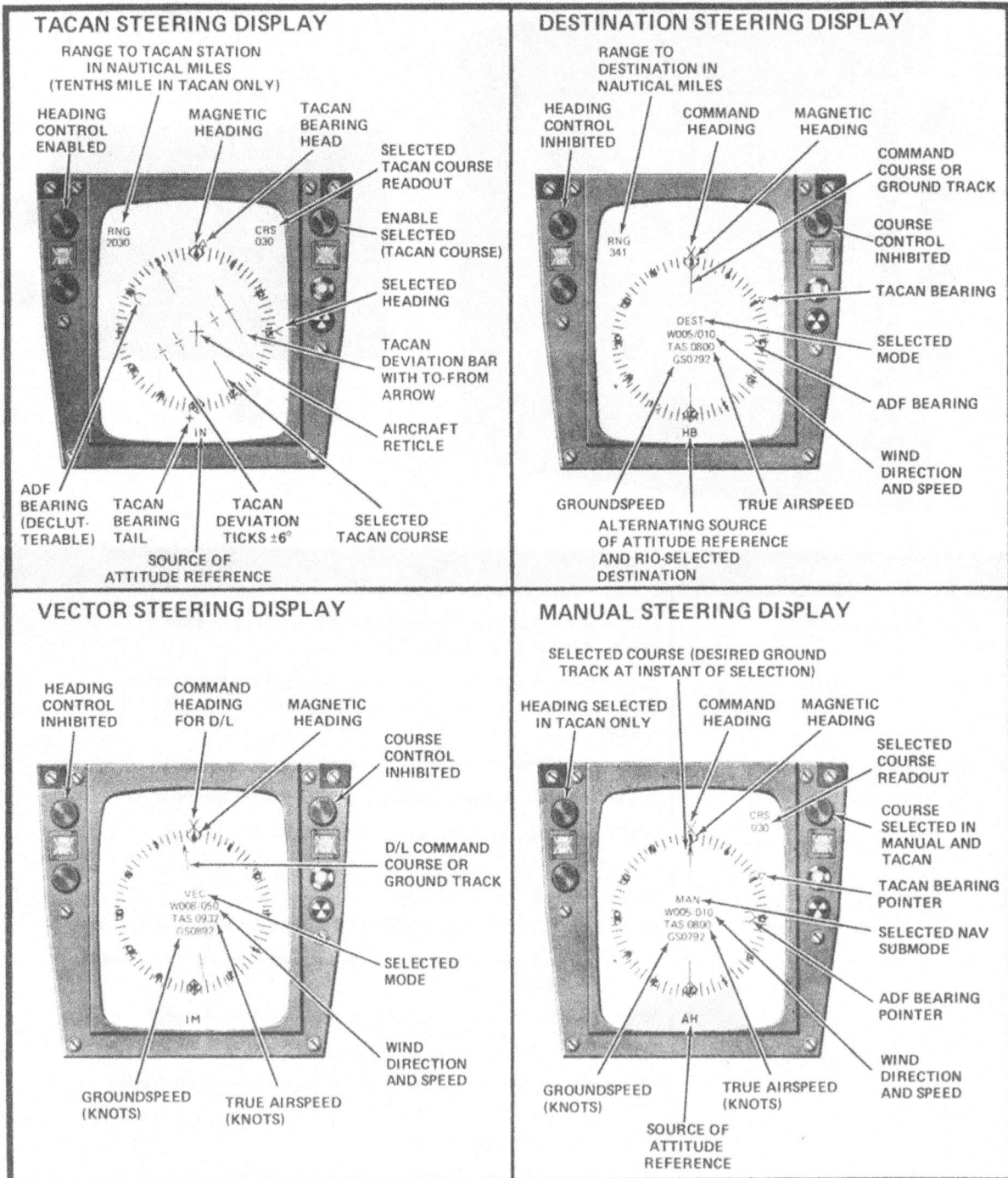

TACAN STEERING DISPLAY

RANGE TO TACAN STATION
IN NAUTICAL MILES
(TENTHS MILE IN TACAN ONLY)

HEADING
CONTROL
ENABLED

MAGNETIC
HEADING

TACAN
BEARING
HEAD

SELECTED
TACAN COURSE
READOUT

ENABLE
SELECTED
(TACAN COURSE)

SELECTED
HEADING

TACAN
DEVIATION BAR
WITH TO-FROM
ARROW

AIRCRAFT
RETICLE

RNG
2030

CRS
030

IN

ADF
BEARING
(DECLUT-
TERABLE)

TACAN
BEARING
TAIL

TACAN
DEVIATION
TICKS ±6°

SELECTED
TACAN COURSE

SOURCE OF
ATTITUDE REFERENCE

DESTINATION STEERING DISPLAY

RANGE TO
DESTINATION IN
NAUTICAL MILES

HEADING
CONTROL
INHIBITED

COMMAND
HEADING

MAGNETIC
HEADING

COMMAND
COURSE OR
GROUND TRACK

COURSE
CONTROL
INHIBITED

TACAN BEARING

SELECTED
MODE

ADF BEARING

WIND
DIRECTION
AND SPEED

RNG
341

DEST
W005/010
TAS 0800
GS0792

HB

GROUNDSPEED

TRUE AIRSPEED

ALTERNATING SOURCE
OF ATTITUDE REFERENCE
AND RIO-SELECTED
DESTINATION

VECTOR STEERING DISPLAY

HEADING
CONTROL
INHIBITED

COMMAND
HEADING
FOR D/L

MAGNETIC
HEADING

COURSE
CONTROL
INHIBITED

D/L COMMAND
COURSE OR
GROUND TRACK

SELECTED
MODE

WIND
DIRECTION
AND SPEED

VEC
W008/050
TAS 0937
GS0892

IM

GROUNDSPEED
(KNOTS)

TRUE AIRSPEED
(KNOTS)

MANUAL STEERING DISPLAY

SELECTED COURSE (DESIRED GROUND
TRACK AT INSTANT OF SELECTION)

HEADING SELECTED
IN TACAN ONLY

COMMAND
HEADING

MAGNETIC
HEADING

SELECTED
COURSE
READOUT

COURSE
SELECTED IN
MANUAL AND
TACAN

TACAN BEARING
POINTER

SELECTED NAV
SUBMODE

ADF BEARING
POINTER

WIND
DIRECTION
AND SPEED

CRS
030

MAN
W005-010
TAS 0800
GS0792

AH

GROUNDSPEED
(KNOTS)

TRUE AIRSPEED
(KNOTS)

SOURCE OF
ATTITUDE
REFERENCE

3-F050-004
182-0

Figure 8-8. MDIG Navigation Formats

TID REPEAT MODE

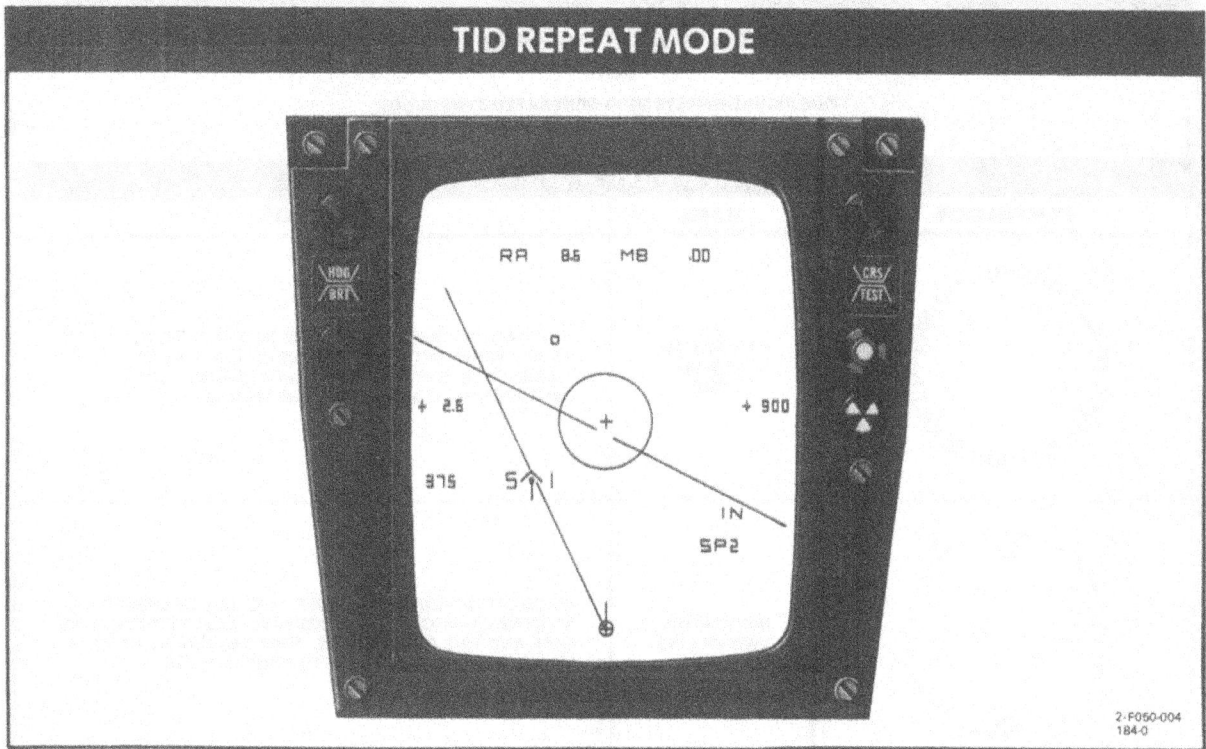

Figure 8-9. TID Repeat Mode

MDIG TEST DISPLAYS

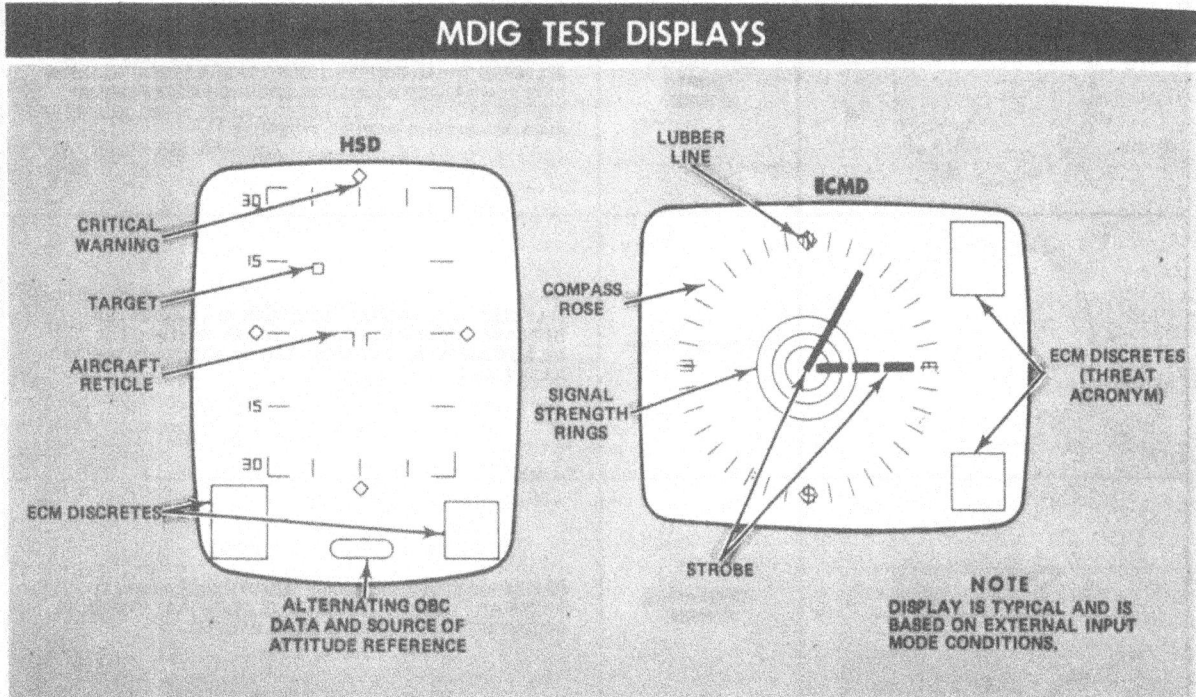

Figure 8-10. MDIG Test Displays

MDIG SYMBOLOGY

NOTE

THIS FIGURE DEPICTS MDIG GENERATED SYMBOLOGY
ONLY. SYMBOLOGY GENERATED DURING TID REPEAT
MODE IS NOT ILLUSTRATED.

NAVIGATION SYMBOLOGY

SYMBOLOGY	NAME	FUNCTION
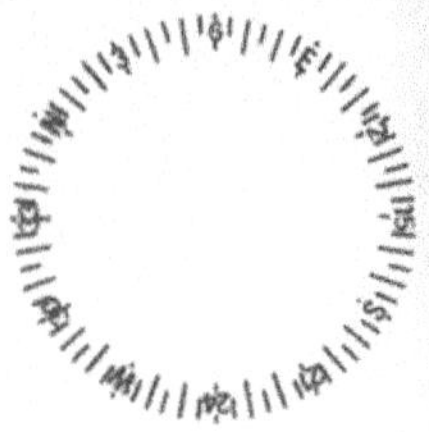	NAVIGATION COMPASS ROSE	NAVIGATION COMPASS ROSE IN 5° INCREMENTS, NUMERICS AT 30° INCREMENTS, AND CARDINAL POINTS AT 90° INCREMENTS. APPEARS IN ALL NAVIGATION MODES. COMPASS ROSE IS MODIFIED IN ECM MODES.
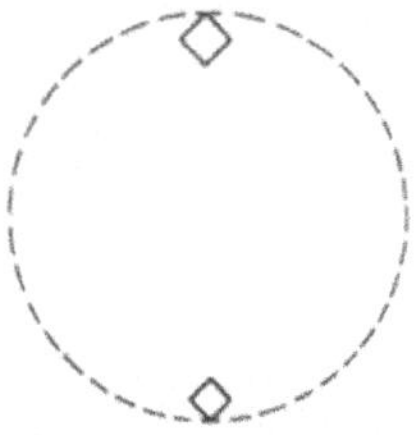	NAVIGATION LUBBER LINE	AIRCRAFT NAVIGATION LUBBER LINE OR ECM LUBBER LINE (WITH NAVIGATION OR ECM COMPASS ROSE) REPRESENTING NOSE AND TAIL OF AIRCRAFT. SHOWS MAGNETIC HEADING OF AIRCRAFT WITH RESPECT TO COMPASS ROSE.
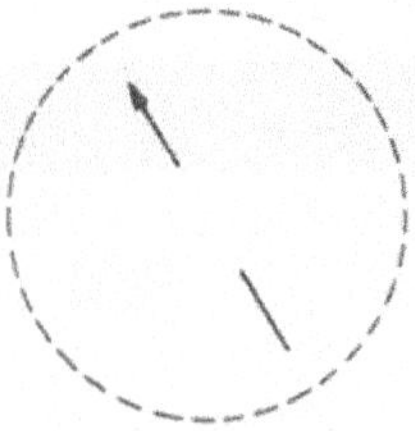	NAVIGATION COURSE SYMBOL	HEAD AND TAIL SYMBOL DEPICTING AIRCRAFT GROUND TRACK OR COMMANDED COURSE DURING DEST STEERING; SELECTED TACAN COURSE DURING TACAN STEERING; DATA LINK COMMANDED COURSE OR GROUND TRACK DURING VECTOR STEERING; OR SELECTED COURSE USING COURSE CONTROL DURING MANUAL STEERING
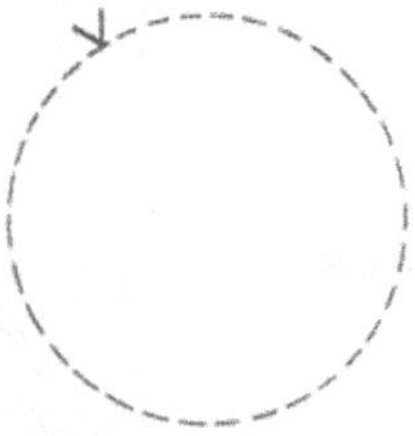	HEADING SYMBOL	SELECTED OR COMMAND HEADING IN ALL NAVIGATION DISPLAYS (MANUALLY CONTROLLED BY HEADING CONTROL OR AUTOMATICALLY CONTROLLED BY WCS COMPUTER OR DATA LINK).
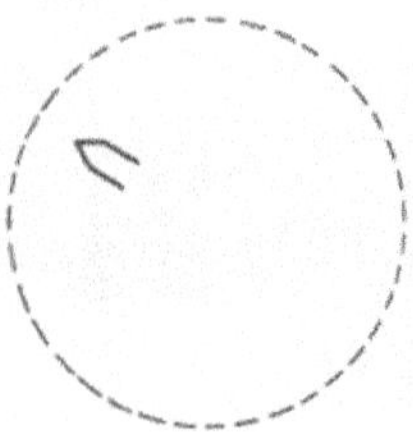	ADF BEARING SYMBOL	ADF BEARING SYMBOL SHOWS DIRECTION OF NEAREST AUTOMATIC DIRECTION FINDING STATION (ADF SYMBOL IS DECLUTTERABLE ON ECM DISPLAY PANEL).

1-F050-004
231-1

Figure 8-11. MDIG Symbology (Sheet 1 of 4)

SYMBOLOGY	NAME	FUNCTION
	FIXED AIRCRAFT SYMBOL	CENTERED IN NAVIGATION COMPASS ROSE AND DEPICTS AIRCRAFT WINGS AND ELEVATORS (ONLY IN TACAN MODE).
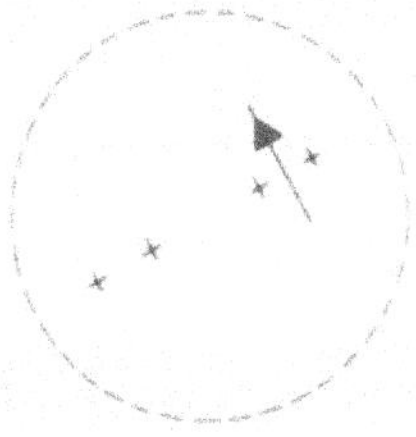	TACAN DEVIATION BAR AND DEVIATION TICKS	DEVIATION BAR SYMBOL SHOWS TACAN STATION DIRECTION WITH DEVIATION TICKS (DEVIATION BAR MAY BE THE SAME DIRECTION AS THE COURSE ARROWHEAD TO STATION, OR 180° FROM THE ARROWHEAD FROM STATION). TICKS ARE 6° APART.
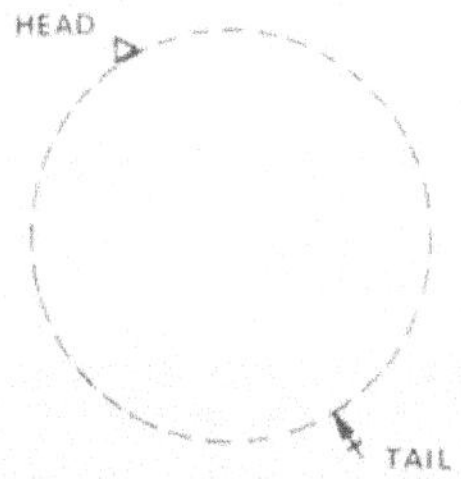	TACAN BEARING SYMBOL	HEAD AND TAIL TACAN BEARING SYMBOL SHOWS DIRECTION OF SELECTED TACAN STATION. ARROWHEAD SHOWS COURSE TO STATION.

ELECTRONIC COUNTERMEASURES SYMBOLOGY		
SYMBOLOGY	**NAME**	**FUNCTION**
	ECM COMPASS ROSE	ECM COMPASS ROSE IN 10° INCREMENTS (VERSUS 5° FOR NAVIGATION) AND WITH CARDINAL POINTS AT 90° INCREMENTS.
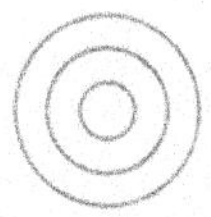	ECM CIRCLES	PROVIDES REFERENCE TO GAGE RECEIVED ECM SOURCE SIGNAL STRENGTH. THE STRONGER THE SIGNAL, THE LONGER THE STROBE.
	ECM STROBES	THREE TYPES OF CODED STROBES INDICATE TYPE AND APPROXIMATE RELATIVE BEARING AND STRENGTH TO ECM SOURCE.

3-F050-004
231-2

Figure 8-11. MDIG Symbology (Sheet 2 of 4)

MULTIPLE DISPLAY INDICATOR GROUP PART BLOCKS

SYMBOLOGY	NAME	FUNCTION
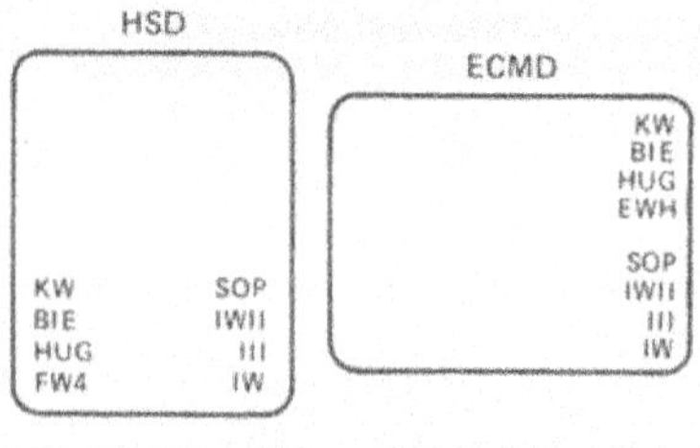 HSD / ECMD (KW BIE HUG EWH SOP IWII III IW) (KW SOP / BIE IWII / HUG III / FW4 IW)	HSD DATA BLOCK 1 (LOWER RIGHT) HSD DATA BLOCK 2 (LOWER RIGHT) ECMD DATA BLOCK 1 (LOWER RIGHT) ECMD DATA BLOCK 2 (UPPER RIGHT)	ECM DISCRETES (THREAT ACRONYMS) ON BOTTOM OF HSD AND ON RIGHT SIDE AND TOP AND BOTTOM OF ECMD. DATA BLOCKS 1 AND 2 DISPLAYABLE IN ALL MODES. REFER TO NAVAIR 01-F14AAA-1A OR TACTICAL MANUAL, NWP-55-5-F14, FOR SPECIFIC INTERPRETATION OF ACRONYMS.
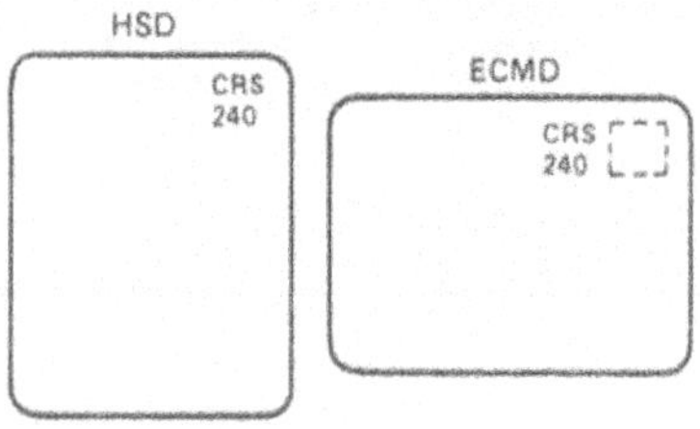 HSD (CRS 240) / ECMD (CRS 240)	DATA BLOCK 3	HSD BLOCK 3, MANUAL OR COMMANDED SELECTED COURSE. ON ECMD, TO LEFT OF DATA BLOCK 2 (MAN AND TACAN MODE ONLY).
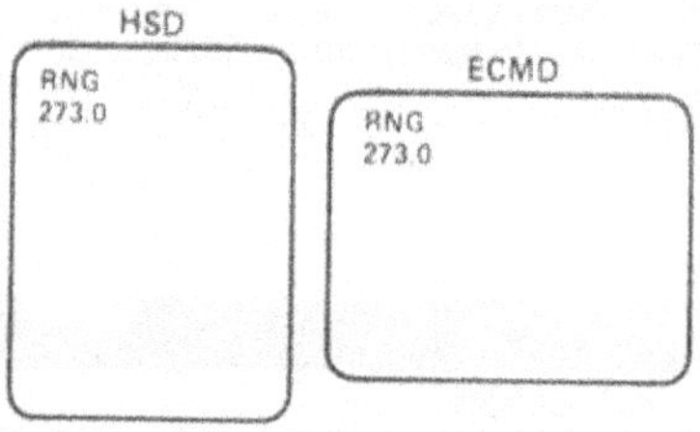 HSD (RNG 273.0) / ECMD (RNG 273.0)	DATA BLOCK 4	RANGE TO DESTINATION IN MILES OR RANGE TO TACAN STATION IN TENTHS OF MILES (TACAN AND DEST MODE ONLY).
HSD (IN)	DATA BLOCK 5 (HSD ONLY)	PREFLIGHT ALTERNATING DISPLAY OF SYSTEM FAILURES AS DETER-MINED BY OBC AND SOURCE OF ATTITUDE REFERENCE. THE SOURCE OF ATTITUDE REFERENCE REFERS TO SYSTEM ACTUALLY DRIVING DISPLAY, NOT THE SYSTEM SELECTED BY THE NAV MODE SWITCH. REFER TO SECTION IX OF THIS MANUAL FOR DESCRIPTION OF THE OBC ACRONYMS THAT MAY BE DISPLAYED. THE SOURCE OF ATTITUDE REFERENCE ACRONYM MAY BE: IN – INERTIAL NAVIGATION MODE IM – IMU/AIRMASS MODE AH – AHRS/AIRMASS MODE IN FLIGHT ALTERNATING DISPLAY (1-SECOND RATE) OF SOURCE OF ATTITUDE REFERENCE AND THE RIO-SELECTED DESTINATION STEERING. THE SOURCE OF ATTITUDE REFER-ENCE IS THE SAME AS PREFLIGHT. THE DISPLAY OF SELECTED DESTINATION STEERING IS DETERMINED BY THE POSITION OF DEST STEERING SWITCH ON THE NAVIGATION CONTROL AND DATA READOUT PANEL.
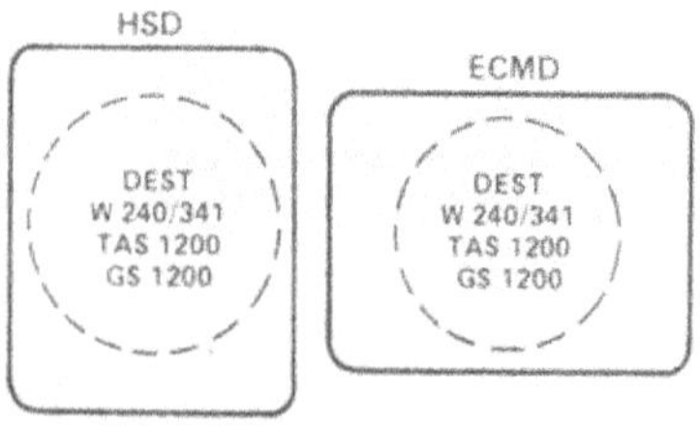 HSD (DEST W 240/341 TAS 1200 GS 1200) / ECMD (DEST W 240/341 TAS 1200 GS 1200)	DATA BLOCK 6	1st LINE – SELECTED MODE OF DEST, VEC, OR MAN 2nd LINE – WIND DIRECTION (TRUE) AND WIND SPEED (KNOTS) CK SPEED ON AND LINE 3rd LINE – TRUE AIRSPEED (KNOTS) 4th LINE – GROUNDSPEED (KNOTS)

3-F080-004
231-3

Figure 8-11. MDIG Symbology (Sheet 3 of 4)

ECMD TEST PATTERN SYMBOLOGY		
SYMBOLOGY	NAME	FUNCTION
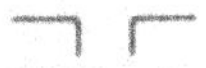	IR A/C RETICLE SYMBOL	DEPICTS OWN AIRCRAFT WINGS (IN IR MODE ONLY)
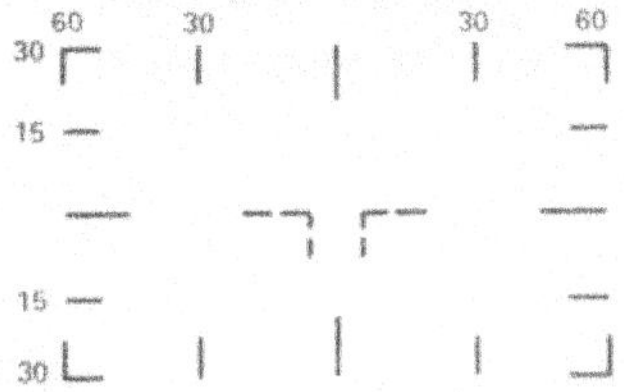	IR CALIBRATION SYMBOLOGY	IR ALPHA NUMERICS AND SYMBOL CALIBRATION RELATING TO AZIMUTH AND ELEVATION ANGLE REFERENCED TO TAIL BORESIGHT.
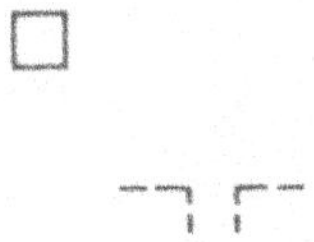	IR TARGET	IR TARGET POSITION RELATIVE TO TAIL OF AIRCRAFT IN AZIMUTH AND ELEVATION.
	IR LIMIT LINE	INDICATES LOWER LIMIT OF IR TARGET DETECTION.
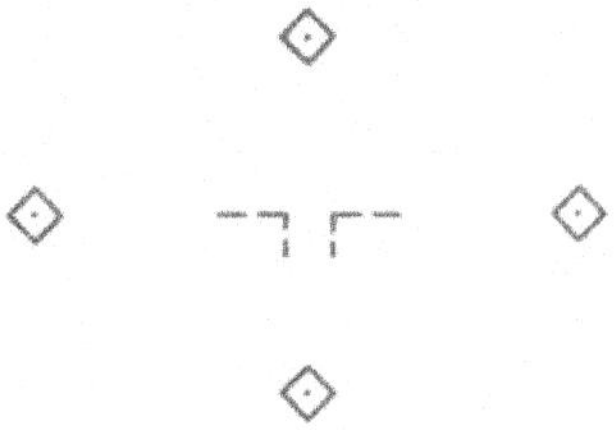	CRITICAL WARNING	CRITICAL WARNING DIAMONDS SHOWN WHEN BEING TRACKED — BLINKS ON HOSTILE IR WEAPON LAUNCH.

0-F050-004
231-4

Figure 8-11. MDIG Symbology (Sheet 4 of 4)

PART 10 — AIR-TO-AIR WEAPONS TACTICAL FUNCTIONS

Note

For a complete discussion of Air-To-Air Tactical Functions refer to NAVAIR 01-F14AAA-1A.

BANNER-TOWED TARGET EQUIPMENT

The F-14A aerial banner towed target equipment consists of a tow adapter, a standard Navy or Air Force 7 1/2 x 40 foot or 6 x 30 foot aerial banner target and approximately 1500 feet of 11/64 inch armored cable towline fitted at both ends with a MK-8 tow ring.

The tow adapter is installed on the hinge point assembly of the tail hook by ground crew personnel. Pilot action is not required for banner hookup. The banner is released in flight or on deck by lowering the tail hook.

Refer to Section III, Part 4 for banner towed target procedures and Section I, Part 4 for banner towing restrictions.

PART 16 — TARPS COMPUTER SUBSYSTEM AND DISPLAYS

Note

Except for the following, all weapons system descriptions
are covered in NAVAIR 01-F14AAA-1A.

TARPS TID SYMBOLOGY

The reconnaissance tactical situation is displayed for the
RIO on the TID using standard AWG-9 symbology.
Additional TARPS alphanumerics are used to display
time- and range-to-go to the target, and range and time remaining to camera OFF. The TID acronyms are unique
to TARPS (see figure 8-12). Figure 8-13 shows a typical
TID reconnaissance display.

TARPS MODE ENTRY

The TARPS-configured aircraft will fly to the target area
using the basic navigation system in the fighter configuration. Upon reaching the target area, the flightcrew must
perform the following steps to obtain the TARPS HUD/
TID symbology and CAP panel entry:

PILOT
1. Selects A/G display on PDCP.
2. Selects either destination or manual steering on
   the PDCP.

RIO
1. Select CAP NAV CATEGORY and depress
   function button No. 5.

Reconnaissance Reference Point Entry

Reconnaissance reference points are entered into the
AWG-9 computer memory bank by the RIO via the CAP
with the CATEGORY switch in TAC DATA. They are
displayed on the TID for navigation and tactical evaluation. The maximum number of inserted reference points
is eight: three waypoints, a fixed point, a surface target,
home base hostile area, and a defended point. Initial
point (IP) is used by the program and therefore cannot be
used as a reference point. Data entered includes latitude,
longitude, target length, target altitude, and command
heading. The heading entered is the desired target
crossing angle used by the pilot for steering. The altitude
entered is the target mean sea level altitude. The target
length is entered via the SPD push-tile in even tenths
of a mile. Figure 8-14 shows the TARPS CAP entry
matrix.

ACRONYM	DEFINITION	CREW ACTION
IRW	IRLS switch to NFOV and Vg/H exceeds 0.357.	Set IRLS switch to WFOV. NFOV is inadequate for Vg/H requirement.
POD	TARPS system failure. One or more failure (amber) lights illuminated on CPS. If no TARPS pod on aircraft, acronym will flash for 60 seconds and then go dim.	Select RESET position with SYSTEM switch on CPS to remove flashing POD acronym.
MAP XX	Normal indication when in mapping mode. Digits under acronym indicate lines remaining; number decreases by one as each line is completed.	None. To exit MAP mode, enter zero value for either IP, ALT (separation distance) or SPD (number of map lines).
TARP	Pilot has not selected A/G display on PDCP.	Pilot must select A/G display on PDCP.

Figure 8-12. TARPS TID Acronyms

TARPS TID DISPLAYS

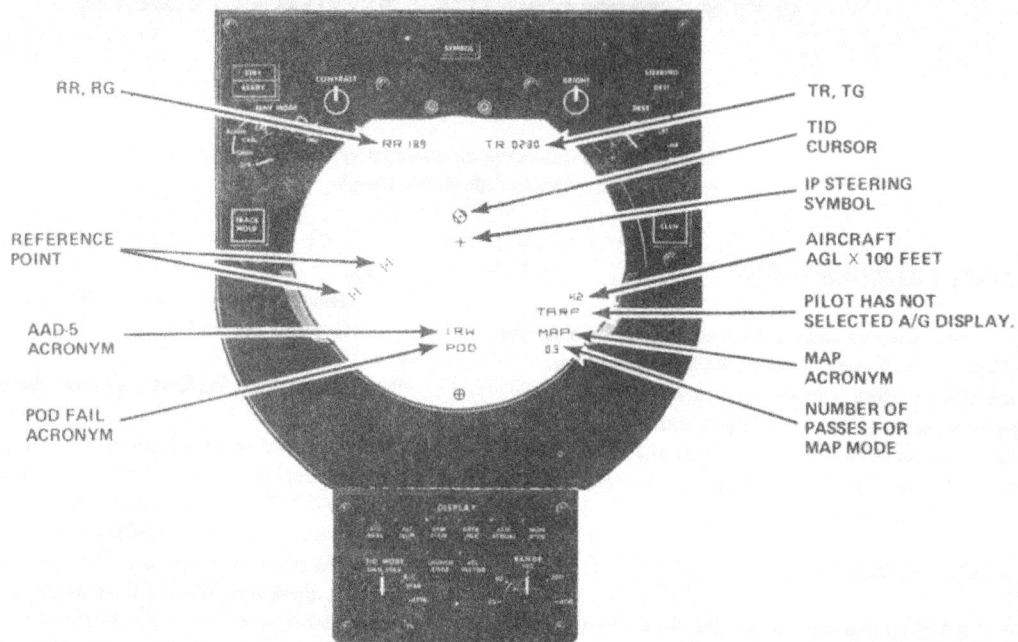

SYMBOL	DESCRIPTION
RG	RANGE TO GO TO TARGET
TG	TIME TO GO TO TARGET
RR	RANGE REMAINING TO CAMERA OFF (LENGTH OF TARGET DEPENDENT)
TR	TIME REMAINING TO CAMERA OFF (LENGTH OF TARGET DEPENDENT)
TID CURSOR	USED TO HOOK TARGET VIA HAND CONTROL; ALTERNATE TO CAP HOOK.
REFERENCE POINT	RECCE TARGET RELATIVE TO OWN AIRCRAFT.
POD	POD/CPS FAILURE INDICATION; FLASHES BRIGHT FOR 60 SECONDS, THEN DIMS UNTIL RESET ON CPS.
IRW	Vg/H IS ABOVE OPERATIONAL VALUE OF AAD-5 NARROW FIELD OF VIEW; RIO SELECTS WFOV ON CPS.
MAP XX	DISPLAYED IN MAP MODE ONLY WHEN (IP ALT) AND (IP SPD) VALUES ARE PRESENT. WHEN MAP MODE IS SELECTED, MAPPING RUNS WILL BE INITIATED ON ANY TARGET HOOKED, EXCEPT IN MANUAL SUBMODE OR TARGET OF OPPORTUNITY (NO HOOK).
IP	DYNAMIC STEERING POINT RELATIVE TO TARGET AND OWN AIRCRAFT.
XXX	OWN AIRCRAFT AGL TO NEAREST 100 FEET
TARP	PILOT HAS NOT SELECTED A/G DISPLAY WITH RIO IN TARPS MODE.

0-F50-268-0

Figure 8-13. TARPS TID Displays

PREFIX	DISPLAY INDICATION	DESCRIPTION	RANGE VALUE	UNITS
LAT/1	LN or LS	Target position north or south of equator.	00° 00.0' to 90°	0.1 minute
SPD/3	LR	Readout/update target length	0 to 409.4 nmi	even tenths of a nmi
ALT/4	AL	Readout/update target altitude	±9,999 feet	1 foot
RNG/5	Prior to target: RG TG Ater target: RR TR	• Great circle range from ownship to target. • Time remaining before target intercept. • After target, range and time to end of run.	0 to 2,048 nmi 0 to 512 minutes	0.1 nmi 1 second (TG XXX XX) min sec
LONG/6	LE or LW	Target longitude position east or west of prime meridian.	000° 00.0' to 180°	0.1 minute
HDG/8	MH/CH	Readout/update command ground track over target.	000° to 360°	1°
IP ● ALT/4 ● N+E IP ● ALT/4 ● S—W	SD	Readout/update separation distance between map lines preceded by a (+) for right or a (–) for left map lines.	±97,218 feet	1 foot
IP ● SPD/3	NL	Readout/update number of lines for MAP mode. **Note** MAP mode will be initiated when both IP speed (LR) and altitude (SD) values have been entered.	1 to 99 lines	1 line

Figure 8-14. TARPS CAP Entry Matrix

Note

Entries of odd tenths will be rounded
to the next lowest even digit. If an odd
tenth of target length is required, enter
the next higher even tenth. (For example,
LR desired 0.3 nmi, enter 0.4 nmi.)

IN-FLIGHT ENTRY OF RECONNAISSANCE REFERENCE POINTS

In order for the RIO to enter reconnaissance reference
point data in flight with TARPS selected and not disturb
pilot steering on the HUD, the following procedures are
required:

1. Hook target desired for pilot steering via CAP or
   HCU.

2. Select MAN position on DEST switch and deselect
   target (step 1) using TID cursor or CAP with following results:

   a. HSD steering is caged.

   b. Pilot steering via HUD reticle is still valid to
      target previously selected in step 1.

3. Hook target file desired for change and enter data.
   (IP cannot be used as a reference point.)

4. Reselect target previously deselected in step 2.

5. TID DEST switch . . . . AS DESIRED

Note

If a change of HUD target steering is
desired from step 1, MAN position must
be deselected.

PILOT OPERATION OF SENSORS

A sensor operating button is provided on the pilot's
control stick. With SYSTEM switch on CPS set to RDY,
and any or all sensor selector switches in the standby/
ready position, the activated sensor can be cycled by the
pilot pressing the bomb button on the control stick. This
is the only TARPS control capability provided to the
pilot. Each camera will cycle at its proper rate for Vg/H
and the IRLS will run continuously at the proper speed
until the pilot releases the bomb button.

Hud Symbology

TARPS HUD symbology (figure 8-15) and changes from
the current aircraft are summarized below:

- Heading — Same

- Velocity Vector — Same

- Movable Reticle — Command Error Indicator
  (Azimuth Steering)

- Bomb Fall Line — Command Ground Track Line
  (CGTL)

- Pitch Ladder — Same

- Diamond — Target Designator

- Aircraft Reticle

- Altitude — Same

Additionally, the following symbols can be decluttered
by the pilot as required during TARPS runs when the
air-to-ground mode is selected:

- Heading

- Pitch Ladder

- Aircraft Reticle

Pilot Steering

Aircraft steering is displayed on the HUD using TARPS
symbology (figure 8-16). The target designator diamond
is used to indicate target position. As the RIO hooks
each successive target, either destination or manual
steering may be selected. If a target is not hooked, the
designator diamond is superimposed on the velocity
vector symbol. If a target is hooked, steering is accomplished by noting the direction the pipper is displaced from the velocity vector. The aircraft should
be banked in that direction. The pipper will move
with bank angle until the two symbols align
(maximum 45° bank angle). Under visual flight conditions steering may be disregarded to approach the
target, then referenced for precise corrections at
close range.

TARPS HUD SYMBOLOGY

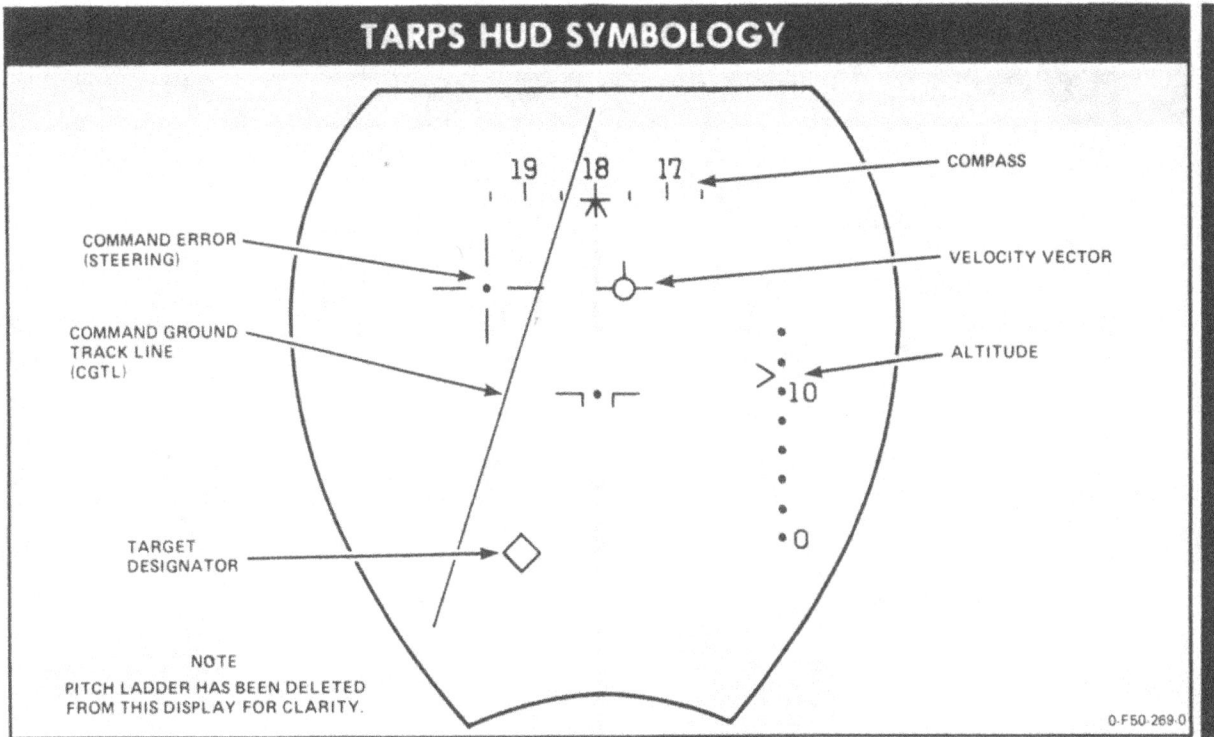

Figure 8-15. TARPS — HUD Symbology

WARNING

Following steering too closely can result
in pilot fixation to the exclusion of safe
altitude control.

DESTINATION STEERING

Destination steering provides steering back to the commanded ground track line (CGTL); not directly to target.
This CGTL is determined from the heading (target
crossing angle) stored for each reference point. To
obtain destination steering, pilot must have DEST steering
selected on PDCP, RIO must have desired heading stored
via CAP, and have target hooked. If ECMD steering to IP
is desired, RIO DEST switch must be in IP position. If
ECMD steering to target is desired select appropriate

waypoint on DEST switch. The bomb fall line is used to
indicate CGTL. Inputs used to compute the CGTL are:
drift angle, aircraft heading, and stored command heading.

MANUAL STEERING

Manual (direct steering) is provided to any reference point
stored by selecting manual steering on the PDCP. This
will provide direct (point-to-point) steering from present
own aircraft position to the selected target. To follow
either type of steering, the velocity vector on the HUD
must be flown to and maintained over the command error
indicator (pipper).

Navigation System Updates

Navigation system updates can be performed via the three
normal methods; VIS FIX, TACAN FIX, and RDR FIX.
The TARPS mode provides an additional update
capability via the HUD.

TARPS HUD DISPLAYS

HOOKED TARGET (DESTINATION STEERING SELECTED)

HOOKED TARGET (MANUAL STEERING OR TARGET OF OPPORTUNITY)

NOTE

PITCH LADDER HAS BEEN DELETED
FROM THESE DISPLAYS FOR CLARITY.

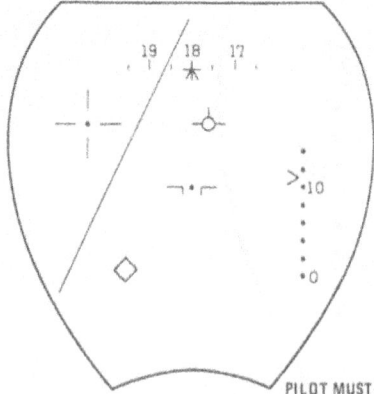

PILOT MUST FLY LEFT TO
CENTER CGTL. WHEN
PILOT FLIES VELOCITY
VECTOR TO OVERLAP
COMMAND ERROR,
INDICATOR CGTL WILL
CENTER AND AIRCRAFT
WILL CROSS TARGET ON
COMMANDED HEADING.

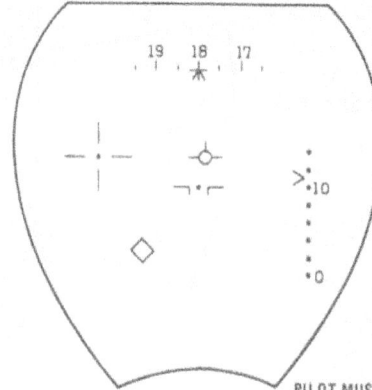

PILOT MUST FLY LEFT TO
PLACE VELOCITY
VECTOR OVER COMMAND
ERROR INDICATOR. THIS
GIVES DIRECT STEERING
TO TARGET.

HUD SLEW DISPLAY

PREHOOK OR AUTO UNHOOK (END OF RUN)

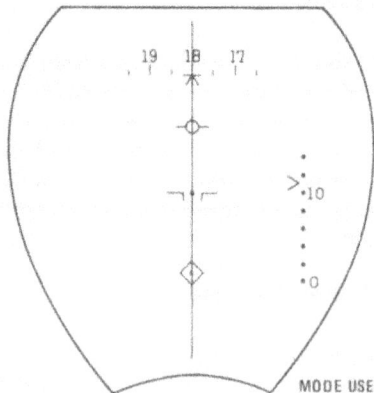

MODE USED FOR
NAVIGATION/TARGET
UPDATE AND TARGET
DESIGNATE FUNCTIONS.

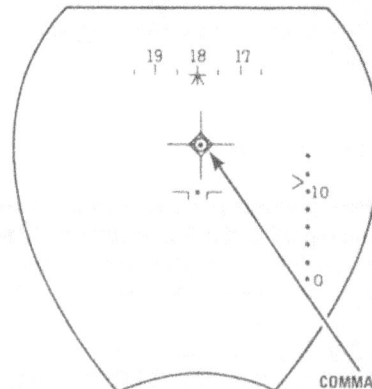

COMMAND STEERING
ERROR/TARGET
SYMBOLS OVERLAYED
ON VELOCITY VECTOR

0-F50-270-0

Figure 8-16. TARPS HUD Displays

SURF TGT file can be used for TARPS
HUD updates, but not for VIS, TACAN,
or RDR FIX updates.

If the navigation system is operating accurately, the HUD
diamond will be superimposed over the hooked target. If
the pilot determines that the system is not accurate
(diamond significantly off hooked target), he moves the
target designate switch (outboard of throttle quadrant)
forward, undesignating the target. This action cages the
CGTL in azimuth and positions the designate diamond
on the CGTL midway between the center and bottom of
the HUD FOV. The pilot can then fly the aircraft so as to
align the CGTL with the target, and slew the diamond
via the designate switch up or down to superimpose it
over the target. Moving the designate switch forward
again will supply the system with the new updated
position of the target. If the RIO selects NAV category
on the CAP and presses function pushbutton no. 3 (not
labeled), he can read delta latitude/longitude displayed
on the TID. If system update is desired, he presses FIX
ENABLE.

Note

Accurate target altitude determines the
quality of the HUD navigation update.
Additionally, the quality of an update increases with greater lookdown angles.
Low grazing angles may induce navigation error.

TARGETS OF OPPORTUNITY

The pilot, using the target designate procedure, can
selectively reconnoiter targets of opportunity. Direct
(manual) steering will be provided to intercept the re-
connoitered target. Before the pilot designates a target
of opportunity, it is necessary for the RIO to be un-
hooked and have entered an estimated target altitude in
the hostile area altitude file.

MAPPING MODE

Mapping mode software will be initialized when IP
altitude and IP speed contain valid data (non zero) and a
target file is selected that has a command heading and

target length entered. The following steps are required
of the RIO to enter the mapping mode:

1. Select TAC data category on CAP.

2. Press IP function pushbutton.

3. Select speed and enter the number of mapping
   lines required.

4. Select altitude and enter the offset distance between
   passes (in feet). If the map is to be to the right of
   the first line, press the N + E pushtile, then ENTER.
   If the map is to be to the left of the first line, press
   the S − W pushtile, then ENTER.

5. Select reconnaissance reference point to be mapped
   by hooking it on CAP or the TID. This stored
   target must have a command heading stored as well
   as target length. (Length of map legs is in even
   tenths of a mile.)

The MAP acronym will appear with the associated
number of lines requested on the TID (lower right-
hand corner). The number of lines will decrease by
one as each pass is completed. Map steering is
basically the same as for a single target with a stored
CGTL. The primary difference is that steering
commands a $90°$ and a $270°$ turn following each
mapping line and then updates steering for each
consecutive pass. At camera off (TR=00 on the
TID) completing the first leg, (RIO must turn
camera off with FILM switch) the target designator,
steering reticle and velocity vector will coincide.
After approximately 8 seconds, the HUD steering
reticle will update, commanding the $90°/270°$ turn
to set up for the next pass; the TID MAP acronym
will blink for 8 seconds and stop; the pass number
will decrease by one; and the IP symbol will then
move to the proper location for the next pass. On
the last pass, the TID MAP acronym will go off
and the HUD symbology will freeze. At this time,
IP altitude and speed entries will be zeroed by the
computer and the MAP acronym will be removed.

ALTITUDE (AGL) MECHANIZATION

AGL information for initial F-14/TARPS software calcu-
lation of Vg/H uses the following sources in the order
given below:

1. APN-194 radar altimeter — This altitude source is
   used whenever the F-14 is below 5,000 feet and the
   APN-194 is operating properly.

2. AWG-9 radar altitude — Altitude will be calculated using a 55° lookdown angle, earth stabilized antenna (TARPS mode). This souce will be used if above 5,000 feet or APN-194 is inoperative.

3. Ownship system altitude (hooked target) — Used whenever the APN-194 and AWG-9 derived altitude are not available. For target stored in file (hooked), AGL is calculated as system barometric altitude minus target altitude.

4. Ownship system altitude (no-hooked target) — For non-stored or no-hooked targets, AGL is calculated as barometric system altitude minus hostile area altitude is chosen that represents the average terrain in the area of interest and inserted into the hostile area altitude file before flight.

Note

Computed AGL is presented on the right side, center of the TID, TID in hundreds of feet, that is 42 = 4,200 feet AGL.

The CPS provides a manual V/H analog output via the velocity and altitude thumbwheels for use by the TARPS sensors when primary sources of V/H (digital data from the aircraft computer) are not available. Manual V/H may be selected at any time by the RIO and should be used instead of steps 3 or 4 above when doubt exists as to the quality of the inputs.

The RIO can determine which altitude information is being used by selecting Flycatcher 7-02767. The following readouts will be displayed in the last four digits:

XXXX0000 — Radar Altimeter

XXXX0001 — AWG-9

XXXX0002 — Barometric minus target altitude

XXXX0003 — Barometric minus hostile area altitude

SECTION IX — FLIGHTCREW COORDINATION

TABLE OF CONTENTS

PART 1 — FLIGHTCREW COORDINATION

PILOT AND RIO RESPONSIBILITIES

FLIGHTCREW RESPONSIBILITIES

The duties of the pilot-RIO team are necessarily integrated, and each must support and contribute to the performance of the other. In this section, specific responsibilities are delineated; however, in the event of aircraft system malfunction, emergency, or unfamiliar circumstances where assistance is desired, cooperation and initiative would become paramount. The pilot is the aircraft commander, whereas the pilot or RIO who is designated mission commander is responsible for the successful completion of any mission assigned to his aircraft. The RIO should constitute an extension of the pilot's observation facilities. By intercommunication, the RIO should anticipate, rather than await, developments. A challenge and reply system will be used between pilot and RIO when using the checklist. The RIO will normally be responsible for all communications except in ACM.

SPECIFIC RESPONSIBILITIES

Flight Planning

PILOT

The pilot is responsible for the preparation of required charts, flight logs, and navigation computations including

fuel planning, checking weather and NOTAMS, and for filing required flight plans.

RIO

The RIO is responsible for the preparation of charts, flight logs, navigation computations including fuel planning, checking NOTAMS, obtaining weather for filing purposes, and for completing required flight plans.

Briefing

Accomplish those tasks delineated in the preceding paragraph.

PILOT OR RIO

The pilot or RIO designated as flight leader is responsible for briefing all crewmembers on all aspects of the mission to be flown.

Preflight

PILOT

The pilot is responsible for accepting and preflighting the aircraft assigned in accordance with this manual and appropriate preflight checklists contained in NAVAIR 01-F14AAA-1B.

RIO

The RIO will be capable of, and proficient in, performing a complete aircraft preflight, including all armament, in accordance with this manual and appropriate preflight checklists contained in NAVAIR 01-F14AAA-1B.

Prestart

PILOT

The pilot will execute prestart checks prescribed in NAVAIR 01-F14AAA-1B and, when external power is applied and checks requiring external power are completed, will inform the RIO "PRESTART CHECKS COMPLETED - READY TO START."

RIO

The RIO will execute prestart checks prescribed in NAVAIR 01-F14AAA-1B and, when external power is applied, will inform the pilot "PRESTART CHECKS COMPLETE."

Starting

PILOT

The pilot will start engines as prescribed in engine start-pilot paragraph in section III, part 2, and will keep the RIO informed of any unusual occurrences.

RIO

The RIO will remain alert for any emergency signal from the ground crew and will inform the pilot if such signals are not observed.

Poststart

PILOT

At completion of the emergency generator check, the pilot will inform the RIO "EMERGENCY GENERATOR CHECK COMPLETE." The pilot will complete all poststart checks prescribed in NAVAIR 01-F14AAA-1B and stand by for OBC.

RIO

At completion of the emergency generator check, the RIO will perform the poststart checks prescribed in NAVAIR 01-F14AAA-1B. When onboard checkout and (INS) is complete, RIO informs the pilot "READY TO TAXI."

Pre-Takeoff

PILOT

The pilot will execute pretakeoff, instrument, and takeoff checklists prescribed in NAVAIR 01-F14AAA-1B and as posted in the aircraft. The pilot will report to the RIO takeoff checklist items, using the challenge-reply method. The pilot will receive the "READY FOR TAKEOFF" report from the RIO and advise him of type and configuration takeoff planned, prior to rolling or catapulting. The pilot will report "ROLLING" or "SALUTING", as appropriate, to the RIO.

RIO

The RIO will execute pretakeoff checklists prescribed in NAVAIR 01-F14AAA-1B; will initiate, using the challenge-reply method, the posted TAKEOFF checklist in the aircraft; and, at completion of the TAKEOFF checklist, RIO informs the pilot "READY FOR TAKEOFF."

Takeoff and Departure

PILOT

The pilot shall ensure that the intercom remains in the HOT MIKE position for normal flight operations and will report "GEAR UP" and "FLAPS UP" to the RIO insofar as safety permits. The RIO should be advised of any unusual occurrences during takeoff that may affect safety of flight. The pilot or RIO will request, copy, and acknowledge all clearances.

RIO

Where departures are made in actual instrument conditions, the RIO will monitor the published clearance departure procedures and inform the pilot of any deviation from the prescribed flightpath. The RIO will copy all clearances received and at all times be prepared to provide the pilot with clearance information of navigational information derived from these instruments. Built-in test checks will not be conducted during instrument climbouts.

Inflight (General)

PILOT

The pilot will inform the RIO of any unusual occurrences and will ensure that the aircraft is operated within prescribed operating limitations at all times. The pilot or RIO will normally request, copy, and acknowledge all clearances.

RIO

The RIO will inform the pilot of the weapon system status and assist the pilot in normal or emergency situations. During descent, the RIO will inform the pilot 1000 feet prior to reaching the intended level-off altitude.

Intercept

PILOT

The pilot will maneuver or coordinate aircraft maneuvers with, or as directed by, the RIO, observing normal operating limitations. The pilot will inform the RIO of weapons status, weapons selected and armed, and when the target is sighted visually. The pilot will monitor aircraft position from initial vector through breakaway by pigeons information or navigational display.

RIO

The RIO will handle all communications from initial vector through breakaway, excluding missile-away transmissions; provide the pilot with descriptive commentary, including weapon status and target aspect, if available, and direct and coordinate aircraft maneuvers with the pilot, as necessary, to complete the intercept.

Instrument Approaches

PILOT

The pilot is responsible for the safe control of the aircraft, the decision to commence the approach with the existing weather, and the selection of the type of approach to be made. The pilot, before commencing any penetration, will report to the RIO the completion of each item of the instrument checklist. In addition, the pilot will challenge the RIO instrument penetration checklist, as to approach plate availability and corrected altimeter setting.

RIO

The RIO will monitor aircraft instruments and appropriate approach plate during holding, penetration, and approach and shall be ready to provide the pilot with any required information. He shall be particularly alert to advise the pilot of deviations from the course of minimum altitudes prescribed on the approach plate. Built-in test checks will not be conducted in actual instrument conditions. The RIO will inform the pilot of the status of the radar and will do nothing to cause the display to be lost. During penetrations and/or descents (VFR or IFR), the RIO will report to the pilot the aircraft descent through each 5,000 feet of altitude above 5,000 feet and each 1,000 feet of altitude loss below 5,000 feet, until, on reaching the desired altitude, the RIO will report when altitude error exceeds 10% of actual altitude or ±300 feet.

Landing

PILOT

The pilot will utilize the LANDING checklist and report each item to the RIO prior to reporting "GEAR DOWN, HOOK DOWN" to the final controller, tower, or Pri-Fly. The pilot will receive a "READY TO LAND" report from the RIO.

RIO

In the landing pattern, the pilot shall read and the RIO acknowledge the posted LANDING checklist. The RIO shall visually check the flap position and landing gear position by looking through the opening on the left side of the instrument panel. The RIO will report "READY TO LAND" to the pilot. Built-in test checks shall not be conducted while in the landing pattern.

Postflight

PILOT

The pilot will inform the RIO of any unusual occurrences on the landing roll or arrestment. The pilot will report flap and wing position to the RIO when clear of the runway or landing area, and report when the wing is actuated. The pilot will inform the RIO when shutting down engines. The pilot will conduct a postflight inspection of the aircraft.

RIO

The RIO will challenge the pilot on flap position if the report is not received. When informed by the pilot that the wing has been actuated, the RIO will visually verify wing and spoiler positioning. The RIO will complete the built-in test checks remaining and secure the rear cockpit for shutdown, then notify the pilot "READY FOR SHUTDOWN." The RIO will assist the pilot in conducting a postflight inspection of the aircraft.

Debriefing

The pilot and RIO will complete the Yellow Sheet and all required debriefing forms.

Note

The RIO will vacate the aircraft first and after he is on the ground, flight-deck, or hangar-deck, the pilot will exit. This is particularly important during shipboard operations.

PROCEDURES, TECHNIQUES, AND CHECKLISTS

Even though some of the procedures, techniques, and checklist are specifically designed for the pilot or RIO, the entire contents of the Flight Manual and the Pocket Checklist should be thoroughly read, understood, discussed, and agreed upon collectively by the pilot-RIO team. Discrepancies in existing procedures, or the need for additional procedures, should be brought to the attention of the NATOPS evaluator and/or instructor. Most of the procedures (individual and coordinated) are covered in this manual and are grouped under flight phases and/or categories. Aircraft systems description, with their individual operating criteria, is covered in section I. Classified systems descriptions and procedures, and some limitations information, are covered in the classified supplement. The Pocket Checklist contains the pilot's and RIO's check items for Preflight, Prestart, Start, Poststart, Takeoff, Built-In Test (BIT), Instrument and Descent, and Postflight procedures.

PART 2 — AIRCRAFT SELF-TEST

AIRCRAFT SELF-TEST INTRODUCTION

There are six self-test procedures in the weapon system; four in the airframe and avionics systems, and two that deal specifically with the AN/AWG-9. The self-test procedures are presented in the following table.

NAME	ORIGINATOR	PURPOSE
MASTER TEST checks	PILOT	Selectable tests of instruments, fuel system, warning system (lights), Mach lever, wing sweep, ECM and AOA.
Onboard checkout (OBC)	PILOT and RIO	Tests various avionics, flight controls, actuators, AICS, and computers.
Onboard checkout continuous monitor (CM)	AUTOMATIC	Monitors various avionics functions for in-flight or on-deck failures. Works on 2-second cycle time.
AWG-9 continuous monitor	AUTOMATIC	Monitors various AWG-9 functions for in-flight or on-deck failures. Operates on 2-second cycle time.
AWG-9 BIT	RIO	Tests AWG-9 and missile functions.
Unit self-test	PILOT and RIO	Test incorporated in various components independent of other tests.

MASTER TEST CHECKS

MASTER TEST checks are initiated by the pilot through the MASTER TEST panel (figure 9-1) on the right outboard console. These tests check the operational status of specific aircraft systems basic to safety of flight and mission success. The OBC, MACH LEV, WG SWP, FLT GR UP, and FLT GR DN positions are used on the deck only, and are prevented from inadvertent use in flight by the weight-on-wheels safety switches. The remaining tests, except for emergency generator, which also requires combined hydraulic pressure can be done whenever electrical power and cooling air are available. For details of specific aircraft system tests, refer to the applicable system description.

WARNING

During ground operations, once the OBC position is selected, do not deselect OBC until the program has completed the entire cycle. When the disable signal, which inhibits throttle movement is removed, the APC will run throught its BIT test and advance the throttles greater than 80%.

Note

- Before starting the test, the MASTER RESET button on the left vertical console must be depressed to turn off any caution or advisory lights associated with the air data computer.
- Cycling CSDC circuit breakers while the MASTER TEST switch is in the OBC position can cause uncommanded AFCS and/or AICS OBC.
- In the LTS position, the MASTER CAUTION light will flash unless there is a circuit failure within the caution advisory indicator, in which case the lights will remain on.

Master Test Switch Operation

The master test check is made by pulling the knob up, rotating to the desired position, and depressing it. After the test is completed, the master TEST switch must be pulled up to deenergize the system.

System status and test results are indicated on the cockpit instruments: GO-NO GO lights on the master test panel; warning, caution, and advisory lights in both cockpits; and displays on the HSD, VDI, HUD, and TID.

Note

If INST test is selected for more than 30 seconds with engines running, opening the interconnect valve may allow fuel to fill the vent tank and overflow.

MASTER TEST PANEL

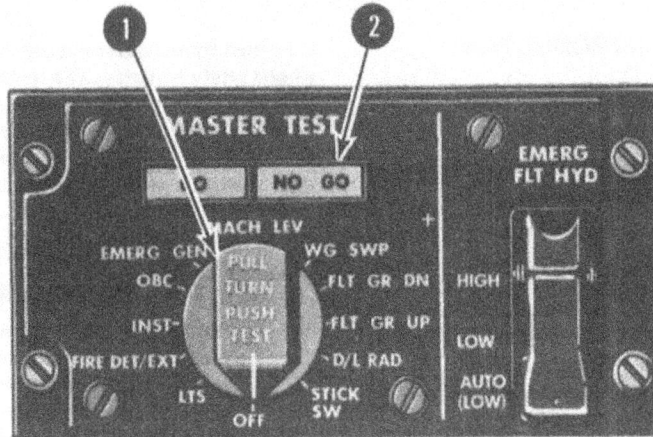

1-F060-004
120-0

NOMENCLATURE		FUNCTION
① MASTER TEST switch	OFF -	Disables test functions.
	LTS -	Turns on caution, warning, and advisory lights; emergency stores jettison button; GO and NO GO lights; landing gear and hook transition lights; approach indexer; FIRE warning lights; and ACM panel lights. D/L power switch must be ON to check DDI lights.
	FIRE DET/EXT (before engine start)	L and R FIRE warning lights illuminate. If a circuit problem exists, the corresponding FIRE light will not illuminate. Simultaneously, the fire extinguishing system initiates a self-test. If tests pass, the GO light illuminates, if the NO-GO light illuminates or if both or neither GO or NO-GO lights illuminate, a failure exists in the system.
	INST -	Lights the FUEL LOW and OXY LOW caution lights in both cockpits, and displays the following pilot cockpit indications.

Before engine start:

RPM . . . . 80%±2%
TIT . . . . 1300°±20°C (initiates engine over temperature alarm at 1215°±15°C.
FF . . . . 4300±100 pounds per hour
FUEL QTY . . . . . . 2000±200 pounds (both cockpits)
LOX . . . . 2 liters
WING SWEEP . . . . 45°±2.5°
AOA . . . . 18±.5 units
L and R OVSP/
 VALVE . . . . ON
BINGO . . . . ON (if Bingo set ≥2000)

Figure 9-1. Master Test Panel (Sheet 1 of 2)

NOMENCLATURE	FUNCTION
	After engine start:
	Symbology on the VDIG and HSD are determined by display mode selected.
OBC - (after engine start)	Enables Class II A checks during AUTO SAT mode or through selection of SPL by RIO on computer address panel and initiation of OBC BIT. Test acronyms available on HSD and TID for the AFCS, AICS, CADC, and auto throttle, through the computer signal data converter.
EMERG GEN - (after engine start)	Activates automatic transfer feature of generator and checks tie contactors. GO lights indicate satisfactory check. If the NO GO light remains illuminated, a malfunction is indicated.
MACH LEV - (before engine start)	Air data computer simulates predetermined signal that checks Mach lever control unit and actuators of left and right engine. GO light indicates satisfactory check.
WG SWP - (after engine start)	Air data computer simulates that circuit to the wing sweep system (wings do not move). Requires wings in oversweep.
FLT GR DN - (Throttles must be midway between IDLE and MIL.)	Initiates ground check of auto throttle interlocks. Auto throttle mode switch should remain in AUTO position if selected and AUTO THROT advisory light should illuminate for 10 seconds when AUTO MODE is deselected or master test deengaged.
FLT GR UP - (after engine start)	Permits checking external fuel tank pressurization. GO light indicates required pressure. WING/EXT TRANS switch must be in AUTO position and DUMP switch set to OFF.
	Engine RPM above idle (approximately 75%) may be required to provide sufficient bleed air pressure for satisfactory check.
D/L RAD - (after engine start)	Tests the data link converter. Test is displayed on DDI indicators, the ECMD, HSD, and VDIG. Inhibits tactical control messages during test sequence. Symbology displayed is determined by the display mode selected.
STICK SW - (after engine start)	Checks left and right spoilers and 1-inch stick switches (left and right) for yaw SAS gain control (>1 inch stick throw, yaw SAS gain tripled).
② GO-NO GO lights	
	GO - Indicates a valid test
Note	NO GO - Indicates unsatisfactory test
Functional only in LTS, FIRE DET/EXT, INST, EMER GEN, MACH LEV, FLT GR UP, and STICK SW positions.	In aircraft BUNO 159825 and subsequent and aircraft incorporating AFC 400 the GO-NO GO lights indicate operation of automatic rudder interconnect with MASTER TEST switch in INST position.

Figure 9-1. Master Test Panel (Sheet 2 of 2)

The GO-NO GO indicator lights on the MASTER TEST panel will illuminate only in the LTS, FIRE DET/EXT, EMER GEN, MACH LEV, and FLT GR UP positions. In the LTS test position, only the bulbs in the GO-NO GO indicators are checked. In the EMER GEN, MACH LEV, FIRE DET/EXT, and FLT GR UP positions, a GO light indicates a valid test and a NO GO light indicates an unsatisfactory test. The STICK SW position utilizes only the GO light; therefore, a valid test in STICK SW is indicated by a GO light but the lack of a light indicates an unsatisfactory test.

Electrical power for the master test panel comes from the left main dc bus through the MASTER TEST circuit breaker (8H5) on the main dc circuit breaker panel. When operating on aircraft power or when external electrical power is connected to the aircraft, cooling air must be supplied to all avionic equipment before a test is initiated.

ONBOARD CHECKOUT (OBC)

The OBC system checks the operational status of approximately 85% of the non-AWG-9 avionics equipment. It provides fault isolation to the weapon replaceable assembly (WRA) without the use of ground support equipment. The system is capable of monitoring various avionic systems to detect failures or to initiate test signals that simulate a response, which the OBC system monitors. The OBC system uses the AWG-9 computer and the computer signal data converter (CSDC) (figure 9-2) to process test information, maintain a record of all failures, and display test information on the TID and HSD (TID repeat). The record of failures is maintained during the course of the flight and may be displayed at any time, including postflight operations by maintenance personnel. The record, however, must be erased prior to the next mission by the flightcrew. However, if OBC is run, systems that have passed will clear previous fail acronyms from the maintenance file. Systems that have not passed will display any existing acronyms that are in the maintenance file.

The type of test information displayed is dependent on which OBC test routine is selected. The command-activated BIT (CAB) and maintenance readout (MR) routines are stored on the magnetic tape module (MTM), and the continuous monitoring (CM) is permanently stored in non-destructive readout (NDRO).

Note

In flight, the maintenance function (OBC FILE CLEAR pushbutton) does not function.

Some of the avionic equipment that the OBC does not check are the ADF, UHF, Auxiliary UHF receiver, speech security system KY-28, barometric altimeter, vertical velocity indicator, standby compass, parts of the electrical power supply, ICS, antenna assemblies, and DLS during a SAT align.

Onboard Checkout Built-in Test

To facilitate testing, the avionic equipment has built-in test features, associated output lines that provide GO-NO GO indications, and command-initiated BIT circuits of equipment that have both continuous and command-initiated BIT features. The OBC program, under the control of the WCS computer, sends BIT commands to the avionic equipment under test and subsequently monitors the response from the BIT GO-NO GO features of the equipment. This equipment, upon completion of a particular test, outputs GO-NO GO voltage levels and an equipment identification code on its BIT lines. The CSDC converts this raw data into digitally-formatted data words for the WCS computer. A history file of all NO GO responses is maintained by the WCS computer and all NO GO responses are displayed to the WRA level on the TID. (See figures 9-3 and 9-4.)

Due to hardware inhibits, DLS tests are not valid with the data link MODE switch in the CAINS/WAYPT position. DLS tests can only be accomplished in a NON SAT mode with the data link MODE switch in the TAC position.

There are two ways to complete OBC; simultaneous alignment and test (SAT) and a non SAT. The differences between the two are in the methods for entering and exiting the routines. The systems under test in OBC are presented on the TID under their respective classes: Class I (in-flight only), Class IIA and Class IIB (preflight only), and Class III (inflight and preflight). (See figure 9-5.) SAT OBC can only be performed on the ground when preflight systems are checked. Systems not tested glow bright and steady. Systems under test are flashing and those which have been tested are dim and steady.

Note

Weight-on-wheels or TAS < 70 knots inhibits tests that radiate power on the deck.

Simultaneous Alignment and Test (SAT)

The SAT permits concurrent On-Board Checkout and platform alignment. The method of entry requires the RIO to set the NAV MODE switch to ALIGN prior to the completion of the auto BIT sequence 2 (figure 9-6). At the completion of the auto BIT sequence 2, simultaneous alignment and OBC will automatically begin. If BIT sequence 2 is degraded or failed, auto SAT will not be entered. A program restart will cause align to be read in. However, OBC will not run unless manually selected, since auto SAT requires a successful completion of auto BIT 2.

Note

- Only SAT mode displays will appear between the align display and the OBC display. The acronym S2 and the status of the last sequence 2 run will appear upon entrance to SAT mode. A √ will appear if sequence 2 passed. A D will appear if sequence 2 ran to completion but did not pass. An X will appear if sequence 2 did not run to completion. The RAMP acronym will appear if an AICS failure is detected by OBC.

- A successful auto BIT 2 is necessary to initialize certain radar constants. If an X appears above the S2 in the SAT displays, another power up auto BIT 2 is necessary.

For a manual SAT, SMAL, or OBC may be initiated first, but the second mode initiation causes the SAT entry. OBC and alignment can still be performed independently. To obtain a complete OBC, the pilot must also select OBC on the MASTER TEST panel. If the pilot should fail to select OBC, POBC DIS (figure 9-4) appears on the TID and only a Class 11B and III OBC will run.

The pilot's test panel will be continually monitored to check if he subsequently selects OBC. If he makes the selection while there is still 6 seconds of test time remaining, Class IIA OBC will be run.

Non-SAT OBC

A non-SAT OBC can be run by the pilot selecting OBC on the MASTER TEST panel and the RIO selecting OBC BIT in SPL category (figure 9-6). This initiates the OBC routine only. If the pilot does not select OBC, only a Class IIB and III will run and POBC DIS will appear on the TID (figure 9-4).

Once initialization is complete, approximately 95 to 200 seconds later (after the built-in tests have been completed), the acronyms of the aircraft systems that have not passed their BIT tests appear on the TID, followed by a TEST COMP legend. Also, a DIS POBC legend will appear when

the OBC BIT testing is complete. The DIS POBC legend is deleted when the pilot deselects OBC with the master TEST switch. Additionally, a CHAL IFF legend appears during OBC testing. This legend requires that the RIO challenge the APX-76 by depressing the IFF DISPLAY pushbutton on the DDD panel. When challenged, the legend will be deleted from the display.

Class IIA is enabled by the pilot via his MASTER TEST panel with weight-on-wheels, TAS <70 knots, pilot's OBC discrete, throttles at IDLE, and handbrake set. Class IIB is enabled by weight-on-wheels only. The weight-on-wheels BITS are constantly checked for change in status to ensure proper class testing. Class I tests are performed in flight only because they radiate power. Class II tests are designated for preflights because a failure of these systems constitute flight safety hazards. Class III tests do not radiate energy and are not considered safety of flight components.

The fault display following Class I OBC contains a CIA acronym followed by a four-digit number representing the time, in milliseconds, the AWG-15 was in self-test. If an AWG-15 failure was detected, the appropriate WRA acronyms will be displayed following the test time display. AWG-15 self-test is terminated if a failure is detected in the control indicator (CI) or the power switch unit (PSU). A CI or PSU acronym and the test time can be used to perform a limited degraded mode assessment of remaining system capabilities.

Note

After completion of the last test cycle, the MASTER TEST switch must be set to OFF. If left in the OBC position, the weight-on-wheels interlock prevents the CSDC from commanding AFCS, APC, AICS, and CADC self-test.

Maintenance Readout

The maintenance readout (MR) routine cycles through the failure history file and displays failure acronyms at the WRA level on the TID, two-thirds page format. Upon depression of the MAINT DISPL pushbutton on the computer address panel, the routine will branch directly to the MR routine. When the routine is completed, the words TEST COMP are displayed under the last acronym and the routine terminates.

The maintenance readout routine has the capability to clear the WCS computer and the CSDC failure history file upon command from the RIO. This feature is enabled on the ground only. To clear the file, the RIO calls up MAINT DISPL, as before, and then depresses the OBC DISPL

F-14A TOMCAT

THIS PAGE INTENTIONALY LEFT BLANK.

OBC BLOCK DIAGRAM

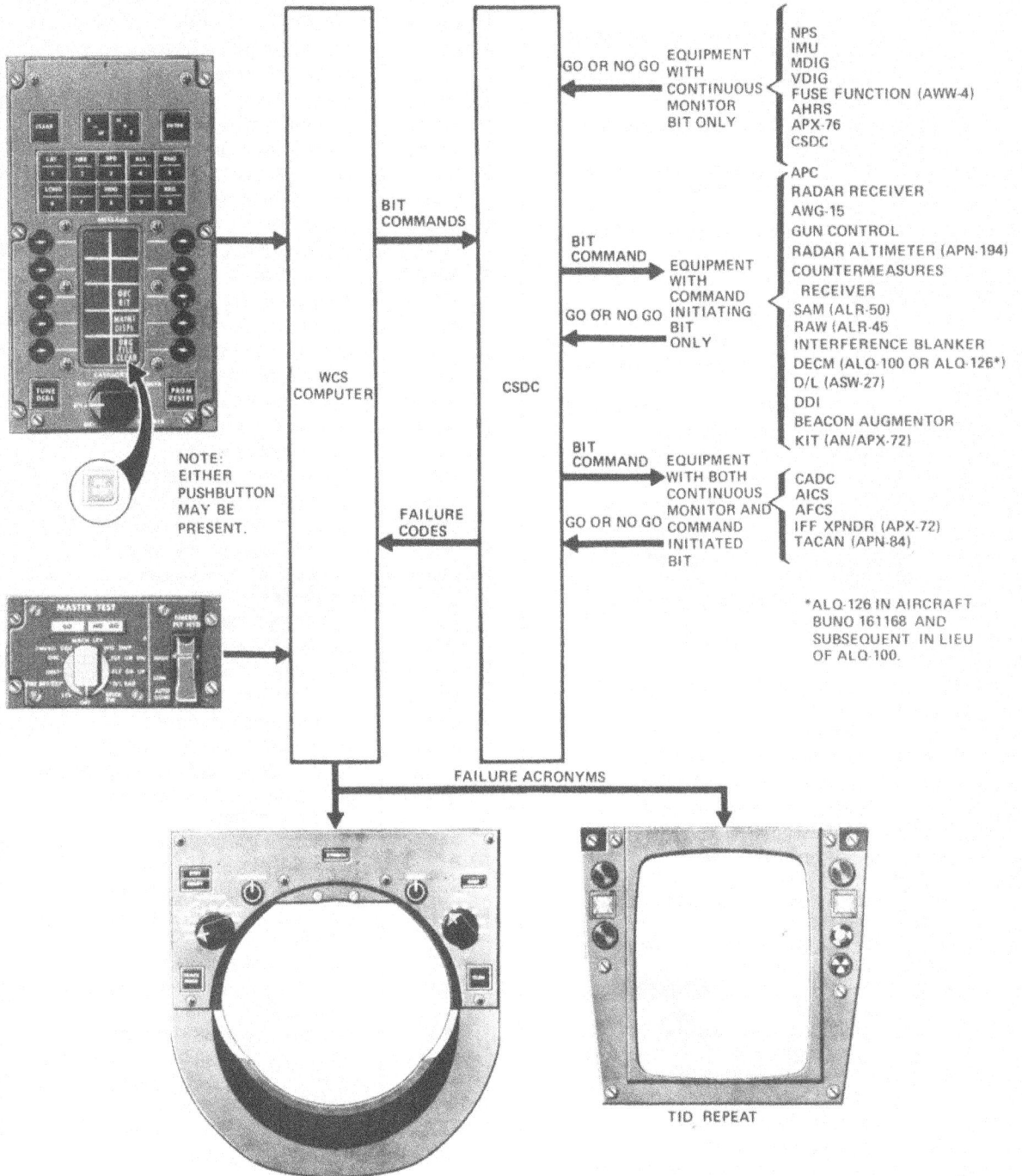

NOTE:
EITHER
PUSHBUTTON
MAY BE
PRESENT.

WCS
COMPUTER

CSDC

BIT
COMMANDS

GO OR NO GO | EQUIPMENT
WITH
CONTINUOUS
MONITOR
BIT ONLY

NPS
IMU
MDIG
VDIG
FUSE FUNCTION (AWW-4)
AHRS
APX-76
CSDC

BIT
COMMAND | EQUIPMENT
WITH
COMMAND
INITIATING
GO OR NO GO | BIT
ONLY

APC
RADAR RECEIVER
AWG-15
GUN CONTROL
RADAR ALTIMETER (APN-194)
COUNTERMEASURES
 RECEIVER
SAM (ALR-50)
RAW (ALR-45
INTERFERENCE BLANKER
DECM (ALQ-100 OR ALQ-126*)
D/L (ASW-27)
DDI
BEACON AUGMENTOR
KIT (AN/APX-72)

BIT
COMMAND | EQUIPMENT
WITH BOTH
CONTINUOUS
MONITOR AND
GO OR NO GO | COMMAND
INITIATED
BIT

CADC
AICS
AFCS
IFF XPNDR (APX-72)
TACAN (APN-84)

FAILURE
CODES

*ALQ-126 IN AIRCRAFT
 BUNO 161168 AND
 SUBSEQUENT IN LIEU
 OF ALQ-100.

FAILURE ACRONYMS

TID REPEAT

8-F-050-004
194-0

Figure 9-2. OBC Block Diagram

ACRONYM	SUBSYSTEM WRA LEVEL	ACRONYM	SUBSYSTEM WRA LEVEL
AFC AM	AUTO FLIGHT CONTROL SYSTEM ACCELEROMETER	CIA A1	CIACS A DECODER STATION NO. 1 RAILS
AFC PA	AUTO FLIGHT CONTROL SYSTEM PITCH ACTUATOR	CIA A3	CIACS A DECODER STATION NO. 3 RAILS
AFC PC	AUTO FLIGHT CONTROL SYSTEM PITCH COMPUTER	CIA A4	CIACS A DECODER STATION NO. 4 RAILS
		CIA A5	CIACS A DECODER STATION NO. 5 RAILS
AFC PS	AUTO FLIGHT CONTROL SYSTEM PITCH SENSOR	CIA A6	CIACS A DECODER STATION NO. 6 RAILS
AFC RA	AUTO FLIGHT CONTROL SYSTEM ROLL ACTUATOR	CIA A8	CIACS A DECODER STATION NO. 8
		CIA CI	CIACS CONTROL INDICATOR
AFC RC	AUTO FLIGHT CONTROL SYSTEM ROLL COMPUTER OR LATERAL ARI COMPUTER	CIA B2	CIACS B DECODER STATION NO. 2
		CIA B3	CIACS B DECODER STATION NO. 3 AND 6
AFC RS	AUTO FLIGHT CONTROL SYSTEM ROLL SENSOR	CIA B4	CIACS B DECODER STATION NO. 4 AND 5
AFC YA	AUTO FLIGHT CONTROL SYSTEM YAW ACTUATOR	CIA B7	CIACS B DECODER STATION NO. 7
		CIA PSU	CIACS PWR SWITCH UNIT
AFC YC	AUTO FLIGHT CONTROL SYSTEM YAW COMPUTER OR RUDDER ARI COMPUTER	CIA CPA	CIACS POWER A
		CIA CPB	CIACS POWER B
AFC YS	AUTO FLIGHT CONTROL SYSTEM YAW SENSOR	CSD	CSDC FAILURE
		CSI	SSI FAILURE (SYNCHRONIZER 3, 4, OR 5)
AHR	AHRS HEADING AND VERTICAL FAILURE	DDI	DIGITAL DATA INDICATOR
AICL A1	AICS LEFT INLET ACTUATOR	DIG C	VDIG CONVERTER-VDI
AICL A2	AICS LEFT INLET ACTUATOR	DIG CH	VDIG CONVERTER-HUD
AICL A3	AICS LEFT INLET ACTUATOR	DIG IH	VDIG CONVERTER-HUD
AICL A4	AICS LEFT BLEED DOOR ACTUATOR	DIG I	VDIG INDICATOR-VDI
AICL P	AICS LEFT INLET PROGRAMMER	DLS	UHF D/L TRANSCEIVER AN/ASW-27B
AICL S1	AICS LEFT INLET SENSOR (STATIC PRESSURE)	DSM E	MDIG-ECM INDICATOR FAILURE
		DSM H	MDIG-HSD INDICATOR FAILURE
AICL S2	AICS LEFT INLET SENSOR (TOTAL PRESSURE)	DSM P	MDIG PROCESSOR FAILURE
AICL S3	AICS LEFT INLET SENSOR (TOTAL TEMPERATURE	ECM	DEFENSIVE ECM (ALQ-100 OR ALQ-126 FAILURE)
AICL S4	AICS LEFT INLET SENSOR (ANGLE OF ATTACK)	FSE	FUZE FUNCTION
		GCS	GUN CONTROL UNIT
AICR A1	AICS RIGHT INLET ACTUATOR	IFA	COMPUTER IFF (APX-72)
AICR A2	AICS RIGHT INLET ACTUATOR	IFB	INTERFERENCE BLANKER
AICR A3	AICS RIGHT INLET ACTUATOR	IFI CH	IFF CHALLENGE (APX-78)
AICR A4	AICS RIGHT BLEED DOOR ACTUATOR	IFI RT	IFF INTERROGATOR AN/APX-76 RT
AICR P	AICS RIGHT INLET PROGRAMMER	IFI SW	IFF INTERROGATOR AN/APX-76 SW/AMP
AICR S1	AICS RIGHT INLET SENSOR (STATIC PRESSURE)	IFI SY	IFF INTERROGATOR AN/APX-76 SYNCHRONIZER
AICR S2	AICS RIGHT INLET SENSOR (TOTAL PRESSURE)	IFN	COMPUTER IFF (APX-76)
		IFX	IFF TRANSPONDER AN/APX-72
AICR S3	AICS RIGHT INLET SENSOR (TOTAL TEMPERATURE)	IMU	INERTIAL MEASUREMENT UNIT
AICR S4	AICS RIGHT INLET SENSOR (ANGLE OF ATTACK)	NPS	INERTIAL NAVIGATION SYSTEM POWER SUPPLY
APC A	APPROACH POWER CONTROLLER ACCELEROMETER	OBC	OBC FAILURE (CSDC)
		RAW A	RADAR RECEIVER (AN/ALR-45 OR APR-25) ANALYZER
APC C	APPROACH POWER CONTROLLER COMPUTER	RDA	RADAR ALTIMETER CONTROL
		SAM	RADAR RECEIVER (APR-27 OR ALR-50)
BAG	BEACON AUGMENTOR	TCN	TACAN, CONTINUOUS MONITOR/ COMMANDED
CAA	CSDC ANALOG FAILURE	WOW	WEIGHT-ON-WHEELS FAILURE
CAD	CENTRAL AIR DATA COMPUTER (CADC)	WRA	ILLEGAL FAILED BIT

Figure 9-3. OBC and Maintenance Readout Display Acronyms

OBC DISPLAY

AWG-9
CONTINUOUS
MONITORING
DISPLAY (2
CHARACTERS)

OBC
CONTINUOUS
MONITORING
DISPLAY
(3 CHARACTERS)

HO
CRO

CONTINUOUS MONITORING

OBC MASK
ENABLE

NOTE
RAMP
ACRONYM
SHOWS AICS
DETECTED
FAILURE.

LN 3057 LW11 5067
00517 40004064
01
52   RAMP

IMU
AICL A3
DIG CH
ROR
DDI
RAW R
OLS
IFI SW
TEST COMP
DIS POBC

CM +

RH

NAV STATUS

MAINTENANCE DISPLAY

ALIGNMENT
TICKS

DISPLAY
READOUTS
AND
FLYCATCHER

LN 3057 LW11 5067
00517 40004064
01
52

ECM AFC RAW TCN
IFX AIC ROR DDI
CIR APC   GCS
BRG CRO   DLS
               SAM
               IFB
POBC DIS
CHRL IFF

SAT DISPLAY — COARSE ALIGN

OBC
MASK
DISABLED

LN 3057 LW11 5067
00517 60004064

IMU
SAM
DIG CH
ROR
CM − DDI
RAW
OLS
IFI SW
TEST COMP
DIS POBC

RH

NAV STATUS

**OBC DISPLAY — FAULT DISPLAY
(NOT IN SAT MODE)**

9-F50-196-0

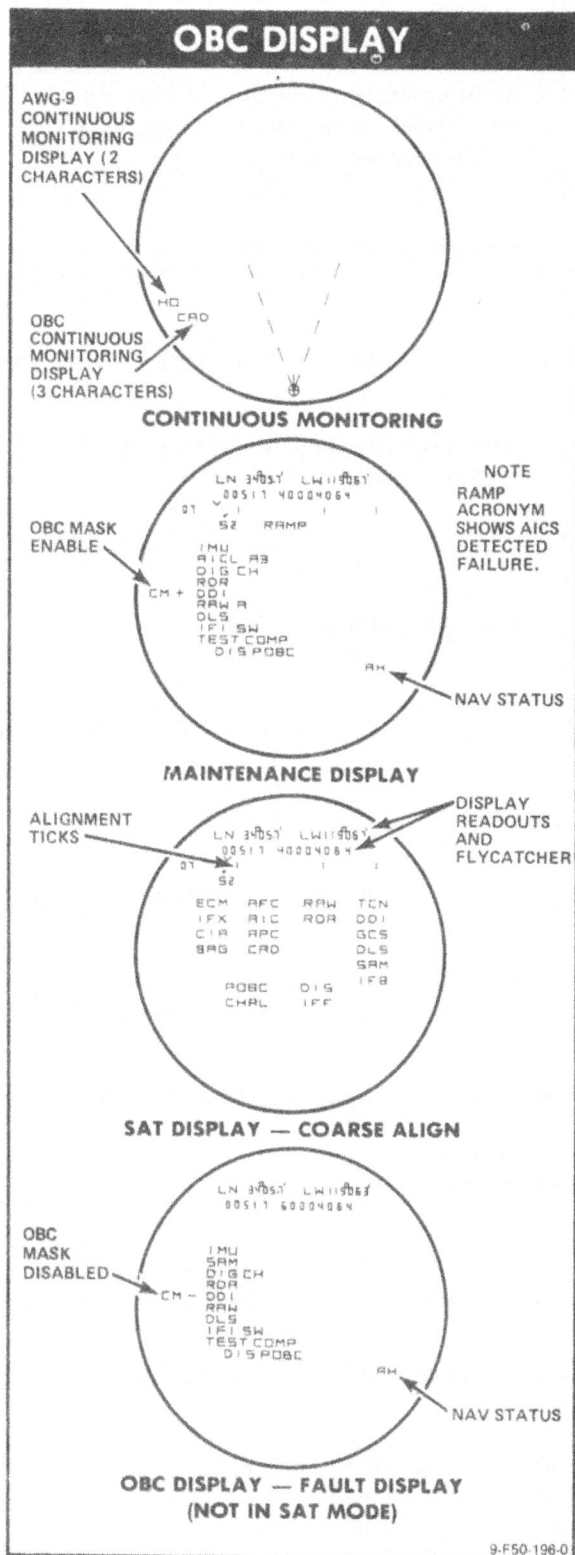

Figure 9-4. OBC Display

function pushbutton. The word CLEARED is then displayed below the maintenance readout display (figure 9-4).

OBC Masking

Masking is the AWG-9 software method of inhibiting known WRA faults from producing OBC system acronyms in CM (except the CSI and CSD acronyms).

To set the OBC fault mask:

1. Perform OBC BIT or OBC maintenance display and wait until TEST COMP is displayed on the TID.

2. Enter CLEAR 7 ENTER on the AWG~9 keyboard.

Note

To enable CM mask during SAT, SPL category must be selected.

3. The TID will display CM+ at the 9-o'clock position and acronyms will be masked from CM display.

To clear OBC fault mask:

1. Perform OBC BIT or OBC maintenance display and wait until TEST COMP is displayed on the TID.

2. Enter CLEAR 77 ENTER on the AWG-9 keyboard.

3. The display CM− will appear on the TID in the 9-o'clock position, indicating the CM mask is cleared.

4. Mask will be cleared by running OBC or by initiating AWG-9 OBC file clear.

OBC Continuous Monitoring

The continuous monitoring (CM) routine is an integral part of the AWG-9 tactical program. It displays system failure information on the TID during normal tactical processing. The routine is permanently stored in nondestructive readout file (NDRO) and is cycled every 2 seconds.

If an aircraft failure occurs, the appropriate three-letter acronym (figure 9-7) for this failure is displayed on the lower left quadrant of the TID below the two-letter AWG-9 acronym (figure 9-4). If multiple failures occur, the appropriate acronyms are automatically displayed at a 2 second rate.

CLASS	FLIGHT STATUS	EQUIPMENT MONITORED
I	INFLIGHT (WEIGHT-OFF-WHEELS, TAS > 70 KNOTS)	DECM (AN/ALQ-100 OR AN/ALQ-126), IFF TRANSPONDER (AN/APX-72), CIACS, AND BEACON AUGMENTOR
IIA	PREFLIGHT ONLY (WEIGHT-ON-WHEELS, HAND BRAKE SET, TAS < 70 KNOTS, AND PILOT'S OBC DISCRETE) THROTTLE AT IDLE	AICS, AFCS, CADC, AND APC
IIB	PREFLIGHT ONLY (WEIGHT-ON-WHEELS)	COUNTERMEASURES RECEIVER (ALR-45), AND RADAR ALTIMETER (APN-194)
III	PREFLIGHT OR INFLIGHT	TACAN (ARN-84), DDI, GUN CONTROL UNIT, D/L (ASW-27), RADAR RECEIVER (ALR-50), AND INTERFERENCE BLANKER

Figure 9-5. OBC Classifications

METHOD	ENTER	EXIT	RERUN
SAT OBC - Automatic	Set NAV MODE switch to ALIGN prior to completion of auto BIT sequence 2.	Depress PRGM RESTRT pushbutton and set NAV MODE switch to any position but ALIGN.	Depress PRGM RESTRT pushbutton and set NAV MODE switch ALIGN; depress OBC.
SAT OBC - Manual	SMAL or OBC-initiated second mode entry causes SAT after auto BIT sequence 2.	Manual selection of BIT sequence 2. Weight-off-wheels automatically aborts SAT. OBC will continue to Class IIB or Class III.	Depress OBC in SPL category.
Non SAT OBC	Depress OBC in SPL category.	Rotate category switch out of SPL.	Depress OBC in SPL category. Depress PRGM RESTRT pushbutton.
Note			
Rerunning OBC by PRGM RESTRT pushbutton will require that OBC be depressed in the SPL category. If this is not done, only initialization of the alignment will occur and OBC will not return. Preferred method is to depress OBC in SPL category.			

Figure 9-6. SAT and Non SAT OBC Operations

ACRONYM	DEFINITION	EXPLANATION
AFC	Automatic Flight Control	With caution lights — See PCL.
AM		Yaw accelerometer fail.
PA		Pitch actuator position does not agree with command.
PC		Pitch computer — For aircraft with ARI alpha probe and AFC 400/488, ignore PC with autopilot caution light. For aircraft without ARI alpha probe and AFC 400/488, ignore PC when no caution light is illuminated. Caution lights are valid failures.
PS		Pitch sensor failure.
RA		Roll actuator position does not agree with command.
RC		Roll Computer — For aircraft without ARI alpha probe and AFC 400/488, ignore RC when no caution light is illuminated.
YC		Yaw computer — For aircraft with ARI alpha probe and AFC 400/488, check that circuit breaker is engaged. ALPHA COMP/PEDAL SHAKER cb (7C8).
YS		Yaw sensor failure.
AHR	Attitude Heading Reference System	AHRS failure. Computer will select best NAV mode available. Select COMP mode for flux valve MAG HDG.
AICL or AICR	Air Inlet Control Left or Right	Air INLET RAMPS switch is stow without other acronyms. With other acronyms call up maintenance file.
P		Programmer fail. Without INLET light computer uses normal values. Operational mode no flight restriction.
A1		No. 1 ramp actuator position does not agree with command.
A2		No. 2 ramp actuator position does not agree with command.
A3		No. 3 ramp actuator position does not agree with command.

Figure 9-7. OBC Fail Acronyms (Sheet 1 of 5)

ACRONYM	DEFINITION	EXPLANATION
A4		MCB may remain on at greater than 0.35 Mach.
AIC		Static Pressure.
S1		With INLET light, fail safe mode.
		Without INLET light, fail operational. No flight restriction.
S2		Total pressure.
		With INLET light, fail safe mode.
S3		Total temperature.
		Without INLET light, fail operational mode. No flight restriction.
S4		Angle-of-Attack.
		Without INLET light, fail operational mode. No flight restriction.
APC	Approach Power Compensator	Auto throttle inoperative.
		System will shift to BOOST automatically.
A		APC accelerometer. No associated light. Auto throttle inoperative. APC not authorized for landing.
C		APC computer fail.
		Auto throttle inoperative.
BAG	Beacon Augmenter	System not on. Run manual BIT.
		Degraded precision approach (ACL) and/or ground vectoring.
CAD	Central Air Data Computer	Check caution/advisory lights.
		Run CADC flycatchers.

Figure 9-7. OBC Fail Acronyms (Sheet 2 of 5)

ACRONYM	DEFINITION	EXPLANATION
CIA	Coded Integrated Armament System	Call up Maintenance File. See AWG-15 BIT. Valid after inflight OBC. Degraded AWG-15 operation. CIA A1, A3, A4, A5, A6, A8, B3, and B4 degraded station operation. C1A B2 and B7 are valid only if d op tank MXU 611 is installed.
C1		Control indicator failure.
PSU		Power switching unit failure.
CIA CPA	CIACS Power A	Neither normal weapons release nor gun are available. (Ignore if COATS not installed). EMER JETT available.
CPB	CIACS Power B	Weapons, gun and jettison modes are available. (Ignore if COATS not installed).
CSD	Computer Signal Data Converter	See CSD in troubleshooting guide.
CSI	Standard Serial Interface Failure	See CSI in troubleshooting guide.
DDI	Digital Data Indicator	Data link not on. Run manual BIT.
DIG	Vertical Display Indicator Group	Evaluate displays for flight. Acronyms may indicate degraded mode but useable. Reset BIT indicators.
C	Vertical Display Converter	Evaluate displays.
CH	HUD Converter	Evaluate displays.
I	Vertical Display Indicator	Evaluate displays.
IH	HUD Indicator	Evaluate displays.
DLS	Data Link System	Data link off. Run manual test. Run D/L Rad test.
DSM	Multiple Display Group	Evaluate displays for mission requirements.
E	ECM Indicator	Defensive ECM may not be available.

Figure 9-7. OBC Fail Acronyms (Sheet 3 of 5)

ACRONYM	DEFINITION	EXPLANATION
H	HSD Indicator	Horizontal display may be degraded.
P	HSD/ECMD Processor	HSD/ECMD may not be available.
ECM	Electronic Countermeasures	ALQ-100/126 failure. Defensive ECM may not be available. Run manual BIT.
FSE	Electrical Fuses	Electric bomb fuses cannot be set.
GCS	Gun Control System	Trigger shorted. See AWG-15 BIT BUNO 158978 and subsequent, normal after trigger activation.
IFA	IFF Computer APX-72	IFF failure. Run IFF manual test.
IFB	Interference Blanker	Possible interference between TACAN, Radar altimeter, IFF, DECM, AWG-9.
IFI	Air to Air Interrogator APX-76	APX-76 not on/installed. APX-76 failure.
IFI CH	Challenge	Air-to-Air Interrogate not available.
RT	RT unit	
SW	Switch Amplifier	
SY	Synchronizer Unit	
IFN	APX-76 Computer	APX-76 computer failure/not installed.
IFX	IFF Transponder APX-72	IFF failure. Master switch to NORM. Select test for each mode and observe test light.
IMU	Inertial Measuring Unit	IMU failure. Degraded missile and navigation capability.
NPS	Navigation Power Supply	Possible IMU failure.

Figure 9-7. OBC Fail Acronyms (Sheet 4 of 5)

ACRONYM	DEFINITION	EXPLANATION
RAMP	Air Inlet Control System	OBC detected failure of the air inlet control system. Refer to maintenance file for AICS related acronym.
RAW	Radar Warning (ALR-45)	Verify system on/installed. Run manual test.
RDA	Radar Altimeter	Verify system on. Run manual test.
SAM	Surface-To-Air Missile (ALR-50)	Verify system on/installed. Run manual test.
TCN	TACAN	Verify System On. Run manual test. Evaluate TACAN display on BDHI.
WOW	Weight on/off Wheels	Failure of weight on/off wheels switches. Refer to WOW in emergency section.

Note

Before correcting a malfunction, ensure available flycatcher/CM data is recorded.

Figure 9-7. OBC Fail Acronyms (Sheet 5 of 5)

F-14A TOMCAT

THIS PAGE INTENTIONALY LEFT BLANK.

Note

A CSD acronym appearing on the TID
may indicate failure of the primary
attitude reference. When an acronym
appears, the VDI and the standby
attitude indicator should be cross-checked
to determine if the primary reference
system is degraded. If the VDI is
erroneous, the standby attitude indicator should be used as the primary
reference instrument and instrument
flight should be avoided.

BUILT-IN TEST (BIT)

The built-in test (BIT) is designed to assist the flightcrew
in determining if the system is operational and capable of
performing an assigned mission.

BIT Capabilities

The BIT has the following four major capabilities:

- Fault detection (FD) – uses eight computer-controlled
  and RIO-initiated test sequences to detect system
  failures in flight or on the deck.

- Fault isolation (FI) – enables the operator to isolate
  a detected system failure by indicating decision
  points (DP) and suspect WRAs.

- Degraded modes assessment (DMA) – provides a pass,
  fail, or degraded evaluation of the operational modes
  at the completion of sequence BIT 3. BIT 2
  sequence will also post the DMA of the last computed sequence BIT 3 if one has been run since the
  AWG-9 power up.

- Continuous monitoring (CM) – automatically provides the RIO with a warning when system failures
  occur during tactical modes.

Types of Tests

These BIT capabilities are incorporated into the four types
of test within BIT: confidence test, detailed radar tests,
continuous monitoring tests, and special tests. The confidence tests and the detailed radar tests include keyboard
digital entries to enable partial or complete rerun of the
tests. BIT sequence 1 uses digital entry to perform the
dynamic portions of the display tests without the need for
additional MTM read-ins. BIT sequence 3 does not have
this digital entry capability. BIT sequence 2 and 4 read in

from the MTM with digital entries. BIT sequence 5, 6,
7, and 8 can be partially or completely rerun by digital
entry without additional MTM tape read-in.

CONFIDENCE TESTS

The confidence tests provide the RIO with a rapid check of
all essential system functions utilizing the fault detection,
fault isolation, and DMA capabilities of the system. The
confidence tests provide a subsystem-by-subsystem check
of the AWG-9. The RIO may perform these tests before
takeoff in order to verify that the system is capable of
performing the assigned mission. After a mission, the
confidence tests may be repeated by the RIO to determine
if the WCS is still operating properly and can be scheduled
for another mission. These postflight confidence tests may
be performed before or after landing. The confidence tests
consist of four test sequences:

- SEQUENCE 1 - Displays test (DISPL 1)

- SEQUENCE 2 - Computer test (CMPTR 2)

- SEQUENCE 3 – AWG-9 confidence tests
  (AMCS CONF 3)

- SEQUENCE 4 – Missile auxiliaries subsystem and
  missile-on-aircraft test (MAS-MOAT).

These tests are initiated when the RIO selects the BIT
mode, using the CATEGORY select switch on the computer address panel, and depresses the appropriate
MESSAGE select pushbutton. These four test sequences
can be run in any order convenient to the RIO but
sequence 2 and sequence 3 should be run prior to
sequences 5 through 8. Refer to the following section on
magnetic tape memory for the optimum sequence to perform BIT.

DETAILED RADAR TESTS

The detailed radar tests provide the RIO with an extensive
test of the radar subsystem. The RIO may choose to perform these tests based on the results of the confidence tests.
Sequence 3 of the confidence tests performs a rapid check
of the radar subsystem. If a fault is detected during this
rapid check, but cannot be isolated, the sequence 3 display
may recommend detailed radar tests be performed. The
detailed radar tests consist of four test sequences:

- SEQUENCE 5 – Receiver test (RCVR 5)

- SEQUENCE 6 – Transmitter test (XMTR 6)

- SEQUENCE 7 - Radar antenna servo test (ANT)

- SEQUENCE 8 - Single target track set (STT 8)

Continuous Monitoring Tests

The continuous monitoring (CM) fault display is in the lower left quadrant of the TID and consists of two alpha symbols for a radar or MAS function that failed. Each CM fault detected is displayed for 2 seconds, and then the CM program is continued. Basic AWG-9 related CM symbology is shown in figure 9-8.

In the program, a test monitors missile-separated and missile-not-separated flags with a display of MA on the TID if both flags are not the same except during the LTE cycle.

Note

When performing BIT sequences, refer to section III, part 2, for standard operating procedures.

SPECIAL TESTS

Special tests are provided in the program primarily to aid the maintenance crew. All special tests are entered via the SPL TEST selection of the BIT category, then NBR XX, then ENTER. The special tests provide the operational interface with AIM-7 and AIM-9 armament test sets, individual processing of computer inputs and outputs, and computer memory address inspection. Tests 90, 91, 94, 97, 98, and 99 cannot be entered when airborne. Test 101 will function the same as 100, if airborne. Special tests are as follows:

● NBR-90 (SPAM)	SIP-SOP, PIP-POP, AI-AO, and memory inspection
● NBR-91 AWM-71 test set	Tests CWI noise and modulation
● NBR-94	MATS/MITS interface test set
● NBR-97 (AIM-7 SMTS)	AIM-7 missile station test set (GAC)

CONTINUOUS MONITOR DISPL	FAULT DESCRIPTION
CS	COMPUTER NON DESTRUCTIVE READOUT CHECKSUM ERROR INDICATING COMPUTER MEMORY FAILURE PROGRAM FUNCTIONS MAY BE UNRELIABLE.
HO	CLUTTER TERM SATURATED. DEGRADED OR NO DOPPLER PROCESSING CAPABILITY.
MA	MISSILE AUXILIARY MALFUNCTION FAULT IS DISPLAYED CONTINUOUSLY. MISSILE IRREVERSIBLES NOT IN NORMAL READY-TO-LAUNCH STATE IF DISPLAYED FOR 2 OUT OF 6 SECONDS (NORMAL WITH TNG SELECTED).
PM	LOW FLOW IN THE AWG-9 COOLANT LOOP TO THE TRANSMITTER.
RP	RADAR POWER FAULT INDICATION.
SA	RADAR MASTER OSCILLATOR FAILED TO TUNE SEMIACTIVE CHANNEL BECAUSE OF DECODER CIRCUIT FAILURE.
T	BIT RADOME HORN TARGET TURNED ON.
XM	TRANSMITTER PEAK POWER BELOW MINIMUM ACCEPTABLE LEVEL OR TRANSMITTER IS NOT SELECTED ON.
XO	NO PHASE LOCK ON TRANSMITTER CHANNEL SELECTED. NO RECEIVE CAPABILITY.

Figure 9-8. AWG-9 Acronyms

- NBR-98　　　　　　　　AIM-9 SEAM test set
  (AIM-9 SEAM)

- NBR-99　　　　　　　　AIM-7 missile station
  (AIM-7 MSTS)　　　　　test set (Navy)

- NBR-100　　　　　　　Recalls transmitter over-
  　　　　　　　　　　　loads and/or power
  　　　　　　　　　　　faults

- NBR-101　　　　　　　Clears SPL TEST 100
  　　　　　　　　　　　on the deck (WOW)

Note

- Sacred cells are cleared upon entering
  Special Test 101 on the deck; displays
  remain until next entry.

If a radar transmitter fault is detected, the continuous
monitor (CM) routine stores the transmitter status in a
sacred cell, which maintenance personnel can read using
Special Test 100 (ST 100).

Transmitter fault monitoring is called at an 8-msec rate.
Also, only detected faults will be written in the sacred
cell. This will ensure a cumulative record of faults during
a mission.

MAGNETIC TAPE MEMORY (MTM) FAILURE INDICATION

If the magnetic tape unit fails to run and respond with an
enabled signal when commanded to run, an E will be displayed instead of a B, M, N, I, or X. To provide the proper
perspective for the display of the different acronyms, the
operational logic for each is briefly described.

The magnetic tape memory is a long single strip (figure 9-9).
There are two identical copies of the data on the tape. The
first copy is the M copy; the second, the B copy. As shown
in figure 9-9, the individual program files are sequentially
located on each copy. Data cannot be used directly from
the tape; it must be loaded into the destructive readout
(DRO) memory first. This is done by running the tape

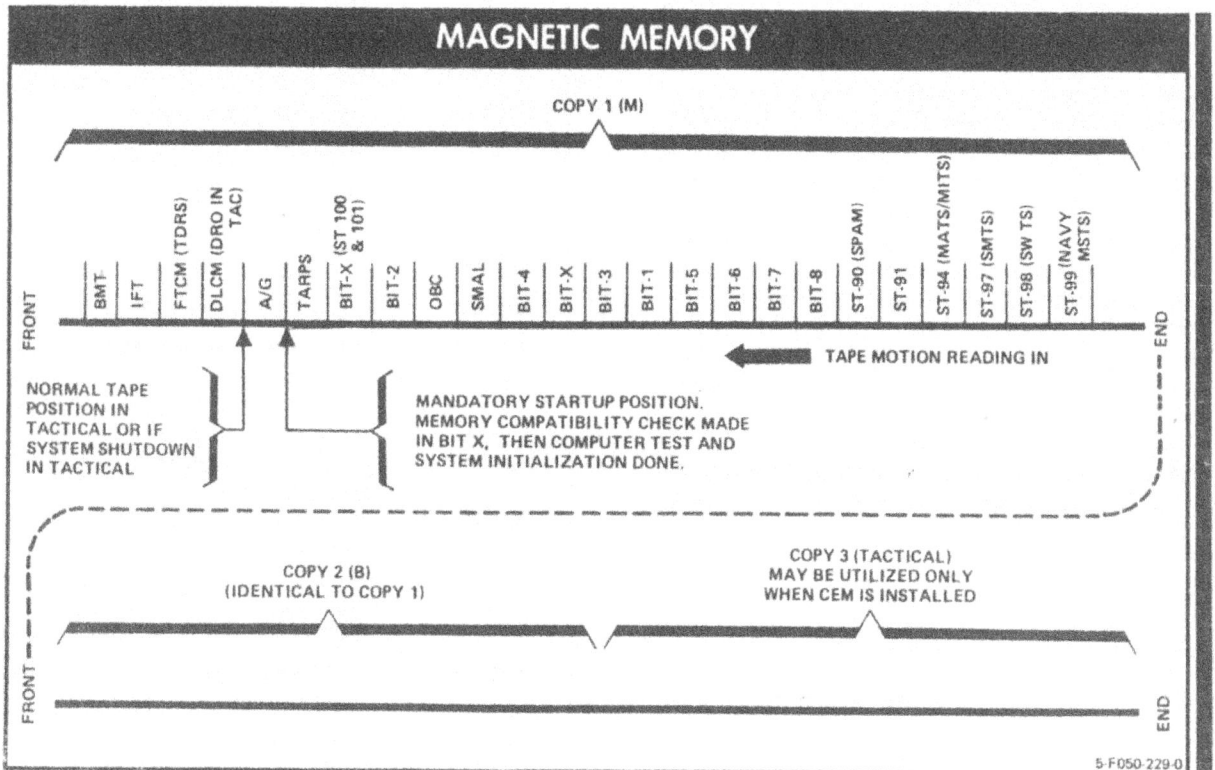

Figure 9-9. Magnetic Memory

through a tape read head similar to a home recorder. While the tape is searching for the right file and loading the file into the DRO, it displays an M or B, depending on which copy is being used.

Depending on where maintenance left the tape, either copy may be used by the system, and will continue to be used, until a problem is encountered. After a block of data is read in, the program checks the sum and, if it is wrong (such as might be caused by a weak recording or loss of oxide from the tape surface), it tries again. If after five tries it still gets the wrong sum, it assumes the copy is defective and runs to the other copy. When it does this, it displays the letter of the new copy and runs the data in. If a problem is encountered with the new copy, it will go back to the original copy and try (it does not remember that a copy is defective).

If, when commanded to run, the MTM fails to run, an E will be displayed on the TID. This means that: (1) no new data can be run in; (2) the MTM is still being commanded to run and, if the trouble is of a transitory nature, may recover; and (3) the present system status can be maintained if power is not recycled. If an E appears, even momentarily, it should be reported to maintenance.

Two separate computer programs are loaded into the computer memories. Their programs must be compatible. Since these memories are loaded into two different units, there is a possibility of having one tape configuration in one unit and another tape configuration in the other unit. To prevent this situation, the computer checks a number from one memory against a number in the other memory when BIT executive file is read in. If they do not match, the program displays an X and goes into an idle loop (same as a holding pattern), awaiting further instructions. The only command it will accept is PRGM RESTRT, which causes the program to proceed as though nothing were wrong. However, there is no way of telling what the results might be. The incompatibilities may be few or many; they might show up in the displays or subtly interfere with a missile launch. When the program displays an X, request maintenance action. When the FTC load is in TAC DRO and no data link capability exists, an I will be displayed on the TID.

Note

With the reordered BIT sequence 1 thru 4, the complete set of confidence checks should be run in the following order to minimize time; 2, 4, 3 and 1.

FLYCATCHER

Flycatcher is a computer routine that allows the RIO to examine the contents of specific CADC, CSDC, or WCS memory locations. The difference between CADC, CSDC, or WCS is the numerical designation of the different memory locations (figure 9-10). The flycatcher routine is entered using the computer address panel. The specific memory location and its contents, which are in octal numbers are displayed on the TID. (See figure 9-11.) Figure 9-12 shows the interpretation of the displayed flycatcher data.

LOCATION	DATA IN LOCATION
7-00107	Transmitter Power
7-00157	Radar Overloads
7-00166	Transmitter Overloads (When difficulty is encountered entering Special Test 100 or 101)
7-00426	MSL Directional Problems
7-00427	MSL Type and Station Loading
7-00434	Missile Select and AIM-9 Scan and Slave Data
7-00435	MSL Power and Tuning AIM-54 and AIM-7
7-00444	MSL Parameters
7-00445	MSL MOAT Parameters
7-00504	Navigation Alignment Data
7-00507	BIT Lift Accelerometer
7-00517	Navigation Modes
7-00537	INS Wander Angle
7-00555	Weight-on-Wheels Check
7-00646	Missile Station Select Indicators
7-00647	Rapid Lock-on Problems, Missile Gate Aspect Select, and Weapon Type Select
7-00650	AIM-54 Ready, AIM-7 Tuned ACM Launch, Training Select, and SEAM LOCK indicators
7-16735	AVC 2102 incorporated (Should read XXXXXXX1)
7-17273	Automatic Sequence 2 and DMA Initialization problems (Should read 7XX77777)
7-77776	AWG-9 Checksum
71-00101	CSDC Self Test Loop
71-00750	CSDC Fail Word
71-11777	CSDC Checksum

Figure 9-10. Typical Flycatcher Locations

FLYCATCHER DISPLAY

OCTAL ADDRESS
ENTERED (NAV ALIGN
DATA FLYCATCHER)

ADDRESS CONTENTS
(OCTAL CODED)

RS   450   MH 175
00517 00000000

+00.0

730

DIGIT 1
OCTAL CODE: 1

DIGIT 4
OCTAL CODE:
4 AND 1

DIGIT 8
OCTAL CODE:
4, 2 AND 1

RS 77 H 17
00150 10050007

READ

LEFT   RIGHT

+00.0

730

3 FS0 221-0

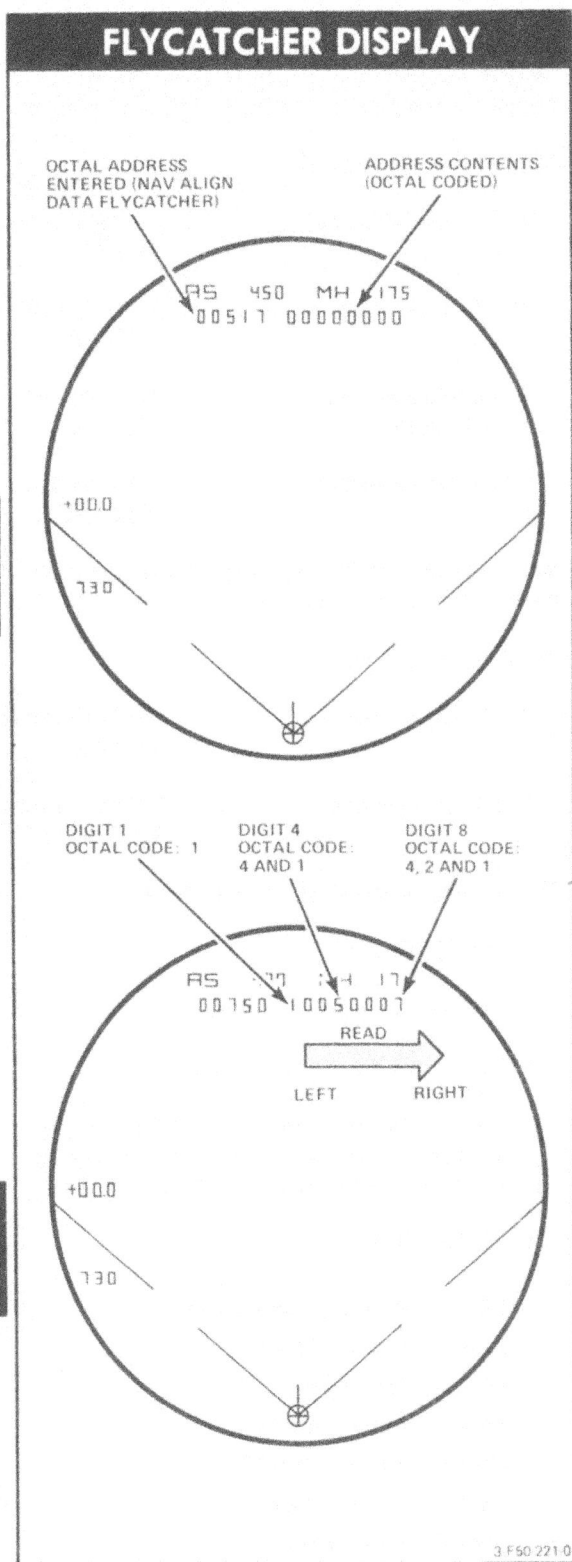

Figure 9-11. Flycatcher Display

OCTAL DISPLAYED	INTERPRETATION
1	INDICATES CONDITION FOR 1 EXISTS
2	INDICATES CONDITION FOR 2 EXISTS
3	INDICATES CONDITION FOR 1 AND 2 EXISTS
4	INDICATES CONDITION FOR 4 EXISTS
5	INDICATES CONDITION FOR 1 AND 4 EXISTS
6	INDICATES CONDITION FOR 2 AND 4 EXISTS
7	INDICATES CONDITION FOR 1, 2, AND 4 EXISTS

Note

The octal numbers describe the existence of three possible situations or some composite of the three.

Figure 9-12. Decoding Flycatcher

CSDC Flycatcher

The CSDC flycatcher routine is selected by entering 71, and then entering a five-digit octal number. The five-digit number is the octal address of a specific memory location under interrogation. Once entered, the TID displays the contents of the memory location in octal form as the WCS interrogates the CSDC memory location. Refer to NAVAIR 01-F14AAA-1B for pertinent CSDC flycatcher locations.

CADC Flycatcher
(Buno 159825 and subsequent and aircraft incorporating AFC 417 and AVC 1992)

The CADC flycatcher routine provides a rapid inflight and flight line identification of CADC related malfunctions. Use of CADC flycatcher will allow sounder judgments to be made following CADC failures. CADC flycatcher routines are selected in the same manner as the CSDC flycatcher routine. Figure 9-13 shows the major failure functions for each of the three words that can be called up. Refer to NAVAIR 01-F14AAA-1B for details indicated in the CADC flycatcher locations.

WARNING/ CAUTION/ ADVISORY	CAP ADDRESS	MAJOR FAILURE FUNCTION
CADC	7100037	Mach Lever, ECS, Input Discretes, Sum Check
CADC, GLOVE VANE, WING SWEEP	7100027	Wing Sweep, Glove Vane, Tail Auth, ECS, RIO Altimeter, Total Temperature, CSDC Output
CADC, FLAP, MACH TRIM	7100025	Flaps, Autopilot, Pilot Altitude, Angle of Attack, Qc Monitor

Figure 9-13. CADC Flycatcher

Note

If the CADC failure flag is reset, the flycatcher will not retain the failure indications.

WCS Flycatcher

The WCS flycatcher routine is selected in the same manner as for CSDC, except the access code is 7 followed by the five-digit octal address. Entering a 9 will incrementally advance the address one location.

Entering 9X will decrement the address by one location. Multiple depressions of the ENTER button following 9 or 9X will increment or decrement the address location, once for each depression.

Flycatcher Exit

To exit a flycatcher routine, depress:

- CLEAR
- 7
- 0
- ENTER

KEYBOARD DIGITAL ENTRY TESTS

The capability is provided to rerun all or selected portions of a sequence (1, 2, 4, 5, 6, 7, and 8) only after the test

sequence has been run to its completion. This capability has been incorporated to aid the maintenance crews. Digital entry tests are initiated by the following sequence.

On computer address panel:

1. CLEAR pushbutton . . . . DEPRESS AND RELEASE

2. NBR pushbutton . . . . DEPRESS AND RELEASE

3. Enter number via digit . . . . DEPRESS AND pushbuttons RELEASE

4. ENTER pushbutton . . . . DEPRESS AND RELEASE

NBR 63, 71, 81, and 86 digital entry tests may be terminated by the following sequence:

On computer address panel:

1. CLEAR pushbutton . . . . DEPRESS AND RELEASE

2. ENTER pushbutton . . . . DEPRESS AND RELEASE

The following digital entry tests are available:

- SEQUENCE 1

  NBR-11 – Dynamic displays test

- SEQUENCE 2 (NAV MODE switch OFF)

  The provision for digital entires (DE 2X) in sequence 2 is made for the benefit of maintenance personnel and is intended for use only on the deck.

- SEQUENCE 4

  NBR-40 – Repeat sequence 4 MAS only.
  NBR-41 – Station 1 MOAT*
  NBR-43 – Station 3 MOAT*
  NBR-44 – Station 4 MOAT*
  NBR-45 – Station 5 MOAT*
  NBR-46 – Station 6 MOAT*
  NBR-48 – Station 8 MOAT*
  NBR-49 – MOAT all stations with AIM-54 ID's

*With AIM-54 ID present

- **SEQUENCE 5**

    NBR-50 - Repeat sequence 5.

    NBR-51 - Doppler filter display with FMR off

    NBR-52 - Doppler filter display with FMR on

- **SEQUENCE 6**

    NBR-60 - Repeat sequence 6

    NBR-61 - Repeats all sequence 6 tests affected by channel selection on the operator-selected channel. The noise plot and the pulse doppler power test are performed at one-third transmitter duty cycle.

    NBR-62 - Repeats all sequence 6 tests affected by channel selection on the operator-selected channel. The noise plot and the pulse doppler power test are performed at one-half transmitter duty cycle.

    NBR-63 - Displays transmitter peak power on the RIO-selected mode (PDS or PS) and selected channel. Channel may be changed without reentering NBR 63.

    NBR-64 - Decodes CM detected transmitter problem. Displays DPs and faulty WRA (applied fault isolation to octal (XM) display at ST 100).

- **SEQUENCE 7**

    NBR-70 - Repeat sequence 7.

    NBR-71 - Radar antenna scan pattern display with STAB in or STAB OUT

- **SEQUENCE 8**

    NBR-80 - Repeat sequence 8.

    NBR-81 - Radome target horn calibration tests (BIT adjustment)

    NBR-82 - Track loop test on lobing frequency 1 only

NBR-83 - Track loop test on lobing frequency 2 only

NBR-84 - Track loop test on lobing frequency 3 only

NBR-85 - Repeat PSTT tests only.

NBR-86 - Transmitter oscillator leakage test

Note

The continuous iterations of NBR's 63, 71, 81, and 86 may be stopped by pressing CLEAR and then ENTER on the computer address panel.

BIT OPERATION

The BIT (figure 9-14) can be exercised by the RIO while in flight or on the ground. The confidence tests, with the exception of MOATS, take approximately 5 minutes and provide a good indication of system condition. An additional 30 seconds is required for each AIM-54A missile onboard.

Note

When a higher priority mode such as DFM is selected, BIT is automatically exited.

BIT Readout

BIT results are displayed on the TID. Each of the confidence test sequences and each of the detailed radar test sequences has two basic displays: a fault-detection display and a fault-isolation display. Continuous monitoring displays appear only when a failure occurs during tactical modes.

To initialize the desired BIT, RIO selects the appropriate sequence and enters it via the computer address panel. Tape read-in on the TID is indicated by the appearance of an M or B. At the completion of read-in, the BIT test result box will appear with a blinking sequence number. During the sequence test, the number will appear in the upper left-hand corner of the BIT test result box. The sequence number will blink until the test is complete.

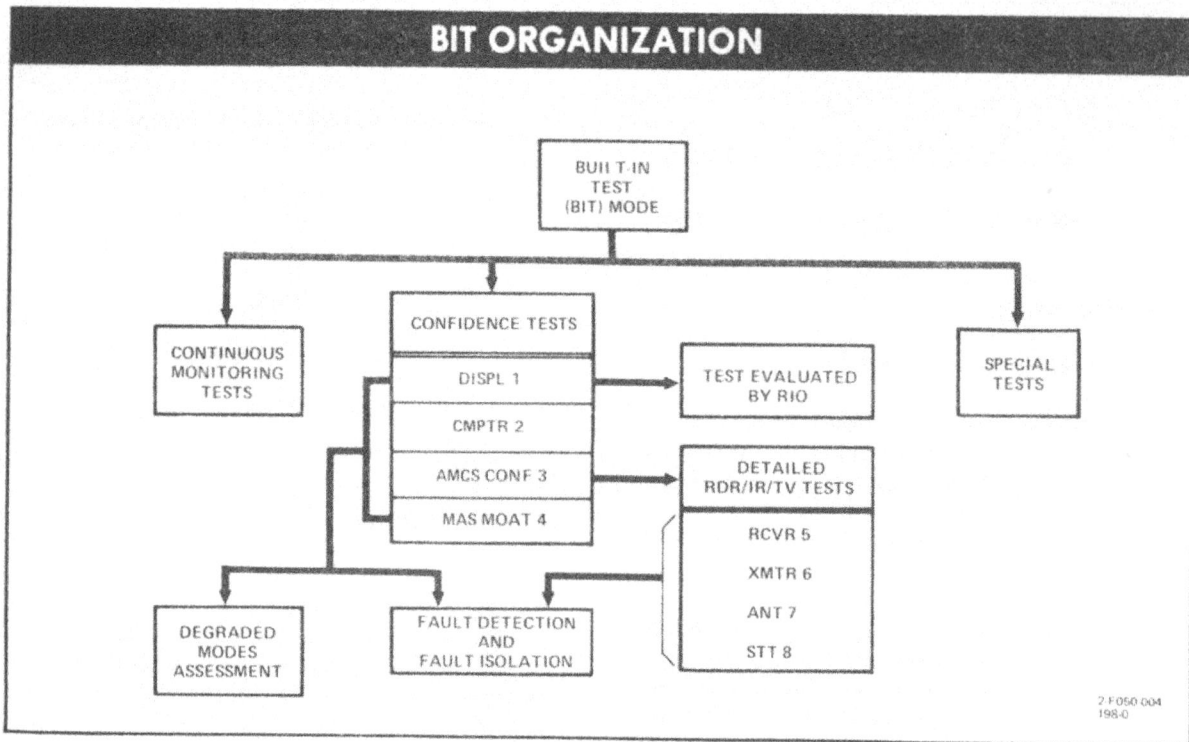

Figure 9-14. BIT Organization

BIT Fault Detection Displays

A typical BIT fault detection display is shown in figure 9-15. In all the BIT fault detection displays (except BIT sequence 4), the square BIT box appears at the top of center of the TID. The sequence number appears in the upper left-hand corner of the BIT box.

When the sequence number is blinking, the test is in progress. If the test cannot proceed because a manual switch or control setting is required, a symbol blinking above the BIT box (below the box in BIT sequence 4) indicates which switch is not properly set. Proper settings are as defined in the test procedure for each sequence.

The WCS has been designed to require a minimum number of these manual switch settings while executing the confidence tests. If the required switch setting is not made within 28 seconds, the computer may continue and execute an abbreviated version of the test sequence. At the completion of the BIT sequence, the test sequence number remains steady and the test sequence is evaluated by a symbol in the lower left corner of the BIT box (left center of the box in BIT sequence 4). A pass (√) or a discrepancy (D) symbol will appear. An exception is sequence 1, the displays test, in which no (√) will appear because it is

evaluated by the RIO. The discrepancy (D) symbol may appear if there is a C and D power fault. When the test sequence number goes steady, the RIO knows the fault detection test is complete and progression to the next test or fault isolation is possible. The RIO will note the discrepancies prior to the start of the next sequence.

If, at any time during execution of a BIT sequence (except auto sequence 2) incoming data link is detected, the blinking DL acronym will appear at the lower left portion of the TID. At this time, the RIO may elect to exit BIT and receive data link. If a power transient should occur during the execution of a BIT sequence, the blinking symbol PTI (power transient interrupt) will be displayed to the right of the BIT box. This advises the RIO that a power transient has occurred and that the data displayed may not be entirely correct.

Note

If the aircraft was in motion (pitch or roll) during sequence 3, sequence 4 (MAS or MOAT), or sequence 8 tests, an aircraft-in-motion symbol (⊘) may appear in the BIT box. When this happens, it indicates portions of the test were either affected by the motion and/or were bypassed. The test should be rerun, with the aircraft stabilized, to perform a complete test.

BIT DISPLAYS

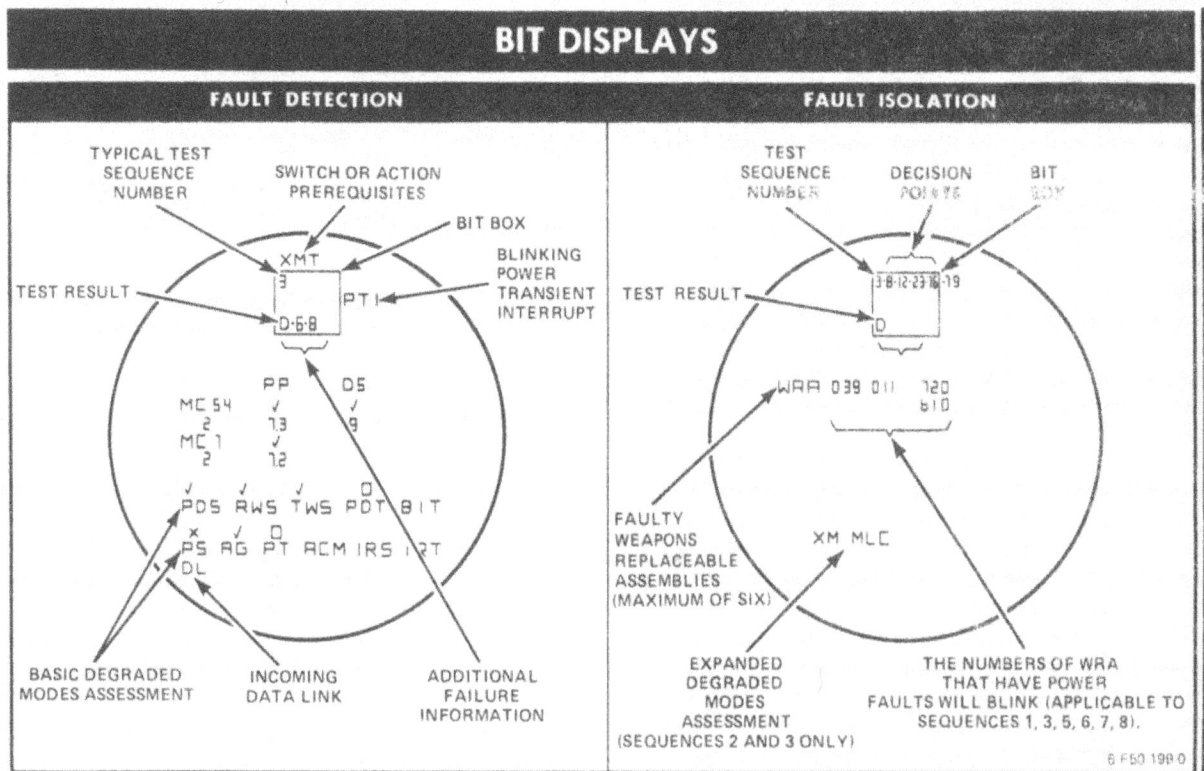

Figure 9-15. BIT Displays

Degraded Mode Assessment

The fault displays associated with sequences 2 and 3 also include the basic degraded modes assessment (DMA). An acronym for each mode is displayed on the TID and a pass (√), fail (X), or degraded (□) evaluation is presented with each acronym when sufficient information becomes available to the computer. The basic DMA is cumulative and carries and updates evaluation throughout sequences. Sequence 1 displays test consists of operator-evaluated patterns and contains no DMA display. Sequence 4 has an abbreviated DMA for AWG-9 modes related to missile launch capability.

Note

A mode with no fault detected, but not fully evaluated, will have no evaluation associated with it. In other words, a pass will be inhibited.

The operator may at any time call up the latest evaluated basic DMA display by depressing the SPL TEST (NBR) pushbutton. The symbols that appear on the displays and the corresponding modes or function named for the basic DMA are listed as follows:

- PDS — Pulse Doppler Search

- RWS — Range While Search

- TWS — Track While Scan

- PDT — Pulse Doppler Single Target Track

- BIT — Built-In Test

- PS — Pulse Search

- AG — Air-to-Ground

- PT — Pulse Single Target Track

- ACM — Pulse Dog Fight Modes

BIT Fault Isolation Displays

A typical BIT fault-isolation display is shown in figure 9-15. The fault-isolation display for a test sequence is

initiated by depressing the FAULT DISPL pushbutton on
the computer address panel when a fault has been detected.
The fault-isolation displays contain a BIT box near the top
of the TID. The sequence number is in the upper left
corner of the BIT box, and the test evaluation is in the
lower left corner. In the BIT box, and immediately to the
right of the test sequence number, will appear a list of
decision points. These decision points represent the
specific locations of failed decisions within the computer
program of the test sequence. Displayed to the right of
the acronym WRA are the unit numbers responsible for
the failure of the test sequence. The relationship of unit
number to unit common name is shown in figure 9-16.
Sequence 4 is an exception to this description. Refer to
the sequence 4 detailed discussion for locations information.

The fault isolation display for sequence 4 MAS/MOAT is
the only exception to the above display description. A
typical sequence and fault isolation display is shown in
figure 9-20. The fault display is called up in the same way
as all other BIT sequences. The sequence and fault isolation display contains a BIT box near the bottom of the
TID. The sequence number is to the left of the BIT box,
the test evaluation is in the center, and degraded weapon
status is to the right. Weapon stations affected by MAS/
MOAT fault are displayed vertically on the left side of the
display. First through fourth maintenance action choices
are displayed horizontally from left to right beside the
affected weapon station, if applicable. Test failures are on
the right side in form of acronyms and maintenance decision (DP) numbers. An R in front of a DP number indicates a MOAT readiness failure, while S indicates a WSC
system failure affecting MOAT. When MAS/MOAT is repeated through a digital entry and a horizontal line appears
across the TID, real failures are displayed above the line,
while indeterminate failures will be displayed below.

Expanded Degraded Mode Assessment

Also included in sequence 2 and 3 fault-isolation displays
is an expanded DMA presentation. The expanded DMA
presentation is a list of acronyms, near the bottom of the
display, that represent submodes or specific functions that
are degraded or have failed. This expanded DMA display
does not evaluate the functions with pass, fail, or degraded
symbols; it displays only the acronyms corresponding to
the failed or degraded submodes or functions. The expanded DMA symbols which may appear are as follows:

- SEQUENCE 2

    DL - Data Link IFU failure

    IFF - Identification Friend or Foe IFU failure

R/R - Range and Range Rate in the IFU failure

SAL - Semiactive Launch

SSI - Standard Serial Interface

- SEQUENCE 3

    ANT - Antenna failed two-bar and four-bar
            tests, or antenna has failed the $\pm5^{\circ}$
            pointing test.

    ANT 2 - Antenna failed two-bar test.

    ANT 4 - Antenna failed four-bar test.

    CWI - Continuous wave illuminator power
            is below acceptable level to support
            AIM-7 inflight, the transmitter
            flood antenna switch was not enabled,
            or the CW transmitter dumped.

    JET - Jam Exceeds Theshold interrupt
            failed.

    MLC - Main Lobe Clutter notch fails to
            take out clutter.

    MRL - Manual Rapid Lockon (indicates
            the antenna failed to reverse scan
            direction within 300 milliseconds
            during $\pm10^{\circ}$ scan).

    PA - Paramp failure

    PC - Pulse Compression failure

    SAL - Semiactive Launch decoder com-
            mand was true on a search channel
            or false for a semiactive channel.

    VSL - Vertical Scan Lockon antenna
            switching test failed.

    XM - Transmitter was enabled but failed
            to come on, the XMTR position
            was not selected by the RIO, or no
            rf was detected.

SIMPLIFIED WRA NUMBER	COMMON NAME
001	Radar master oscillator
006	Transmitter oscillator waveguide
007	CWI oscillator waveguide
010	Synchronizer
011	Radar transmitter
012	Radar output waveguide
013	Collector power supply
014	Beam power supply
015	Solenoid power supply
021	Radar input waveguide
022	Radar receiver
025	Radar test horn
026	MOAT forward horn
027	Aft wing horn
028	Aft belly horn
029	Aft hook fairing horn
030	Aft center channel horn
031	Radar antenna
039	Doppler processor
042A OR B	Doppler filter
081	Antenna and test controller
083	Low PRF processor
451	Arithmetic and control/DRO
452	NDRO memory
461	Input-output unit (IFU)
462	Computer power supply/Bulk storage memory (MTM)
501	Sensor control panel
505	Computer address panel (CAP)
541	Detail data display (DDD)
560	Hand control unit
580	Tactical information display (TID)
590	Mission recorder
601	Semi-regulated power supply
610	Regulated power supply
710	Missile signal data converter
720	Missile logic timing controller
730	Missile power supply
805	CSDC

Figure 9-16. AN/AWG-9 WRA's

Special Test 100 And 101

Note

Special test 100 and 101 cannot be performed directly after a BIT sequence. Reselecting a BIT category is necessary.

Special test 100 allows ground maintenance personnel to check for the most recent power faults and transmitter problems, which may have occurred on the previous flight. During the time that continuous monitoring (CM) is operating, if a power fault is indicated from any of the units listed below, or, if a transmitter fault is detected in an operate mode, the status of power faults and/or BIT transmitter operate mode are stored in sacred cells.

710/610	001	505
720/610	010	541
730/610	011	580
039	022	501
042A	31E	083
042B	081	610

Special test 100 decodes the contents of those sacred cells, and displays under the PF (figure 9-17) the WRA number of the unit causing the power fault. If the conditions for a transmitter problem exist, the octal contents of location 00166 are displayed under the XM. The symbols WRA, PF, and XM will always be dislayed with or without associated data.

Special test 101 is the same as special test 100, except the sacred cells are cleared if the aircraft is on the ground. If the sacred cells are not cleared after every flight, there is no way of knowing how long ago the data was stored.

To initiate special test 100 or 101, select BIT, then SPL TEST (NBR), then NBR followed by the digits 100 or 101, and then ENTER. Sacred cells are cleared upon entering special test 101 on the ground; displays remain until next entry.

Note

Special test 101 may require two depressions or pushtile to initiate enter.

BIT SEQUENCES

Sequence 1 — Displays Tests

The displays test presents the RIO with standard test patterns on the TID and DDD for his evaluation. Sequence 1 is initiated by selecting BIT with the CATEGORY switch and depressing the DISPL 1 pushbutton on the computer address panel (CAP). The displays test is divided into three tests: direct, static, and dynamic. Holding the DISPL 1 pushbutton in results in the direct display test, which

consists of a 45° diagonal line across the TID screen and a display of decreasing shades of intensity of the DDD video (DDD screen) (figure 9-18, sheets 1 and 2). Release of the DISPL 1 pushbutton initiates the static display test. The normal sequence 1 displays are obtained with the TID MODE switch set to the GND STAB or A/C STAB position. At the same time, the DDD displays a two-bar, ±40° scan pattern. In addition, the following lights on the DDD, TID, CAP, and hand control are illuminated.

DDD	Hand Control
• CCM Modes	• IF power and/or cool indicator
• ANT TRK	• PWR RESET indicator
• RDROT	• WCS power indicator
• JAT	• One of four function lights
• IROT	
• RDR	
• All WCS Modes	

TID	CAP
• LAUNCH ZONE	• All MESSAGE pushbuttons (10)
• VEL VECTOR	
• STBY	
• READY	Not controlled by BIT. May or may not illuminate.
• CLSN	

Eight range-rate video targets are also displayed on the DDD as equally-spaced horizontal lines. If IFT is selected, the IFT target video appears two-thirds of the way down from the center of the DDD, aligned horizontally with the zero closing rate etch mark on the right side of the DDD.

When ATTK is selected with the TID MODE switch, the only lights illuminated are the PDSTT and RDR pushbuttons. The range readout drum (window) indicates 5 nmi. The two-bar, ±40° scan pattern is replaced by a 0° B-scan positioned in azimuth by the AZ CTR control on the sensor control panel. The B-scan shows the minimum launch range marker at 1 nmi, the target range at 3 nmi, and the maximum launch range marker at 5 nmi. The closing rate

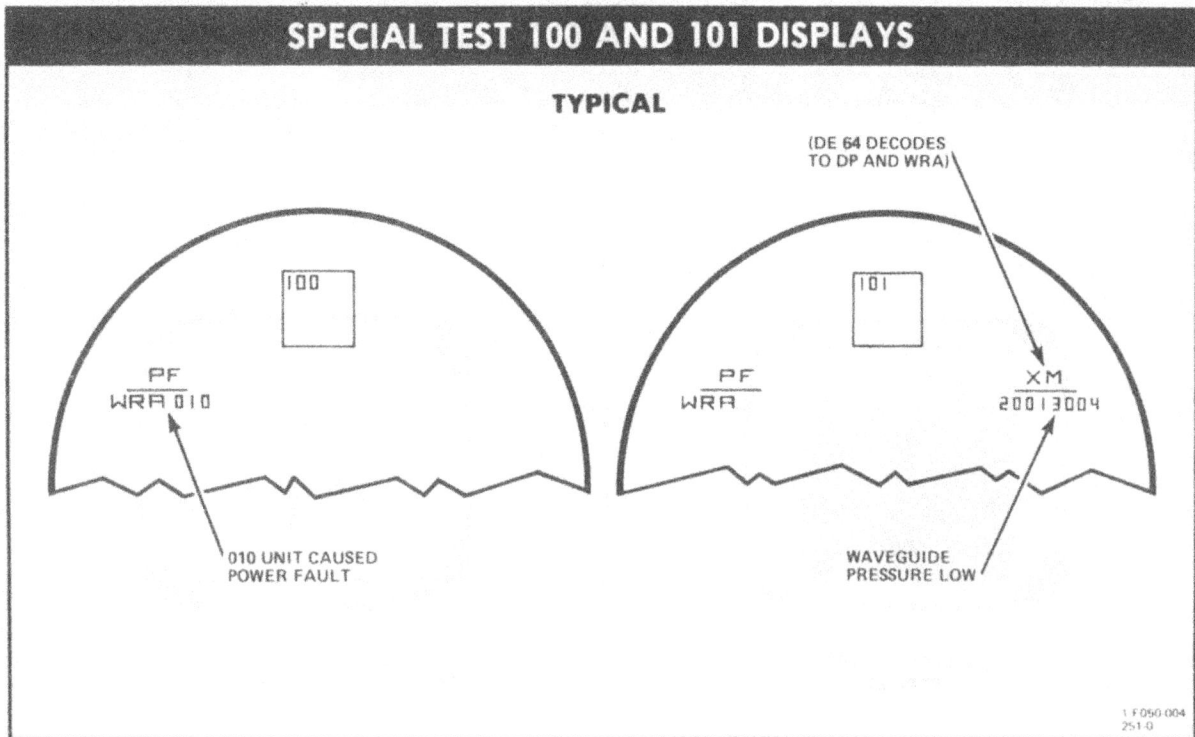

Figure 9-17. Special Test 100 and 101 Displays

symbol is positioned at approximately +220 knots. The eight range-rate video targets are displayed at the left side of the DDD. The acquisition gates are over the target one-third of the way from the top.

These test patterns should be examined by the RIO for the absence of any symbols, for symbol intensity, and for symbol position. During the running of the static test, the RIO should also select HALF ACTION or FULL ACTION on the hand control. Then, moving the hand control, the RIO should ensure that the TID CURSOR can be moved throughout the range of the TID. Upon release of the ACTION switch, the cursor symbols should return to the original position.

It should be noted that the static portion of the displays test gives the RIO an indication that the computer does or does not have the capability of displaying each of the indicated symbols. It is, therefore, more than a displays test because it also tests the computer's ability to generate the symbols it would need in a tactical situation. It should also be pointed out that the computer assists the RIO in the static portion of the displays test by monitoring power

failures that may have occurred in the controls and displays. The presence of a power failure is indicated by the acronym D (indicating a discrepancy) in the BIT Box. If the RIO wishes to know which controls and displays sub-units have suffered a power failure, he depresses the FAULT DISPL pushbutton. This will cause the computer to display the numbers of the WRA that suffered a power failure.

The dynamic test consists of a visual evaluation of the movement of the artificial horizon, ASE circle, steering symbol, launch zone symbols, and the velocity vectors. To enter the dynamic test, the RIO selects CLEAR, NBR, 1, 1, ENTER and selects ATTK on the TID MODE switch. Throughout the dynamic test, the displays on the TID and DDD go through the following movements, at a rate of one step per second.

On the DDD, the following occurs simultaneously:

- The artificial horizon moves in pitch from zero to +15°, +30°, +45°, 0°, −15°, −30°, −45°, and back to 0°.

SEQUENCE 1 DISPLAYS

DIRECT TEST

SHADES OF GREY
TYPICAL)

TID
(45° STROBE)

DYNAMIC TEST

ARTIFICIAL
HORIZON

STEERING
SYMBOL

ASE
CIRCLE

TEST
SEQUENCE
NUMBER

BIT
BOX

ARTIFICIAL
HORIZON

TARGET
DOT

BAR

DOT

VELOCITY
VECTOR

ASE
CIRCLE

STEERING
SYMBOL

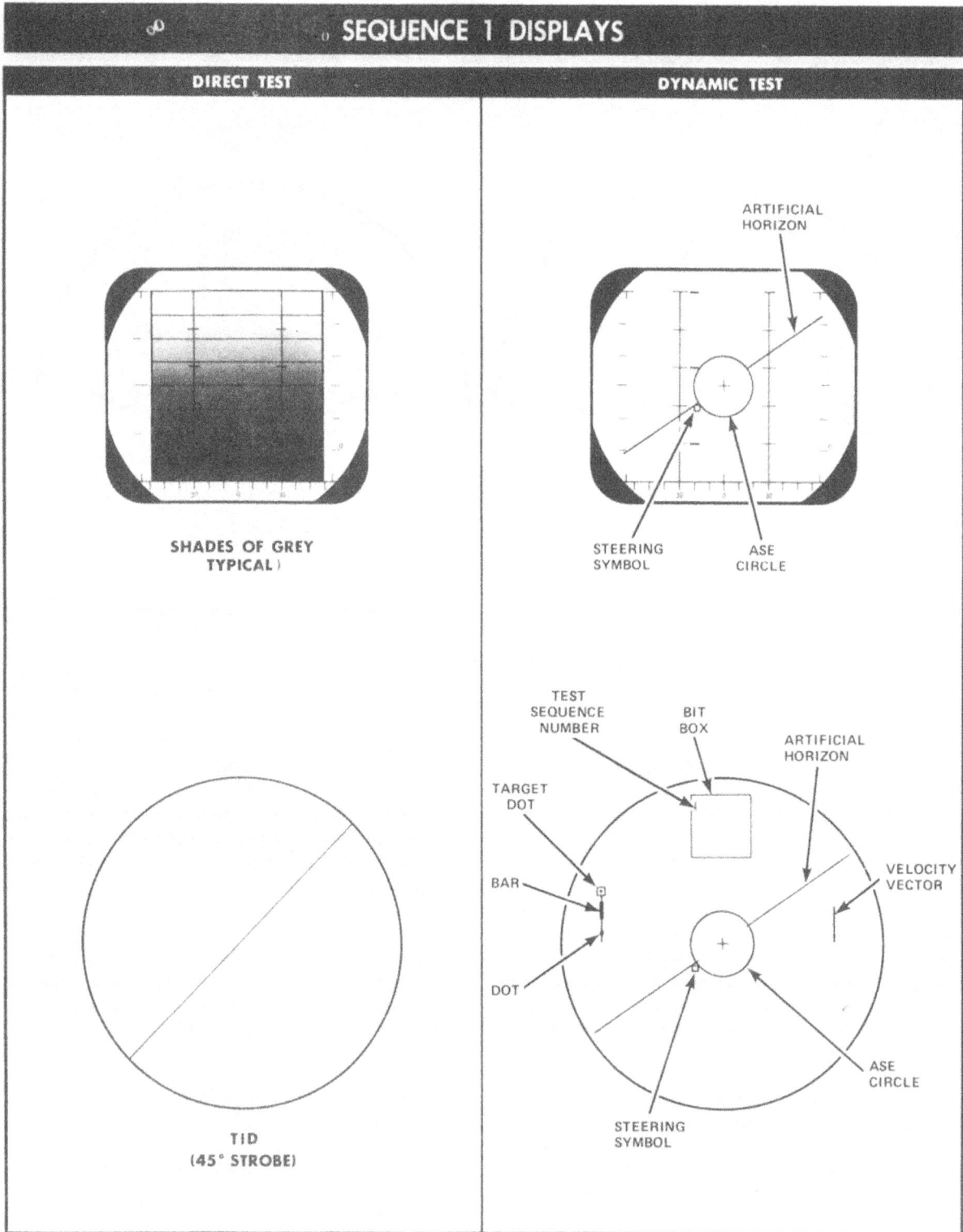

3-F050-004
202-1

Figure 9-18. Sequence 1 Displays (Sheet 1 of 2)

SEQUENCE 1 DISPLAYS

STATIC TEST

ATTK NOT SELECTED	ATTK SELECTED

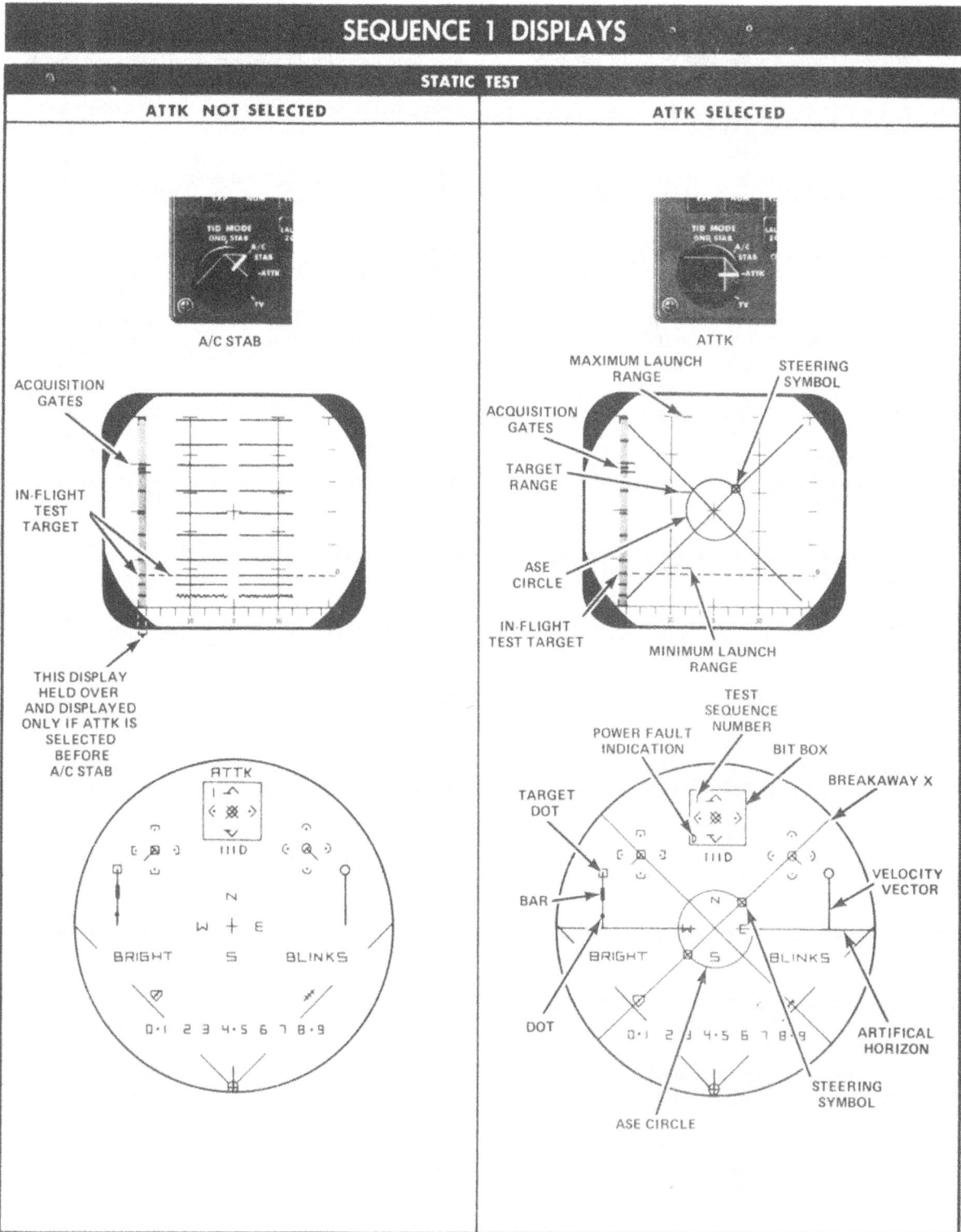

A/C STAB

ATTK

ACQUISITION GATES

IN-FLIGHT TEST TARGET

THIS DISPLAY HELD OVER AND DISPLAYED ONLY IF ATTK IS SELECTED BEFORE A/C STAB

MAXIMUM LAUNCH RANGE

STEERING SYMBOL

ACQUISITION GATES

TARGET RANGE

ASE CIRCLE

IN-FLIGHT TEST TARGET

MINIMUM LAUNCH RANGE

TEST SEQUENCE NUMBER

POWER FAULT INDICATION

BIT BOX

BREAKAWAY X

TARGET DOT

BAR

VELOCITY VECTOR

DOT

ARTIFICAL HORIZON

ASE CIRCLE

STEERING SYMBOL

ATTK
IIID

BRIGHT   BLINKS

N
W + E
S

0·1 2 3 4·5 6 7 8·9

Figure 9-18. Sequence 1 Displays (Sheet 2 of 2)

7-F50-202-2

- The artificial horizon steps in roll from 0° to 15° right wing down, 30°, 45°, back to 0°, 15° left wing down, 30°, 45°, and back to 0°.

- The ASE circle steps from 0.8 inch in diameter to 0.1, 0.3, 0.56, then back to 0.8.

- The steering symbol steps around the ASE circle in a clockwise direction in steps from its position in the upper right quadrant to the lower right, lower left, upper left, and back to the upper right quadrant.

On the TID, the following occur simultaneously:

- The bar marker steps from 1.5 inches above the artificial horizon to 1.0, 0.5, 0, and back to 1.5 inches.

- The dot marker steps from above the artificial horizon to 0.5, 1.0, 1.5 inches and back to the artificial horizon.

- The artificial horizon, ASE, and steering symbol move on the TID at the same rate as on the DDD.

- The ASE circle steps from 2-inches diameter to 0.2, 0.8, 1.4, back to 2-inches diameter.

Note

The events occurring during the dynamic portion of the test are repeated until the flight officer selects another BIT sequence test, or until he rotates the CATEGORY switch to a new position.

Sequence 2 — Computer Test

The check of the digital computer includes the basic elements of the computer subsystem and the interface to the other subsystems. Four WRA are checked in sequence 2; the computer power supply storer (WRA 462), the computer interface unit (WRA 461), and the computer itself (WRA 451 and 452). The computer power supply is not specifically tested by the sequence because, if the unit were faulty, the sequence could not have been initiated. The computer test is performed automatically after the WCS switch has been set to STBY or XMT, or can be initiated manually by selecting BIT with the CATEGORY switch and depressing the CMPTR 2 pushbutton. Sequence 2 displays are shown in figure 9-19.

The automatic sequence 2 normally requires about 25 seconds to run; however, due to a possible delay in tape

read-in, the test might still be in progress when the displays have been timed in. In this case, the BIT box will appear on the TID with flashing 2 in the upper left corner until completion of the sequence 2 test. When sequence 2 testing concludes and the system shows a pass, the symbol S2 with a check mark over it appears on the left side, just above center on the TID. This symbol remains displayed for approximately 15 seconds, after which the system switches to a tactical display. If simultaneous align and test (SAT) is selected, the TID will switch to a SAT display. If the test fails, the fault-isolation display is automatically displayed and held until the RIO depresses the PRGM RESTRT pushbutton on the computer address panel. After auto-sequence 2 has run and passed, the RIO can continue running BIT. If sequence 2 failed, however, the results of the other tests are likely to be invalid.

During an auto sequence 2 if only the M or B is displayed (prior to BIT box), depressing any CAP function switch in any CAP category will result in a skipped auto sequence 2. This will be indicated by an X displayed over S2 in the SAT mode. A power down for greater than 6 seconds and then reselecting standby is required to reinstate auto sequence 2. If power up sequence 2 passes, and the system is not in a SAT mode, a check will be displayed over S2 for approximately 15 seconds, then the program will return to tactical. If sequence 2 CAP function switch is depressed during this 15 seconds, an additional auto sequence 2 will be performed.

After reading in the program, sequence 2 (manually initiated) is executed and a flashing 2 appears in the upper left corner of the BIT box. During the first routine, when performing the central processor test, the mode shown on the DDD is pulse doppler search (PD SRCH pushbutton is illuminated, RDR DISPLAY pushbutton is illuminated, and PD is indicated on the mode readout window). The antenna scan pattern indicated on the DDD and EL meter is a two-bar, ±40°, as established by the BIT executive program (081 indication only). The antenna remains off during the test.

At the completion of the central processor test, the mode readout window indicates XXX. (RDR DISPLAY and PD SRCH mode remains illuminated.) The next four routines contain the interface unit tests and evaluation of the test results. Routine changes may be verified by watching the display of an M or B in the lower left quadrant of the TID. Each time an M or a B appears during sequence 2, a new routine is being read in from the magnetic tape memory.

Completion of the sequence is indicated when the number 2 stops flashing. A pass (√) or fail (D) is displayed in the

SEQUENCE 2 DISPLAYS

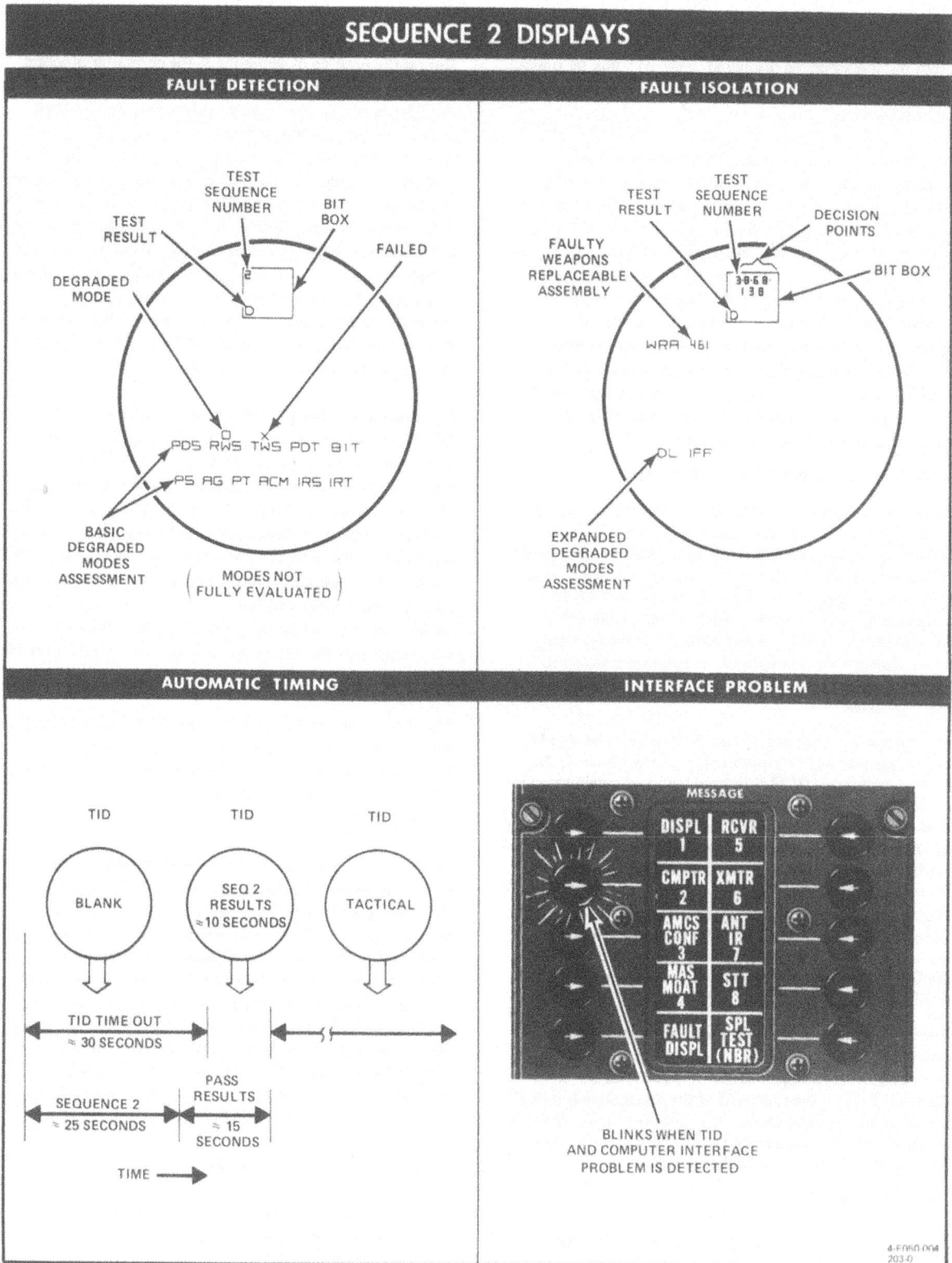

Figure 9-19. Sequence 2 Displays

lower left corner of the BIT box. A typical fault detection display is shown in figure 9-19. If a D appears, the decision point numbers may be displayed in the BIT box by pressing the FAULT DISPL pushbutton on the computer address panel.

At the end of sequence 2, the program determines whether the test was run during power-up or manually selected as part of BIT. If the test was run during power-up, control is returned to the tactical program (if the test passed). If the test failed, the program enters an idle loop and waits for a program restart (the test results remain displayed during the idle loop). If sequence 2 was manually selected, control is transferred to the BIT executive program and the test results displayed while the program waits for another sequence command. If the CATEGORY switch is moved to a position other than BIT after sequence 2 is initiated manually, control is taken away from the BIT executive program.

Automatic Sequence 2 Initialization

Automatic sequence 2 performs certain WCS initializations that are not performed by a manual sequence 2. This initialization does the following tasks: Nominal values are stored for pulse and PC mode zero range bias, a nominal ACM threshold of 2 volts is stored, AIM-7 and AIM-54 A aerodynamic launch order is established, missile active frequency IDs are cleared and AWG-9 DMA (degraded mode assessment) is initialized. If automatic sequence 2 fails or is exited while in progress, the above initializations may not occur.

If sequence 3 is then performed, computed pulse and PC zero range bias will be stored in lieu of the nominal. Similarly, a computed ACM threshold based on a calibrated target will be stored. If sequence 4 is performed, AIM-7 and AIM-54A launch orders will be established and AIM-54A active frequency ID's stored as well as MOAT status by mode. Thus if sequence 3 and sequence 4 are performed following an incomplete power-up sequence 2 initialization, all the missed initialization will be compensated for except DMA initialization. Lack of DMA initialization will prevent passes from being displayed over any mode in the AWG-9 DMA display. Degraded or failed modes, however, would be identified.

Power-up sequence 2 initialization can be verified by observing the check over S2 following power-up or by observing DMA storage location 7-17000 with the flycatcher. 7777 in the right half of this location will indicate initialization has been performed. If power-up sequence 2 initialization has not occurred, a WCS power-down and backup may restore the initialization.

Note

DMA assessment will be incomplete if auto sequence 2 does not run at power up.

Sequence 3 — AWG-9 Confidence Test

During the running of sequence 3, the computer program tests all the WCS subsystems sufficiently to ensure a high confidence of operation. Some of the tests for RADAR confidence include transmitter peak power and receiver detection sensitivity during both pulse and pulse doppler operation. Readout of each individual mode and the test results for that mode will be displayed at the bottom of the TID. The test results will be indicated on a fault detection format (figure 9-20) with a check ($\sqrt{}$) indicating the mode is operative, an X indicating the mode failed the BIT check, or a square ($\square$) indicating the mode is degraded. Additionally, based on what modes are degraded, the computer recommends which other sequences should be executed to get explicit fault isolation information.

Sequence 3 is initiated by selecting BIT with the CATEGORY switch and depressing the AMCS CONF 3 pushbutton on the computer address panel. After sequence 3 has been read into the computer from the magnetic tape memory, a flashing 3 is displayed in the BIT box while the program is being executed. An M or B is displayed throughout much of the sequence because sequence 3 consists of 3 routines. The program loads a routine following the execution of each preceding routine. The modes that are being evaluated for a level of confidence are displayed at the lower center of the TID. Figure 9-20 shows the information displayed following the completion of sequence 3.

The AIM-7 channel tested will be the AIM-7 CWI channel that the AIM-7 was prepped on. If prep was not selected, then CWI is tested on the missile channel displayed. The AIM-54A channel tested will be on the computer-selected channel.

As the test progresses, values for detection sensitivity (DS) and peak power (PP) are displayed on the TID along with the AIM-54A or AIM-7 channel being tested. At the same time, a pass, degrade, or fail assessment is made as shown in figure 9-20. As the mode evaluations are performed, the MODE pushbuttons on the DDD illuminate according to the mode being evaluated. For each mode evaluated, the antenna scan patterns, gate positioning, and lockons that occur change in accordance with mode being tested and are displayed on the DDD. The sequence of the mode evaluation, with an explanation of the events taking place during the test, are discussed in the following paragraphs.

Note

If sequence 3 is not run following initial turnon or any power-down sequence, a back-up ACM threshold will be employed.

SEQUENCE 3 DISPLAYS

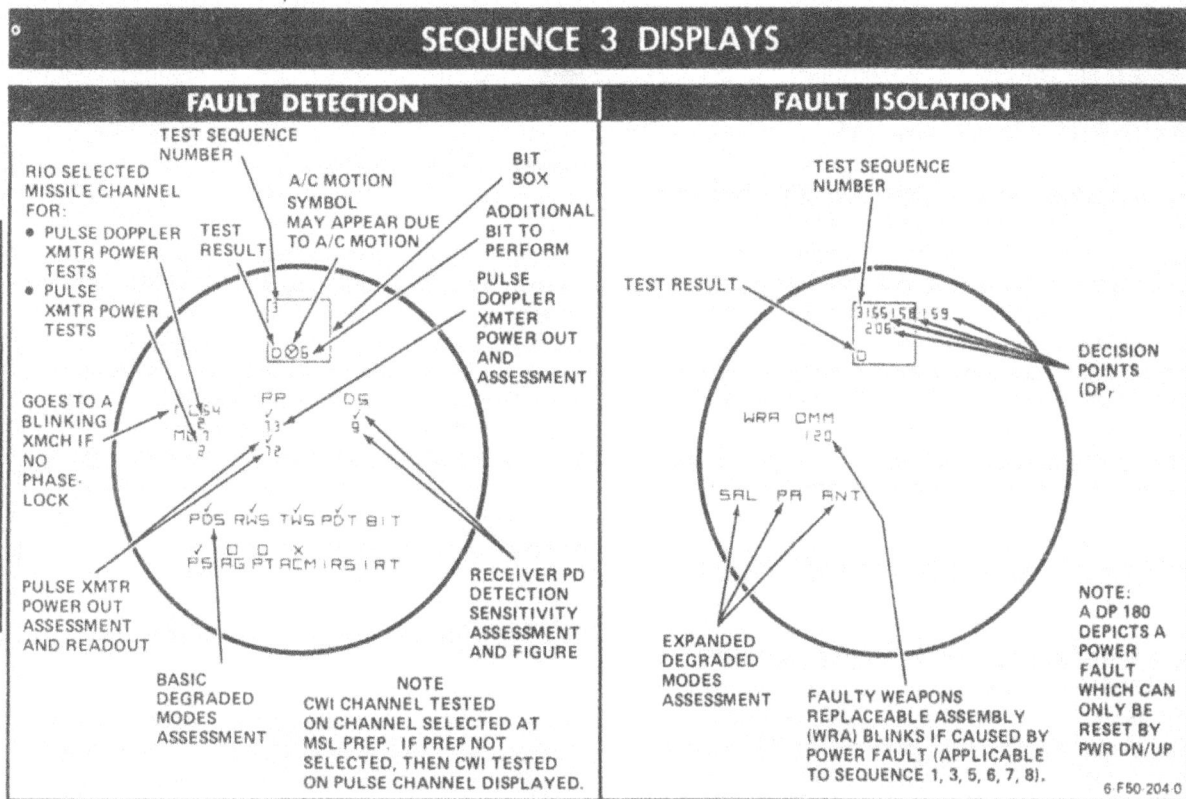

Figure 9-20. Sequence 3 Displays

PULSE DOPPLER SEARCH (PDS)

The initial sequence 3 mode is PDS. The PD SRCH push-button is illuminated and PD is indicated on the mode drum. The DDD scan indication of two bar, ±40°. The antenna now moves to 0° in azimuth and elevation with a 0° B-sweep displayed while the main lobe and side lobe, and modulation target levels are measured. A one-bar, ±65° scan (centered at 0° azimuth and elevation) is then commanded and displayed with JAM intensity noise displayed on the bottom of the DDD. The antenna ±65° scan tests are performed here.

RANGE-WHILE-SEARCH (RWS)

The RWS mode is then commanded (RWS on mode drum and RWS mode pushbutton illuminates and a 0° antenna scan is displayed). Master oscillator phase-lock is checked on the semiactive channel selected by the flight officer. If this channel fails, up to four channels are tried. If phase-lock passes to an alternate search channel, XMCH and the channel number are displayed on the TID in place of MC54.

In this event, the RIO should select a different MC54 channel and rerun the test if AIM-54 missiles are loaded. AIM-54A phase lock is required to perform sequence 4 MOAT. If no channels pass, XMCH is displayed blinking on the TID. The DDD B-scan shows darkness in sub-bands 3 and 4 as the AGC noise tests with the paramp are performed. Sub-band areas 1 and 2 then darken when the BIT JAM intensity tests are performed. Range-rate words now appear in the B-scan as many horizontal lines, while the program is testing for reception of range-rate words. Zero doppler will appear as a lightly shaded square in the middle of the B-scan. The B-scan will then show a sweep up and down during the performance of the pulse doppler sensitivity test in which the target aspect is switching between tail and nose aspects. (A dark sweep-down indicates nose; sweep-up indicates tail.) At the end of this test, the detection sensitivity is displayed under the DS acronym on the TID.

TRACK-WHILE-SCAN AUTOMATIC

The antenna is positioned to 17° in azimuth and elevation. The mode indication on the DDD is TWS AUTO. The speed tracker tests are performed at this time followed by the antenna pointing tests, unless the antenna hydraulics are open, in which case the program will proceed directly to the transmitter test. The antenna gimbal angle positions remains at 17°.

TRACK-WHILE-SCAN MANUAL

The antenna begins to scan in a two-bar, ±40° pattern and the mode indications are switched to TWS MAN. JAM noise is displayed at the bottom of the DDD. The antenna two-bar scan rate, bar spacing, and frame time is tested at this time. The antenna is then switched to a four-bar, ±20° pattern and the four-bar scan rate, bar spacing, and frame time is tested. The antenna next is switched to scan in a one-bar, ±10° supersearch pattern. The supersearch scan time and scan reversal time is verified. The antenna is then commanded to scan zero degrees and the receiver shutter is checked to verify that it opened with the SNIFF command.

PULSE DOPPLER SINGLE-TARGET TRACK (PDSTT)

The mode is switched to PDSTT (PDSTT pushbutton illuminated and PD indicated on the mode drum). The antenna attempts to track the horn target at an elevation angle of –5.2° centered at 0° azimuth. The DDD video shows a B-scan type display with the range-rate gates positioning and then searching repeatedly as the BIT horn target is turned on and off. After antenna on target (AOT) is true, the horn target remains on and the speed tracker is positioned and the gyros caged. The program then samples for fast frequency on target (FFOT) and delayed velocity on target (DVOT). If FFOT is true, the ANT TRK indicator illuminates and if DVOT is true, the RDROT indicator illuminates. Both DDD indicators should illuminate to indicate lockon. The antenna gyros are then uncaged and the velocity tracker and range-rate gate commands are disabled. A test to verify that the antenna does not drift more than 3° within 3 seconds is performed (RDROT and ANT TRK lights will go out).

Transmitter Test

The mode indications during the transmitter test remain as PDSTT indications and the antenna azimuth scan pattern is 0°. The speed tracker VCO frequency is set to a predetermined doppler frequency, at which time a darkening of the B-sweep occurs and BIT LOG DC is measured. The transmitter then turns on and a momentary darkening at zero doppler can be observed. The transmitter central line power (CLP) at one-third, one-half, and two-thirds duty with transmitter peak power for one-third duty is determined. If the mode immediately changes to pulse search

and a flash is observed on the DDD, the antenna hydraulics are open. The PDSTT test was bypassed and the sequence has proceeded to the CWI, low-prf, and pulse compression tests.

Receiver Unblanking Test

The receiver unblanking test checks for degraded receiver blanking. The receiver blanking, which is controlled by the synchronizer, cuts off the return signal path between the antenna and the receiver to prevent damage to the receiver or interference with signal processing when the transmitter is being pulsed. If the blanking is degraded, DP 139 will be displayed in the BIT box and a DMA degrade symbol (□) will be displayed over PDS, RWS, TWS, PDT, PS, AG, PT, and ACM. If one or more of the above modes has failed for another reason, a fail symbol (X) will be displayed over that mode instead of the degrade symbol.

PULSE SINGLE-TARGET TRACK (PSTT)

The antenna is commanded to scan in one-bar, ±10° and the BIT low-prf polar error is measured. The horn target is enabled and the antenna is positioned to –5.2° in elevation and 0° azimuth. The program then monitors for low-prf antenna on target (LAOT) for 1 second, maximum. When LAOT is true, radar full action is commanded and the antenna gyros are caged. When rage on target (ROT) goes true, the RDROT indicator illuminates and the gyros are uncaged for 2 seconds to verify that the antenna does not drift more than 2°. RDROT goes off and when delayed range on target (DROT) goes true, the ANT TRK indicator illuminates. The target, along with the range gates, appears at the top of the DDD. The RDROT and ANT TRK indicators then go out and a one-bar, 0° scan is commanded. The azimuth scan center shifts to about 0° left on the DDD and the elevation meter scan 35° up and then back down. During this time period, the program performs a dogfight-mode lockon, but there is no external indication that it has occurred. The DDD now will go blank. PSTT, pulse, and pulse compression tests are performed on the MC7 channel selected. If this channel does not phase-lock these sections are bypassed and DP 140 will be displayed.

CWI, Low-prf And Pulse Compression Test

The mode switches to pulse search (PULSE on the mode drum and PULSE SRCH pushbutton illuminates). A flash on the DDD is observed when CWI is enabled. Gates are then seen at the bottom of the B-sweep, indicating that the low-prf tests are being performed. The test then proceeds to measure the pulse compression peak power and ratio. The CWI channel tested in sequence 3 will be the one that the AIM-7 missiles tune to after PREP if the PREP switch was turned on.

Note

- With weight-on-wheels, all modes will indicate failures in DMA and no peak power will be displayed if the ground cooling switch is in the CABIN/OBC position.

- If BIT sequence 3 is performed on a cold system, a few 039 WRA may fail the MLC and SLC translation tests. This results in any or all of the following DP: no. 4, 5, 6, 7, 8, and 9. Therefore, to preclude unnecessary maintenance, the tests should be rerun 5 minutes after system turn-on.

Sequence 4 — Missile Auxiliaries Subsystem And Missile On Aircraft Test

MAS

Sequence 4 (figure 9-21) is initiated by selecting BIT with the CATEGORY switch and depressing the MAS MOAT 4 pushbutton on the computer address panel. After sequence 4 has been read in from the MTM, a flashing 4 is displayed in the BIT box while the program is running.

Note

If sequence 4 is run with the MASTER ARM switch in the TNG position, DP 090 and TNG will be displayed;

Functions tested for include:

- MAS BIT missile signal data converter.
- Missile logic timing controller.
- Missile power supply fault indicator.
- SSI loop between the MAS and computer subsystems.
- MAS-MOAT digital test report lines.
- Irreversible safe signal.
- Active-frequency ID.
- Launch-initiate flag output.
- Miscellaneous AIM-54A, AIM-7 and AIM-9 missile parameters, true airspeed and range-at-launch.
- English bias pitch and yaw outputs.

- AIM-7 video injection signal output.
- AIM-54A parameters.
- Missile seekerhead positioning.
- Sequence programmer launch cycle.
- Missile separated flag.
- Launch-cycle-output enable.
- SEAM test for Sidewinder missile.
- Roll and Pitch initial conditions.
- AIM-7 sweep control.

The AIM-7 firing order is resequenced after MAS-BIT is run so that passes are selected first. Failures are placed at the end of the firing order in sequence.

The test circuits used by MAS BIT are cleared at the end of the test, the sequence number stops blinking and fault-detection test results are displayed on the TID. A MAS BIT pass-fail indication and missile-type failures, if they occur, are displayed in the BIT box. Control will remain in the MAS BIT program until MOAT prerequisites have been met, BIT deselection or dogfight selections returns control to tactical, or a function pushbutton other than FAULT DISPL pushbutton is selected and depressed. After the test is completed, the program will idle in a monitoring loop if the WCS switch is in the STBY position, or the MSL PREP switch is in an OFF condition. Settting the WCS switch in the XMIT psoition and the MSL PREP switch in an ON condition would cause a MOAT read-in from magnetic tape.

WARNING

When BIT sequence 4 detects the irreversible missile launch sequence is unsafe to test, the BIT box will flash. This indicates a failure in the WRA 720 sequence programmer which controls all missile timing and other information required by the AIM-7 during the launch sequence. With MSL PREP selected, the WRA 720 can initiate a launch on its own. On the deck with a missile station selected, low probability exists for missile ejection or rocket motor ignition due to multiple aircraft interlocks. It is possible to activate the missile battery which will overvoltage the missile gyros causing an AIM-7 missfire condition. This could overheat the missile gyros and create a danger of fire. Airborne, with a missile station selected, a greater probability exists for missile ejection and/or activation of the missile battery. On deck or airborne, deselect MSL PREP and do not rerun BIT sequence 4. Continue the mission only if operational necessity dictates.

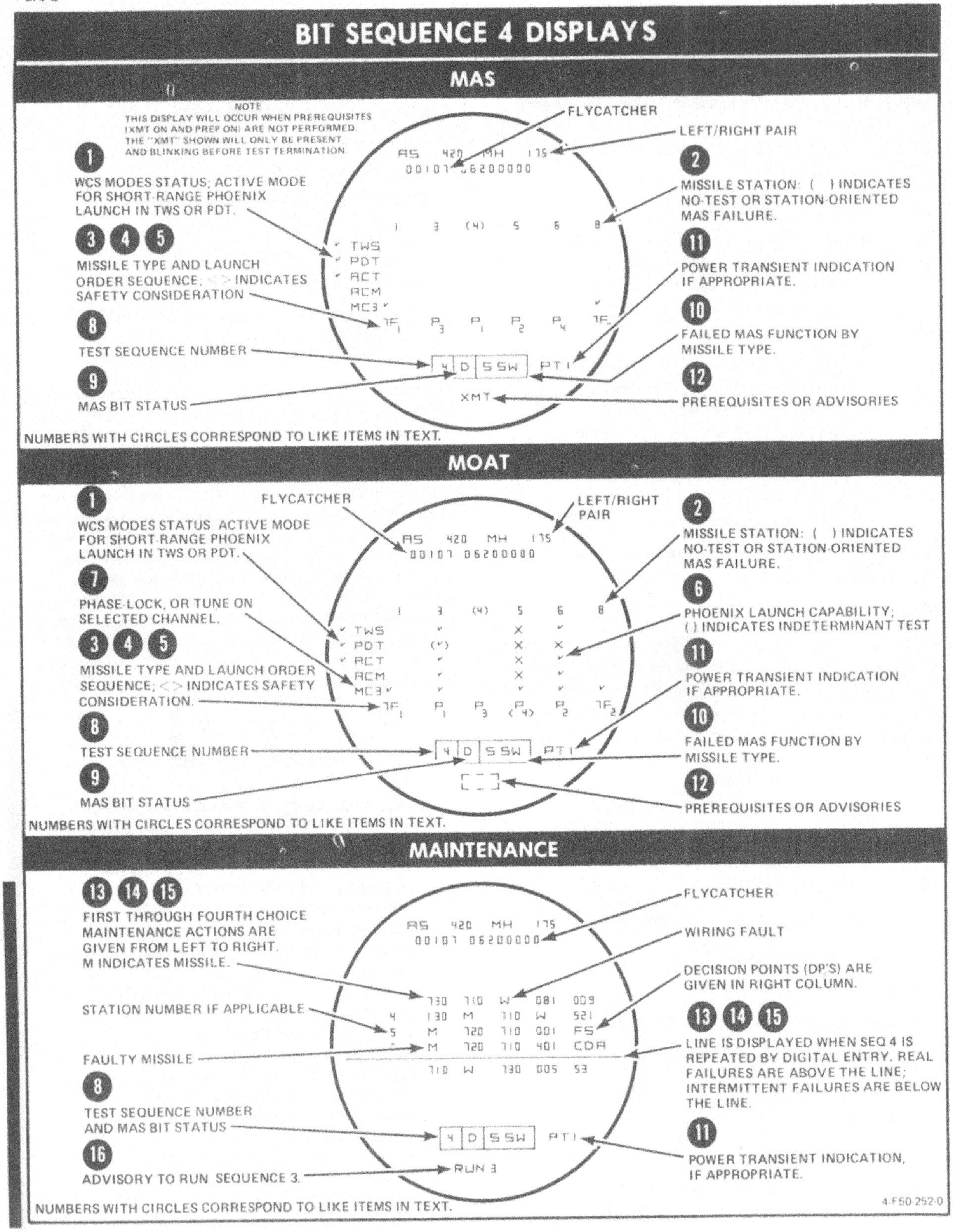

BIT SEQUENCE 4 DISPLAYS

MAS

FLYCATCHER

LEFT/RIGHT PAIR

NOTE
THIS DISPLAY WILL OCCUR WHEN PREREQUISITES (XMT ON AND PREP ON) ARE NOT PERFORMED. THE "XMT" SHOWN WILL ONLY BE PRESENT AND BLINKING BEFORE TEST TERMINATION.

1 WCS MODES STATUS; ACTIVE MODE FOR SHORT-RANGE PHOENIX LAUNCH IN TWS OR PDT.

2 MISSILE STATION: ( ) INDICATES NO-TEST OR STATION-ORIENTED MAS FAILURE.

3 **4** **5** MISSILE TYPE AND LAUNCH ORDER SEQUENCE; < > INDICATES SAFETY CONSIDERATION

11 POWER TRANSIENT INDICATION IF APPROPRIATE.

8 TEST SEQUENCE NUMBER

10 FAILED MAS FUNCTION BY MISSILE TYPE.

9 MAS BIT STATUS

12 PREREQUISITES OR ADVISORIES

NUMBERS WITH CIRCLES CORRESPOND TO LIKE ITEMS IN TEXT.

MOAT

FLYCATCHER

LEFT/RIGHT PAIR

1 WCS MODES STATUS ACTIVE MODE FOR SHORT-RANGE PHOENIX LAUNCH IN TWS OR PDT.

2 MISSILE STATION: ( ) INDICATES NO-TEST OR STATION-ORIENTED MAS FAILURE.

7 PHASE-LOCK, OR TUNE ON SELECTED CHANNEL.

6 PHOENIX LAUNCH CAPABILITY; ( ) INDICATES INDETERMINANT TEST

3 **4** **5** MISSILE TYPE AND LAUNCH ORDER SEQUENCE; < > INDICATES SAFETY CONSIDERATION.

11 POWER TRANSIENT INDICATION IF APPROPRIATE.

8 TEST SEQUENCE NUMBER

10 FAILED MAS FUNCTION BY MISSILE TYPE.

9 MAS BIT STATUS

12 PREREQUISITES OR ADVISORIES

NUMBERS WITH CIRCLES CORRESPOND TO LIKE ITEMS IN TEXT.

MAINTENANCE

FLYCATCHER

WIRING FAULT

13 **14** **15** FIRST THROUGH FOURTH CHOICE MAINTENANCE ACTIONS ARE GIVEN FROM LEFT TO RIGHT. M INDICATES MISSILE.

DECISION POINTS (DP'S) ARE GIVEN IN RIGHT COLUMN.

STATION NUMBER IF APPLICABLE

13 **14** **15** LINE IS DISPLAYED WHEN SEQ 4 IS REPEATED BY DIGITAL ENTRY. REAL FAILURES ARE ABOVE THE LINE; INTERMITTENT FAILURES ARE BELOW THE LINE.

FAULTY MISSILE

8 TEST SEQUENCE NUMBER AND MAS BIT STATUS

11 POWER TRANSIENT INDICATION, IF APPROPRIATE.

16 ADVISORY TO RUN SEQUENCE 3.

4-F50-252-0

NUMBERS WITH CIRCLES CORRESPOND TO LIKE ITEMS IN TEXT.

Figure 9-21. BIT Sequence 4 Displays

MOAT

MOAT functions tested are prelaunch phase lock and velocity slaving, rear receiver and command decoder, transmitter PA output, velocity update-search-acquisition, velocity track, active and ACM mode logic, angle track, and auto-pilot.

Sequence 4 fault detection and fault isolation displays include the following features:

- To validate the tests to be run in MOAT, certain tests are run to assure that all related WCS functions are proper. This is referred to as "MOAT readiness."

- To reduce MOAT failures caused by aircraft motion, adaptive threshold are used for certain tests. These thresholds are increased progressively up to certain limits to allow greater test tolerance when aircraft motion is a factor.

- Statistical data reduction technique is used, so that failures formerly caused by external disturbances, or noise, are identified.

- The lift accelerometer itself is checked, to eliminate it as a cause of MOAT failures.

- Missile phase lock is checked on all 6 missile channels, and the results stored in memory. At launch time, any missiles that failed phase lock on the channel selected are put last in the launch selection order.

  Also, the RIO can check which missiles failed which channel and the corresponding launch order sequence by dialing the channels with the thumbwheel at the end of MOAT.

  AIM-7 missiles are checked for missile tuning on the channel selected at the time of the MOAT, and the results displayed.

- Certain missile messages are sent a second time to reduce missed-message failures.

- MAS seeker-pointing-commands and transmitter on-commands are monitored to separate these from actual missile failures.

Fault isolation revisions provide the following features:

- Test results will be analyzed by comparing the pattern of failures (former DPs) with data stored in the program. This will enable the display of failures by function, WCS modes, station, missile type and MAS function.

- MAS test and MOAT readiness results are included in the analysis to allow separation of missile failures from WCS failures.

- When MOAT is rerun by digital entry, the results of the previous tests are saved and used to separate intermittent failures from hard failures.

Displays added include the following:

- The BIT box and test advisories have been moved to the lower half of the TID.

- MOAT results are displayed by overall WCS mode capability, missile launch capability, WCS mode, missile type, station phase lock by channel (AIM-54A), missile tune by station (AIM-7), MAS status, and launch sequence.

- In the maintenance display, MAS results are displayed in the form of prioritized maintenance actions and DPs. MOAT results are displayed below the MAS display by station number, prioritized WRAs, an M signifying missile fault or W for wiring fault, and an acronym indicating the fault type. When applicable, an advisory is dislayed at the bottom of the display recommending another BIT to run for additional data.

The following item numbers relate to the appropriate number on the MOAT illustrations (figure 9-21):

(1) WCS Mode Status

Display Data

Three-state status for three WCS MODES displayed. Status displayed:

$\sqrt{}$ = pass

X = fail

(blank) = unevaluated

Note

The ACM mode applies to AIM-54A missiles only and not to the AWG-9 ACM mode. AWG-9 mode status is not indicated.

(2)   Missile Stations

Display

- Station number for weapons stations having a missile ID. Maximum stations displayed = 6. Station numbers for maximum load are 1, 3, 4, 5, 6, and 8. Station number will blink during missile test (MOAT).

- MAS station fail indicated by parentheses enclosing station ID number.

- MOAT no-test indicated by parentheses enclosing station ID in the same manner as a MAS station failure.

- Excessive aircraft motion beyond the capacity of the adaptive threshold is indicated by motion symbol ⊘ displayed above the station number.

- Excessive noise that is sensed by the program that precludes successful test evaluation is indicated by an external disturbance symbol ⋈ above the station number.

(3)   Missile Type

Display Data

One missile type symbol for each station with a missile ID (maximum of 6).

The required symbols are:

    P = Phoenix

    S = Sparrow

    7E = Sparrow 7E ID

    7F = Sparrow 7F ID

(4)   Launch Priority

Display Data

Subscript numbers for each missile ID to indicate computer-selected launch sequence for Phoenix and Sparrow missiles on board.

Brackets enclosing a Phoenix missile ID and launch number, for example, [P_1] indicate that missile power cannot be deselected on that station. The launch priority will be number one, and higher number stations loaded with Phoenix missiles will indicate no test; that is, parentheses enclosing station numbers.

Note

This condition of latched missile power prohibits selection and, thus, testing of other stations; however, a missile on a station indicating latched power may be successfully prepared and launched.

(5)   Launch Safety

Display Data

A missile that may not be safe to launch will be indicated by enclosing the missile ID (including launch number) with carats (⟨ ⟩).

(6)   Phoenix Launch Capability

Display Data

Missile mode status of four missile modes on up to six stations. Status of modes displayed are:

    √   = Pass

    X   = Fail

    (√) = Indeterminate

    blank = Unevaluated

(7)   Phoenix Phase Lock/Sparrow Tune

Display Data

Upon completion of MOAT, the RIO has the capability of checking Phoenix phase-lock on all MC54 channels by thumbing through each channel on the DDD. All Phoenix missiles will be evaluated for each channel change. The Sparrow missiles will be evaluated only on the CW channel that they had tuned to, if they were tuned prior to MOAT.

(8) Test Sequence Number

Display

Number 4 blinking indicates MAS BIT is in progress.

(9) MAS BIT Status

Display

Pass or degrade symbol.

$\sqrt{}$ = Pass

D = Degrade (DPs)

(10) Failed MAS Function

Display

Symbol indicates if MAS failures affect Phoenix, Sparrow, or Sidewinder functions. Symbols required are:

P = Phoenix

S = Sparrow

SW = Sidewinder

(11) PTI Indicator

Display

PTI symbol is displayed whenever the power transient monitor routines detects a power transients except when display inhibit is requested by the resident program.

(12) Prerequisites and Advisories

Display

PREREQUISITES. The following control setting symbols are used to request operator action:

XMT = AWG-9 Transmitter On
PREP = AIM-54 Prep

(13) MAS DP/WRA Display

Display

MAS DPs are displayed on the right-hand side of the TID following depression of the fault display button as indicated in figure 9-21. Up to 4 maintenance choices are displayed horizontally across the TID on line with the related DP. If more than 7 failures occur, 7 failures are displayed on the TID initially. Any additional failures stored by the computer can be displayed by re-depressing the fault display button until all faults have been displayed as indicated by return to aircrew display.

Any MAS DPs stored as intermittent DPs are displayed beneath a horizontal bar across the TID as indicated in figure 9-21. The vertical position of the horizontal bar is variable and its position determined by the number of hard DPs or MOAT failures displayed. If there are no intermittent failures, the bar will not be displayed.

(14) MOAT Station Failures

Display

MOAT Station oriented failures and prioritized maintenance actions are displayed in the same manner as MAS failures in Item 13 except failure acronyms are displayed in the right-hand position in lieu of DPs. Additionally, the station number of the failed station is displayed to the left of the maintenance actions. If two different missile functions fail requiring two fail acronyms, two lines of actions are displayed and the station number is repeated. Intermittent MOAT failures are indicated in the same manner as defined for intermittent MAS failures.

(15) MOAT Detected AWG-9 Failures

Display

MOAT detected AWG-9 failures are displayed in the same manner as MOAT missile failures described above except that an AWG-9 system failure number (SXX) is displayed on the right in lieu of missile failure acronyms.

ADVISORIES. The following symbols are used for information only, no operator action required:

Displayed Advisory	When Displayed	Cause
DMLD	MOAT Readiness or MOAT	Dummy Load not IN
XMT[1]	MOAT Readiness or MOAT	Xmtr PP below 1.0 kw
PF[2]	MAS, MOAT Readiness or MOAT	Power Fault
SSI	MAS	SSI Failure
TIME	MOAT Readiness	Prep on but 128 seconds timer not timed out
ANALY-SIS	Pattern Analysis (Fault Isolation)	Test has terminated and results are being analyzed
RUN 3	Displays	Detected MOAT readiness failure needs further definition/ confirmation by sequence 3.

Note

(1) Also a prerequisite during MOAT readiness.
(2) When test is completed, reset power and rerun sequence 4.

Digital Entries For MOAT Rerun

Digital entries for repeat of MOAT are as follows:

NBR-40	Repeat sequence 4 MAS only.
NBR-41	Station 1 MOAT*
NBR-42	MATS/MITS
NBR-43	Station 3 MOAT*

NBR-44	Station 4 MOAT*
NBR-45	Station 5 MOAT*
NBR-46	Station 6 MOAT*
NBR-48	Station 8 MOAT*
NBR-49	MOAT on all stations with AIM-54A ID.

*Station with AIM-54A ID.

General MOAT Utilization Summary

MOAT tests the AIM-54A missile, but since all the weapons system must function correctly to test the missile, MOAT is a very good weapons system test. However, because the entire weapons system is involved, there are a number of problems other than a bad missile that can cause the weapons system to fail a MOAT.

The following guidelines are not in any particular order of importance and are not mandatory, but should be followed whenever time and/or conditions permit.

- Radio frequency interference (RFI) can cause the weapons system to fail MOAT. Separate missile channels should be assigned to each aircraft in a briefing if several aircraft are operating in the same area. A SNIFF check with the AWG-9 or the desired missile channel can be made to look for RFI.

- There are several causes of occasional failures to pass MOAT other than real problems that usually will exist on successive MOATs such as RFI, aircraft maneuvers, or other transient events.

- There are several modes of failures that are unique to a certain missile channel, so a missile may fail MOAT on one channel and pass on another channel.

Sequence 5 — Receiver Test

This test sequence verifies the performance and detects and isolates faults to a WRA within the receiving subsystem. BIT sequence 5 consists of the following tests:

- SSI test
- Overloads/power faults
- Main lobe and side lobe clutter tests
- T.O. channel lock and doppler filter tests

- Doppler processor test

- Receiver gain/processing tests

- Doppler processor/clutter processor tests

- Channel switching test

- Detection sensitivity test

Test results are displayed on the TID during and after the sequence (figure 9-22). These displays indicate the full range of receiving subststem operability from an overall pass (√) or degrade (D) through the individual subtest (pass, degrade, or fail (X)). BIT sequence 5 is primarily intended to be a maintenance tool, but it is often used for system verification by the RIO.

Sequence 5 is initiated by selecting BIT category on the computer address panel and depressing the RCVR 5 message pushbutton. After the sequence has been read into DRO from the MTM, a BIT box is displayed at the top of the TID with a blinking 5 in its upper left corner.

Upon initiation of sequence 5 an M or B appears in the lower left corner of the TID. The BIT box then appears after a short time delay which depends upon the point the MTM tape was situated prior to sequence 5 selection. Approximately 15 seconds after the appearance of the BIT box, the remainder of the sequence 5 TID display appears. The selected missile channel appears under the MC 54. Throughout sequence 5 the DDD indicates PD SRCH mode and shows a vertical bar about three-eights inch wide on the CRT. The next TID indications are the assessment in sequence of XM CH 1 through 8, MC 54 1 through 6, and MC 7 1 through 5, followed by the PA and DS assessments and value. Finally the overall sequence 5 assessment appears in the lower left corner of the BIT box. If the sequence 5 assessment was a pass (√), that test is concluded and the program remains in an idle loop waiting for operator action. If the assessment was a degrade (D), the program also enters an idle loop waiting for operator action such as depression of FAULT DISPL, digital entries, or running another sequence. Depressing FAULT DISPL will display the DPs/WRAs involved, but will not pull the program out of the idle loop.

Initial prerequisite testing is accomplished prior to actual receiving system tests. Thses tests include the SSI test and the overload/power fault test, which ensure that a set of system preformance baselines have been met. If these tests are failed, the sequence is terminated.

When the prerequisites are passed, the main lobe clutter and side lobe clutter tests are initialized. The MCL and SLC VCO's are varied across their frequency spectrum to three distinct points and measured in each of the FRM phases.

The BIT T.O. channel lock and doppler filter tests are then conducted. When a T.O. phase lock on a selected channel is attempted and fails, the testing jumps to the channel switching test. Next, a target aspect indication is checked. Reference levels for subbands 1 through 4 AGC are established, and the false alarm rate thresholds are checked. Comparisons are then made between fast and slow AGC levels. The data processing capability is checked using range rate data. Finally, the subband amplifier gain is measured.

The doppler processor test phase is conducted with a range processing test and clutter processor gain checks. The range processing test is run with FMR enabled and the side lobe clutter verified. The FMR phases jitter SLC to simulate a target in range, and that range is then checked and compared. The clutter processor gain is tested with each subband filter by comparing subband AGC to a 3.2 MHz target.

The receiver gain and receiver processing tests are then conducted. Unblanking in SNIFF mode is verified using an RF target. A blanking test and an inhibiting diode blanking loss test are run by varying the BIT target level by 18 dB and making comparisons. The paramp operation is then verified by varying receiver gain and series attenuation to compare target levels.

The doppler processor and filter processor tests follow. The subband filters are checked in TAIL aspect (subbands 1 through 4) and NOSE aspect (subbands 5 and 6) to respond to the SLC sweep test. At lease 121 out of 128 filters in each subband must respond to a pass. AGC readings are taken and averaged to test the filter ripple.

The channel switching tests are then conducted with T.O. phase lock tests and paramp tests conducted on each channel. If the MC 54 channel and an appropriate missile channel is selected (determined by a channel decoder check) then the AGC is measured for reference. The paramp is then enabled and the AGC is compared to determine a pass or fail. The results of the paramp test on every channel are then tabulated to determine overall paramp condition. A zero or 1 channel fail equals a pass, 2 to 10 channel failures equals a degrade and 11 or more channel failures equal a fail.

The detection sensitivity tests are run to establish overall receiver sensitivity. If the paramp fails on the selected missile channel, it is turned off, detection sensitivity reduced by one, target level increased, and then tested. If the paramp passes on the selected missile channel it is turned on and tested. The target sidebands are then swept through the subbands in TAIL and NOSE aspects as in filter ripple test.

SEQUENCE 5 DISPLAYS

FAULT DETECTION

CHANNEL ON WHICH
DETECTION SENSITIVITY
WAS
EVALUATED

TEST
SEQUENCE
NUMBER

TEST
RESULT

DETECTION*
SENSITIVITY

PARAMP
EVALUATION

TRANSMITTER
CHANNELS

PASSED
PHASE LOCK

FAILED
PHASE
LOCK

PHASE LOCK
SLOW BUT
USABLE

AIM-54A
MISSILE
CHANNELS

PARAMP
FAILED,
PHASE LOCK
PASSED

AIM-7E
AND/OR AIM-7F
MISSILE
CHANNELS

NOTE: THE SLOW (DEGRADED) PHASE LOCK
SYMBOL IS <u>NEVER</u> SHOWN WITH FAILED
PARAMP SYMBOL.

FAULT ISOLATION

TEST
SEQUENCE
NUMBER

FAILED
DECISION
POINTS

TEST
RESULT

FAULTY
WRA

* A PASS, DEGRADE, OR FAIL SYMBOL IS DISPLAYED
ABOVE THE DETECTION SENSITIVITY (DS) ACRONYM
DEPENDING ON THE DS NUMBER. DS 8 OR 9 EQUALS
PASS, DS 6 OR 7 EQUALS DEGRADE, AND DS 5
OR 0 EQUALS FAIL.

2-F050-004
206-0

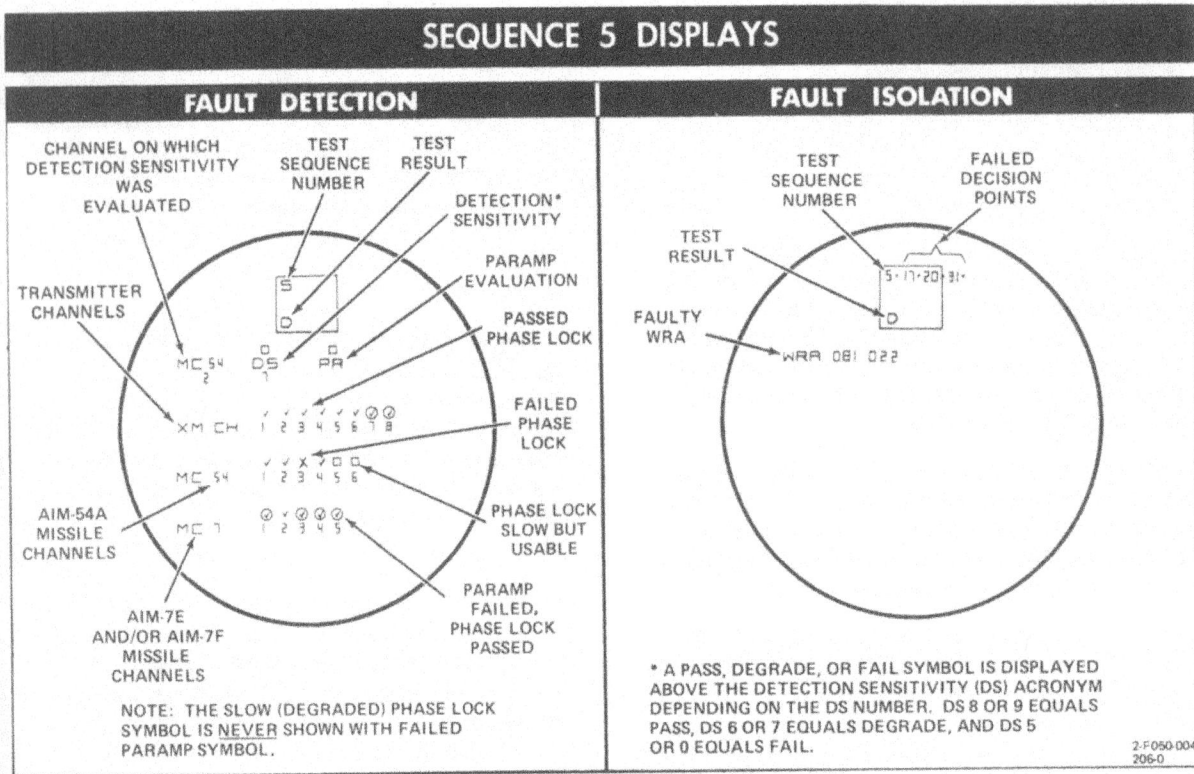

Figure 9-22. Sequence 5 Displays

There are no specific tests within sequence 5 where a failure will terminate the sequence. These failures and their respective DPs indicate a major problem and would cause several other pseudo failures and their DPs. If the sequence were allowed to continue, this would make RIO assessment and flight line maintenance difficult. However, when one of these failures occur, it does not exclude the possibility of failures being present in untested areas of the sequence.

Upon conclusion of sequence 5, with either a pass or degrade, initiation of a digital entry 50 will repeat the sequence. Subsequent depression of the ENTER pushbutton will run the sequence again. Digital entries 51 and 52 will run portions of sequence 5 and, in doing so, will display a different presentation on the TID. In both digital entry 51 and 52 detection sensitivity (DS) will be replaced by target level (T) and the transmitter and missile channels (XM, CH 1 through 8, MC 54 1 through 6, and MC7 105) will be replaced by a vertical row of numbers (6 to 1) representing the subbands, and six 3 digit numbers each representing the number of filters responding in each subband. Digital entry 51 repeats the paramp test on the selected channel and repeats the detection sensitivity test with FMR using slow sweep. Digital entry 52 repeats the paramp test on the selected channel and repeats the detection sensitivity test in range rate search mode with R-dot on high (FMR off) using fast sweep. DP 126 for T.O. phase lock fail is only tested in digital entries 51 and 52, not in the normal sequence 5.

Sequence 6 — Transmitter Test

The test sequence verifies the performance and detects and isolates faults to a WRA within the transmitter. Sequence 6 BIT transmitter test comprises the following tests:

- Synchronizer and prf
- Flood antenna switch
- Transmitter turn-on and low-prf peak power
- CW illuminator power
- High-prf transmitter and receiver blanking test
- Transmitter centrol-line-power
- Transmitter noise
- Central line power versus duty factor
- Transmitter fault isolation

Analog readouts are displayed on the TID during the test. These displays give the RIO a basic graphic indication of the condition of the transmitter. The test result displayed is transmitter peak power.

Sequence 6 is initiated by selecting BIT with the CATEGORY switch and depressing XMRT 6 pushbutton on the computer address panel. A flashing 6 in the upper left corner of the BIT box indicates that sequence 6 is running. The missile channel selected is displayed on the TID along with the acronyms for peak power (PP). If the JAM/JET control on the DDD is not adjusted properly, a blinking JAM will appear above the BIT box. Thirty seconds are allowed to adjust this control, after which time the blinking JAM is removed and the test continued. If the transmitter is not turned on, a blinking XMT will appear above the BIT box and continue to blink for 30 seconds unless another BIT sequence is selected or the CAP CATEGORY switch is rotated out of BIT. The test remains in an idle loop while XMT is blinking. If the transmitter is turned on in time, the test will continue; if not, the test will terminate and display a degrade symbol.

After the transmitter is turned on, the test continues. The next display generated during the test is the transmitter noise plot. A minus 70 dB reference line is displayed. The transmitter is then slewed from 120 kHz closing to 20 kHz opening in approximately 1 kHz increments. The corresponding output voltages are measured at each point and displayed as a series of dots relative to the -70 dB reference line (figure 9-23). The last display generated is the value of PP and its assessment.

When the test ends, the 6 in the BIT box stops blinking and a D is displayed in the lower left corner of the BIT box

if there were any DP; if not, a checkmark is displayed in the BIT box. The FAULT DISPL pushbutton may be depressed to display the failed DP along with the corresponding faulty WRA.

Once the test is completed, the RIO has the option of repeating the entire test by selecting digital entry 60. The repeat test can be performed on the same transmitter channel or any other missile channel (MC 54) selected. If desired, the RIO can repeat only the transmitter power test and the noise for one-third and one-half duty cycle by selecting digital entry 61 or 62 on the channel selected. This can be done for any transmitter channel. Digital entry 63 also allows the RIO to display the transmitter peak power on selected mode (PDS or PS) and a selected channel. It is not necessary to reinitiate the digital entry during channel or mode changes. Digital entry 64 can be used to present the decoded results of CM-indicated transmitter power faults.

Throughtout the sequence 6 test, the mode indication is pulse doppler search (PD on mode drum and PD SRCH pushbutton illuminated). The scan pattern on the DDD is one-bar, ±0° and remains basically unchanged in appearance through the test (wavy lines vertically).

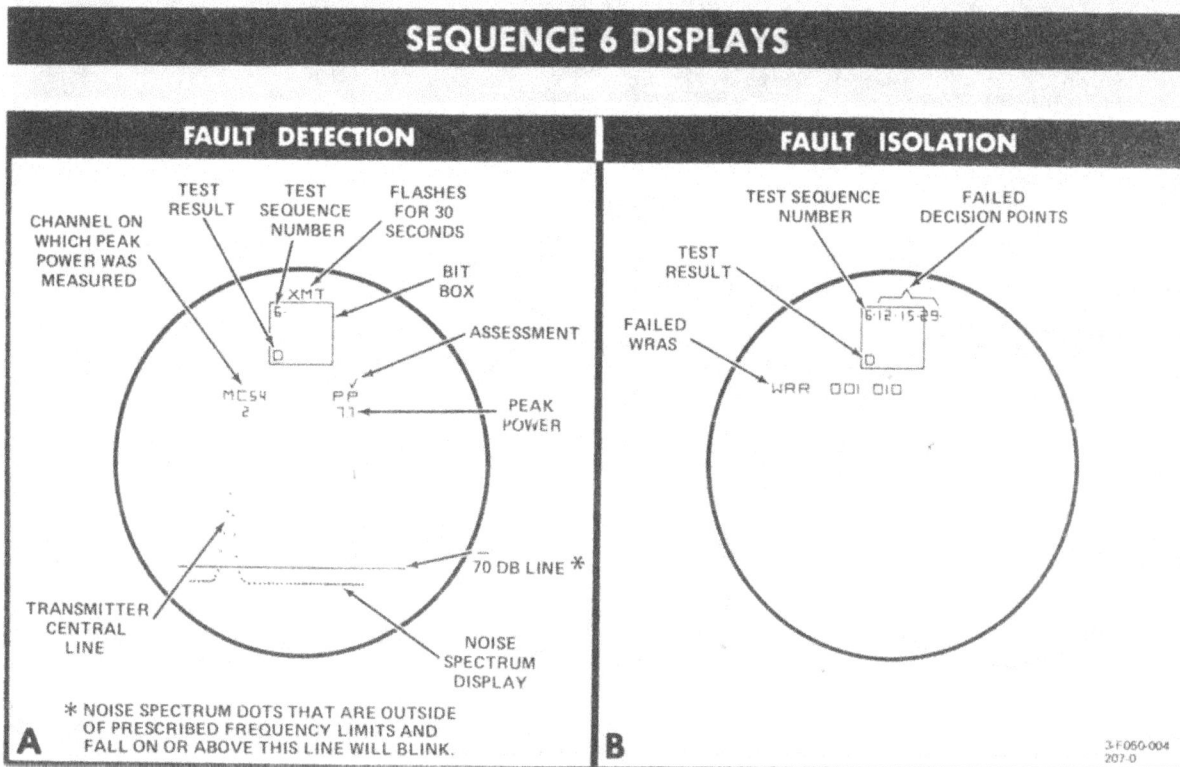

Figure 9-23. Sequence 6 Displays

Sequence 7 — Antenna Servo Tests

Sequence 7 is programmed to test the antenna scan pattern program accuracy and response. Sequence 7 is entered by selecting BIT with CATEGORY switch and depressing the ANT IR 7 MESSAGE pushbutton on the computer address panel. After sequence 7 has been read into the computer from the magnetic tape memory, a flashing 7 is displayed in the upper left corner of the BIT box. At the completion of the sequence, the 7 remains steady and the test results are displayed on the TID (figure 9-24).

Once the test is completed, the RIO may repeat the entire test by selecting digital entry 70. Digital entry 71 will cause the radar antenna scan pattern to be displayed on the TID as shown in figure 9-24. The commanded antenna elevation will appear in degrees as a digital readout on the left side of the TID. The operator can select any combination of antenna azimuth scans and elevation bar steps and examine the resultant radar antenna scan pattern in azimuth versus elevation coordinates. The effects of elevation commands to the antenna can be observed by manually adjusting the antenna elevation control knob on the sensor control panel while the test is in progress. Also, the test can be run with stabilization in or out, as selected on the

sensor control panel. Scan pattern under test should not change when selecting STAB IN or STAB OUT. If it does, the CSDC platform loop is probably at fault.

Note

High aircraft roll rates sometimes cause oscillations on some signals between the CSDC and the AWG-9 computer. These oscillations cause the radar to drop as much as 20° on both ends of the scan. Running BIT Sequence 7 will usually clear the problem.

Sequence 8 — Single Target Track Test

This test verifies the performance of both PSTT and PDSTT modes. Sequence 8 is initiated by selecting BIT with the CATEGORY switch and pressing the STT 8 pushbutton on the computer address panel.

This test sequence is executed in following order:

• Program Standard Serial Interface (SSI)

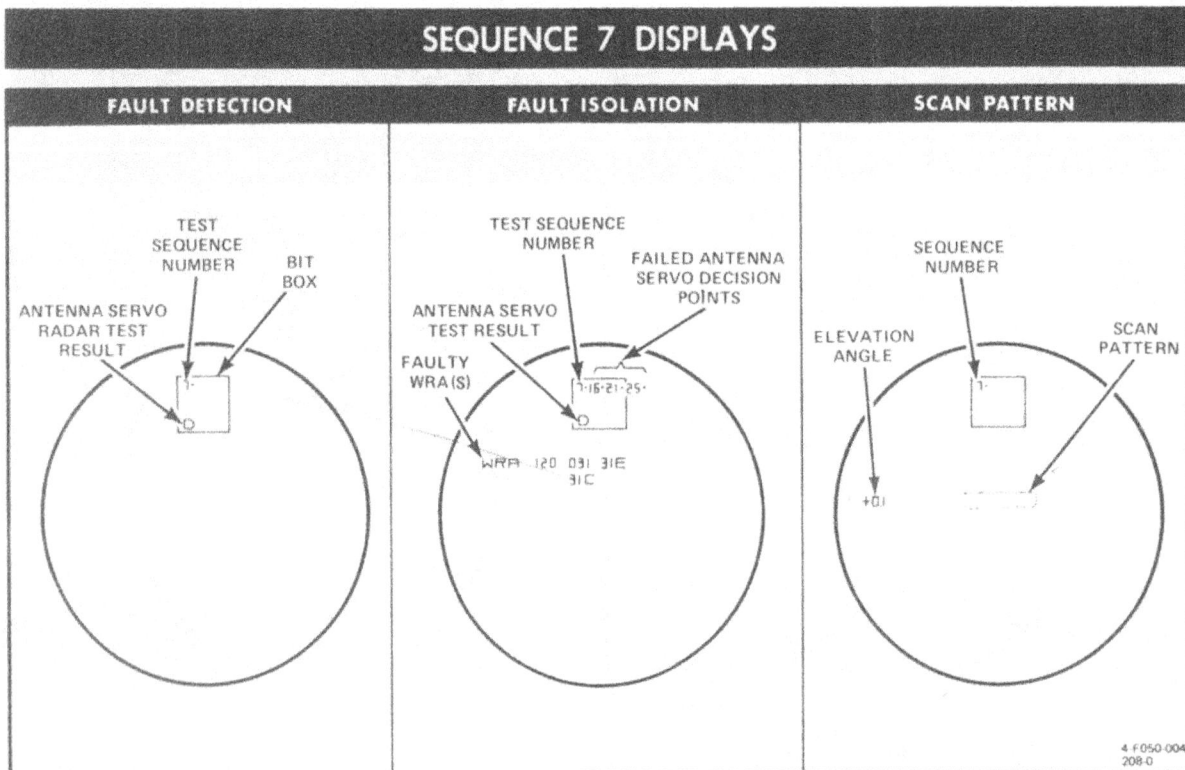

Figure 9-24. Sequence 7 Displays

- Speed Tracker Submit

- PDSTT

- PSTT

The SSI test is performed primarily to establish initial conditions for the test. The speed tracker test verifies proper operation of the radar to ensure it can acquire and track both constant and eclipsing target signals. Sensitivity is tested by decreasing the intensity of a simulated BIT target. PDSTT is also checked with the same BIT target. A doppler JAM track test is also conducted to verify that the system can operate in PDSTT in JAT. An antenna angle rate memory test is performed during the PDSTT to verify antenna drift limits. The PSTT test simulates target acquisition when transferring from PDSTT. A radar-on-target and delayed velocity-on-target condition will be indicated as a successful transition by the illumination on the RDROT and ANT TRK indicator lights on the DDD. Range and range rate are compared by ranging a simulated target from 5 nmi to 2.5 nmi. ACM threshold control and TWS to PSTT transition are tested to assure proper operation. PSTT and PSTT-JAT sensitivity for tracking targets is verified.

When the test is being read in from the magnetic tape memory, the BIT box on the TID and a two-bar, ±40° scan on the DDD will be displayed by the BIT executive routine. As soon as the program has been read in, a flashing 8 is displayed in the upper left corner of the BIT box and the DDD shows an azimuth scan of ±0° and one-bar is shown in the EL meter. The scan center is 0° azimuth and -5.2° elevation. At this point, speed tracker tests are performed. An intermediate-frequency target is generated and then removed during the test of target eclipsing and antenna drift. The RDROT indicator will illuminate when fast frequency-on-target has been detected and the ANT TRK indicator illuminates when delayed velocity-on-target has been detected. These two indicators illuminate and go out as the target eclipsing test is performed. The target level and aspects are then changed to determine the PDSTT sensitivity. The BIT on-target filter tests are also performed at this time.

The RDROT and ANT TRK indicators illuminate, then go out and the antenna is turned on and moves to 0° azimuth and -5.2° elevation (the horn target position). When lock-on is achieved, the RDROT and ANT TRK indicators illuminate again. The PDSTT angle track, horn target comparison, and high-prf tracking loop gain tests are performed at this time. The RDROT indicator goes out and the JAT indicator illuminates indicating that the high-prf jam track test is being performed. The ANT TRK and JAT indicators

then go out and the antenna rate memory test is performed. The antenna sweeps to the -30° position in azimuth on the DDD and the antenna drifts down on the EL meter during the antenna servo fault isolation portion of the test.

Mode transfer to PSTT is then performed. The PSTT mode pushbutton illuminates, PULSE is set on the mode drum, 10 nmi is indicated on the DDD range drum, and RDROT and ANT TRK indicators illuminate. The system tracks the target from 5 nmi to 2.5 nmi. The RDROT and ANT TRK indicators then go out and a target is reacquired at 5 nmi.

The RDROT and ANT TRK indicators illuminate again and the system tracks the horn target (0° azimuth, -5.2° elevation). PSTT lockon from TWS is performed at this time, although no external indication of TWS is given. Low-prf tracking loop gain tests are also performed at this time. The JAT indicator is then illuminated and the RDROT and ANT TRK indicators go out. The low-prf jam track tests are performed at this time and the antenna sweeps to the 30° left position in azimuth on the DDD and drifts down on the EL meter. The DDD then goes black and the TO leakage portion of the test is performed. Sequence 8 displays are shown in figure 9-25.

Sequence 7 And 8 – SRA Fault Isolation

Fault isolation on the radar antenna is performed down to the shop replaceable assembly (SRA) level. This results in some unique WRA displays. These displays may be displayed in combinations with each other or combinations with a WRA. Figure 9-26 shows the display, its meaning, and applicable BIT sequence.

TEST TARGET

It is recommended that the RIO verify proper WCS operation prior to each mission by running the WCS confidence tests (sequences 1 through 4), which require about 5 minutes to perform. In cases where there is insufficient time to perform these checks, the test target can be used to quickly verify that the WCS is capable of detecting, processing, and displaying reasonable-size targets. It is available in, and can be used to check the operation of PD or PULSE tactical modes. To use the test target, the RIO selects the TGT DATA category on the computer address panel and depresses the TEST TGT pushbutton. This activates the radar test horn in front of the radome. Normal mode operation continues. The radar test target will be processed and displayed on the DDD and TID just as would any other newly-detected target in the mode being tested.

SEQUENCE 8 DISPLAYS

FAULT DETECTION

FAULT ISOLATION

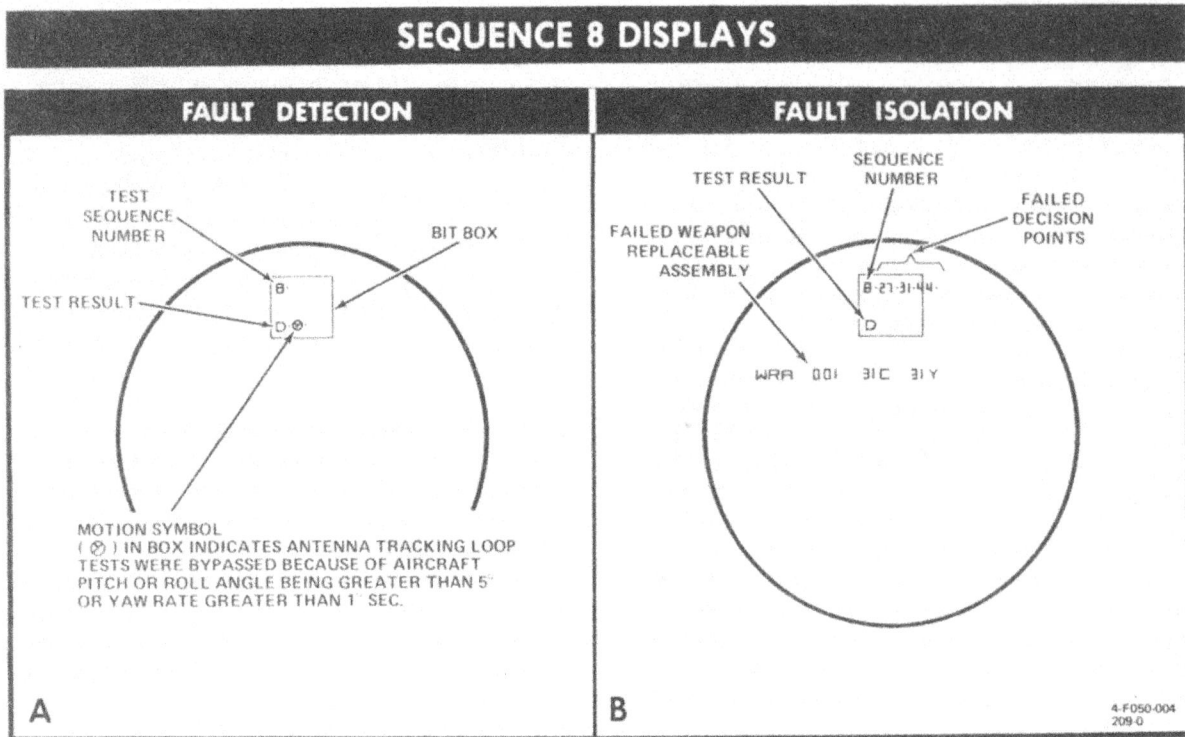

Figure 9-25. Sequence 8 Displays

DISPLAY	MEANING	BIT SEQUENCE 7	8
31A	AZ HYDRAULIC SERVOVALVE	X	
31B	EL HYDRAULIC SERVOVALVE	X	
31C	AZ INSTRUMENT GEARBOX	X	X
31D	EL INSTRUMENT GEARBOX	X	X
31E	ELECTRONICS PACKAGE	X	X
31H	HYDRAULIC POWER SUPPLY	X	
31X	AZ INTEGRATED GYRO		X
31Y	EL INTEGRATING GYRO		X

Figure 9-26. Sequence 7 and 8 SRA Fault Isolation

In addition to testing the operation of the various modes, the test target can also be used for checking many of the WCS controls (such as display controls) and verifying computer functions such as hooking. For example, the RIO can hook the test target (which first appears as an unknown target) on the TID; designate it hostile (noting symbol change); initiate single-target tracking on it (noting operation of ANT TRK and RDROT indicators); enter data pertaining to the target; and even test the track hold function after deselecting the test target.

The PD test targets appear with the following parameters:

ASPECT switch set to:

NOSE . . . . 1800 knots closing at 50 nmi
600 knots closing at 20 nmi

BEAM. . . . . 600 knots closing at 20 nmi
1800 knots closing at 50 nmi

TAIL . . . . 600 knots closing at 20 nmi
600 knots opening at 20 nmi
(opening target may be deleted by altitude return filter)

The pulse test target appears at 2 nmi. All radar test targets appear at -5.2° elevation and 0° azimuth.

Note

When TEST TARGET is selected, a T is displayed on the TID as an AWG-9 CM acronym.

On the deck, perform the following:

1. Set WCS switch to STBY or XMT.

2. Depress TEST TGT pushbutton.

Note

Although TGT DATA category must be selected to achieve or deactivate TEST TGT, the targets retain their status when TGT DATA category is deselected.

3. Select desired RDR mode.

In flight, perform the following:

1. Set WCS switch to XMT.

2. Depress TEST TGT pushbutton.

Note

This procedure supplements BIT; it does not replace BIT.

BIT MOVING TARGETS

The BIT moving-target generator program allows the operator to simulate in-flight target intercept conditions. It provides targets from the radome or MOAT horns to enable the operator to effectively test all AWG-9 radar functions. The operator, by making entries via the computer address panel, has the capability to generate targets at various ranges and closing rates in both high- and low-prf. Initiate BIT moving-target generator as follows:

Note

Ensure that TEST TGT pushbutton is OFF (TEST TGT ON inhibits the BIT moving-target program).

1. Set WCS switch to STBY or XMT.

2. Set CATEGORY switch to SPL.

Note

BIT moving target can be exercised while airborne if the pilot selects TNG.

3. Depress and release MESSAGE pushbutton 7.

Note

Backlighting of function pushbutton indicates that program is reading in or has been read in. Verify that M (magnetic tape read) is removed from lower left quadrant of TID display before proceeding. A normal tactical display appears at end of tape read.

4. The program is executed by making a digital entry in the format NBR BCD for the radome horn or NBR ABCD for the MOAT horn, then depressing ENTER on the computer address panel. Explanation of the format is as follows:

   ● A in the entry selects the MOAT horn. Any digit (such as 9) suffices for an A entry and results in M being displayed in the A position, otherwise position A is blank.

   ● B in the entry selects starting range and closing rate of target. Various possible entries are listed in figure 9-27.

   ● C in the entry selects target level. Various possible target level entries are listed in figure 9-27.

   ● D in the entry selects number of targets to be displayed in pulse Doppler. If a 2 is entered, two targets are generated per entry B. Any other digit will result in a single target. If a pulse doppler jamming target is desired for PDSTT, an entry of digit 7 is required. It is not required to depress ENTER. Depressing CLEAR will remove the jammer. If 7 is depressed, the D position display will blank and stay blank.

5. Once a valid entry code is entered, initial conditions for selected target(s) are set and MT ABCD is displayed below the flycatcher readout on TID figure 9-27.

6. Select SNIFF.

7. Adjust antenna elevation to approximately -5°.

BIT MOVING TARGET GENERATOR ENTRIES AND DISPLAY

ENTRY "B"	STARTING RANGE* (nmi)	RANGE RATE (KNOTS)
0	5	+100
1	10	+200
2	20	+200
3	32	+300
4	60	+300
5	80	+600
6	100	+1200
7	150	+1500

ENTRY "C"	HPRF TARGET LEVEL	LPRF TARGET LEVEL
1	LOWEST	LOW
2		HIGH
3	↓	HIGH
4		HIGH
5	HIGHEST	HIGH

*NO LPRF TARGET GENERATED IF RANGE IS GREATER THAN 33 MILES

TWS WITH RADOME HORN

POSTT WITH MOAT HORN

5-F-50-223-0

Figure 9-27. BIT Moving Target Generator Entries and Displays

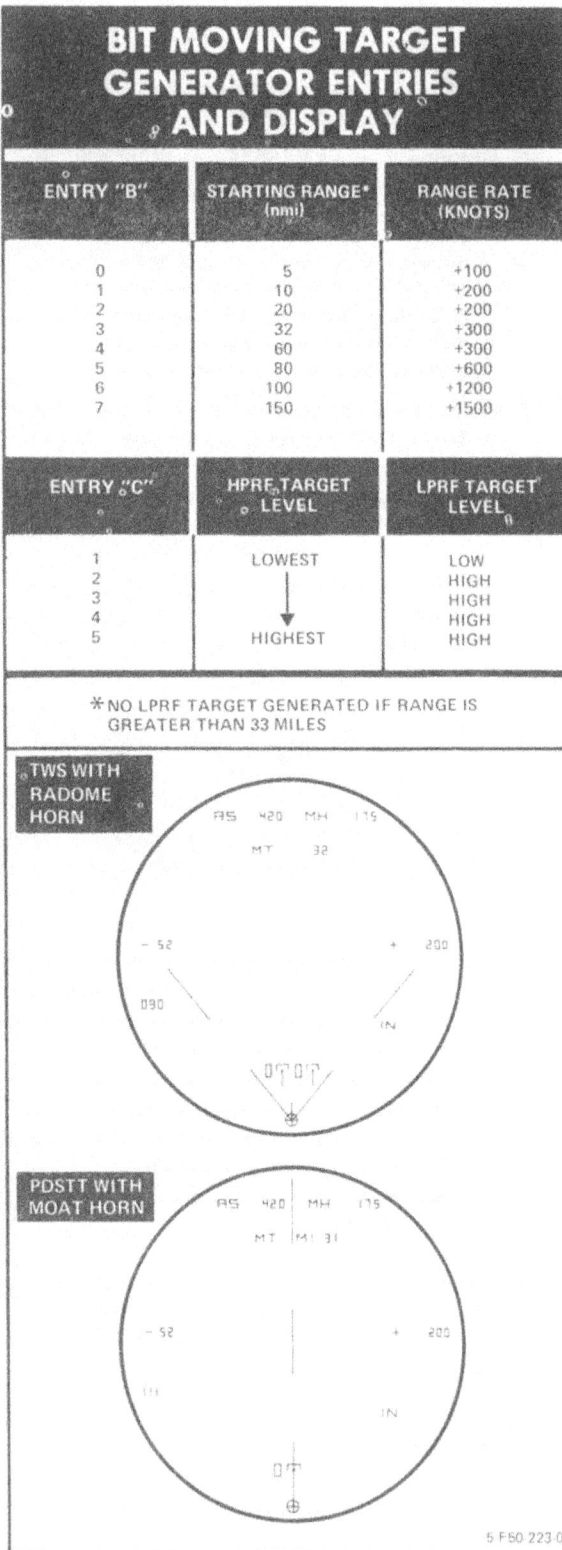

8. Targets displayed on DDD are treated as valid by tactical program.

9. The MOAT horn output target selection is provided for maintenance purposes only. This selection provides the option of a stronger target by manually connecting the radome horn target cable or an external horn to the MOAT horn output at the RMO.

When the WCS is locked on to either the BMT or the test target in PDSTT, the ANT TRK light will blink as it does in the tactical program on an actual target.

Target may be deselected by pressing and releasing CLEAR ENTER.

AWG-15 BUILT-IN TEST (COATS NON-FUNCTIONAL)

In aircraft BUNO 160413 and prior without AFC 520:

Preflight

The status of the AWG-15 is not verified on deck during OBC, however, valid AWG-15 acronyms may be displayed in OBC Continuous Monitoring (CM) and OBC maintenance readout (See figure 9-28). If a time code readout (TCR) is displayed at the end of on-deck OBC, the TCR is not valid.

Aircrews should note on-deck AWG-15 CM and OBC maintenance readout acronyms when stores are onboard. A CIA PSU acronym requires an on-deck abort. Presence of CIA PSU indicates possible failure of both A and B power supplies and no ability to launch or jettison stores.

Inflight

The complete status of the AWG-15 can only be checked during inflight OBC. Verification of the AWG-15 status is determined with acronyms displayed at the completion of OBC in the maintenance readout and in CM. The procedures for running an AWG-15 inflight OBC are:

1. Verify ACM, NEXT LAUNCH, A/G not selected.

Note

If ACM, NEXT LAUNCH, or A/G is selected, it will be interpreted as a failure and terminates the AWG-15 BIT.

2. MASTER TEST — OFF.

		Aircraft BUNO 160413 and prior without AFC 520
LOCATION	ACRONYM	MEANING/ACTIONS
Continuous Monitoring (CM)	CIA	BIT failure in one of the AWG-15 components. **Note** Aircrew should select MAINT DISP to read out failed components.
	CPA	Invalid
	CPB	Invalid
OBC Maintenance Readout	CIA	BIT failure in one of the AWG-15 components.
	CPA	Invalid
	CPB	Invalid
	CI	Control Indicator failure. Caused by a power up anomaly or a failure of CI BIT circuitry. Run airborne OBC to check for validity.
	PSU	Power Switching Unit failure, A or B power supply failure, or power up anomaly in the PSU. Aircrew should ensure the failure is valid and not retained from a previous flight. Abort if external stores are aboard, unless operational necessity dictates otherwise.
	CIA A or B	Station Decoder failure. Run airborne OBC to check validity.

WARNING

Do not cycle AWG-15 circuit breakers or aircraft electrical power to clear an AWG-15 failure either on deck or airborne, as stores may be inadvertently jettisoned.

Figure 9-28. Preflight AWG-15 Acronyms (Non-COATS)

3. MASTER ARM — OFF.

4. RIO initiates OBC and aircrew check for the following:

  ● Station status indicators cycle black, white, checkerboard

  ● HOT TRIG, SEAM LOCK, and COLLISION lights momentarily illuminate (at initiate).

  ● LAUNCH button and FUZE HV lights illuminate (maximum of 4 seconds).

5. Results: (See figure 9-29 for acronym meanings.)

Note

The TCR at the end of OBC is not valid.

  ● If a HOT TRIG light is not present during inflight OBC, the status of the AWG-15 is unknown and troubleshooting procedures should not be attempted. The mission should be aborted with stores aboard.

  ● If the HOT TRIG, SEAM LOCK, COLLISION and LAUNCH button lights illuminate, the trigger and bomb button are not shorted. This indicates that the AWG-15 BIT check has progressed far enough through the test sequence to have checked the trigger and bomb button for shorts.

  ● If the HOT TRIG, SEAM LOCK and COLLISION lights illuminate without a LAUNCH button light illumination, a shorted trigger or bomb button could exist. This indicates that the AWG-15 BIT check has not progressed far enough to check the trigger and bomb button or has detected a shorted trigger.

Aircrews should not attempt to troubleshoot a CIA CI acronym. If operational necessity dictates, aircrews may determine the validity of a CIA PSU acronym or if the trigger is shorted (that is, LAUNCH button did not illuminate) by performing the following procedures:

┌─────────────┐
│ **WARNING** │
└─────────────┘

When operational necessity dictates, troubleshooting procedures may be performed to determine if a malfunction is valid, however, stores jettison, launch or release may result.

1. Proceed to a safe area.

2. RIO select WPN TYPE — OFF.

3. RIO STA SEL — SAFE, TANK JETT 2 and 7 — SAFE.

4. Pilot WPN SELECT — OFF.

5. Pilot MASTER ARM — ON.

6. Pilot master test — OFF.

7. RIO select OBC.

8. Results:

  ● If CIA PSU acronym is cleared, normal operations may be performed. This indicates that the CIA PSU acronym that appeared with MASTER ARM OFF was due to an anomaly.

  ● If CIA PSU acronym recurs, a valid malfunction exists. The aircrew should abort the mission if stores are aboard.

  ● If HOT TRIG, SEAM LOCK, COLLISION and LAUNCH button lights illuminate, the trigger and bomb button are not shorted regardless of acronyms displayed. This indicates that the AWG-15 BIT has progressed far enough through the test sequence to check the trigger and bomb button for shorts. The previous fault and its effect on system performance is unknown. Sidewinder tests, except for 28 V dc decoder power from the PSU, are complete. Since decoder power is redundant, there is a high probability that sidewinder launches will be successful.

		Aircraft BUNO 160413 and prior without AFC 520
LOCATION	**ACRONYM**	**MEANING/ACTIONS**
Continuous Monitoring (CM)	CIA	BIT failure in one of the AWG-15 components. **Note** Aircrew should select MAINT DISP or OBC to readout failed components.
	CPA	Invalid
	CPB	Invalid
OBC Maintenance Readout	CIA	BIT failure in one of the AWG-15 components.
	CPA	Invalid
	CPB	Invalid
	CI	Control Indicator failure. AWG-15 BIT sequence is terminated after CI BIT and the remaining circuits are not tested. The mission should be aborted if stores are aboard.
	PSU	Power Switching Unit failure. AWG-15 BIT sequence is terminated after PSU BIT and the remaining circuits are not tested. The mission should be aborted if stores are aboard. If operational necessity dictates, an OBC with MASTER ARM ON may determine if the malfunction is valid.
	CIA A or B	Station Decoder failure. The probability is that one of the redundant frequency channels of that Station Decoder is inoperative with the remaining channel providing full release and jettison capability. The mission may be continued. **Note** • A — decoders provide the release signal to all missiles on STA 1 and 8 and to the AIM-54 on STA 3, 4, 5, 6. • B — decoders provide the release signal to the external tanks on STA 2 and 7 and to the AIM-7 on STA 3/6 and 4/5.

WARNING

• Do not cycle AWG-15 circuit breakers on aircraft electrical power to clear an AWG-15 failure either on deck or airborne, as stores may be inadvertantly jettisoned.

• With a CIA CI or PSU acronym displayed the total status of the AWG-15 is unknown. When operational necessity dictates, troubleshooting procedures (that is, Re-run OBC with MASTER ARM ON) may be performed to determine if a CIA PSU malfunction is valid, however, stores jettison, launch, or release may result.

Figure 9-29. Inflight AWG-15 Acronyms (Non-COATS)

WARNING

If HOT TRIG, SEAM LOCK, and COLLISION
lights illuminate without a LAUNCH
button illumination, a shorted trigger or
bomb button could exist. This indicates that
the AWG-15 BIT has not progressed far enough
to check the trigger and bomb button or has
detected a shorted trigger or bomb button.
If external stores are aboard, the mission
should be aborted.

If the LAUNCH button does not illuminate (and no external stores aboard), determine if the trigger is shorted for
gun operation only by:

1. Proceed to a safe area.

2. Ensure MASTER ARM – OFF.

3. WPN SELECT – GUN.

WARNING

If GCS appears in CM the trigger is shorted
and MASTER ARM should not be selected.

4. Results:

- If GCS acronym does not appear in CM, the
trigger is not shorted and gun operation may be
performed.

AWG-15 BUILT-IN TEST (COATS FUNCTIONAL)

In aircraft BUNO 160414 and subsequent, and aircraft
incorporating AFC 520:

All aircraft with tactical tape 111D/P7D incorporated
measures the elapsed time when AWG-15 BIT fail, monitor
AWG-15 power supply failure, and transfer failure data
to the AWG-9 computer. Aircraft BUNO 160414 and
subsequent and those incorporating AFC 520 have the
hardware that allows valid status monitoring of the A and
B power supplies and that allows a valid time code readout
(TCR) to be displayed in OBC Continuous Monitoring
(CM) or OBC maintenance readout. OBC includes AWG-15
Coded Integrated Armament Control System (CIACS)
failure isolation through the CIACS operational assessment
timing system (COATS). Detected BIT failures are displayed on the TID by an appropriate acronym, followed
by the BIT failure time in milliseconds. If no failure
occurs, the total time of the test is displayed. The time
status is monitored during the approximately 4 second
period of AWG-15 OBC. WRA acronyms are displayed
following the BIT failure TCR, (figure 9-30).

Note

- The BIT failure TCR is valid only if a
modified (AVC 2234) power switching
unit (PSU) is installed. Aircrews may
determine if AFC 520 and AVC 2234 are
incorporated by running inflight OBC with
NEXT LAUNCH selected and observing
a TCR between 0305 and 0345.

- In aircraft BUNO 160414 and subsequent
and aircraft incorporating AFC 520, the
AWG-15 BIT is a sequential test requiring
a total of 3250 milliseconds. The following
functional testing and test time periods
occur during the BIT cycle:

FUNCTION	START	STOP
Sidewinder Tests	0000	0320
Decoder Frequency Tests	0000	1695
Aux Jett Test	0320	0595
Select Jett Test	0595	0810
Gun Test	1695	1775
AIM-7 Test	1695	1735
AIM-54 Test	1735	1775
Decoder Power Tests	1735	2310
Rocket Test	2310	3250
Emerg Jett 1 Test	2880	3250

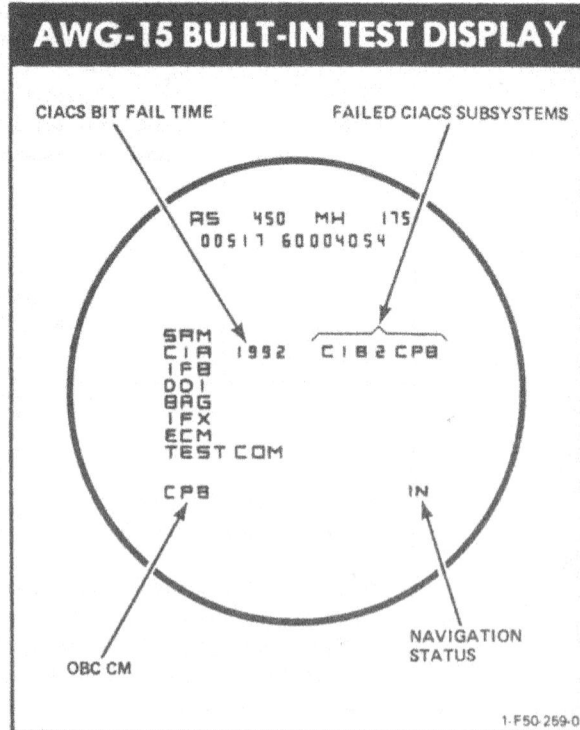

Figure 9-30. AWG-15 Built-In Test Display

To ensure safe operation, detection of a failure terminates the AWG-15 BIT and the appropriate failed component acronym is displayed in the maintenance readout at the conclusion of OBC. Only failures detected in CI or PSU circuitry will terminate AWG-15 BIT. Functional tests sequenced after the displayed failure time will not be tested.

Preflight

The status of the AWG-15 is not checked on deck during OBC, however, valid AWG-15 acronyms may be displayed to the aircrew in OBC CM and OBC maintenance readout (figure 9-31). If a TCR is displayed at the end of on-deck OBC, the TCR is not valid.

Inflight

Complete verification of AWG-15 status can only be checked during inflight OBC. The aircrew assesses the status of the AWG-15 from the acronyms and the TCR displayed at the completion of OBC as well as the acronyms displayed in CM. The procedures for running an AWG-15 inflight OBC are:

1. Verify ACM, NEXT LAUNCH, A/G not selected.

Note

If ACM, NEXT LAUNCH or A/G is selected, it will be interpreted as a failure and terminate the AWG-15 BIT.

2. MASTER TEST – OFF.

3. MASTER ARM – OFF.

4. RIO initiate OBC.

5. Results:

- Note BIT failure TCR displayed if acronyms CI and/or PSU are displayed (figure 9-32).

- Refer to critical failure time testing (figure 9-33) for additional testing and degraded mode assessment determinations.

Note

A TCR is always displayed as a result of AWG-15 OBC BIT. If a CI or PSU acronym is not displayed, the time will be 3900 ms or greater. This indicates successful completion of AWG-15 BIT.

	Aircraft BUNO 160414 and subsequent and aircraft incorporating AFC 520	
LOCATION	**ACRONYM**	**MEANING/ACTIONS**
Continuous Monitoring (CM)	CIA	BIT failure in one of the AWG-15 components. **Note** Aircraft should select MAINT DISP to read out failed components.
	CPA	Failure of the A power supply. No AIM-54, AIM-7, AIM-9, or Gun capability. Emergency jettison available on B power supply.
	CPB	Failure of the B power supply. Loss of the emergency jettison backup power supply.
OBC Maintenance Readout	CIA	BIT failure in one of the AWG-15 components.
	CPA	Failure of the A power supply. No AIM-54, AIM-7, AIM-9, or Gun capability. Emergency jettison available on B power supply. If weapons launch is required, mission should be aborted.
	CPB	Failure of the B power supply. Loss of the emergency jettison backup power supply.
	CI	Control Indicator failure. Caused by a power up anomaly or a failure of the CI BIT circuitry. Run airborne OBC to check validity.
	PSU	Power Switching Unit failure. Caused by a power up anomaly or a failure of the PSU BIT circuitry. Run airborne OBC to check validity.
	CIA A or B	Station Decoder failure. Run an airborne OBC to check validity.

WARNING

Do not cycle AWG-15 circuit breakers or aircraft electrical power to clear an AWG-15 failure either on deck or airborne, as stores may be inadvertently jettisoned.

Note

Aircrew should ensure that any displayed acronym is a current failure and not retained from a previous flight by clearing maintenance readout.

Figure 9-31. Preflight AWG-15 Acronyms (COATS)

Aircraft BUNO 160414 and subsequent and aircraft incorporating AFC 520		
LOCATION	ACRONYM	MEANING/ACTIONS
Continuous Monitoring (CM)	CIA	BIT failure in one of the AWG-15 components. **Note** Aircrew should select MAINT DISP or OBC to readout failed components.
	CPA	Failure of the A power supply. No AIM-54, AIM-7, AIM-9, or Gun capability. Emergency jettison available on B power supply. AWG-15 BIT in Class I OBC will not run. If weapons launch required, abort the mission.
	CPB	Failure of the B power supply. Loss of emergency jettison backup power supply.
OBC Maintenance Readout	CIA	BIT failure in one of the AWG-15 components.
	CPA	Failure of the A power supply. No AIM-54, AIM-7, AIM-9, or Gun capability. Emergency jettison available on B power supply. AWG-15 BIT in Class I OBC will not run. If weapons launch required, abort the mission.
	CPB	Failure of the B power supply. Loss of the emergency jettison backup power supply.
	CI	Refer to Critical Failure Time Testing, figure 9-33.
	PSU	If not in conjunction with CPA or CPB, refer to Critical Failure Time Testing, figure 9-33.
	CIA A or B	Station Decoder failure. The probability is that one of the redundant frequency channels of that Station Decoder is inoperative with the remaining channel providing full release and jettison capability. The mission may be continued. **Note** • A — decoders provide the release signal to all missiles on STA 1 and 8 and to the AIM-54 on STA 3, 4, 5, 6. • B — decoders provide the release signal to the external tanks on STA 2 and 7 and to the AIM-7 on STA 3/6 and 4/5.

WARNING

Do not cycle AWG-15 circuit breakers or aircraft electrical power to clear an AWG-15 failure either on deck or airborne, as stores may be inadvertently jettisoned.

Figure 9-32. Inflight AWG-15 Acronyms (COATS)

OBC DISP	FAIL TIME		POSSIBLE FAILURE	PROCEDURES	MISSION IMPACT
	FROM	TO			
--	0000	0000	No test	None	Unknown AWG-15 status
CI	0000	0025	Shorted Trigger	(1) MASTER ARM — OFF (2) WEAPONS SELECT — GUN If GCS is displayed in CM trigger is shorted. **WARNING** If GCS appears in continuous monitoring, the trigger is shorted and MASTER ARM should not be selected. This indicates that the contacts in the trigger that interface with the gun gas purge switching assembly are shorted and therefore one safety interlock in the gun control system is defeated.	ABORT GUN FIRING
CI	0000	0025	SW Weapon Select	(1) WEAPONS SELECT — SW Monitor VDI for correct display. (2) SW COOL — Ensure ON (3) STEP — SW stations Monitor ACM WEAPONS STATUS INDICATORS to locate faulty stations.	No SW capability if SW not displayed
PSU	0000	0025	SW STATION SELECT or SW COOL	None	Degraded SW capability
CI	0065	0105	Shorted Bomb Button	MASTER ARM — OFF	Crew decision
CI	0305	0345	(NEXT LAUNCH Selected)	**WARNING** If trigger or bomb button is shorted, store separation or gun firing could occur when MASTER ARM is selected and HOT TRIG light illuminated. Do not continue OBC with MASTER ARM selected unless operational necessity absolutely dictates, and then only when the aircraft is in a safe area where accidental firing or release of weapons will not be a hazard to friendly forces.	
CI	1675	1715	Gun Select	(1) WEAPONS SELECT — GUN Monitor VDI for correct weapons display.	No gun capability if G not displayed
CI	1675	1715	AIM-7 Select	(1) WEAPONS SELECT — SP Monitor VDI for correct weapons display.	No AIM-7 capability if SP not displayed
CI	1715	1755	AIM-54 Select	(1) WEAPONS SELECT — PH Monitor VDI for correct weapons display.	No AIM-54 capability if PH not displayed

Figure 9-33. Critical Failure Time Testing

SECTION X — NATOPS EVALUATION

TABLE OF CONTENTS

NATOPS EVALUATION PROGRAM

Concept

The standard operating procedures prescribed in this manual represent the optimum method of operating the aircraft. The NATOPS evaluation is intended to evaluate compliance with NATOPS procedures by observing and grading individuals and units. This evaluation is tailored for compatability with various operational commitments and missions of both Navy and Marine Corps units. The prime objective of the NATOPS evaluation program is to assist the unit commanding officer in improving unit readiness and safety through constructive comment. Maximum benefit from the NATOPS program is achieved only through the vigorous support of the program by commanding officers as well as by flightcrew members.

Implementation

The NATOPS evaluation program shall be carried out in every unit operating naval aircraft. The various categories of flightcrew members desiring to attain and retain qualification in the F-14A shall be evaluated initially in accordance with OPNAV Instruction 3510.9 series, and at least once during the 12 months following initial and subsequent evaluations. Individual and unit NATOPS evaluations will be conducted annually; however, instruction in and observation of adherence to NATOPS procedures must be on a daily basis within each unit to obtain maximum benefits from the program. The NATOPS Coordinators, Evaluators, and Instructors shall administer the program as outlined in OPNAVINST 3510.9 series. Evaluees who receive a grade of Unqualified on a ground or flight evaluation shall be allowed 30 days in which to complete a re-evaluation. A maximum of 60 days may elapse between

the date of the initial ground and flight evaluation and the date that qualification is satisfactorily completed. Flightcrew members possessing a valid F-14A NATOPS Evaluation Report Form, (OPNAV Form 3510-8) are considered qualified in all F-14A model aircraft provided the conditions outlined in Section II, (Indoctrination) are met.

Definitions

The following terms, used throughout this section, are defined below as to their specific meaning within the NATOPS program.

NATOPS EVALUATION

A periodic evaluation of individual flightcrew members standardization consisting of an open-book examination, closed-book examination, oral examination, and flight evaluation.

NATOPS RE-EVALUATION

A partial NATOPS evaluation administered to a flightcrew member who has been placed in an Unqualified status by receiving an Unqualified grade for any ground examination or for the flight evaluations. Only those areas in which an unsatisfactory level was identified need be observed during a re-evaluation.

QUALIFIED

The evaluation term applied to a flightcrew member who is well standardized and who demonstrated highly professional knowledge of and compliance with NATOPS standards and procedures. Momentary deviations from or minor omission in noncritical areas are permitted if prompt and timely remedial action was initiated by the evaluee.

CONDITIONALLY QUALIFIED

The evaluation term applied to a flightcrew member who is satisfactorily standardized, who may have made one or more significant deviations from NATOPS standards and procedure but made no errors in critical areas and no errors jeopardizing mission accomplishment or flight safety.

UNQUALIFIED

The evaluation term applied to a flightcrew member who is not acceptably standardized, who failed to meet minimum standards regarding knowledge of and/or ability to apply NATOPS procedures, or who made one or more significant deviations from NATOPS standards and procedures that could jeopardize mission accomplishment or flight safety.

AREA

A routine of preflight, flight, or postflight.

SUBAREA

A performance subdivision within an area, which is covered and evaluated during an evaluation flight.

CRITICAL AREA AND SUBAREA

Any area or subarea that covers items of significant importance to the overall mission requirements, the marginal performance of which would jeopardize safe conduct of the flight.

GROUND EVALUATION

Prior to commencing the flight evaluation, an evaluee must achieve a minimum grade of Qualified on the open-book and closed-book examinations. The oral examination is also part of the ground evaluation but may be conducted as part of the flight evaluation. To assure a degree of standardization between units, the NATOPS instructors may use the bank of questions contained in this section in preparing portions of the written examinations.

Open-Book Examination

The open-book examination shall consist of, but not be limited to, the question bank. The purpose of the open-book examination portion of the written examination is to evaluate the flightcrew member's knowledge of appropriate publications and the aircraft.

Closed-Book Examination

The closed-book examination may be taken from, but shall not be limited to, the question bank, and shall include questions concerning normal and emergency procedures and aircraft limitations. Questions designated critical will be so marked.

Oral Examination

The questions may be taken from this manual and may be drawn from the experience of the Instructor-Evaluator. Such questions should be direct and positive and should in no way be based solely on opinion.

Emergency

An aircraft component or system failure or condition that requires instantaneous recognition, analysis, and proper action.

Malfunction

An aircraft component or system failure or condition that requires recognition and analysis, but which permits more deliberate action than that required for an emergency.

OFT and WST Procedures Evaluation

A OFT and WST may be used to assist in measuring the flightcrew member's efficiency in the execution of normal operating procedures and his reaction to emergencies and malfunctions. In areas not covered by the OFT and WST facilities, this may be done by placing the flightcrew member in an aircraft and administering appropriate questions.

Grading Instructions

Examination grades shall use a 4.0 scale and be converted to an adjective grade of Qualified or Unqualified.

OPEN-BOOK EXAMINATION

To obtain a grade of Qualified, an evaluee must obtain a minimum score of 3.5.

CLOSED-BOOK EXAMINATION

To obtain a grade of Qualified, an evaluee must obtain a minimum score of 3.3.

ORAL EXAMINATION AND OFT AND WST
PROCEDURE CHECK (IF CONDUCTED)

A grade of Qualified or Unqualified shall be assigned by
the Instructor-Evaluator.

FLIGHT EVALUATION

The flight evaluation may be conducted on any routine
syllabus flight with the exception of flights launched
for FMLP and CARQUAL or ECCM training. Emergencies will not be simulated.

The number of flights required to complete the flight
evaluation should be kept to a minimum; normally one
flight. The areas and subareas to be observed and
graded on a flight evaluation are outlined in the grading
criteria with critical areas marked by an asterisk (*).
Grades on subareas will be assigned in accordance
with the grading criteria. Grades on subareas shall be
combined to arrive at the overall grade for the flight.
If desired, grades of areas shall also be determined in
this manner. At the discretion of the squadron or unit
commander, the evaluation may be conducted in WST,
OFT, or COT.

OPERATIONAL DEPLOYABLE SQUADRONS

Pilots and RIOs assigned to operational deployable
squadrons will normally be checked as a team, with
the flight evaluation being conducted by the checkcrew
flying wing. RIO commentary will be transmitted on
the GCI or CIC control frequency in use.

TRAINING AND EVALUATION SQUADRONS

Units with training or evaluation missions that are
concerned with individual instructor pilot or RIO
standardization rather than with team standardization
may conduct the flight evaluation with the checkcrew-
pilot flying wing or on an individual basis. A pilot may
be individually checked with the Instructor-Evaluator
conducting the flight evaluation from the rear seat.
The RIO may be individually checked by flying with
the Instructor-Evaluator as his pilot.

FLIGHT EVALUATIONS

The areas and subareas in which pilots and RIOs
may be observed and graded for adherence to
standardized operating procedures are outlined in
the following paragraphs.

Note

If desired, units with training missions
may expand the flight evaluation to include evaluation of standardized training
methods and techniques.

(*) The IFR portions of the Flight Evaluation shall be in
accordance with the procedure outlined in the NATOPS
Instrument Flight Manual.

Mission, Planning And Briefing

   a. Flight planning (pilot and RIO)

   b. Briefing (pilot and RIO)

(*)c. Personal flying equipment (pilot and RIO)

Preflight And Line Operations

Inasmuch as preflight and line operations procedures
are graded in detail during the ground evaluation, only
those areas observed on the flight check will be graded.

   a. Aircraft acceptance (pilot and RIO)

   b. Start

   c. Before-taxiing procedures (pilot)

Taxi And Run Up

(*) Takeoff And Transition

   a. ATC clearance (pilot)

   b. Takeoff (pilot)

   c. Transition to climb schedule

Climb And Cruise

   a. Departure (pilot)

   b. Climb and level-off (pilot)

   c. Procedures enroute (pilot)

(*) Approach And Landing

   a. Radar, TACAN (pilot)

   b. Recovery (pilot)

Communications

a. Receiving and transmitting procedures (pilot and RIO)

b. Visual signals (pilot and RIO)

c. IFF and SIF procedures (RIO)

(*) Emergency And Malfunction Procedures

In this area, the pilot and RIO will be evaluated only in the case of actual emergencies, unless evaluation is conducted in the COT, WST, or OFT.

Postflight Procedures

a. Taxi in (pilot)

b. Shutdown (pilot and RIO)

c. Inspection and records (pilot and RIO)

d. Flight debriefing (pilot and RIO)

Mission Evaluation

This area includes missions covered in the NATOPS Flight Manual, F-14A Tactical Manual, and NWP and NWIP for which standardized procedures and techniques have been developed.

RECORDS AND REPORTS

A NATOPS Evaluation Report (OPNAV Form 3510-8) shall be completed for each evaluation and forwarded to the evaluee's Commanding Officer only. This report shall be filed in the individual's flight training record and retained therein for 18 months. In addition, an entry shall be made in the pilot's and RIO's flight log books under "Qualifications and Achievements" as follows:

QUALIFICATION		DATE	SIGNATURE
NATOPS EVAL.	(Air-craft Model) (Crew Posi-tion)	(Date)	(Authenticating signature) (Unit that administered evaluation)

Critique

The critique is the terminal point in the NATOPS evaluation and will be given by the Evaluator-Instructor administering the check. Preparation for the critique involves processing, reconstructing data collected, and oral presentation of the NATOPS Evaluation Report. Deviations from standard operating procedures will be covered in detail using all collected data and worksheets as a guide. Upon completion of the critique, the pilot and RIO will receive the completed copy of the NATOPS Evaluation Report for certification and signature. The completed NATOPS Evaluation report will then be presented to the Unit Commanding Officer.

FLIGHT EVALUATION GRADING CRITERIA

Only those subareas provided or required shall be graded. The grades assigned for a subarea shall be determined by comparing the degree of adherence to standard operating procedures with adjectival ratings listed below. Momentary deviations from standard operating procedures should not be considered as unqualifying provided such deviations do not jeopardize flight safety and the evaluee applies prompt corrective action.

Flight Evaluation Grade Determination

The following procedure shall be used in determining the flight evaluation grade: A grade of Unqualified in any critical area and subarea will result in an overall grade of Unqualified for the flight. Otherwise, flight evaluation (or area) grades shall be determined by assigning the following numerical equivalents to the adjective grade for each subarea. Only the numerals 0, 2, or 4 will be assigned in subareas. No interpolation is allowed.

Unqualified	0.0
Conditionally Qualified	2.0
Qualified	4.0

To determine the numerical grade for each area and the overall grade for the flight, add all the points assigned to the subareas and divide this sum by the number of subareas graded. The adjective grade shall then be determined on the basis of the following scale.

0.0 to 2.19	—	Unqualified
2.2 to 2.99	—	Conditionally Qualified
3.0 to 4.0	—	Qualified

EXAMPLE: (Add subarea numerical equivalents)

$$\frac{4+2+4+2+4}{5} = \frac{16}{5} = 3.20 \text{ or Qualified}$$

Final Grade Determination

The final NATOPS evaluation grade shall be the same as the grade assigned to the flight evaluation. An evaluee who receives an Unqualified on any ground examination or the flight evaluation shall be placed in an Unqualified status until he achieves a grade of Conditionally Qualified or Qualified on a re-evaluation.

APPLICABLE PUBLICATIONS

The NATOPS Flight Manual contains the standard operations criteria for F-14A aircraft. Publications regarding environmental procedures peculiar to shorebased and shipboard operations and tactical missions are listed below:

- F-14 Tactical Manual

- NWP

- NWIP

- NATOPS Air Refueling Manual

- ATCC/CATCC Manual

- Local Air Operations Manual

- Carrier Air Operations Manual

NATOPS EVALUATION QUESTION BANK

The following bank of questions is intended to assist the unit NATOPS Instructor-Evaluator in the preparation of ground examinations and to provide an abbreviated study guide. The questions from the bank may be combined with locally originated questions in the preparation of ground examinations. The closed-book examination will consist of not less than 50 questions nor more than 75 questions. The time limit for the closed-book examination is 1 hour and 30 minutes. The requirements for the open-book examination are the same as those for the closed-book examination, except there is no time limit.

NATOPS EVALUATION QUESTION BANK

1. The aircraft weighs approximately_____ including trapped fuel, oil, gun, pilot, and RIO.

2. The aircraft is _____ in length and has a wingspan of _____ at 20° and _____ in oversweep.

3. The L INLET and R INLET caution lights indicate _____.

4. During normal system operation, the status of AICS ramp control is as follows:

SPEED	Ramp Hydraulic Power	
M<0.35	ON/OFF	Restrained by _____
M 0.35 to 0.5	ON/OFF	Commanded _____
M>0.5	ON/OFF	Programed as a function of _____ .

5. An AICS failure that causes illumination of an INLET and/or RAMP caution light results in the following:

Speed Range	Ramp Resultant
M<0.5	
M 0.5 to 1.2	
M>1.2	

6. During the AICS portion of OBC, simulated variant flight conditions cycle the _____ through their full range of operation in about _____ seconds. This exercises the _____ _____ and insures _____

7. Operation of the L and R AICS is completely independent.

    a. True
    b. False

8. With the gear handle down and one or more ramps not in the stow position, the ramp light will be illuminated.

    a. True
    b. False

9. What are the throttle interlocks at the military power detent?

    a. _____
    b. _____
    c. _____

10. Auto throttle may be a preflight ground-tested on deck either in _____ or _____ . Indications that a malfunction exists in the auto throttle system are _____ _____ or _____ .

11. List oil pressure readings: Normal _____ psi
                                    Minimum at IDLE _____ psi

12. The OIL PRESS light illuminates at _____ psi. Allowable oil pressure gage fluctuation is _____ psi.

13. List nozzle position indicator readings corresponding to the full open and closed positions.

_____ OPEN
_____ CLOSED

14. What interlocks must be satisfied to activate the nozzle to the full open position to reduce residual thrust?

    a. _____
    b. _____

15. Minimum RPM for ground start of the TF-30 P-414 is _____ % RPM.

16. Maximum allowable turbine inlet temperatures for ground starting the TF-30 and P-414 is _____ °C.

17. The starting temperature limits are the same for both ground and air starts.

    a. True
    b. False

18. The maximum engine operating TIT for a 45-minute MAX time period is _____ °C.

19. Maximum transient TIT (2 minutes) is _____ °C.

20. Zero or negative -g flight is limited to a maximum of _____ seconds in military power or less and _____ seconds in afterburner in order not to _____ .

21. The engine Mach levers control:

    a. Maximum and minimum engine RPM
    b. Maximum engine RPM
    c. Minimum engine RPM

22. If the throttle boost system fails, the throttles automatically revert to manual mode, and the throttle mode switch returns to MAN.

    a. True
    b. False

23. What pilot action is required to reset the boost mode of throttle control subsequent to reversion to the manual mode?

24. If the auto throttles are disengaged by any means, the AUTO THROT light illuminates for a 10-second duration.

    a. True
    b. False

25. Engine N$_2$ RPM must be above _____% to supply sufficient power for normal operation of the engine ignition system.

26. Do not attempt a crossbleed start with engine RPM greater than _____%.

27. Normal airstarts are initiated at _____% minimum stabilized RPM.

28. Failure of the afterburner hydraulic pump will have what effect on engine operation?

29. The INLET ICE caution light illuminates when _____.

30. In the AUTO position, pitot probe heat is available only with weight off wheels.

    a.   True
    b.   False

31. The L or R FIRE warning lights illuminate when the respective entire sensing loop is heated approximately _____°F or when any 6-inch section is heated to approximately _____ °F.

32. Which of the following would result in illumination of the FUEL PRESS caution light?

    a.   Failure of a motive flow pump.
    b.   Failure of a main fuel pump stage.

33. Failure of a motive flow fuel pump will have what effect on the engine and fuel system operation?

34. Failure of the second stage on the main engine fuel pump will have what effect on engine operation?

35. The engine boost pumps are powered by _____ .

36. The loss of an engine boost pump will have what effect on operation of both engines?

37. Selecting either the AFT or FWD position with the fuel FEED switch performs what functions in the fuel system?

    a. _____   c. _____
    b. _____   d. _____

38. The L/R FUEL LOW lights illuminate with approximately _____ pounds remaining in individual feed group.

39. Automatic shutoff of wing and drop tanks when empty occurs with WING/EXT TRANS switch in either AUTO or ORIDE.

     a. True
     b. False

40. The BINGO caution light illuminates when _____ .

41. Is vent tank fuel quantity included in the fuel totalized and AFT and L indicator readings?

42. When should the FEED switch be activated to the FWD or AFT position?

43. What medium is used to actuate the feed tank interconnect valve, wing motive flow shut-off valves, and fuel dump valve?

44. Wing fuel is transferred by:

     a. Engine bleed air
     b. Motive flow fuel

45. The fuel thermistors in the outboard section of the wing tanks perform what function?

46. The fuel thermistors in fuel cell No. 2 and 5 perform these functions when either is uncovered:

     a. _____   d. _____
     b. _____   e. _____
     c. _____

47. All fuel entering the vent tank is vented overboard through the vent mast in the tailhook attachment fairing.

     a. True
     b. False

48. Fuel transfer from the external drop tanks is accomplished by _____

49. External fuel transfer can be checked on the deck by _____ or _____ .

50. Fuel dump is prohibited with speed brakes open and/or afterburner operation.

     a. True
     b. False

51. When the fuel dump circuit is activated, wing and external drop tank transfer is automatically initiated.

     a. True
     b. False

52. Is it possible to refuel in flight and accomplish total fuel transfer without electrical power or a combined hydraulic system? If not, why?

53. How may the refueling probe be extended with combined hydraulic system failure?

54. On engine start with the generator switch in normal the generator is automatically excited and the generator control unit brings it on the line when engine RPM is approximately _____ %.

55. If the thermal cutout decouples the drive clutch to either main generator in flight, the IDG may be recoupled (reset) a maximum of three times.

     a. True
     b. False

56. With a generator caution light, the TEST position of the generator switch is selected; the generator light goes OUT; this indicates a failure in the power distribution circuits.

     a. True
     b. False

57. Failure of either ac generator automatically connects the left and right main ac buses to the operative generator. The cockpit indicator will be a _____ caution light.

58. The emergency generator is powered by _____ .

59. If the emergency generator switch is in NORM, it will come on the line automatically when _____ _____ .

60. With a generator failure, the cockpit indication when the buses tie together will be: _____ _____ .

61. When the emergency generator is energized with normal hydraulic pressure only the essential dc 1 bus will be energized.

     a. True
     b. False

62. If the emergency generator is operating under reduced hydraulic pressure only the essential dc 1 bus will be energized.

     a. True
     b. False

63. When operating on the emergency generator the cockpit lighting available consists of _____ and _____ .

64. A single engine-driven pump on the left engine powers the combined hydraulic system and a single engine-driven pump on the right engine powers the flight hydraulic system.

   a. True
   b. False

65. If the pilot extinguishes MASTER CAUTION light after a failure of one main hydraulic system, failure of the other system will or will not illuminate the MASTER CAUTION light.

66. With the left engine shut down in flight and 0 windmill RPM, the combined hydraulic system can be powered by _____ .

67. The hydraulic transfer pump will supply approximately_____psi on the failed side with 3000 psi on the other side.

68. With total loss of fluid from either main hydraulic system, the hydraulic transfer pump will: _____
_____

69. The cockpit handpump will charge the brake accumulator in flight if _____
_____ .

70. Loss of all hydraulic fluid from the flight hydraulic system will mean loss of power to the right inlet ramps.

   a. True
   b. False

71. With loss of the combined hydraulic system, (combined system pressure 0) the main flaps are powered by_____and the auxiliary flaps are _____ .

72. When the gear is blown down, the nosewheel steering and normal brakes will operate normally after touchdown.

   a. True
   b. False

73. The outboard spoiler module uses a combined systems fluid.

   a. True
   b. False

74. Outboard spoiler operation is inhibited with wings aft of 57° and the spoiler module is _____ with wings aft of 62° .

75. The outboard spoiler module thermal cut out is inhibited when _____
    _____ .

76. The ON-OFF flag in the spoiler window of the hydraulic indicator indicates:

    a. The outboard spoiler module is energized.
    b. The outboard spoiler system is pressurized.

77. With loss of the combined hydraulic system (combined system pressure 0) the inboard spoilers will:
    _____ .

78. The backup flight control module powers the _____ and the _____ .

79. With the backup flight control module switch in AUTO, the module is automatically energized when
    _____ .

80. The backup flight control module switch has three positions, AUTO, _____ and _____
    _____ .

81. The backup flight control module operates in the high-speed mode when _____
    _____ .

82. Operational status of the backup flight control module is indicated in the cockpit by _____
    _____ .

83. Failure of either the combined or flight hydraulic system will have what effect on wing sweep?

84. On the wing sweep indicator there are three position indicators. These show _____ , _____
    and _____ wing sweep position.

85. The aircraft is being operated with the wing aft of the forward limit. The wing sweep control mode indicator
    reads MAN. If speed is now increased beyond where the wing sweep angle and forward limit coincide, the control
    mode indicator will read _____ and the wings will _____ .

86. The most forward wing sweep angle allowed in bomb mode is _____ .

87. The emergency wingsweep mode is a manual method of positioning the wings. This method incorporates locks
    every_____from 20° to 68° to prevent random wing movement in this mode.

88. Illumination of the WING SWEEP advisory light means:

89. Illumination of the WING SWEEP warning light means:

90. Transient failures in the CADC may be reset by:

91. The CADC is self-tested in _____ .

92. List the caution, advisory, and warning lights activated by the CADC directly or via the AFCS:

    a. _____   e. _____   i. _____
    b. _____   f. _____   _____
    c. _____   g. _____   _____
    d. _____   h. _____

93. Maneuver flaps can be lowered at any wing sweep angle between 20° and _____ .

94. The maneuvering flap thumbwheel will lower the main flaps _____ , the auxiliary flaps _____ , and the slats _____ . Use of the maneuvering devices does or does not put more restrictive g limitations on the aircraft _____ .

95. The glove vanes can be extended above 1.1 Mach by _____ .

96. The glove vanes automatically begin extension at _____ Mach.

97. What is the meaning of?

    a. FLAP caution light:

    b. GLOVE VANE caution light:

    c. REDUCE SPEED warning light:

98. Power for emergency extensions of the landing gear is supplied by: _____
_____ .

99. The minimum bottle pressure for accomplishing emergency extension of the landing gear is _____ psi but a minimum preflight bottle pressure should be _____ psi at 70°F.

100. The wing spoilers are or are not actuated for lateral trim control.

101. Full rudder throw of ± _____° corresponds to ± _____ inches of rudder pedal travel.

102. Control surface authority of the stability augmentation system (SAS) is:

    Stabilizer ± _____°
    Differential Stabilizer ± _____°
    Rudder ± _____°

103. The gear handle is down and the three gear position indicators show the gear down, but the transition light is illuminated. What does this indicate and what action should be taken?

104. The ANTI SKID SPOILER BK switch is OFF and the BRAKE light is illuminated. This would indicate:

    a. _____
    b. _____

105. The BRAKE light (ANTI SKID SPOILER BK switch OFF) operates only when the brakes are depressed or the parking handle is pulled.

    a. True
    b. False

106. The procedure for lowering the launch bar is: _____.

107. Nosewheel steering cannot be engaged until the weight is on wheels.

    a. True
    b. False

108. Regardless of nosewheel position, the nosewheel assumes the position commanded by the rudder pedals when nosewheel steering is engaged.

    a. True
    b. False

109. Illumination of the BLEED DUCT light indicates _____ and corrective action would be
_____ .

110. The ram air door can be opened only if the _____ or _____ buttons are depressed on the ECS control panel.

111. The ram air door automatically closes with selection of L ENG, R ENG or BOTH ENG on the ECS control panel.

    a. True
    b. False

112. The ram air door requires _____ seconds to go full open and vice versa.

113. The RIO has a low cockpit pressure caution light (CABIN PRESS) which illuminates if _____ or
_____ .

114. The OXY LOW light indicates _____ . This low quantity should supply each flightcrew member with _____ hours of oxygen at 20,000 feet (8000 feet cabin altitude).

115. Pulling the emergency oxygen actuator releases gaseous oxygen charged to _____ psi and will provide approximately a _____ minute supply.

116. Windshield rain removal is accomplished by blowing 390°F air over the outside of the windshield. If the temperature sensor detects an overtemperature condition, the WSHLD HOT advisory light will illuminate and
_____
_____ .

117. Maximum allowable headwind for the open canopy is _____ knots.

118. When the canopy is jettisoned, the sill locks are released by _____
_____ .

119. The canopy pneumatic reservoir must be serviced by a ground servicing unit.

    a. True
    b. False

120. The pilot can tell the position of the command ejection lever by _____ .

121. The RIO can eject both himself and the pilot with EJECT CMD handle set to PILOT.

    a. True
    b. False

122. The pilot can eject both himself and the RIO, with the EJECT CMD handle set to MCO.

    a. True
    b. False

123. In the event the canopy does not separate from the aircraft when either flightcrew member has initiated ejection using the primary firing handle, actuating the alternate firing handle will allow "through the canopy" ejection.

    a. True
    b. False

124. There are _____ safety pins per ejection seat.

125. Command ejection by either flightcrew member will eject the RIO in _____ seconds and the pilot _____ seconds later.

126. The time-release mechanism is set for _____ feet ± _____ .

127. All exterior lighting controls except for the _____ light are located on the MASTER LIGHT panel on the pilot's console, and the exterior lights master switch on the outboard throttle.

128. When the wings are swept aft of _____°, the _____ position lights are disabled and the position lights are operable.

129. When the ANTI COLLISION light switch is ON, the position lights flasher switch is disabled.

    a. True
    b. False

130. A proper indicator lights test has the MASTER CAUTION light on steady.

    a. True
    b. False

131. The RIO can monitor SW tones by selection of _____ position on his ICS panel.

132. The standby attitude indicator is capable of providing reliable attitude information within _____ for up to _____ minutes after a complete loss of power.

133. On deck, the allowable error between the primary, reset mode, and standby readings is _____ feet and both modes should agree within _____ feet at field elevation.

134. With the altimeter in standby, what is the position error for the following conditions:

	PILOT	RIO
45,000 lb 116 KIAS LANDING CONFIGURATION (Non DLC)	_____	_____
50,000 lb 123 KIAS LANDING CONFIGURATION (Non DLC)	_____	_____
50,000 lb 0.8 IMN CLEAN CONFIGURATION	_____	_____
50,000 lb 0.95 IMN CLEAN CONFIGURATION	_____	_____

135. The angle-of-attack indicator is checked during _____ and the indexer during _____ . Proper indications are:

      INDICATOR — _____

      INDEXER   — _____

136. In the landing configuration, 15 UNITS AOA is equivalent in airspeed for:

      45,000 lb (DLC NOT ENGAGED) = _____ KIAS
      45,000 lb (DLC ENGAGED/NEUTRAL) = _____ KIAS
      50,000 lb (DLC NOT ENGAGED) = _____ KIAS

137. With an airspeed indicator failure, list the angle of attack to fly for the following conditions: (DRAG INDEX 8)

      CATAPULT _____
      CLIMB (MRT) SL _____ to combat ceiling _____
      CRUISE at OPT ALT _____
      ENDURANCE at OPT ALT _____

138. ACM jettison requires the MASTER ARM power switch ON.

    a. True
    b. False

139. Selective jettison can be completely controlled by either flightcrew member.

    a. True
    b. False

140. In the emergency jettison mode, the weight-on-wheels interlock is bypassed.

    a. True
    b. False

141. Emergency jettison mode will jettison Sidewinders.

    a. True
    b. False

142. Sidewinder is jettisoned by firing the motor and safing the warhead.

    a. True
    b. False

143. The pretaxi (weight on wheels) OBC master test is a complete check of the AWG-15 system.

    a. True
    b. False

144. Selection of any pulse dogfight mode (DFM) automatically provides stab out aircraft reference.

    a. True
    b. False

145. The RIO must clear maintenance display prior to running OBC for current test results.

    a. True
    b. False

146. Class I OBC tests can be performed only _____ .

147. With track files established in TWS, the HUD and VDI provides the pilot complete steering information to the centroid of the targets.

    a. True
    b. False

148. The navigation system may be updated by three methods, they are:

    a. _____
    b. _____
    c. _____
      _____

149. In TACAN BIT, the range and bearing on the HSD and BDHI should indicate _____ nmi and _____ °.

150. The target designator (diamond) is valid to $\pm$_____° off the nose.

151. With MASTER ARM OFF, the HUD and VDI armament legend will appear with _____

152. The PLM button must be held depressed for at least 2 seconds to acquire a radar lock.

    a. True
    b. False

*153. A duty cycle reduction will occur upon selection of AIM-7, with an AIM-54A in flight, assuming the AWG-9 has a valid PDSTT track file.

    a. True
    b. False

154. To acquire an attack presentation, the _____ button must be selected on the pilot's panel.

*Confidential when filled in.

*155. Only the AIM-7E missile can be tuned to a different channel prior to LTE by simply selecting another number on the missile channel thumbwheel.

    a. True
    b. False

156. The AWG-9 aligns the IMU; the CSDC maintains that alignment.

    a. True
    b. False

157. COOLING AIR light refers to air cooling out of tolerance while AWG-9 COND light indicates _____ _____ cooling out of tolerance.

158. The TID is oriented to _____ north, with selection of GND STAB on the TID mode switch.

159. RWS sensor targets can be hooked by the RIO for information readout.

    a. True
    b. False

160. Which of the following R/C presentations are available to the pilot:

    a. IRS
    b. PS
    c. PDS
    d. TWS

161. A _____ acronym indicates a failure of the AWG-15 primary power supplies, thus preventing normal separation of stores in any launch mode.

162. It is possible for the radar antenna drive to fail and the _____ will continue to display the antenna scan commanded on the sensor control panel.

163. The erase control continues to function in all STT modes.

    a. True
    b. False

*Confidential when filled in.

164. The upper limit of VSL HI is _____ ; the lower limit of VSL LO is _____ .

165. The LIQ COOLING switch in the RIO's cockpit controls liquid coolant to the _____
system and _____ missiles.

166. Hostile area altitude entry is set in _____ pseudo file to properly reject altitude line return.

167. Selecting PLM will break an existing radar track.

   a. True
   b. False

168. MLC can be displayed on the DDD with the MLC switch in the _____ position.

*169. It is possible to fire an AIM-54A with TWS MAN selected.

   a. True
   b. False

170. In sequence 5, the evaluation of search and missile channels with a circled check mark above the channel number
indicates that channel has _____
_____ .

171. In sequence 7, fault isolation, decision points in the BIT box would indicate: _____
_____ .

172. The RIO can read the NDRO tactical program number and its revision directly from the TID in BIT sequence
_____ .

173. Maximum range displayed on the HUD is _____ nmi.

174. Maximum range displayed on the VDI is _____ nmi.

175. Wind is automatically computed by the AWG-9 in the IMU/AM mode:

   a. True
   b. False

176. List the navigation files available for command steering on the HSD and ECMD:

   a. _____   e. _____
   b. _____   f. _____
   c. _____   g. _____
   d. _____

*Confidential when filled in.

177. A wind of 35 knots and 050° relative to the duty runway represents a headwind component of _____ knots and crosswind component of _____ knots.

178. CM masking is initiated upon completion of OBC or with maintenance file displayed when the RIO selects
_____ .

179. A CM mask may be cleared by:

    a. _____
    b. _____
    c. _____

180. A blinking LAR indicates

    a. TGT at MIN RANGE
    b. TGT at OPT RANGE
    c. TGT at MAX RANGE
    d. A and B
    e. All of the above

181. List the five modes of AIM-54 operation:

    a. _____
    b. _____
    c. _____
    d. _____
    e. _____

182. What does a station six checkerboard on deck mean with a Phoenix onboard the station: _____
_____

*183. A Phoenix normal active launch requires AWG-9 illumination.

    a. True
    b. False

*184. The Phoenix requires an AWG-9 transmitter to complete MSL PREP.

    a. True
    b. False

*185. The Phoenix can be launched in the ACM active mode with a PDSTT.

    a. True
    b. False

*186. AIM-54 normal active mode is automatically selected if the launch range is less than the active seekerhead range in TWS or PDSTT.

    a. True
    b. False

187. The Phoenix active seekerhead range is determined by the position of the _____ switch.

*Confidential when filled in.

*188. The frequency range of the AIM-54 active mode is _____ to _____ GHz.

*189. An AIM-7 CW launch is possible in which of the following transmitter channels?

    a. SEARCH 1-4/5-8
    b. MC-7/1-6
    c. MC-54/1-6
    d. All of the above.

*190. The CWI operates in the _____ to _____ GHz frequency band and consists of _____ operator selectable channels.

191. During an AIM-7F tune cycle, an AIM-7F PD launch is possible by selection of the _____ switch.

192. A transition from PDSTT to PSTT with a CW AIM-7 in flight will cause CW illumination to transmit through the CW flood antenna.

    a. True
    b. False

193. After completion of AIM-7 tune, the CW channel can be changed by selecting the desired MSL CHAN and recycling the _____ switch.

194. A flashing BIT box during MAS sequence 4 means _____

_____ .

*195. AIM-9 SEAM acquisition from a radar track is available out to _____ degrees.

196. An AIM-9 SEAM lock acquired from an STT will be lost if the radar breaks lock.

    a. True
    b. False

197. The RTGS diamond indicates a _____ foot solution.

198. Without a radar lock, the RTGS pipper indicates a _____ foot solution.

199. Hydraulic power to drive the gun comes from the _____ system.

200. The gun can be fired with a CPA acronym in CM.

    a. True
    b. False

201. At 1,000 feet, the width of the RTGS diamond equals _____ feet.

202. Pilot selection of the COMP mode commands the INS to use the manually entered magnetic variation value.

    a. True
    b. False

203. With an AHRS advisory light the INS automatically uses the manually entered magnetic variation.

    a. True
    b. False

204. Failure of AUTO sequence 2 to run to completion will inhibit SAT.

    a. True
    b. False

*Confidential when filled in.

F-14A TOMCAT

THIS PAGE INTENTIONALY LEFT BLANK.

SECTION XI — PERFORMANCE DATA

For aircraft performance data and charts
refer to NAVAIR 01-F14AAA-1.1.

F-14A TOMCAT

THIS PAGE INTENTIONALY LEFT BLANK.

FOLDOUT ILLUSTRATIONS

TABLE OF CONTENTS

ANTICOLLISION LIGHT   FORMATION LIGHT

                             DECM RECEIVER ANTENNA
                                                                    INLET CONTROL PROGRAMMER (RIG
                                                                                          GEN
                                                                                          CON
                                                                              TRANS/RECT
                                                               IFF/APX-72   CADC
                                                               UHF DATA LINK   HEAT EXCHANGER
ALQ-126 *
MID- AND
HIGH-BAND   STABILIZER ACTUATOR
ANTENNA

FUEL
DUMP
ARRESTING HOOK
                                                                                    TURBI
               ARRESTING HOOK LIFT   AIM-7/E AND F (SPARROW)
               (RETRACT) CYLINDER
                                                       AIM-7/E AND F (SPARROW)
               ARRESTING HOOK DASHPOT
                                                                                  LOW FUE
                    ENGINE TF30-P-414   UHF BAND BLADE   LEVEL S
                                         ANTENNA   ENGINE
                                         IFF AN-APX-72 ACCESSORIES
                                         UHF DATA LINK AIM-7/E AND F
                                                       (SPARROW) AT
                                                       CENTERLINE

                                               VIEW LOOKING INBOARD RIGHT

        *ALQ-126 IN AIRCRAFT BUNO 161168 AND
        SUBSEQUENT IN LIEU OF ALQ-100.

                              WINDSHIELD RAIN
                              REMOVAL   MARTIN BAKER
                                                         EJECTION SEATS
                       UHF ADF ANTENNA   AIC SENSORS   MK-GRU7A (2)
                       ARA-50   INLET CONTROL PROGRAMMER (LEFT SIDE)
                                        MISSION RECORDER
                                        TACTICAL INFO   ANTENNA
                       INERTIAL NAVIGATION ANGLE-OF-ATTACK   DISPLAY   TACAN ANTENNA IFF/APX-72- UHF D
                       UNIT AN/ASN-92   TRANSMITTER   UHF COMM   EMERGENCY
           AWG-9   UHF DATA LINK GENERATOR   BLEED
           RADAR ANTENNA   ALPHA   ARC-124
     AWG-9   COMPUTER
     TARGET   PITOT STATIC PROBE
     HORN   APR-69
                         ARN-84   FUEL
                                         CANOPY JETTISON HANDLE
                                           AIC PROBE
ALPHA
PROBE
        ARA-63
        ANTENNA   APN   DECODER
        INFRARED SEEKER/TV   194   ALR-50   APX-76   CONTACTOR
        OPTICAL UNIT   TOTAL ANTENNA   APX-72   OUTBOARD
ALQ-100 OR   HYBRID   TEMP   SPOILER
ALQ-126* ANTENNA   COUPLER   PROBE RAIN   AMMUNITION   ARA-50
(FORWARD   ALQ-100 OR   REPELLANT DRUM
LOOKING)   ARA-63   ALQ-126 M-61A1 GUN CONTAINER EXTERNAL POWER
         APN-194 RECEIVER   GUN CONTROL   RECEPTACLE
         ANTENNA ALQ-100 OR ALQ-126* APN-154 BOARDING
                  ANTENNA (AFT LOOKING) TRANSMITTER LADDER   **VIEW LOOKING INBOARD LEFT SIDE**
                                                  (OPEN
                                                  POSITION)

INLET CONTROL PROGRAMMER (RIGHT SIDE)
GENERATOR CONTROL UNIT
TRANS/RECT
CADC
HEAT EXCHANGER
DETAIL DATA DISPLAY
RIO'S EYE
TACTICAL INFORMATION DISPLAY
PILOT'S EYE
PITOT STATIC PROBE
TOTAL TEMPERATURE PROBE
VERTICAL DISPLAY INDICATOR GROUP
REFUELING PROBE (EXTENDED)
UHF/ADF ANTENNA
SPARROW BORESIGHT FLOOD ANTENNA
WINDSHIELD TEMPERATURE CONTROLLER
IFF INTERROGATOR ANTENNA
LOX
ARC-51
ALPHA PROBE

TURBINE
AIM-7/E AND F (SPARROW)
(SPARROW)
CANOPY ACTUATOR BOTTLE
WATER SEPARATOR
LOW FUEL LEVEL SENSING
HEAT EXCHANGER
EMERGENCY RAM AIR INLET
ALQ-100 OR ALQ-126° RECEIVER/ TRANSMITTER
LOX
AIC PROBE
LAUNCH BAR
DESICCANT TANKS (2)
JULIET-28
GYRO
FIRE DETECTION CONTROL UNITS
GROUND REFUELING
CANOPY JETTISON HANDLE
IR SEEKER/TV OPTICAL UNIT
ACLS BEACON ANTENNA
RADAR APN-194 ANTENNA
ANTICOLLISION LIGHT
AN/APN-154 ANTENNA

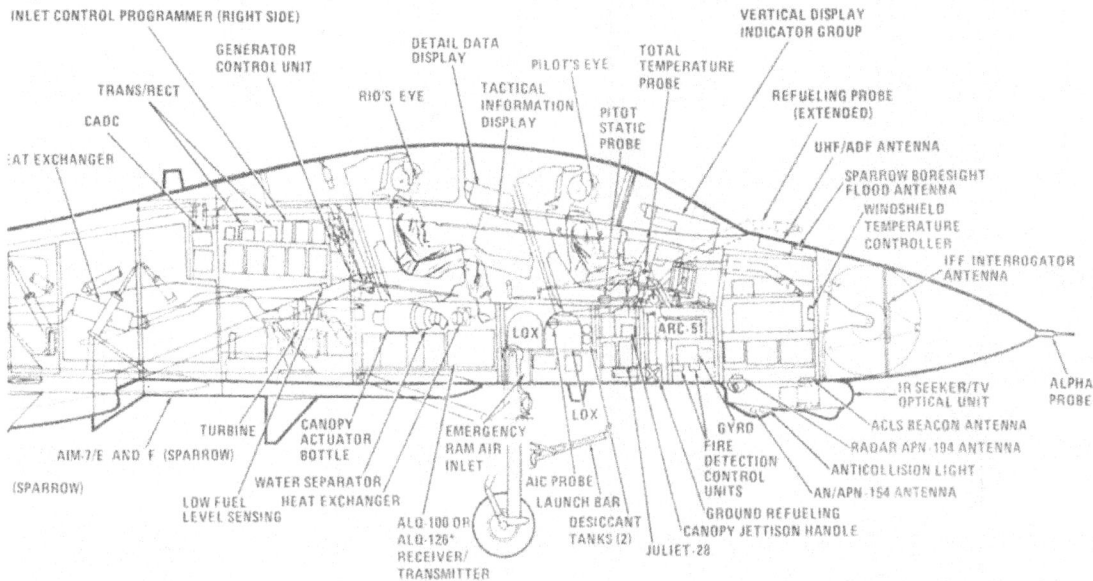

NBOARD RIGHT SIDE

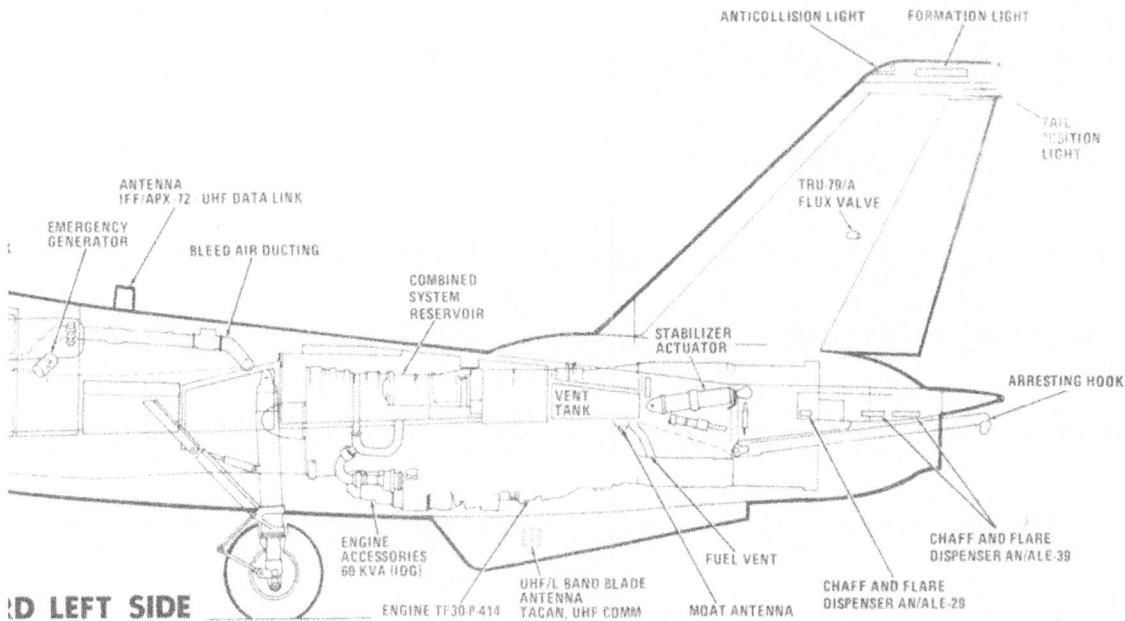

ANTICOLLISION LIGHT
FORMATION LIGHT
TAIL POSITION LIGHT
TRU-79/A FLUX VALVE

ANTENNA IFF/APX-72 UHF DATA LINK
EMERGENCY GENERATOR
BLEED AIR DUCTING
COMBINED SYSTEM RESERVOIR
STABILIZER ACTUATOR
VENT TANK
ARRESTING HOOK

ENGINE ACCESSORIES 60 KVA (IOG)
UHF/L BAND BLADE ANTENNA TACAN, UHF COMM
ENGINE TF30-P-414
FUEL VENT
MOAT ANTENNA
CHAFF AND FLARE DISPENSER AN/ALE-39
CHAFF AND FLARE DISPENSER AN/ALE-29

RD LEFT SIDE

7-F50-019-1

AUXILIARY
FLAP
ACTUATOR

PRIMARY HEAT
EXCHANGER

AIC
ANGLE-OF-ATTACK
PROBE

ALR-45 ANTENNA

HYDRAULIC
COOLER

PITOT-STATIC
PROBE

VARIABLE
INLET ACTR

REFUELING PROBE
(EXTENDED)

TOTAL TEMPERATURE
PROBE (AIC)

RELAY
BOXES

BLEED AIR
DUCTING

IFF
INTERROGATOR
ANTENNA

LOX
BOTTLES

HEAT
EXCHANGER

WATER
SEPARATOR

DUMP
VALVE

INLET

WING
SWEEP
HYDRAULIC
VALVE

AWG-9 TARGET HORN

CADC

LEFT
RUDDER

SPOIL

LEFT STICK
LEFT STICK

MANEUVER
FLAP SERVO
WING SWEEP
SERVO

AFT STICK

DIRECT
FEEL UNIT

ALPHA
PROBE

AWG-9
ANTENNA

SPARROW
BORESIGHT
FLOOD ANTENNA

FUEL QTY
JUNCT BOX

FUEL

FLAP
SLAT
GEAR-
BOX

LEFT RUDDER

ANGLE-OF-ATTACK
PROBE

AMMUNITION
DRUM

BRAKE
MODULE

DUMP
VALVE

COMBINED
PRESSURE
MODULE

PITOT-STATIC
PROBE

AIC
ANGLE-OF-ATTACK
PROBE

RELAY
BOXES

INLET

HYDRAULIC
COOLER

M-61A1 GUN

ALR-45 ANTENNA

WHEEL
BRAKE
ACCUMULATORS

VARIABLE
INLET
ACTUATOR

WING SWEEP
ACTUATOR

SECONDARY
HEAT
EXCHANGER

WING
PIVOT

ALQ-126
HIGH-BAND
ANTENNA
(LEFT AND
RIGHT)

GLOVE
VANE
EXTENDED

WING GLOVE
POSITION
LIGHT

ALQ-126
MID-BAND
ANTENNA
(LEFT AND
RIGHT)

DECM HIGH BAND
TRANSMITTER
ANTENNA
(LEFT AND RIGHT)

DECM HIGH BAND
RCVR ANTENNA
(LEFT AND RIGHT)

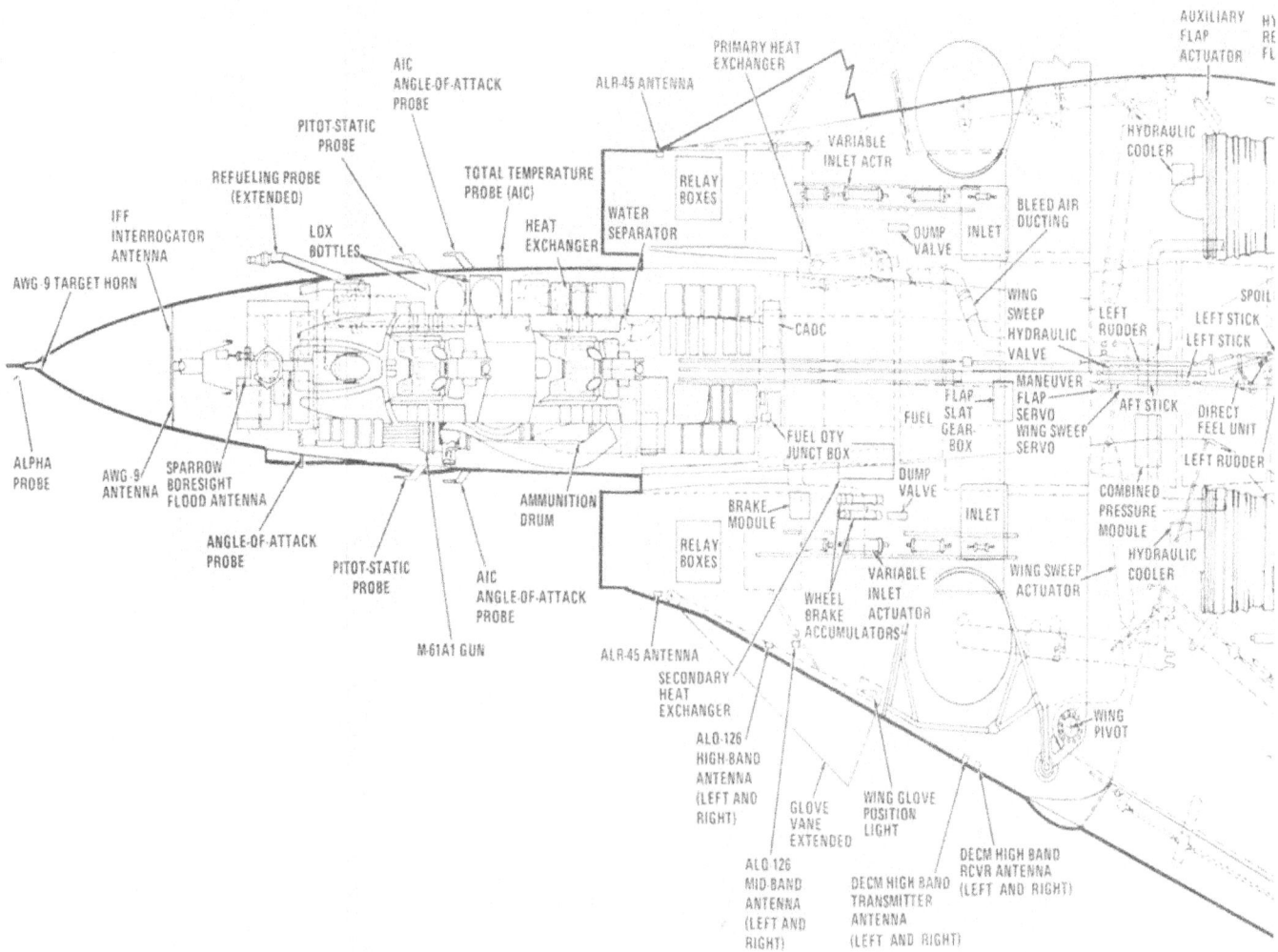

VIEW LOOKING DOWN

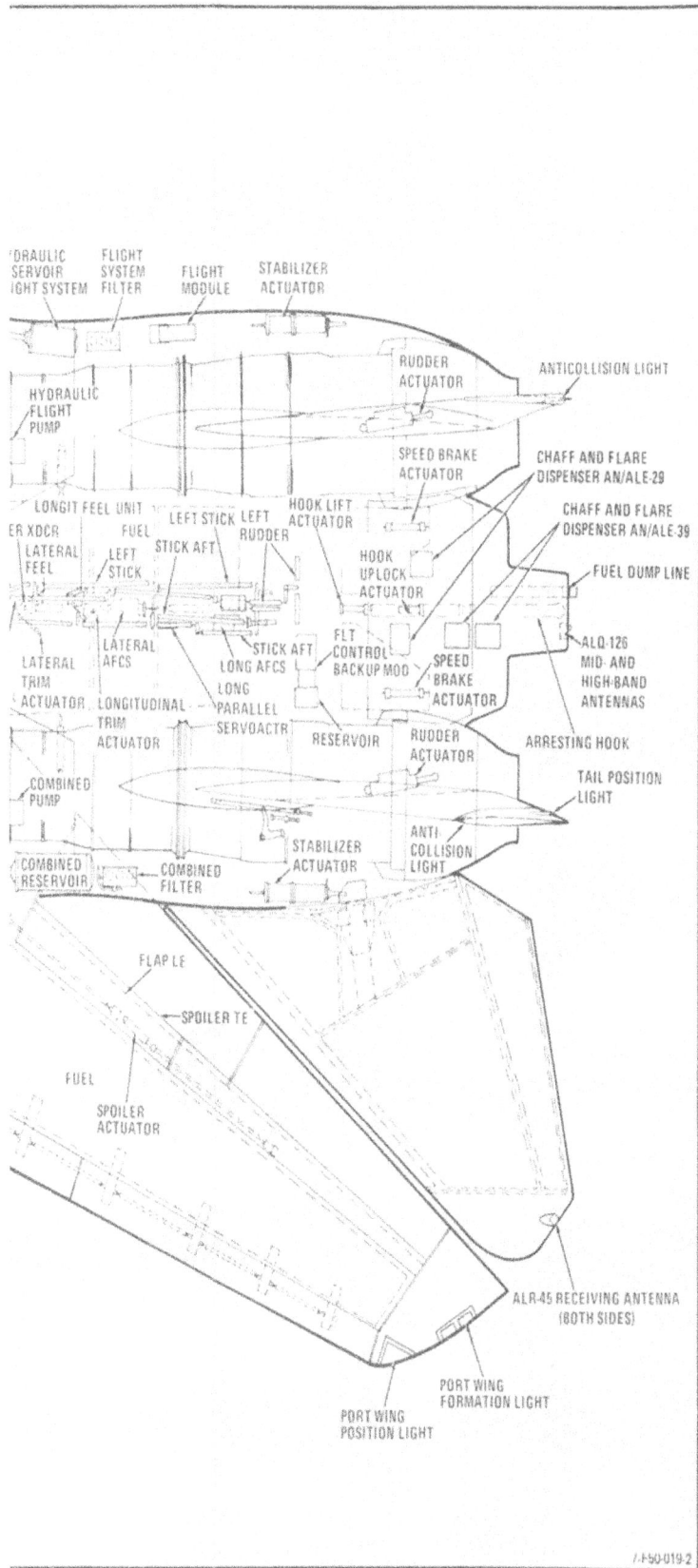

DRAULIC
SERVOIR
IGHT SYSTEM

FLIGHT
SYSTEM
FILTER

FLIGHT
MODULE

STABILIZER
ACTUATOR

HYDRAULIC
FLIGHT
PUMP

RUDDER
ACTUATOR

ANTICOLLISION LIGHT

SPEED BRAKE
ACTUATOR

CHAFF AND FLARE
DISPENSER AN/ALE-29

LONGIT FEEL UNIT

HOOK LIFT
ACTUATOR

CHAFF AND FLARE
DISPENSER AN/ALE-39

ER XDCR
LATERAL
FEEL

LEFT STICK LEFT
RUDDER

LEFT
STICK

FUEL

STICK AFT

HOOK
UPLOCK
ACTUATOR

FUEL DUMP LINE

LATERAL
AFCS

STICK AFT
LONG AFCS

FLT
CONTROL
BACKUP MOD

ALQ-126
MID- AND
HIGH-BAND
ANTENNAS

LATERAL
TRIM
ACTUATOR

LONGITUDINAL
TRIM
ACTUATOR

LONG
PARALLEL
SERVOACTR

SPEED
BRAKE
ACTUATOR

RESERVOIR

RUDDER
ACTUATOR

ARRESTING HOOK

COMBINED
PUMP

TAIL POSITION
LIGHT

COMBINED
RESERVOIR

COMBINED
FILTER

STABILIZER
ACTUATOR

ANTI-
COLLISION
LIGHT

FLAP LE

SPOILER TE

FUEL

SPOILER
ACTUATOR

ALR-45 RECEIVING ANTENNA
(BOTH SIDES)

PORT WING
FORMATION LIGHT

PORT WING
POSITION LIGHT

7-F50-019-2

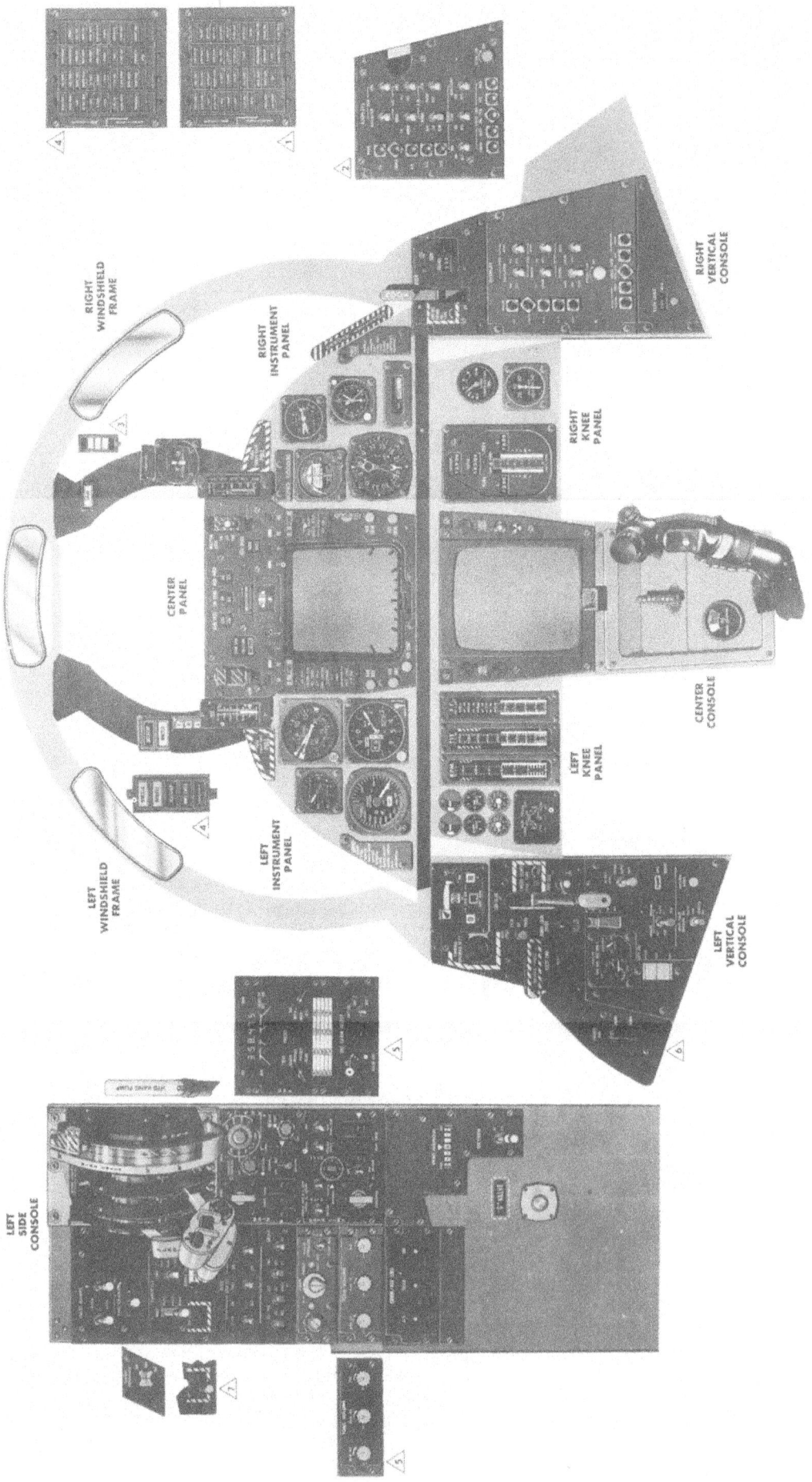

RIGHT WINDSHIELD FRAME

RIGHT INSTRUMENT PANEL

RIGHT VERTICAL CONSOLE

RIGHT KNEE PANEL

CENTER PANEL

LEFT KNEE PANEL

CENTER CONSOLE

LEFT WINDSHIELD FRAME

LEFT INSTRUMENT PANEL

LEFT VERTICAL CONSOLE

LEFT SIDE CONSOLE

NAVAIR 01-F14AAA-1

PILOT'S INSTRUMENT
PANEL AND CONSOLES

AIRCRAFT BUNO 159468 AND PRIOR

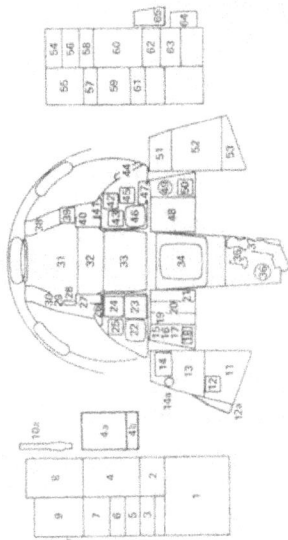

EFFECTIVITY

1. AIRCRAFT BUNO 158612 THRU 159001
2. AIRCRAFT BUNO 158631 AND SUBSEQUENT AND AIRCRAFT INCORPORATING AFC 35
3. AIRCRAFT BUNO 160876 AND SUBSEQUENT
4. AIRCRAFT BUNO 159002 AND SUBSEQUENT AND AIRCRAFT INCORPORATING AFC 181
5. AIRCRAFT BUNO 160378 AND SUBSEQUENT AND AIRCRAFT INCORPORATING AFC 599
6. AIRCRAFT BUNO 159567 AND SUBSEQUENT AND AIRCRAFT INCORPORATING AFC 529
7. AIRCRAFT BUNO 160378 AND SUBSEQUENT AND AIRCRAFT INCORPORATING AFC 334

RIGHT SIDE CONSOLE

LEFT SIDE CONSOLE
1. G VALVE PUSHBUTTON
2. OXYGEN VENT AIRFLOW CONTROL PANEL
3. COMM/NAV COMMAND CONTROL PANEL
4. INTEGRATED CONTROL PANEL
4a. UHF (AN/ARC 159)
4b. UHF COMM SELECT PANEL
5. TONE VOLUME CONTROL PANEL
6. ICS CONTROL PANEL
7. AFCS CONTROL PANEL
8. THROTTLE QUADRANT
9. INLET RAMPS/THROTTLE CONTROL PANEL
10. TARGET DESIGNATE SWITCH
10a. HYDRAULIC HAND PUMP

LEFT VERTICAL CONSOLE
11. FUEL MANAGEMENT PANEL
12. CONTROL SURFACE POSITION INDICATOR
12a. LAUNCH BAR ABORT PANEL
13. LANDING GEAR CONTROL PANEL
14. WHEELS FLAPS POSITION INDICATOR
14a. EMER STORES JETTISON BUTTON

LEFT KNEE PANEL
15. ENGINE PRESSURE RATIO INDICATOR
16. EXHAUST NOZZLE POSITION INDICATOR
17. OIL PRESSURE INDICATOR
18. HYDRAULIC PRESSURE INDICATOR
19. ELECTRICAL TACHOMETER INDICATOR (RPM)
20. THERMOCOUPLE TEMPERATURE INDICATOR (TIT)
21. RATE OF FLOW INDICATOR (FF)

LEFT INSTRUMENT PANEL
22. RADAR ALTIMETER
23. SERVOPNEUMATIC ALTIMETER
24. AIRSPEED MACH INDICATOR
25. VERTICAL VELOCITY INDICATOR
26. LEFT ENGINE FUEL SHUT OFF HANDLE
27. ANGLE OF ATTACK INDICATOR

LEFT FRONT WINDSHIELD FRAME
28. APPROACH INDEXER
29. WHEELS WARNING LIGHT
29a. BRAKES WARNING LIGHT
30. ACLS/AP CAUTION LIGHT
30a. NWS ENGA CAUTION LIGHT

CENTER PANEL
31. HEADS UP DISPLAY
32. AIR COMBAT MANEUVER PANEL
33. VERTICAL DISPLAY INDICATOR (VDI)
34. HORIZONTAL SITUATION DISPLAY INDICATOR (HSI)
35. PEDAL ADJUST HANDLE
36. BRAKE PRESSURE INDICATOR
37. CONTROL STICK

RIGHT FRONT WINDSHIELD FRAME
38. ECM WARNING LIGHTS
39. STANDBY COMPASS

RIGHT INSTRUMENT PANEL
40. WING SWEEP INDICATOR
41. RIGHT ENGINE FUEL SHUT OFF HANDLE
42. ACCELEROMETER
43. STANDBY ATTITUDE INDICATOR
44. CANOPY JETTISON HANDLE
45. CLOCK
46. BEARING DISTANCE HEADING INDICATOR (BDHI)
47. UHF REMOTE INDICATOR

RIGHT KNEE PANEL
48. FUEL QUANTITY INDICATOR
49. LIQUID OXYGEN QUANTITY INDICATOR
50. CABIN PRESSURE ALTIMETER

RIGHT VERTICAL CONSOLE
51. ARRESTING HOOK PANEL
52. DISPLAYS CONTROL PANEL
53. ELEVATION LEAD PANEL

RIGHT SIDE CONSOLE
54. COMPASS CONTROL PANEL
55. CAUTION–ADVISORY INDICATOR
56. TACAN CONTROL PANEL
57. MASTER GENERATOR CONTROL PANEL
58. ARA-63 CONTROL PANEL
59. AIR CONDITIONING CONTROL PANEL
60. MASTER LIGHT CONTROL PANEL
61. EXTERNAL ENVIRONMENTAL CONTROL PANEL
62. MASTER TEST PANEL
63. HYDRAULIC TRANSFER PUMP SWITCH
64. DEFOG CONTROL LEVER
65. WINDSHIELD DEFOG SWITCH

FO-3
(Reverse blank)

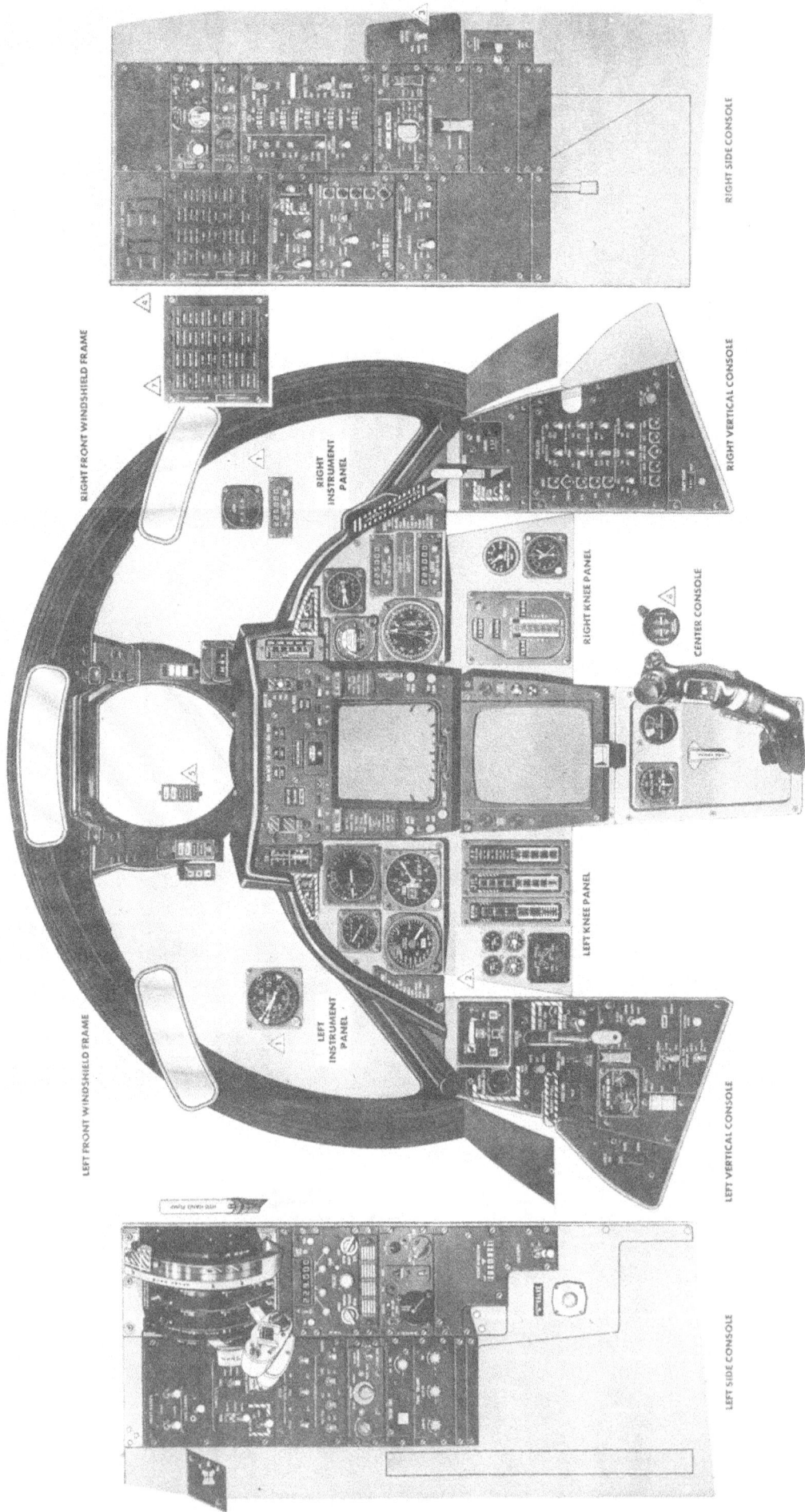

RIGHT SIDE CONSOLE

RIGHT FRONT WINDSHIELD FRAME

RIGHT INSTRUMENT PANEL

RIGHT VERTICAL CONSOLE

RIGHT KNEE PANEL

CENTER CONSOLE

LEFT KNEE PANEL

LEFT FRONT WINDSHIELD FRAME

LEFT INSTRUMENT PANEL

LEFT VERTICAL CONSOLE

LEFT SIDE CONSOLE

AIRCRAFT BUNO 159588 AND SUBSEQUENT

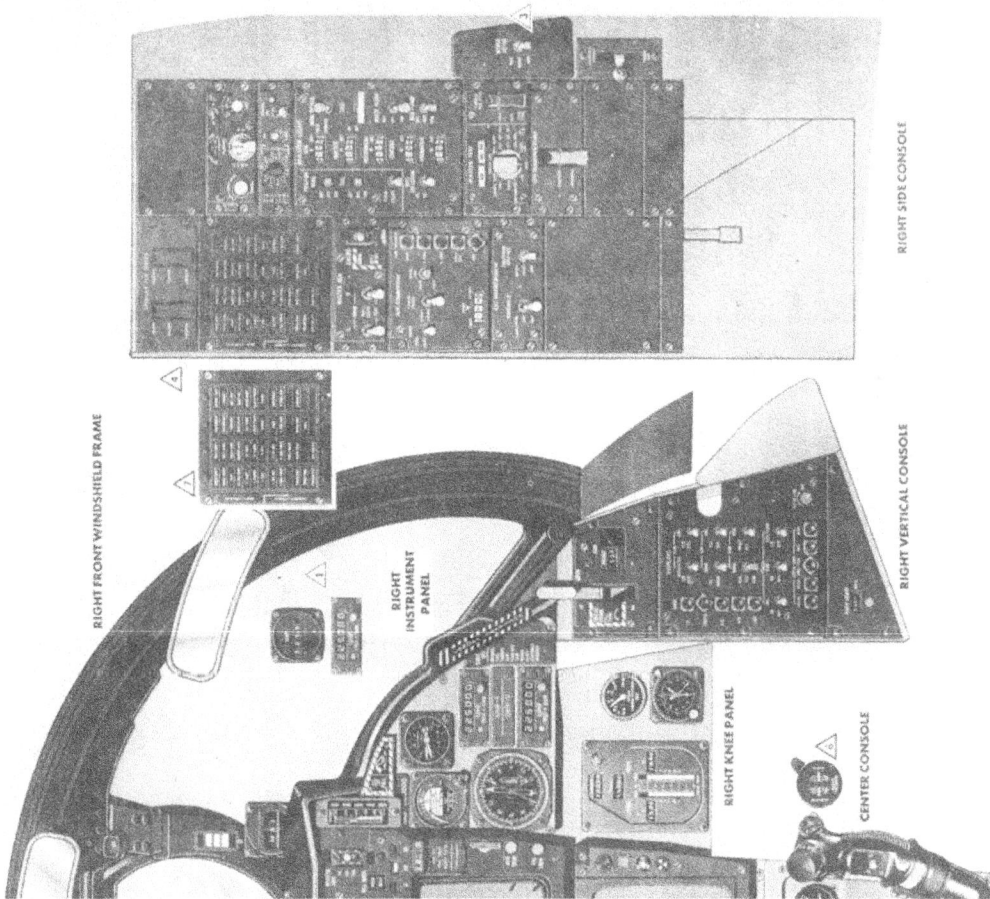

LEFT SIDE CONSOLE
1. G VALVE PUSHBUTTON
2. OXYGEN/VENT AIRFLOW CONTROL PANEL
3. TONE VOLUME CONTROL PANEL
4. TACAN CMD PANEL
5. TACAN CONTROL PANEL
6. ICS CONTROL PANEL
7. ARC-159 UHF CONTROL PANEL
8. AFCS CONTROL PANEL
9. THROTTLE QUADRANT
10. HYDRAULIC HANDPUMP
11. INLET RAMPS, THROTTLE CONTROL PANEL
12. TARGET DESIGNATE SWITCH

LEFT VERTICAL CONSOLE
13. FUEL MANAGEMENT PANEL
14. CONTROL SURFACE POSITION INDICATOR
15. LAUNCH BAR ABORT PANEL
16. LANDING GEAR CONTROL PANEL
17. WHEELS-FLAPS POSITION INDICATOR
18. EMER STORES JETTISON BUTTON

LEFT KNEE PANEL
19. HYDRAULIC PRESSURE INDICATOR
20. OIL PRESSURE INDICATOR
21. EXHAUST NOZZLE POSITION INDICATOR △
22. ENGINE PRESSURE RATIO INDICATOR
23. ELECTRICAL TACHOMETER INDICATOR (RPM)
24. TURBINE INLET TEMPERATURE INDICATOR (TIT)
25. RATE OF FLOW INDICATOR (FF)

LEFT INSTRUMENT PANEL
26. RADAR ALTIMETER
27. SERVOPNEUMATIC ALTIMETER △
28. AIRSPEED MACH INDICATOR
29. VERTICAL VELOCITY INDICATOR
30. LEFT ENGINE FUEL SHUT OFF HANDLE
31. ANGLE-OF-ATTACK INDICATOR

LEFT FRONT WINDSHIELD FRAME
32. APPROACH INDEXER
33. WHEELS WARNING / BRAKES
WARNING / ACLS / APC CAUTION / NWS
ENGA, CAUTION / AUTO THROT, △
CAUTION LIGHTS

CENTER PANEL
33a. HEADS-UP DISPLAY
34. AIR COMBAT MANEUVER PANEL
35. VERTICAL DISPLAY INDICATOR (VDI)
36. HORIZONTAL SITUATION DISPLAY
INDICATOR (HSD)
37. CABIN PRESSURE ALTIMETER
38. PEDAL ADJUST HANDLE
39. BRAKE PRESSURE INDICATOR △
40. CONTROL STICK

RIGHT FRONT WINDSHIELD FRAME
41. ECM WARNING LIGHT
42. STANDBY COMPASS

RIGHT INSTRUMENT PANEL
43. WING SWEEP INDICATOR
44. RIGHT ENGINE FUEL SHUT OFF HANDLE

45. ACCELEROMETER
46. STANDBY ATTITUDE INDICATOR
47. BEARING DISTANCE HEADING
INDICATOR (BDHI)
48. UHF INDICATOR
49. UHF REMOTE INDICATOR △
50. CANOPY JETTISON HANDLE

RIGHT KNEE PANEL
51. FUEL QUANTITY INDICATOR
52. LIQUID OXYGEN QUANTITY INDICATOR
53. CLOCK

RIGHT VERTICAL CONSOLE
54. ARRESTING HOOK PANEL
55. DISPLAYS CONTROL PANEL
56. ELEVATION LEAD PANEL

RIGHT SIDE CONSOLE
57. SPOILER FAILURE OVERRIDE PANEL △
58. COMPASS CONTROL PANEL
59. ARA-63 CONTROL PANEL
60. CAUTION-ADVISORY INDICATOR
61. MASTER GENERATOR CONTROL PANEL
62. AIR CONDITIONING CONTROL PANEL
63. MASTER LIGHT CONTROL PANEL
64. EXTERNAL ENVIRONMENTAL CONTROL PANEL
65. MASTER TEST PANEL
66. HYDRAULIC TRANSFER PUMP SWITCH
67. WINDSHIELD DEFOG CONTROL LEVER △
68. CANOPY DEFOG 'CABIN AIR LEVER
69. DATA STOWAGE COMPARTMENT

EFFECTIVITY

△ AIRCRAFT BUNO 159637 AND EARLIER AIRCRAFT.

△ AIRCRAFT BUNO 159635 AND SUBSEQUENT DO NOT
HAVE ENGINE PRESSURE RATIO INDICATOR.

△ AIRCRAFT BUNO 160696 AND EARLIER AIRCRAFT ONLY.

△ AIRCRAFT BUNO 160920 AND SUBSEQUENT AND
AIRCRAFT INCORPORATING AFC 573.

△ AIRCRAFT BUNO 160937 AND EARLIER AIRCRAFT AND
AIRCRAFT NOT INCORPORATING AFC 508.

△ AIRCRAFT BUNO 161168 AND SUBSEQUENT AND
AIRCRAFT INCORPORATING AFC 622.

△ AIRCRAFT BUNO 161282 AND SUBSEQUENT AND
AIRCRAFT INCORPORATING AFC 629.

RIGHT FRONT WINDSHIELD FRAME

RIGHT
INSTRUMENT
PANEL

RIGHT KNEE PANEL

CENTER CONSOLE

RIGHT SIDE CONSOLE

RIGHT VERTICAL CONSOLE

11-F9x509-2

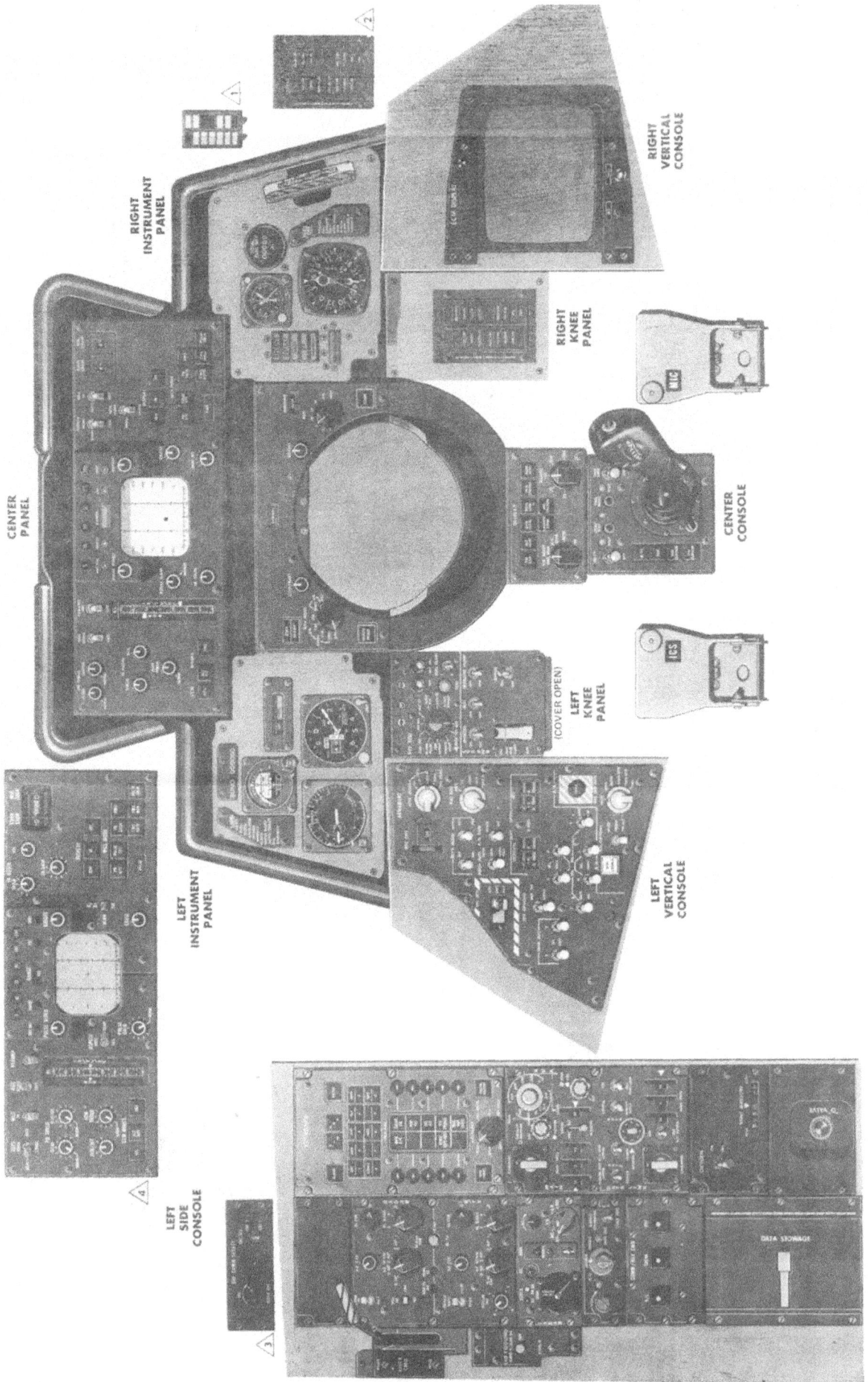

RIGHT
INSTRUMENT
PANEL

RIGHT
VERTICAL
CONSOLE

RIGHT
KNEE
PANEL

CENTER
PANEL

CENTER
CONSOLE

(COVER OPEN)
LEFT KNEE
PANEL

ICS

LEFT INSTRUMENT PANEL

LEFT
VERTICAL
CONSOLE

DATA STORAGE

LEFT SIDE
CONSOLE

EFFECTIVITY

1. AIRCRAFT BUNO 158978 AND SUBSEQUENT

2. AIRCRAFT BUNO 158612 AND SUBSEQUENT

3. AIRCRAFT BUNO 158978 AND SUBSEQUENT AND AIRCRAFT INCORPORATING AFC 599

4. AIRCRAFT BUNO 158978 AND SUBSEQUENT

LEFT SIDE CONSOLE
1. G VALVE PUSHBUTTON
2. OXYGEN VENT AIRFLOW CONTROL PANEL
3. COMM/NAV COMMAND PANEL
4. ICS CONTROL PANEL
5. INTEGRATED CONTROL PANEL
6. TACAN CONTROL PANEL
7. LIQUID COOLING CONTROL PANEL
8. COMPUTER ADDRESS PANEL
9. RADAR IR/TV CONTROL PANEL
9a. UHF COMM SELECT PANEL
10. EJECT COMMAND PANEL

LEFT VERTICAL CONSOLE
11. ARMAMENT PANEL

LEFT KNEE PANEL
12. SYSTEM TEST – SYSTEM POWER PANEL

LEFT INSTRUMENT PANEL
13. SERVOPNEUMATIC ALTIMETER
14. AIRSPEED MACH INDICATOR
15. UHF REMOTE INDICATOR
16. STANDBY ATTITUDE INDICATOR

CENTER PANEL
17. DETAIL DATA DISPLAY PANEL (DDD)

CENTER CONSOLE
18. NAVIGATION CONTROL AND DATA READOUT
19. TACTICAL INFORMATION DISPLAY (TID)
20. TACTICAL INFORMATION CONTROL PANEL
21. HAND CONTROL UNIT

LEFT AND RIGHT FOOT WELLS
22. MIC FOOT BUTTON
23. ICS FOOT BUTTON

RIGHT INSTRUMENT PANEL
24. FUEL QUANTITY TOTALIZER
25. CLOCK
26. THREAT ADVISORY LIGHTS
27. CANOPY JETTISON HANDLE
28. BEARING DISTANCE HEADING INDICATOR (BDHI)

RIGHT KNEE PANEL
29. CAUTION-ADVISORY PANEL

RIGHT VERTICAL CONSOLE
30. MULTIPLE DISPLAY INDICATOR

RIGHT SIDE CONSOLE
31. DIGITAL DATA INDICATOR (DDI)
32. ECM DISPLAY CONTROL PANEL
33. DATA LINK REPLY AND INTERIOR LIGHT CONTROL PANEL
34. ECM CONTROL PANEL
35. DECM CONTROL PANEL
36. DEFOG CONTROL LEVER
37. IFF TRANSPONDER CONTROL PANEL
38. CHAFF/FLARE DISPENSE PANEL
39. AA1 CONTROL PANEL
40. AN/ALE 29A PROGRAMMER
41. IFF ANTENNA ANT AND TEST PANEL
42. RADAR BEACON CONTROL PANEL
43. KY 28 CONTROL PANEL
44. ELECTRICAL POWER SYSTEM TEST PANEL

RIGHT SIDE CONSOLE

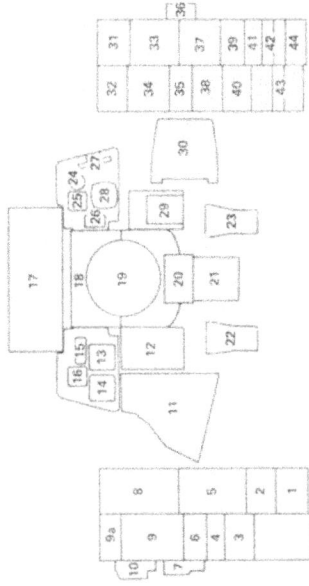

RIGHT VERTICAL CONSOLE

RIGHT INSTRUMENT PANEL

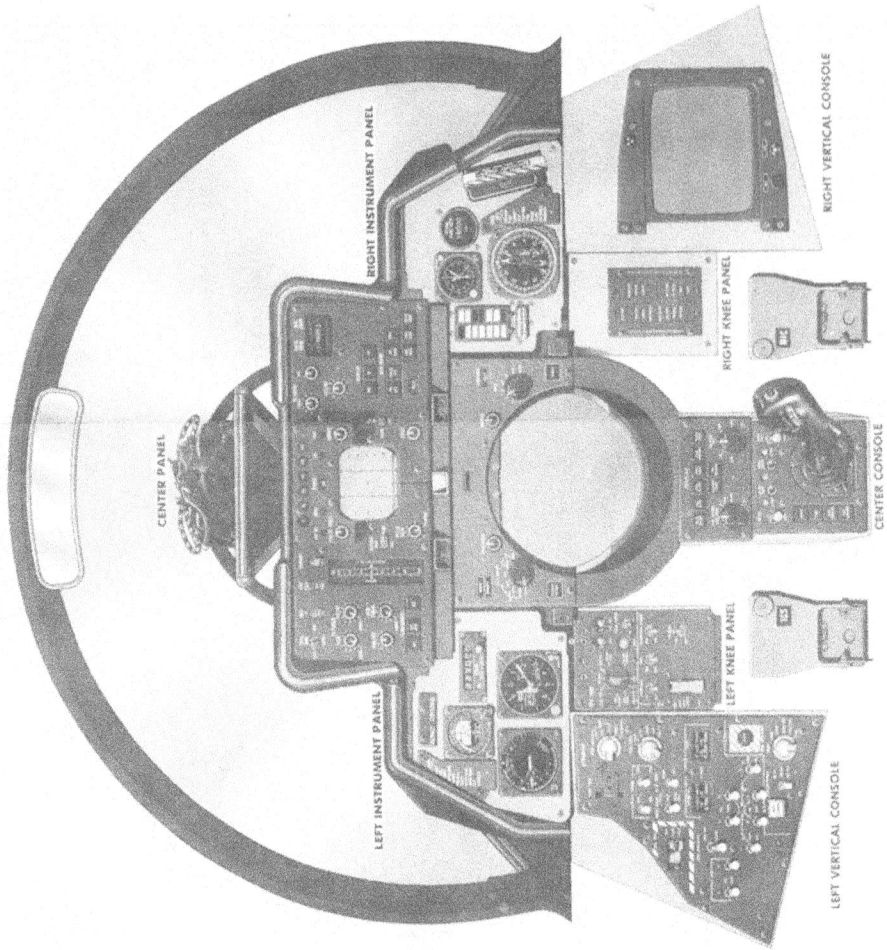

RIGHT INSTRUMENT PANEL

RIGHT VERTICAL CONSOLE

RIGHT KNEE PANEL

CENTER CONSOLE

CENTER PANEL

LEFT KNEE PANEL

LEFT INSTRUMENT PANEL

LEFT VERTICAL CONSOLE

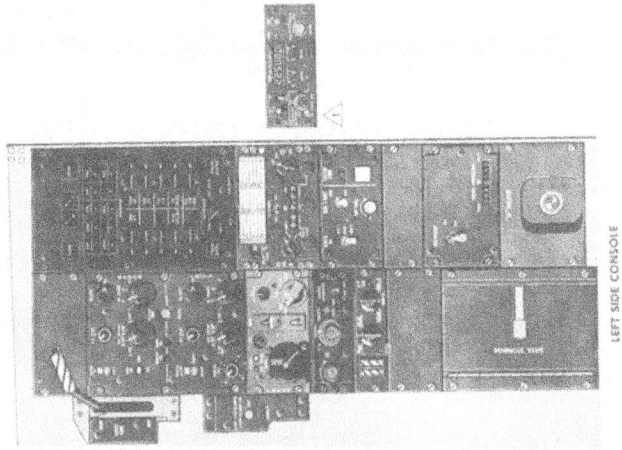

LEFT SIDE CONSOLE

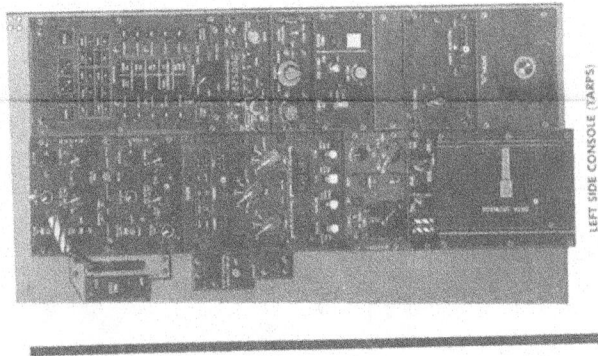

LEFT SIDE CONSOLE (TARPS)

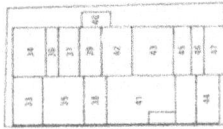

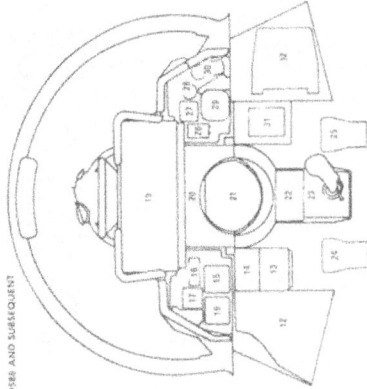

AIRCRAFT BUNO 159588 AND SUBSEQUENT

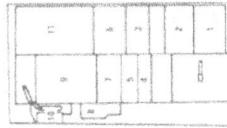

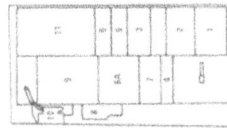

TARPS

SIDE CONSOLE

LEFT SIDE CONSOLE
1. G VALVE PUSHBUTTON
2. OXYGEN-VENT AIRFLOW CONTROL PANEL
3. XMTR SEL / ANT SEL / TACAN CMD
4. KY-28 CONTROL PANEL
5. ICS CONTROL PANEL
6. ARC 159A (V)S CONTROL PANEL
6A. CONTROLLER PROCESSOR SIGNAL (CPS) UNIT
7. TACAN CONTROL PANEL
8. LIQUID COOLING CONTROL PANEL
9. RADAR IR /TV CONTROL PANEL
10. EJECT COMMAND LEVER
11. COMPUTER ADDRESS PANEL

LEFT VERTICAL CONSOLE
12. ARMAMENT PANEL

LEFT KNEE PANEL
13. SYSTEM POWER PANEL
14. SYSTEM TEST PANEL

LEFT INSTRUMENT PANEL
15. SERVOPNEUMATIC ALTIMETER
16. AIRSPEED MACH INDICATOR
17. STANDBY ATTITUDE INDICATOR
18. UHF REMOTE INDICATOR

CENTER PANEL
19. DETAIL DATA DISPLAY PANEL (DDD)

CENTER CONSOLE
20. NAVIGATION CONTROL AND DATA READOUT
21. TACTICAL INFORMATION DISPLAY (TID)
22. TACTICAL INFORMATION CONTROL PANEL
23. HAND CONTROL UNIT

LEFT AND RIGHT FOOT WELLS
24. ICS FOOT BUTTON
25. MIC FOOT BUTTON

RIGHT INSTRUMENT PANEL
26. THREAT ADVISORY LIGHTS
27. CLOCK
28. FUEL QUANTITY TOTALIZER
29. BEARING DISTANCE HEADING INDICATOR (BDHI)
30. CANOPY JETTISON HANDLE

RIGHT KNEE PANEL
31. CAUTION-ADVISORY PANEL

RIGHT VERTICAL CONSOLE
32. MULTIPLE DISPLAY INDICATOR

RIGHT SIDE CONSOLE
33. ECM DISPLAY CONTROL PANEL
34. DIGITAL DATA INDICATOR (DDI)
35. ECM CONTROL PANEL
36. DATA LINK CONTROL PANEL
37. DATA LINK REPLY AND ANT CONTROL PANEL
38. DECM CONTROL PANEL
39. AA1 CONTROL PANEL
40. DEFOG CONTROL LEVER
41. AN/ALE-39 PROGRAMMER AND CONTROL
42. INTERIOR LIGHT CONTROL PANEL
43. IFF TRANSPONDER CONTROL PANEL
44. MID COMPRESSION BYPASS TEST PANEL
45. IFF ANTENNA CONTROL AND TEST PANEL
46. RADAR BEACON CONTROL PANEL
47. ELECTRICAL POWER SYSTEM TEST PANEL

EFFECTIVITY

1 AIRCRAFT BUNO 159825 AND SUBSEQUENT

2 AIRCRAFT BUNO 160921 AND SUBSEQUENT AND AIRCRAFT INCORPORATING AFC 592

3 TARPS AIRCRAFT ONLY

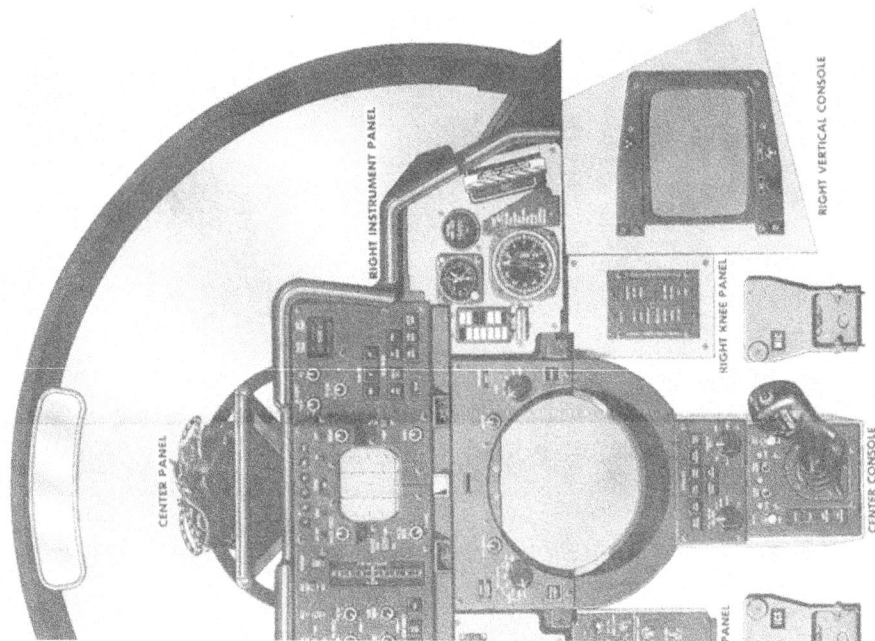

CENTER PANEL

RIGHT INSTRUMENT PANEL

RIGHT VERTICAL CONSOLE

RIGHT KNEE PANEL

CENTER CONSOLE

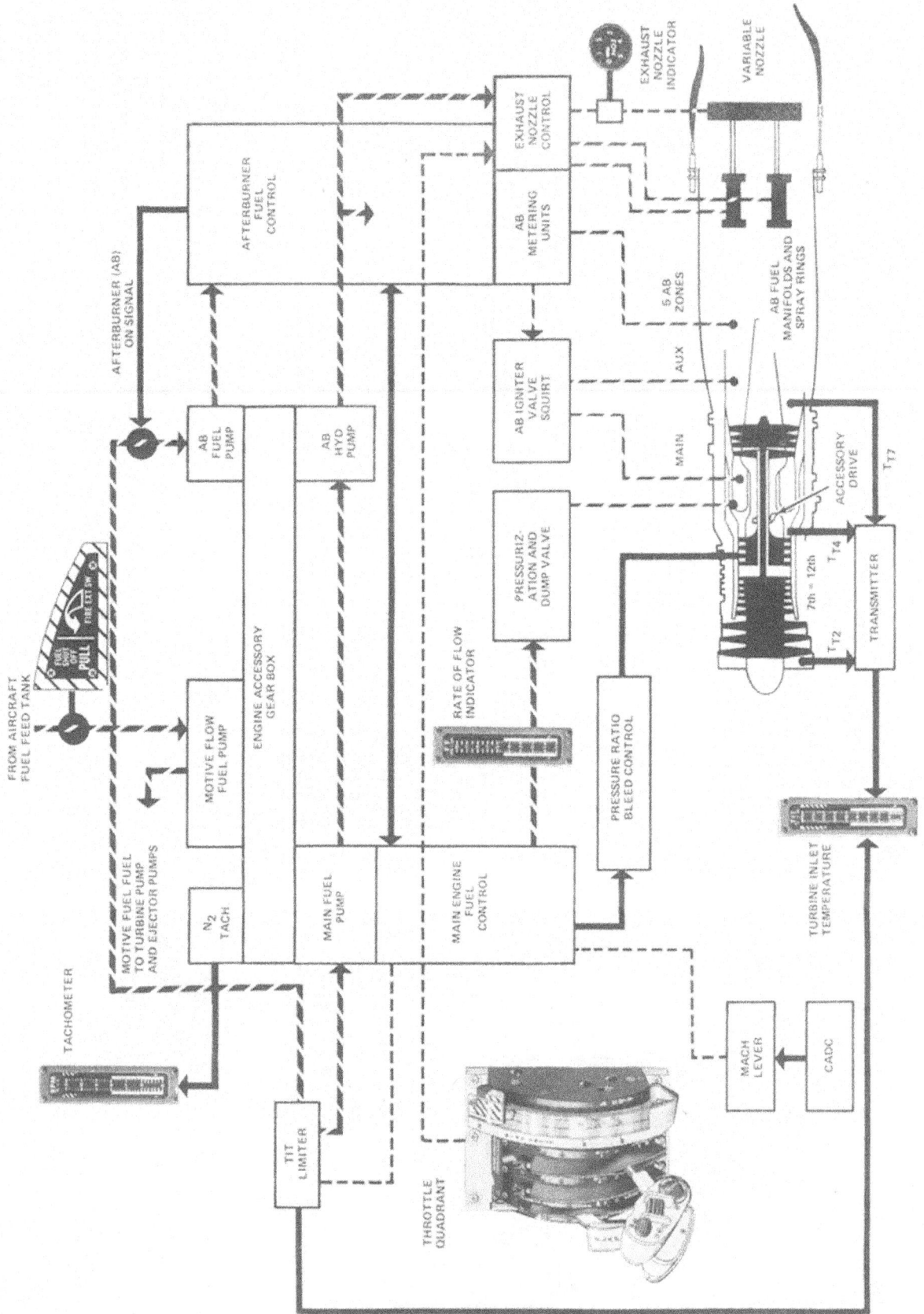

ENGINE CONTROL SYSTEM

5-F60-004-0

F-14A TOMCAT

THIS PAGE INTENTIONALY LEFT BLANK.

LEGEND

CHECK VALVE		MOTIVE FLOW
SCAVENGE EJECTOR PUMP	WATER DRAIN VALVE	PRECHECK SENSING
		TRANSFER
TRANSFER EJECTOR PUMP	ORIFICE	GRAVITY TRANSFER
	INERTIA CHECK VALVE	FEED
DEFUELING FLOAT VALVE		FUELING
	FLAPPER CHECK VALVE	DEFUELING
BOX BEAM FILTER		DUMP
		ECS AIR
		VENT
		FUELING AND TRANSFER
		VENT AND TRANSFER
		SIPHONING

AIR PRESSURE REGULATOR (25 PSI)

ZERO-L

AIR PRESS SWITCH

AIR PRESSURE REGULATOR (2 PSI)

AIR FILTER

DEFUELING-TRANSFER SELECTOR VALVE

REFUELING/TRANSFER SHUTOFF VALVE

LEFT BOX BEAM TANK

FUEL CELL NO. 3

INTERC... VALVE

PILOT VALVE

TURBINE PUMP

PENDULUM INLET

AIR EJECTOR

VENT SOLENOID VALVE

FUEL FEED VALVE

4.5 TO 9.5 PSI FUEL PRESS SWITCH

FUEL SHUTOFF HANDLE

ENGINE FEED ISOLATION VALVE

MOTIVE FLOW PUMP

LEFT ENGINE

LEFT WING TANK

HIGH LEVEL PILOT VALVE

TRANSFER SHUTOFF VALVE

REFUELING SHUTOFF VALVE

ENGINE FEED CROSSFLOW VALVE

AIR VENT VALVE

PILOT VALVE

DUMP VALVE

FUSELAGE MOTIVE FLOW SHUTOFF VALVE

AIR VENT VALVE

FUEL CELL NO. 5

LEFT AUXILIARY TANK

HIGH LEVEL PILOT VALVE

VENT BOX

SIGHT GAGE

VENT PRESSURE VALVE

OVERBOARD VENT

27-PSI RELIEF

REFUELING/TRANS SHUTOFF VALVE

LOW-LEVEL PILOT VALVE (AIR)

LOW-LEVEL PILOT VALVE (FUEL)

NEG "G" CHECK VALVES (2)

VENT PLENUM

FUEL CELL NO. 7

HIGH-LEVEL PILOT VALVE

VENT INLET CHECK VALVE

FUEL CELL NO. 6

FUEL CELL NO. 8

RAM AIR INLET

VENT TANK

VENT RELIEF VALVE (2)

OVERBOARD VENT

STATIC PORT

FLAME ARRESTER

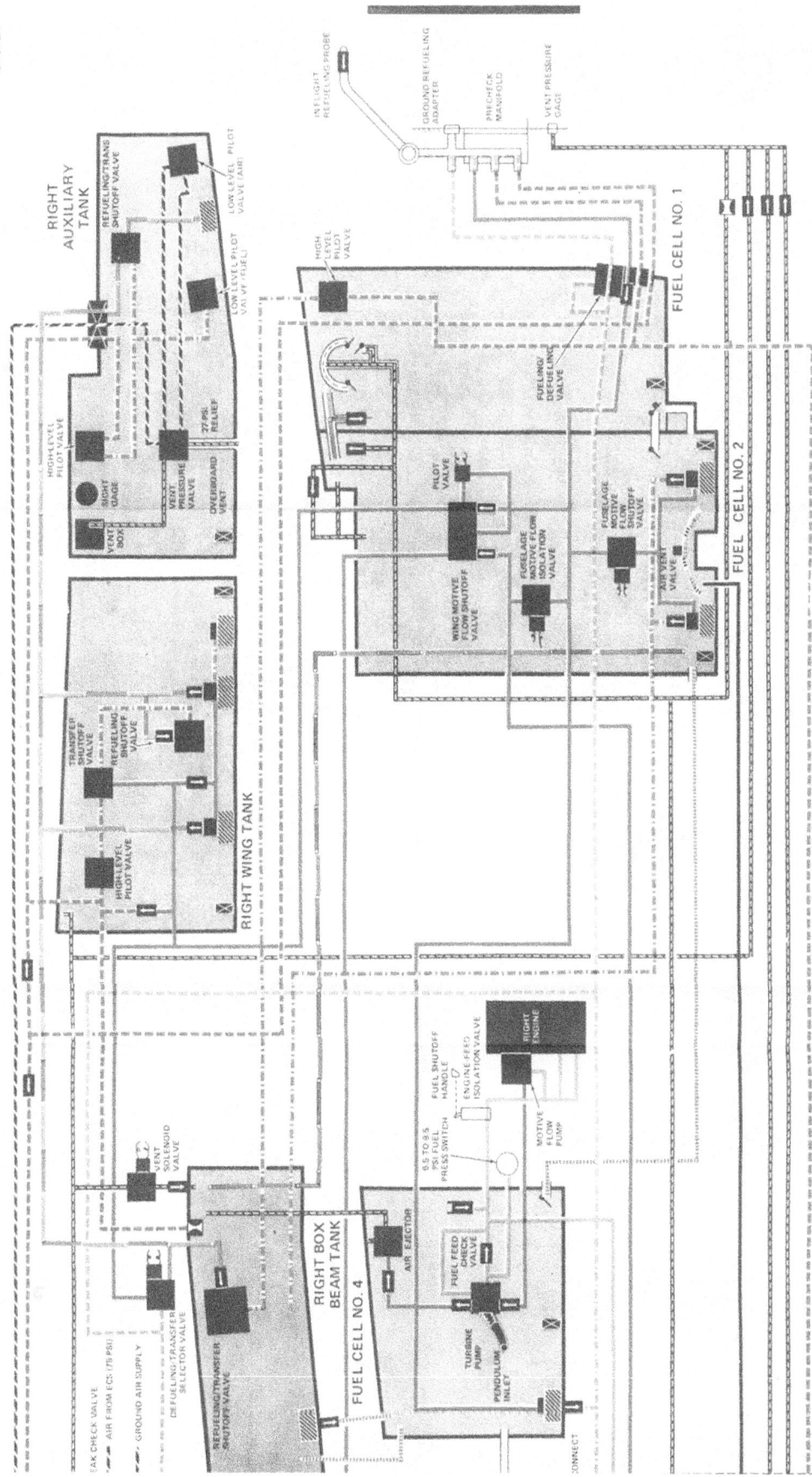

NAVAIR 01-F14AAA-1

RIGHT AUXILIARY TANK

REFUELING/TRANS SHUTOFF VALVE
LOW-LEVEL PILOT VALVE (AIR)
LOW-LEVEL PILOT VALVE (FUEL)

HIGH-LEVEL PILOT VALVE

SIGHT GAGE

VENT BOX

27 PSI RELIEF
VENT PRESSURE VALVE
OVERBOARD VENT

TRANSFER SHUTOFF VALVE
REFUELING SHUTOFF VALVE

HIGH-LEVEL PILOT VALVE

RIGHT WING TANK

IN-FLIGHT REFUELING PROBE
GROUND REFUELING ADAPTER
PRECHECK MANIFOLD
VENT PRESSURE GAGE

HIGH LEVEL PILOT VALVE

FUEL CELL NO. 1

FUELING/DEFUELING VALVE

FUEL CELL NO. 2

PILOT VALVE

FUSELAGE MOTIVE FLOW ISOLATION VALVE

FUSELAGE MOTIVE FLOW SHUTOFF VALVE

AIR VENT VALVE

WING MOTIVE FLOW SHUTOFF VALVE

VENT SOLENOID VALVE

RIGHT BOX BEAM TANK

DEFUELING/TRANSFER SELECTOR VALVE

REFUELING/TRANSFER SHUTOFF VALVE

FUEL CELL NO. 4

AIR EJECTOR

FUEL FEED CHECK VALVE

TURBINE PUMP

PENDULUM INLET

6.5 TO 8.5 PSI FUEL PRESS SWITCH

FUEL SHUTOFF HANDLE

ENGINE-FEED ISOLATION VALVE

RIGHT ENGINE

MOTIVE FLOW PUMP

LEAK CHECK VALVE
AIR FROM ECS (75 PSI)
GROUND AIR SUPPLY

CONNECT

FWD AND RIGHT FUEL SYSTEM

FO-8
(Reverse blank)

DC LEFT MAIN BUS (28 V DC)

INTRPT FREE DC BUS FDR NO. 1

INTRPT FREE DC BUS (28 V DC)

POWER FROM AWG-9 SYSTEM

AWG-9 DC BUS (28 V DC)

INTRPT FREE DC BUS FDR NO. 2

DC RIGHT MAIN BUS (28 V DC)

L MAIN DC POWER CONTACTOR

R MAIN DC POWER CONTACTOR

LEFT TRANSFORMER RECTIFIER (100 AMPERE)

TRANS/RECT

RIGHT TRANSFORMER RECTIFIER (100 AMPERE)

L MAIN XFMR/RECT

R MAIN XFMR RECT

AC LEFT MAIN BUS (3 φ) 115 V AC

INST. BUS FDR (NOTE 14)

INST BUS FDR (NOTE 13)

AC RIGHT MAIN BUS (3 φ) 115 V AC

AC ESS BUS NO. 2 FDR PH C

AC ESSENTIAL NO. 2 PH B & C BUS 115 V AC

AC ESS BUS NO. 2 FDR PH B

115 V AC INSTRUMENT BUS (φ B)

AC MONITOR BUS RELAY

AC ESS BUS NO. 2 FDR PH A

PILOT AC ESSENTIAL CIRCUIT BREAKER PANEL

AC ESSENTIAL NO. 2 PH A BUS 115 V AC

AC ESS BUS NO. 2 FDR PH A

AC MONITOR BUS (3 φ) 115 V AC

26 V AC BUS FDR

26 V AC TRANSFORMER

26 V AC INSTRUMENT BUS

26 V AC NAVIGATION BUS

GENERATOR CONTROL UNIT

EMERGENCY GENERATOR DC

EMERGENCY GENERATOR AC

AC ESSENTIAL NO. 2 TRANSFER RELAY

AC ESSENTIAL NO. 1 TRANSFER RELAY

AC ESSENTIAL NO. 1 BUS (3 φ) 115 V AC

HYDRAULIC MOTOR

FROM COMBINED SYSTEM HYDRAULIC PRESSURE

L MAIN AC POWER CONTACTOR

R MAIN AC POWER CONTACTOR

LEFT GENERATOR CONTROL UNIT

AC GENERATOR

LEFT ENGINE CSD

AC EXTERNAL POWER CONTACTOR

GROUND POWER MONITOR

EXTERNAL POWER RECEPTACLE

RIGHT ENGINE CSD

AC GENERATOR

RIGHT GENERATOR CONTROL UNIT

L GEN

MASTER GEN

RESET

R GEN

NAVAIR 01-F14AAA-1

EFFECTIVITY

△	AIRCRAFT BUNO 159978 AND SUBSEQUENT
△	AIRCRAFT BUNO 159825 AND SUBSEQUENT AND AIRCRAFT INCORPORATING AFC 365.
△	AIRCRAFT INCORPORATING AFC 181.
△	AIRCRAFT BUNO 158620 THROUGH 159636.
△	AIRCRAFT BUNO 159637 AND SUBSEQUENT
△	AIRCRAFT BUNO 159825 AND SUBSEQUENT AND AIRCRAFT INCORPORATING AFC 400.
△	AIRCRAFT BUNO 159430 AND SUBSEQUENT AND AIRCRAFT INCORPORATING AVC 1585.
△	AIRCRAFT INCORPORATING AFC 181.
△	AIRCRAFT BUNO 158620 THROUGH 159468 INCORPORATING AFC 338 BUT NOT INCORPORATING AFC 363.
△	AIRCRAFT INCORPORATING AFC 529
△	AIRCRAFT BUNO 159588 AND AIRCRAFT INCORPORATING AFC 363.
△	AIRCRAFT BUNO 158620 THROUGH 159468 AND SUBSEQUENT 159589 AND SUBSEQUENT NOT INCORPORATING AFC 363.
△	AIRCRAFT BUNO 161270 AND SUBSEQUENT AND AIRCRAFT INCORPORATING AFC 610.
△	AIRCRAFT NOT INCORPORATING AFC 610.
△	ON AIRCRAFT NOT INCORPORATED ON DC ESS NO. 2 BUS.
△	AIRCRAFT BUNO 161168 AND SUBSEQUENT.
△	ALQ-126 DECM ON AIRCRAFT BUNO 161168 AND SUBSEQUENT
△	AIRCRAFT BUNO 161279 AND SUBSEQUENT AND AIRCRAFT INCORPORATING AFC 618.
△	AIRCRAFT BUNO 161279 AND SUBSEQUENT AND AIRCRAFT INCORPORATING AFC 622.
△	AIRCRAFT BUNO 161279 AND SUBSEQUENT AND AIRCRAFT INCORPORATING AFC 618.
△	TARPS AIRCRAFT

9150-0120

PILOT

AC ESSENTIAL (LEFT KNEE) CIRCUIT BREAKER PANEL

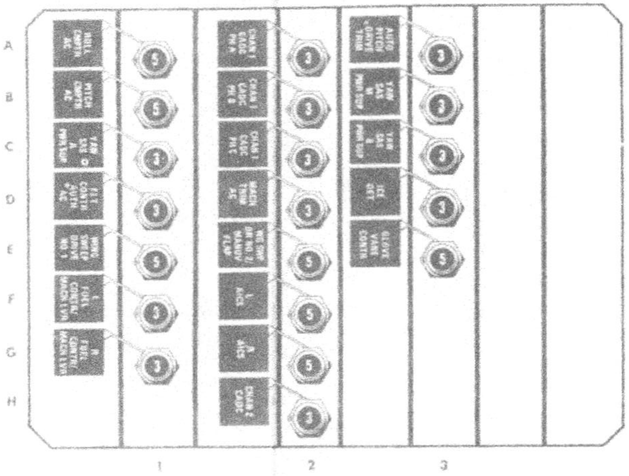

AC LEFT MAIN CIRCUIT BREAKER PANEL (LEFT AFT) (PANEL NO. 1 — UPPER SEGMENT)

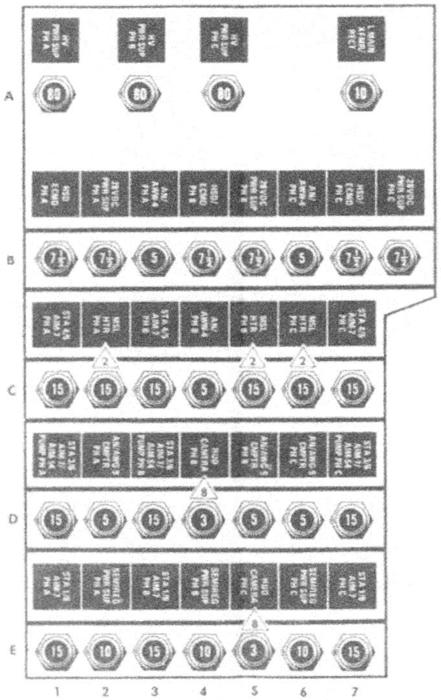

AC LEFT MAIN CIRCUIT BREAKER PANEL (LEFT AFT) (PANEL NO. 1 — LOWER SEGMENT)

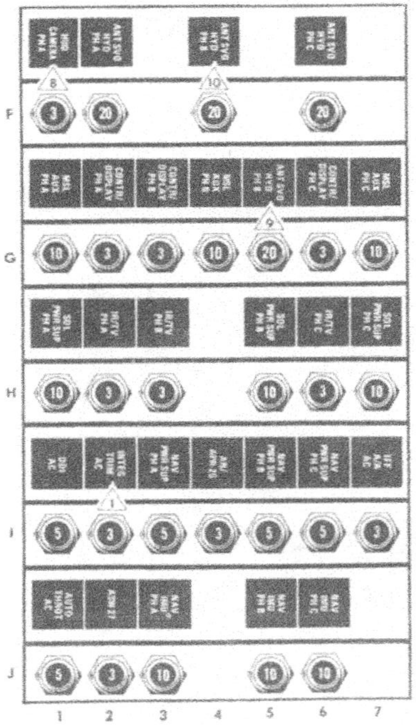

EFFECTIVITY

1. AIRCRAFT BUNO 159825 AND SUBSEQUENT AND AIRCRAFT INCORPORATING AFC 365.
2. AIRCRAFT BUNO 161168 AND SUBSEQUENT, ALQ 100 PH A, B, C CIRCUIT BREAKERS (PANEL NO. 2) ARE 15 AMPERES LABELED MSL HTR, PH A, B, C, MSL HTR PHA, B, C CIRCUIT BREAKERS (PANEL NO. 1, UPPER SEGMENT) ARE 10 AMPERES LABELED ALQ 126 PH A, B, C.
3. AS/ING/BARO ALTMDC CIRCUIT BREAKER ON AIRCRAFT BUNO 158620 THROUGH 159636.
4. AIRCRAFT BUNO 159825 AND SUBSEQUENT AND AIRCRAFT INCORPORATING AFC 400.
5. AIRCRAFT BUNO 158620 THROUGH 159636.
6. AIRCRAFT BUNO 161142 AND SUBSEQUENT AND AIRCRAFT INCORPORATING AFC 610.
7. AIRCRAFT NOT INCORPORATING AFC 610.
8. AIRCRAFT BUNO 161270 AND SUBSEQUENT AND AIRCRAFT INCORPORATING AFC 618.
9. TARPS AIRCRAFT ONLY.
10. RECON POD ON TARPS AIRCRAFT.

NOTE

PANEL NO. 1
ROW A –
THE COLUMN NUMBERING
IS DIFFERENT THAN
ROWS B THROUGH J.

AC ESSENTIAL NO. 1 CIRCUIT BREAKER PANEL (RIGHT SIDE) (PANEL NO. 5)

AC ESSENTIAL NO. 2 PHASE C CIRCUIT BREAKER PANEL (LEFT) (PANEL NO. 4)

AC ESSENTIAL NO. 2 PHASE A CIRCUIT BREAKER PANEL (LEFT) (PANEL NO. 3)

AC RIGHT MAIN CIRCUIT BREAKER PANEL (LEFT AFT) (PANEL NO. 2)

RIO

NOTE
IN AIRCRAFT BUNO 160887 AND SUBSEQUENT AND AIRCRAFT INCORPORATING AFC-624, THE THREE OUTBD SPOILER PUMP CIRCUIT BREAKERS ARE REPLACED BY ONE CB92.

PANEL NO. 2
ROWS A THROUGH F.
THE COLUMN NUMBERING IS DIFFERENT THAN ROWS 0 THROUGH I.

RIO

DC ESSENTIAL NO. 2 CIRCUIT BREAKER PANEL (RIGHT AFT) (PANEL NO. 7)

DC ESSENTIAL NO. 1 CIRCUIT BREAKER PANEL (RIGHT SIDE) (PANEL NO. 6)

PILOT

DC ESSENTIAL (RIGHT KNEE) CIRCUIT BREAKER PANEL

EFFECTIVITY

△1 ALE-29 SEQ 1 & 2 SQUIBS CIRCUIT BREAKER ON AIRCRAFT 158620 THROUGH 158637.

△2 ALE-29 CHAFF FLARE DISP CIRCUIT BREAKER ON AIRCRAFT BUNO 158620 THROUGH 158637.

△3 AN/APR-27 CIRCUIT BREAKER ON AIRCRAFT BUNO 158637 AND PRIOR.

△4 THREE-AMPERE CIRCUIT BREAKER ON AIRCRAFT BUNO 159444 AND PRIOR NOT INCORPORATING AFC 337.

△5 AIRCRAFT BUNO 159430 AND SUBSEQUENT AND AIRCRAFT INCORPORATING AVC 1585.

△6 A/S IND/BARO ALTM DC CIRCUIT BREAKER ON AIRCRAFT BUNO 159620 THROUGH 159636.

△7 AN/ARA-50 CIRCUIT BREAKER ON AIRCRAFT BUNO 159588 AND AIRCRAFT INCORPORATING AFC 363.

△8 AIRCRAFT BUNO 159620 THROUGH 159608 INCORPORATING AFC 338, BUT NOT INCORPORATING AFC 363.

△9 AIRCRAFT BUNO 159002 AND SUBSEQUENT AND AIRCRAFT INCORPORATING AFC 181.

△10 AIRCRAFT BUNO 159826 AND SUBSEQUENT AND AIRCRAFT INCORPORATING AFC 365.

△11 ICS MCO CIRCUIT BREAKER ON AIRCRAFT BUNO 159620 THROUGH 158637.

△12 BARO ALTM DC CIRCUIT BREAKER ON AIRCRAFT NOT INCORPORATING AFC 610.

△13 AIRCRAFT BUNO 161166 AND SUBSEQUENT AND AIRCRAFT INCORPORATING AFC 622.

△14 AIRCRAFT BUNO 161279 AND SUBSEQUENT AND AIRCRAFT INCORPORATING AFC 618.

△15 ON AIRCRAFT PRIOR TO BUNO 161168 CIRCUIT BREAKER IS PLACARDED ECM DESTR.

△16 TARPS AIRCRAFT ONLY

DC COCKPIT CIRCUIT
BREAKER PANELS

RIO

DC MAIN CIRCUIT BREAKER PANEL
(RIGHT AFT) (PANEL NO. 8-LOWER SEGMENT)

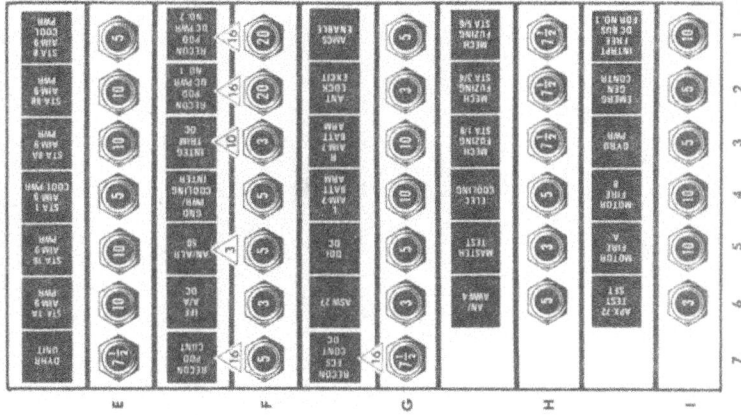

DC MAIN CIRCUIT BREAKER PANEL
(RIGHT AFT) (PANEL NO. 8-UPPER SEGMENT)

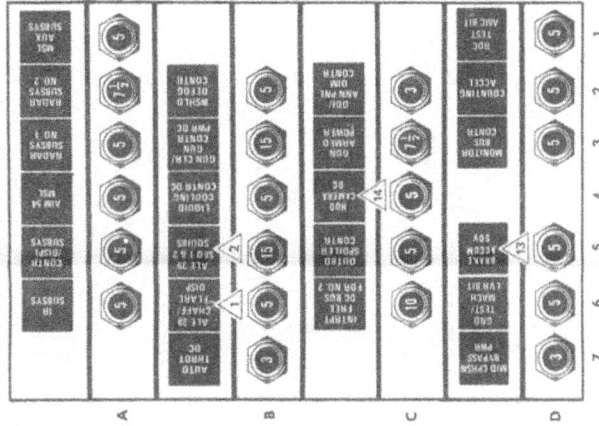

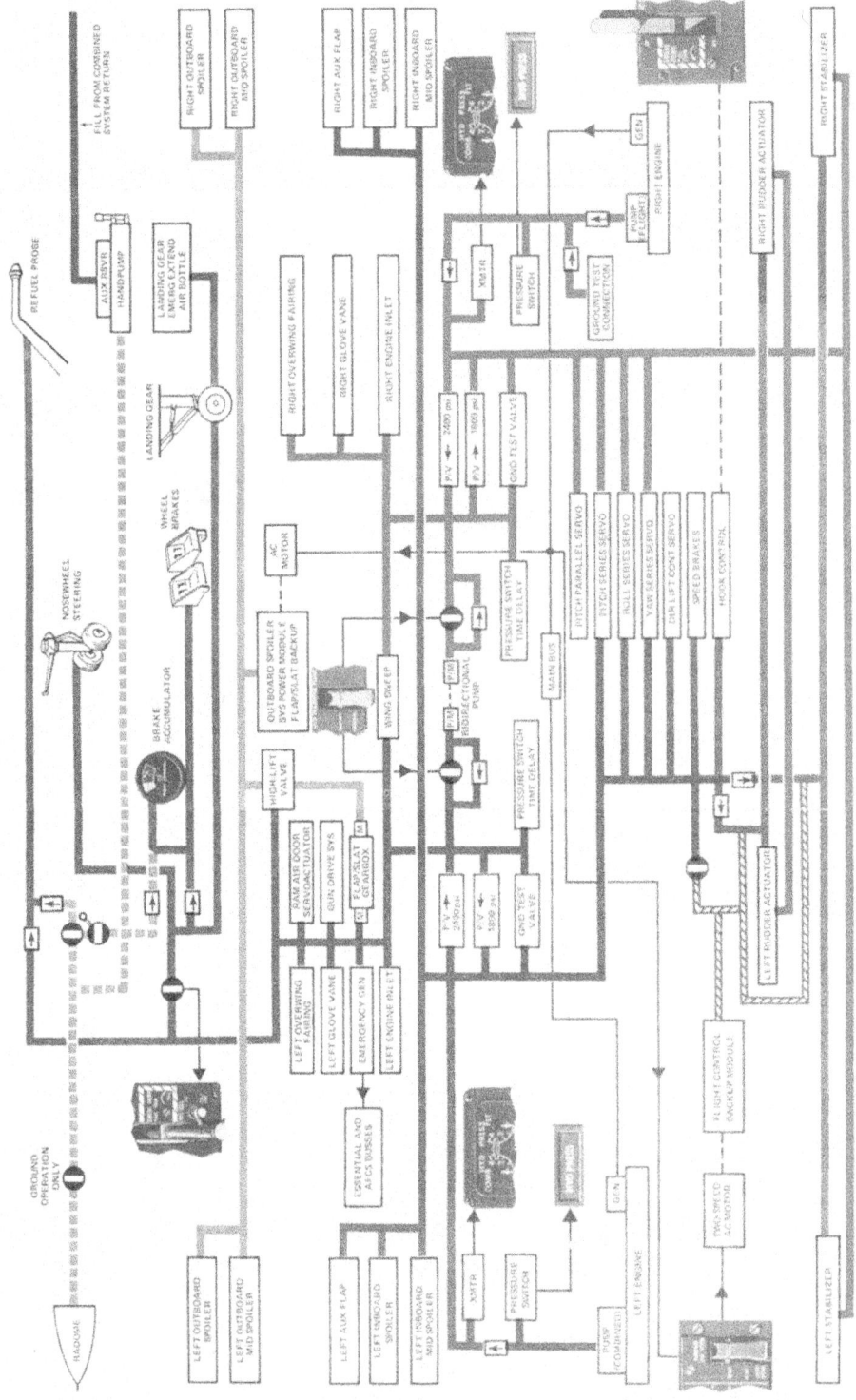

HYDRAULIC SYSTEM
(PRESSURE ONLY)

NOTE

ISOLATION LOGIC*

• HYDRAULIC ISOLATION SWITCH
  FLIGHT-COMBINED SYSTEM SHUTOFF TO
  LANDING GEAR
  COMPONENTS BARRING AUTO ISOLATE
  NOSEWHEEL STEERING
  WHEEL BRAKES

• 1.0: LANDING (WEIGHT ON WHEELS)
  COMBINED SYSTEM AVAILABLE TO ALL
  COMPONENTS BARRING AUTO ISOLATE
  AUTO ISOLATION
  WEIGHT ON WHEELS — OFF (EXCEPT FOR
  NOSEWHEEL STEERING WITH GEAR
  EMERG DOWN)
  IN FLIGHT WITH FLIGHT HYDRAULIC
  FAILURE — ON

*ON AIRCRAFT BUNO 159421 AND SUBSEQUENT
AND AIRCRAFT INCORPORATING AFC 336
THE AUTOMATIC ISOLATION CIRCUIT HAS BEEN
DELETED, EXCEPT FOR THE LANDING-HOOK
SUBSYSTEM

P/M		PUMP-MOTOR
P/V		PRIORITY VALVE
		FLIGHT HYDRAULIC SYSTEM
		COMBINED HYDRAULIC SYSTEM
		SPOILER SYSTEM POWER SOURCE
		COMBINED BACKUP POWER SOURCE
		HANDPUMP
		ELECTRICAL
		MECHANICAL CONNECTION
		CHECK VALVE (ARROW SHOWING FLOW DIRECTION)

REFUEL PROBE

FILL FROM COMBINED
SYSTEM RETURN

RIGHT OUTBOARD SPOILER
RIGHT OUTBOARD MID SPOILER

RIGHT AUX FLAP
RIGHT INBOARD SPOILER
RIGHT INBOARD MID SPOILER

AUX RSVR
HANDPUMP

LANDING GEAR EMERG EXTEND
AIR BOTTLE

RIGHT OVERWING FAIRING
RIGHT GLOVE VANE
RIGHT ENGINE INLET

RIGHT ENGINE
PUMP ENGINE
PUMP FLIGHT

NOSEWHEEL STEERING

LANDING GEAR

WHEEL BRAKES

BRAKE ACCUMULATOR

RIGHT RUDDER ACTUATOR

RIGHT STABILIZER

P/V 2400 psi
P/V 1000 psi
GND TEST VALVE

PRESSURE SWITCH

GROUND TEST CONNECTION

XMTR

AC MOTOR

OUTBOARD SPOILER SYS POWER MODULE
FLAP/SLAT BACKUP

WING SWEEP

PRESSURE SWITCH TIME DELAY

MAIN BUS

PITCH PARALLEL SERVO
PITCH SERIES SERVO
ROLL SERIES SERVO
YAW SERIES SERVO
DIR LAT CONT SERVO
SPEED BRAKES
HOOK CONTROL

HIGH LIFT VALVE

RAM AIR DOOR SERVOACTUATOR
GUN DRIVE SYS
FLAP/SLAT GEARBOX

BIDIRECTIONAL PUMP

PRESSURE SWITCH TIME DELAY

P/V 2450 psi
P/V 1650 psi
GND TEST VALVE

LEFT OVERWING FAIRING
LEFT GLOVE VANE
EMERGENCY GEN
LEFT ENGINE INLET

ESSENTIAL AND AICS BUSSES

FLIGHT CONTROL BACKUP MODULE

LEFT RUDDER ACTUATOR

GROUND OPERATION ONLY

RADOME

LEFT OUTBOARD SPOILER
LEFT OUTBOARD MID SPOILER

LEFT AUX FLAP
LEFT INBOARD SPOILER
LEFT INBOARD MID SPOILER

XMTR
PRESSURE SWITCH

PUMP (COMBINED)

GEN

PUMP FLD ACCESSORY

LEFT ENGINE

LEFT STABILIZER

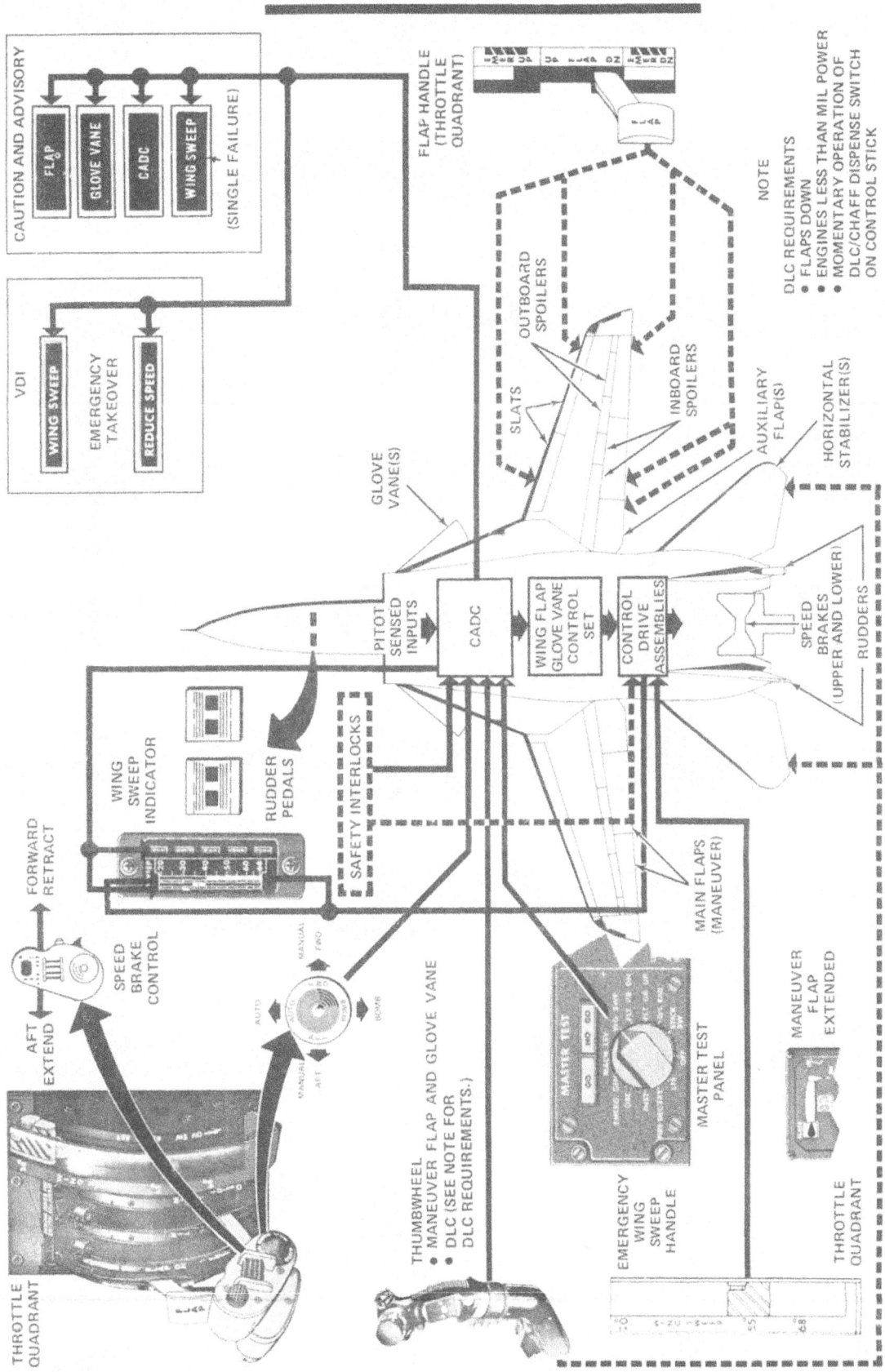

CAUTION AND ADVISORY

FLAP

GLOVE VANE

CADC

WING SWEEP

(SINGLE FAILURE)

VDI

WING SWEEP

EMERGENCY TAKEOVER

REDUCE SPEED

FLAP HANDLE
(THROTTLE
QUADRANT)

NOTE

DLC REQUIREMENTS
- FLAPS DOWN
- ENGINES LESS THAN MIL POWER
- MOMENTARY OPERATION OF
  DLC/CHAFF DISPENSE SWITCH
  ON CONTROL STICK

OUTBOARD
SPOILERS

INBOARD
SPOILERS

SLATS

AUXILIARY
FLAP(S)

HORIZONTAL
STABILIZER(S)

GLOVE
VANE(S)

PITOT
SENSED INPUTS

CADC

WING FLAP
GLOVE VANE
CONTROL
SET

CONTROL
DRIVE
ASSEMBLIES

SPEED
BRAKES
(UPPER AND LOWER)

RUDDERS

SAFETY INTERLOCKS

RUDDER
PEDALS

WING
SWEEP
INDICATOR

FORWARD
RETRACT

AFT
EXTEND

SPEED
BRAKE
CONTROL

THROTTLE
QUADRANT

AUTO

MANUAL
FWD

MANUAL
AFT

BOMB

THUMBWHEEL
MANEUVER FLAP AND GLOVE VANE
- DLC (SEE NOTE FOR
  DLC REQUIREMENTS.)

MAIN FLAPS
(MANEUVER)

MASTER TEST

MASTER TEST
PANEL

EMERGENCY
WING SWEEP
HANDLE

MANEUVER
FLAP
EXTENDED

THROTTLE
QUADRANT

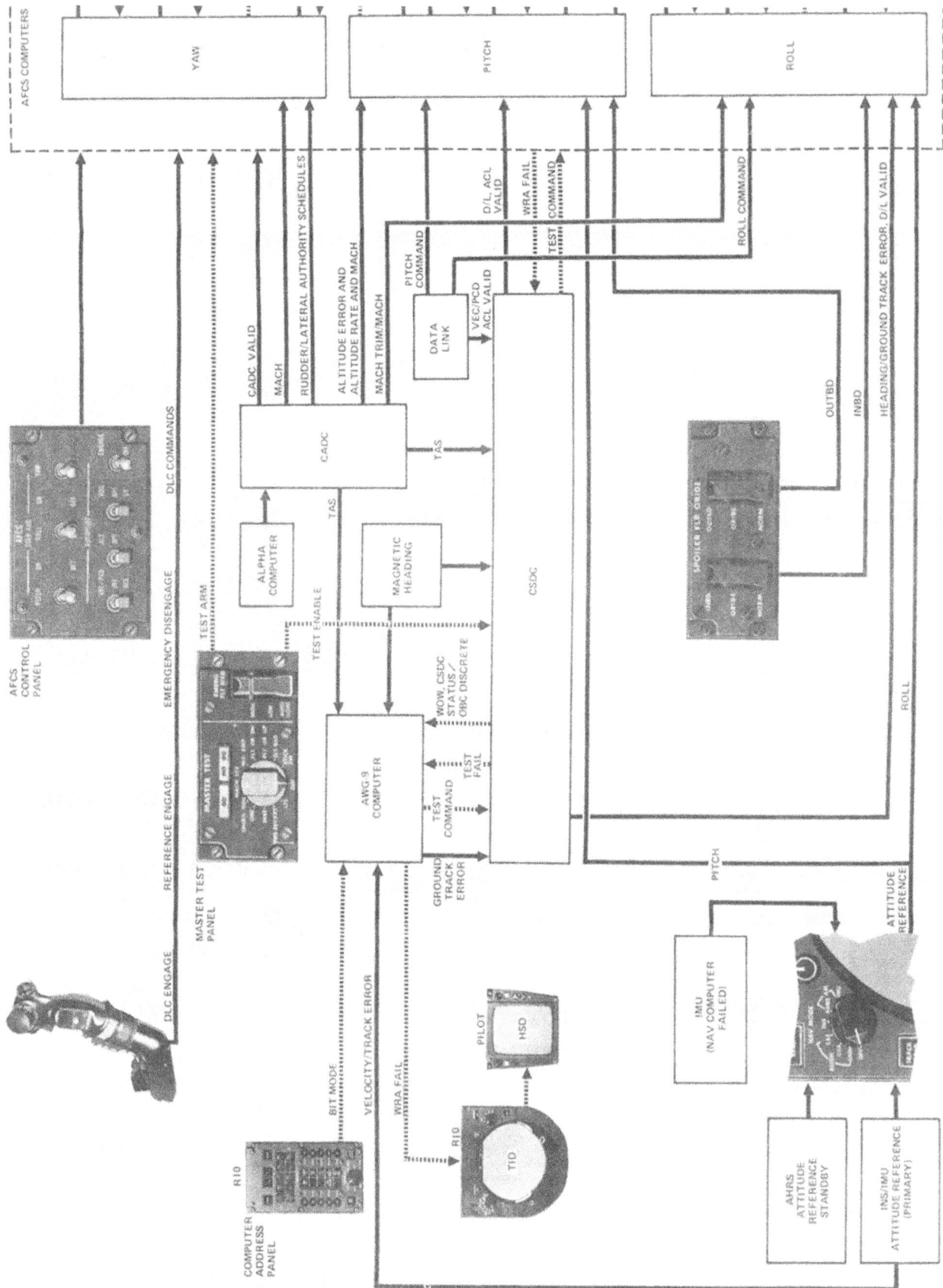

STABILIZERS

PARALLEL ACTUATOR

FLIGHT HYDRAULIC

RUDDERS

OUTBOARD SPOILER ACTUATORS

AUTO PITCH TRIM ACTUATOR

INBOARD SPOILER ACTUATORS

DIFFERENTIAL STABILIZERS

STABILIZERS

COMBINED HYDRAULIC

DUAL SERIES ACTUATOR

FLIGHT HYDRAULIC

COMBINED HYDRAULIC

FLIGHT HYDRAULIC

DUAL SERIES ACTUATOR

DLC TRIM

PITCH MACH TRIM

COMBINED HYDRAULIC

COMBINED HYDRAULIC

DUAL SERIES ACTUATOR

FLIGHT HYDRAULIC

ACLS/AP

SPOILERS   AUTO PILOT

PITCH STAB 1   PITCH STAB 2

YAW STAB OP   YAW STAB OUT   MZ TAIL AUTH

BACK AUTH

ROLL STAB 1   ROLL STAB 2

BACK TRIM

ROLL STOP

YAW STOP

TRIPLE LATERAL ACCELEROMETER

TRIPLE RATE GYRO

LATERAL STICK POSITION TRANSDUCERS

DUAL RATE GYRO

LATERAL STICK POSITION TRANSDUCERS

DUAL RATE GYRO

LATERAL STICK AUTHORITY COMMAND

RUDDER PEDAL AUTHORITY COMMAND

FAIL

YAW SAS/ROLL ARI COMMANDS

FAIL

ACL PITCH COMMAND

SPOILER COMMAND

AUTO TRIM

PITCH SAS/AFCS COMMANDS

SPOILER COMMAND AND ROLL SAS

FAIL

ROLL SAS/LATERAL ARI/AFCS COMMANDS

AFCS COMPUTERS

YAW

PITCH

ROLL

ENVIRONMENTAL CONTROL SYSTEM

NAVAIR 01-F14AAA-1

FO-15
(Reverse blank)

10 590 0230

HIGH PRESSURE ENGINE BLEED AIR

ENGINE SHUTOFF
TO ENGINE STARTER
ENGINE PNEUMATIC START CONNECTION
SHROUD CONTAINING LEAK DETECTION

BLED DUCT
TO ENGINE STARTER
ENGINE SHUTOFF
1100°F 400 PSI
HIGH-PRESSURE ENGINE BLEED AIR

CLOSES MODULATING, REGULATING, SHUTOFF VALVE; OPENS SAFETY VALVE
TO ENGINE SHUTOFF VALVES
TO CABIN, SUIT, ANTI-ICE CONTROLLER

SENSOR AND VENT SHROUD

DUAL PRESSURE REGULATOR AND SHUTOFF 80/90 PSI NORMAL
PRIMARY HEAT EXCHANGER
RAM AIR OVERBOARD
COOLING FAN
400°F
RAM AIR IN

400° MANIFOLD MODULATING AND SHUTOFF
SWITCH, OVERTEMP 475°F
TO CABIN, SUIT ANTI-ICE CONTROLLER

MACH 0.25
28 V DC
CADC SWITCH
WEIGHT ON-WHEELS SWITCH
FLIGHT/GROUND SWITCH

FUEL POD PRESSURE SYSTEM

AUTO (CLOSES AT 200°F)
MAN (CLOSES AT 250°F)
TO COCKPIT HOT AIR REGULATING VALVE

PRESSURE SUIT/VENT REGULATOR
120°F
SUIT HOT AIR MODULATING
SWITCH, SUIT AIR, OVERTEMP

WING SEAL SYSTEM

CABIN, SUIT, ANTI-ICE MODULATING
COMPRESSOR/TURBINE INLET MODULATING, REGULATING, AND SHUTOFF
550°F
TURBINE COMPRESSOR
475°F
RAM AIR OVERBOARD
COOLING FAN
SECONDARY HEAT EXCHANGER
RAM AIR IN

TO MISSILE COOLING SYSTEM
TO MISSILE COOLING SYSTEM
SENSOR, MANIFOLD TEMP 400°F

SUIT SENSOR

PRESSURE/VENTILATION SUIT SYSTEM
ANTI-G SUIT PRESSURE SYSTEM
CANOPY REGULATING VALVE
CANOPY SEAL PRESSURE SYSTEM

HEAT EXCHANGER AIR/COOLANT
TO GUN ACTUATOR
SOLENOID
GUN AND AMMO SHUTOFF
AMMO
AMMO COMPARTMENT
GUN GAS EXHAUST
HOOD
GUN COMPARTMENT

GROUND AIR DIVERTED
PRESS SW
TO ESSENTIAL ELECTRONICS
GROUND COOLING INLET
TO NON-ESSENTIAL ELECTRONICS
SERVICE HEAT EXCH

TEMP SENSOR 35°F/0°F
WATER SEPARATOR

(RIO CAUTION PANEL)
CABIN PRESS

CABIN TEMPERATURE SENSOR
CABIN LOW PRESSURE SWITCH

CABIN, SERVO AIR FLOW CONTROL
CABIN, SUIT, ANTI-ICE CONTROLLER

LEGEND
VENTURI TUBE
CONTROL PNEUMATIC
CONDITIONED AIR
RAM AIR
BLEED AIR
ELECTRICAL
PRESSURE SWITCH
CHECK VALVE
QUICK-DISCONNECT
FILL PORT
MODULATING OR SHUTOFF VALVE
MODULATING VALVE (SERVO AIR)
REGULATING VALVE

OVERTEMP SENSOR
ELECTRIC WINDSHIELD DEFOG AND CONTROL (105° AUTO)
TO RAIN REPELLANT AND ANTI-ICE SYSTEM

CABIN PRESSURE REGULATOR
CABIN PRESSURE CONTROL
CABIN SAFETY VALVE
GAGE PORT
COCKPIT GROUND PRESSURIZATION PORT
COCKPIT HOT AIR MODULATING

DIFFUSERS (2 EACH)
200°F
250°F
SERVO AIR PRESSURE REGULATOR

TO EQUIPMENT PRESSURIZATION SYSTEM
SERVO AIR TO OPERATE MODULATING VALVES
GROUND SERVO AIR CONN
HOT AIR MODULATING (64°/82°)
EMERGENCY RAM AIR
COOLING AIR MODULATING (62°/42°) (30K)
TO AIR-COOLED ELECTRONICS

AWG-9/AIM-54 AND AVIONIC
EQUIPMENT COOLING

FROM TURBINE COMPRESSOR

TO ESSENTIAL ELECTRONIC GROUND POWER RELAY

TO NONESSENTIAL ELECTRONIC GROUND POWER RELAY

GROUND AIR DIVERTER

TO AMMO AND GUN

FROM 400 F MANIFOLD

FROM 400 F MANIFOLD

AIM-54

AWG-9

LOW FLOW SIGNAL

RADAR MISSILE

REGENERATOR (SPACE PROVISIONS)

CONTROLLER

70 / 104 / 115 F OVERHEAT

SERVICE HEAT EXCHANGER

GROUND COOLING INLET

COOLING AIR MODULATING VALVE (64 /42 F) (30,000)

AWG-9 AIR-COOLED ELECTRONICS

CSDC
RACK III
RACK II
RACK I
WCS POWER SUPPLY
TRANSMITTER SYNCHRONIZER

INS POWER SUPPLY
UHF RECEIVER TRANSMITTER
CSDC POWER SUPPLY
TRANSFORMER RECTIFIER 100 AMPERE (2)
IMU
WCS CONTROL SYNCHRONIZER
TACAN RECEIVER TRANSMITTER

MULTIPLE DISPLAY PROCESSOR
VDIG CONVERTER
DECM AN/ALQ-100 OR AN/ALQ-126
IFF TRANSPONDER
CADC

104 F OVERHEAT

ACTUATOR

RAM AIR

AWG-9 OFF /ON 88 /110 F

70 /160 40 F

EXPANSION TANK

FLUID RELIEF 45 PSIG

THERMAL SWITCH

PHOENIX MISSILE PALLETS

FROM MISSILES

FLUID RELIEF 45 PSIG

THERMAL SWITCH

40 /85 F

TO MISSILES

TYPICAL MISSILE

AWG-9 LIQUID-COOLED ELECTRONICS

POWER SUPPLY
POWER SUPPLY
RADAR TRANSMITTER
OSCILLATOR RADAR REQUIRED
ANTENNA RADAR HYDRAULIC

30 PSIA REGULATOR

DEHYDRATOR

VENT

WAVEGUIDE TRANSMITTER ANTENNA

WCS LIQUID-COOLED AND PRESSURIZED EQUIPMENT

EFFECTIVITY

1. Aircraft BUNO 160887 and subsequent and aircraft incorporating AFC 580.

2. Aircraft BUNO 160917 and subsequent and aircraft incorporating AFC 597 and AAC 733.

3. Aircraft BUNO 161168 and subsequent have ALQ-126 in lieu of ALQ-100.

LEGEND

↑	CHECK VALVE
	QUICK DISCONNECT
	ELECTRICAL
	FILTER
	COOLANT PUMP
	REGULATING VALVE
TT	PRESSURE SWITCH

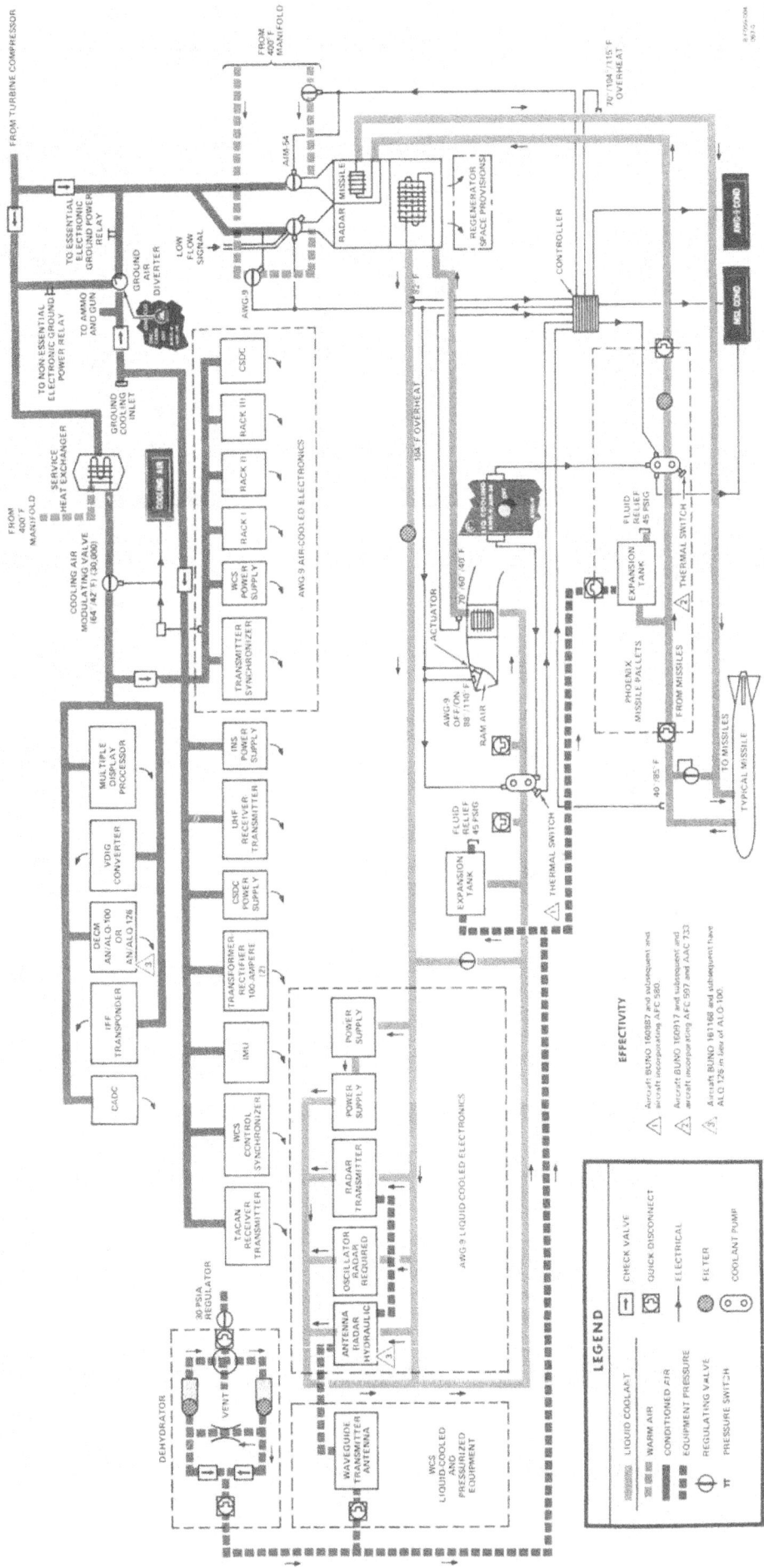

LIQUID COOLANT
WARM AIR
CONDITIONED AIR
EQUIPMENT PRESSURE

FO-16
(Reverse blank)

CANOPY PNEUMATIC AND
PYROTECHNIC JETTISON SYSTEMS

CANOPY HOOKS

ADJUSTMENT SCREW

SILL LOCKS

EXPANDABLE DETONATING CORD TUBING

HOLD-DOWN BOLT

FRANGIBLE HOLD-DOWN BOLT

EXPANDABLE DETONATING CORD TUBING

HOLD-DOWN STRAP

STRAP

HOLD DOWN BOLT

FRANGIBLE BOLT

EXPANDABLE DETONATING CORD TUBING

CANOPY JETTISON HANDLE

CANOPY JETTISON INITIATOR

CANOPY CONTROL HANDLE (PILOT)

CANOPY JETTISON HANDLE

CANOPY JETTISON INITIATOR

MANIFOLD

TO CANOPY SEPARATION CHARGE (BOTH SIDES)

CONNECTING CABLE

CANOPY CONTROL HANDLE (RIO)

CANOPY SEPARATION CHARGE (EXPANDED SHIELDED DETONATING CORD)

TO RIGHT FUSELAGE EXTERNAL CANOPY JETTISON HANDLE

TIMER VALVE

JETTISON INITIATOR

EXTERNAL EMERGENCY JETTISON

RESCUE

(BOTH SIDES FUSELAGE)

EXTERNAL CANOPY CONTROL (LEFT SIDE ONLY)

AUX UNLOCK

NORM OPEN

HOLD

NORM CLOSE

PWR CLOSE

CANOPY CONTROL CABLES

CANOPY LOCK

N₂ STORAGE BOTTLE

CONTROL VALVE

PUSH PULL CABLE TO LOCKING CAM

NEUTRAL RETURN MECHANISM

CANOPY JETTISON GAS GENERATOR

CANOPY UNLOCK

MANIFOLD

CANOPY ACTUATOR

SHIELDED MILD DETONATING CORD

PNEUMATIC CYLINDER FOR CANOPY ACTUATOR

FLEXIBLE CONFINDED DETONATION CORD

CANOPY OPEN

TO CANOPY SEPARATION CHARGE

UNLOCK

SHUTTLE VALVE

LOCK

AUXILIARY BOTTLE GAGE

CANOPY OPEN

UNLOCK

AUXILIARY UNLOCK VALVE

CABLE FROM COCKPIT CANOPY LEVER

PRESSURE REDUCER

N₂ STORAGE BOTTLE

CANOPY ACTUATOR OPEN POSITION

CANOPY LOCKING ACTUATOR

TIMER VALVE

NORMAL OPEN POSITION

25°

F-14A TOMCAT

THIS PAGE INTENTIONALY LEFT BLANK.

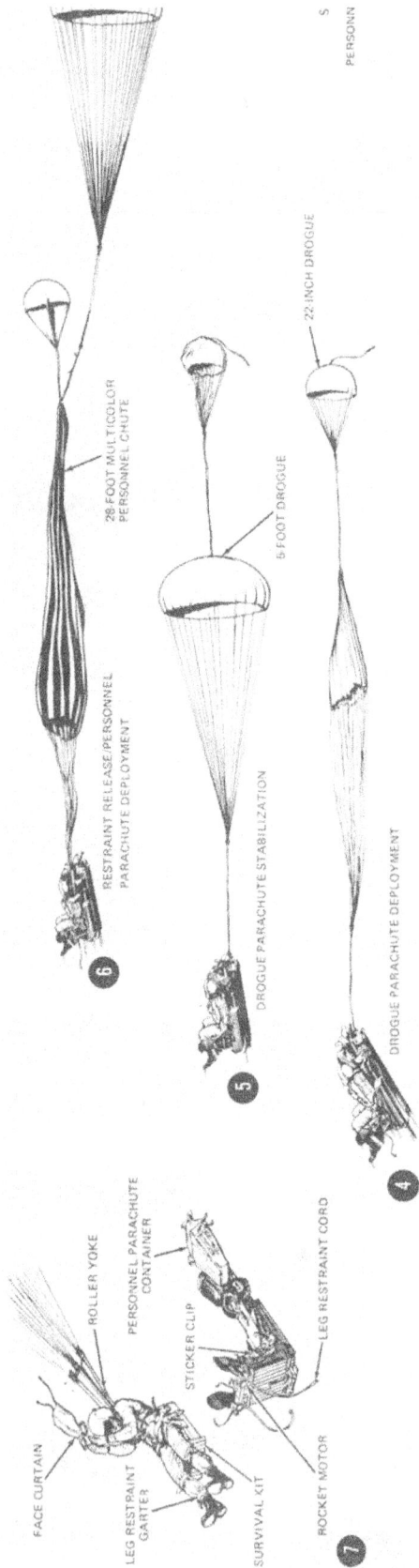

28 FOOT MULTICOLOR PERSONNEL CHUTE

22-INCH DROGUE

5-FOOT DROGUE

RESTRAINT RELEASE/PERSONNEL PARACHUTE DEPLOYMENT

DROGUE PARACHUTE STABILIZATION

DROGUE PARACHUTE DEPLOYMENT

DROGUE GUN FIRING

ROCKET THRUSTING CATAPULT FIRES

SAFE AND ARM FIRING LANYARDS

FACE CURTAIN

LEG RESTRAINT GARTER

SURVIVAL KIT

ROCKET MOTOR

ROLLER YOKE

PERSONNEL PARACHUTE CONTAINER

STICKER CLIP

LEG RESTRAINT CORD

WARNING

THE FOUR LINE RELEASE SHOULD NOT BE ACTIVATED IF DAMAGE TO THE CANOPY OR BROKEN SUSPENSION LINES ARE OBSERVED AFTER DEPLOYMENT OF A FULL CANOPY.

7. LINE STRETCH OF MAIN PARACHUTE PULLS OCCUPANT, SURVIVAL KIT, AND LOWER RESTRAINT HARNESS FREE OF STICKER CLIPS. SEAT FALLS FREE. OCCUPANT DISCARDS FACE CURTAIN AND CONTINUES NORMAL PARACHUTE DESCENT.

8. ON PARACHUTE ASSEMBLIES EQUIPPED WITH ACES 383 (FOUR LINE RELEASE) CAPABILITIES, THE FOLLOWING PROCEDURES APPLY DURING DESCENT.

1. AFTER VISUAL INSPECTION OF PARACHUTE AND DECIDING TO ACTIVATE FOUR LINE RELEASE, THE FLIGHTCREWMAN SHOULD

6. ABOVE 13,000-1,500 FEET, BAROSTAT SECURES TIME RELEASE ESCAPEMENT MECHANISM UNTIL DESCENT TO LOWER ALTITUDE. DROGUE PARACHUTE RETENTION SHACKLE REMAINS LOCKED TO SEAT BY RESTRAINT SCISSOR. SEAT AND OCCUPANT DESCEND THROUGH HIGHER ALTITUDES. BELOW 13,000-1,500 FEET BAROSTAT FREES TIME RELEASE ESCAPEMENT MECHANISM. TIME RELEASE MECHANISM SUBSEQUENTLY RELEASES DROGUE SHACKLE RESTRAINT SCISSOR. OCCUPANT'S SHOULDER HARNESS RESTRAINTS, LAP BELT, LEG RESTRAINTS, PARACHUTE CONTAINER, AND SURVIVAL KIT. RELEASE OF SHACKLE PERMITS CONTINUED PULL OF DROGUE PARACHUTES ON LINK LINES TO RELEASE FACE CURTAIN RESTRAINT AND MAIN PERSONNEL PARACHUTE. OCCUPANT REMAINS ATTACHED TO SEAT BY STICKER CLIP RETENTION OF LOWER RESTRAINT HARNESS ON SEAT BUCKET.

1. EJECTION INITIATED. SHOULDER HARNESS RETRACTED. CANOPY HOOKS CUT AND CANOPY BALLISTICALLY JETTISONED. SAFE AND ARM DEVICE FIRED. (RIO'S SEAT TRIGGERED. PILOT'S SEAT DELAYED 0.4 SECOND.)

2. CATAPULT FIRES. LEG RESTRAINT PULLED TIGHT. DROGUE GUN AND TIME-RELEASE MECHANISM TRIPPED. EMERGENCY OXYGEN BOTTLE ACTIVATED. IFF AND ECM DESTRUCT SWITCH ACTUATED. PERSONNEL SERVICES DISCONNECTED. ROCKET MOTOR GAS GENERATOR LANYARD PULLED AS SEAT GOES UP RAIL. ROCKET MOTOR IGNITED.

3. DROGUE GUN FIRES 0.5 SECOND AFTER EJECTION.

4. DUPLEX DROGUE PARACHUTE SYSTEM DEPLOYED.

5. DUPLEX DROGUE PARACHUTE SYSTEM STABILIZES AND DECELERATES SEAT.

28 FOOT MULTICOLOR PERSONNEL CHUTE

22 INCH DROGUE

5 FOOT DROGUE

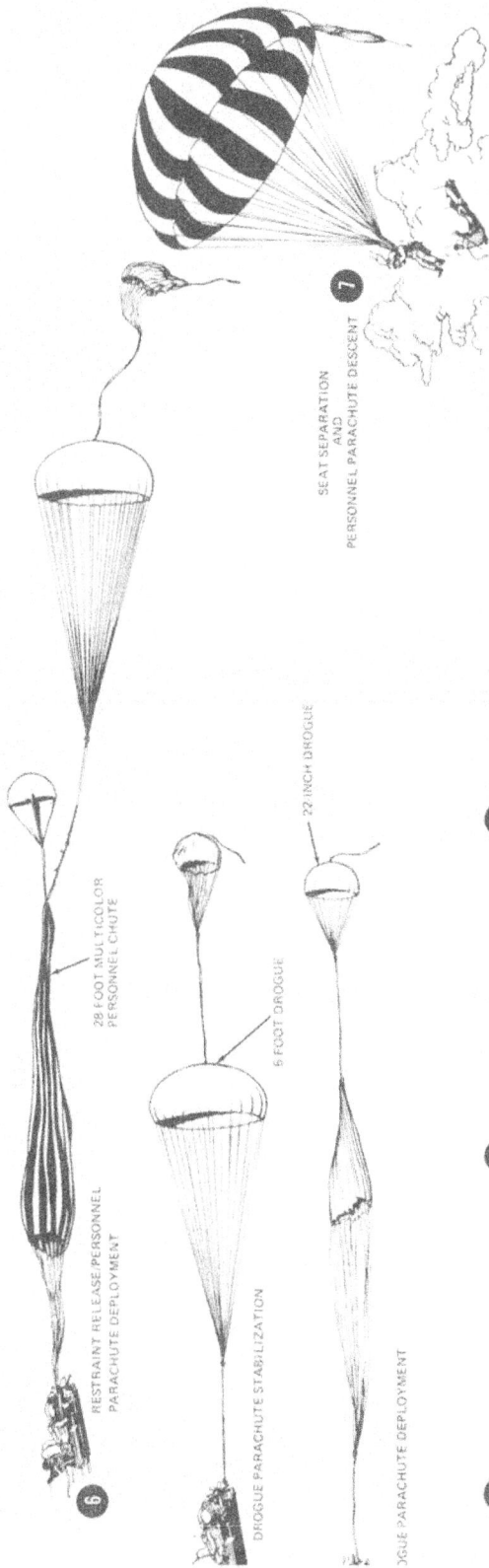

RESTRAINT RELEASE PERSONNEL PARACHUTE DEPLOYMENT

DROGUE PARACHUTE STABILIZATION

DROGUE PARACHUTE DEPLOYMENT

SEAT SEPARATION AND PERSONNEL PARACHUTE DESCENT

(6) RESTRAINT RELEASE PERSONNEL PARACHUTE DEPLOYMENT

(7) SEAT SEPARATION AND PERSONNEL PARACHUTE DESCENT

(1) EJECTION INITIATED. SHOULDER HARNESS RETRACTED, CANOPY HOOKS CUT AND CANOPY BALLISTICALLY JETTISONED, SAFE AND ARM DEVICE FIRED, RIO'S SEAT TRIGGERED, PILOT'S SEAT DELAYED 0.4 SECOND.

(2) CATAPULT FIRES, LEG RESTRAINT PULLED TIGHT, DROGUE GUN AND TIME-RELEASE MECHANISM TRIPPED, EMERGENCY OXYGEN BOTTLE ACTIVATED, IFF AND ECM DESTRUCT SWITCH ACTUATED. PERSONNEL SERVICES DISCONNECTED. ROCKET MOTOR GAS GENERATOR LANYARD PULLED AS SEAT GOES UP RAIL, ROCKET MOTOR IGNITED.

(3) DROGUE GUN FIRES 0.5 SECOND AFTER EJECTION

(4) DUPLEX DROGUE PARACHUTE SYSTEM DEPLOYED

(5) DUPLEX DROGUE PARACHUTE SYSTEM STABILIZES AND DECELERATES SEAT.

(6) ABOVE 13,000/1500 FEET, BAROSTAT SECURES TIME RELEASE ESCAPEMENT MECHANISM UNTIL DESCENT TO LOWER ALTITUDE. DROGUE PARACHUTE RETENTION SHACKLE REMAINS LOCKED TO SEAT BY RESTRAINT SCISSOR, SEAT AND OCCUPANT DESCEND THROUGH HIGHER ALTITUDES ON DROGUE PARACHUTES ONLY. BELOW 13,000/1500 FEET BAROSTAT FREES TIME RELEASE ESCAPEMENT MECHANISM. TIME RELEASE MECHANISM SUBSEQUENTLY RELEASES DROGUE SHACKLE RESTRAINT SCISSOR, OCCUPANT'S SHOULDER HARNESS RESTRAINTS, LAP BELT, LEG RESTRAINTS. PARACHUTE CONTAINER, AND SURVIVAL KIT. RELEASE OF SHACKLE PERMITS CONTINUED PULL OF DROGUE PARACHUTES ON LINK LINES TO RELEASE FACE CURTAIN RESTRAINT AND MAIN PERSONNEL PARACHUTE. OCCUPANT REMAINS ATTACHED TO SEAT BY STICKER CLIP RETENTION OF LOWER RESTRAINT HARNESS ON SEAT BUCKET.

(7) LINE STRETCH OF MAIN PARACHUTE PULLS OCCUPANT, SURVIVAL KIT, AND LOWER RESTRAINT HARNESS FREE OF STICKER CLIPS. SEAT FALLS FREE OF OCCUPANT DISCARDS FACE CURTAIN AND CONTINUES NORMAL PARACHUTE DESCENT.

(8) ON PARACHUTE ASSEMBLIES EQUIPPED WITH ACSE 383 (FOUR LINE RELEASE) CAPABILITIES, THE FOLLOWING PROCEDURES APPLY DURING DESCENT.

WARNING

THE FOUR LINE RELEASE SHOULD NOT BE ACTIVATED IF DAMAGE TO THE CANOPY OR BROKEN SUSPENSION LINES ARE OBSERVED AFTER DEPLOYMENT OF A FULL CANOPY.

1. AFTER VISUAL INSPECTION OF PARACHUTE AND DECIDING TO ACTIVATE FOUR LINE RELEASE, THE FLIGHTCREWMAN SHOULD GRASP RELEASE LANYARD LOOPS ON INSIDE OF REAR RISERS AND BREAK RELEASE TIES BY SHARP PULL (APPROXIMATELY 20 POUNDS FORCE). THIS ACTION FREES REAR FOUR SUSPENSION LINES FROM CONNECTOR LINKS, ALLOWING CANOPY TO FORM LOBE IN REAR CENTER AND PERMIT A STEADY ESCAPE OF AIR, WHICH ELIMINATES OSCILLATION AND GIVES MINIMAL DIRECTIONAL CONTROL OF CANOPY (TO RIGHT OR LEFT) BY PULLING ON RESPECTIVE RELEASE LANYARD.

2. IN PREPARATION FOR LANDING, FLIGHTCREWMAN SHOULD ATTEMPT TO DETERMINE SURFACE WIND DIRECTION BY ANY MEANS POSSIBLE (SMOKE, DUST, ETC) AND TURN CANOPY INTO WIND IF POSSIBLE BEFORE LANDING. OVER LAND, THIS GIVES ADVANTAGE OF MISSING OBSTACLES AND CHOOSING LANDING AREA. OVER WATER, THIS MINIMIZES CHANCE OF LINE ENTANGLEMENT DUE TO FORWARD TRAVEL OF CANOPY AWAY FROM FLIGHTCREWMAN.

21-050-004
07-9

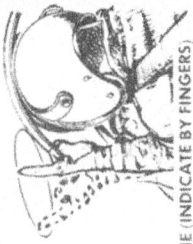

START/WINDMILL ENGINE (INDICATE BY FINGERS)
PILOT EXTENDS FINGERS TO INDICATE WHICH ENGINE IS READY FOR START. (1 PORT, 2 STARBOARD). IF ALL CLEAR, SIGNALMAN RESPONDS WITH SIMILAR GESTURE POINTING AT PROPER ENGINE WHILE ROTATING OTHER HAND IN CLOCKWISE MOTION.

OPEN SPEED BRAKES
EXTEND ARMS AT WAIST WITH PALMS TOGETHER. KEEP WRISTS TOGETHER AND OPEN PALMS.

TURN RIGHT
PULL DESIRED WING AROUND WITH REGULAR "COME AHEAD" AND POINT AT OPPOSITE BRAKE.

NOSEWHEEL STEERING
RIGHT INDEX FINGER POINTING TO RIGHT SIDE OF NOSE FOR RIGHT TURN AND VICE VERSA FOR LEFT TURN. OPPOSITE HAND POINTING TO NOSEGEAR.

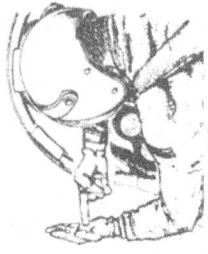

INSERT/PULL EXTERNAL AIR
PILOT INSERTS AND PULLS INDEX FINGER TO AND FROM OPEN PALM. SIGNALMAN RESPONDS WITH SAME SIGNAL.

OBC
HANDS HELD CHEST HIGH TO FORM A "T".

EXTEND IFR PROBE
FIST CLENCHED TO CHEST, THEN ARM EXTENDED TO SIDE.

CUT ENGINES
HAND DRAWN ACROSS NECK IN "THROAT CUTTING" MOTION.

LOWER LAUNCH BAR
ONE ARM BENT UPWARD, ELBOW IN PALM OF OTHER HAND. EXTEND ARM FROM VERTICAL TO HORIZONTAL.

EXTERNAL LIGHTS CHECK
HOLD THE INDEX AND MIDDLE FINGER IN A "V" SIGNAL POINTING TOWARDS THE EYES.

LOWER SPOILER
ARMS EXTENDED HORIZONTAL DIRECTLY AHEAD WITH PALMS SHOWING. DROP PALMS FROM WRIST.

HOT BRAKE
MAKE RAPID FANNING MOTION WITH ONE HAND IN FRONT OF FACE. POINT TO WHEEL WITH OTHER HAND.

KNEEL/UNKNEEL (EXTEND) AIRCRAFT
EXTEND ARM TO SIDE AT SHOULDER LEVEL BRING OTHER ARM DOWN PALMS TOGETHER MOVE ARMS APART TO SIGNAL EXTEND.

RAISE TAILHOOK
RIGHT THUMB JERKED UP TO MEET HORIZONTAL LEFT HAND.

INSERT/PULL ELECTRICAL POWER
PILOT INSERTS AND PULLS INDEX AND MIDDLE FINGER TO AND FROM OPEN PALM. SIGNALMAN RESPONDS WITH SAME SIGNAL.

BACKUP MODULE CHECK/ FLIGHT CONTROLS CYCLE CHECK
PILOT ROTATES CLENCHED FIST IN A HORIZONTAL PLANE. SIGNAL MAN REPEATS SIGNAL. IF ALL CONTROL SURFACES ARE CLEAR OF PERSONNEL/EQUIPMENT.

DLC CHECK/ RAISE SPOILERS
ARMS EXTENDED HORIZONTAL AND DIRECTLY AHEAD. HANDS POP UP FROM WRIST TO SHOW PALMS.

HOLD
ARM RAISED OVER HEAD WITH FIST CLENCHED.

LOWER TAILHOOK
LOWER RIGHT FIST SUDDENLY, THUMB EXTENDED DOWNWARD, TO MEET HORIZONTAL PALM OF LEFT HAND HELD IN FRONT OF BODY.

FLAPS DOWN
ARMS IN CLOSE TO BODY. HANDS FLAT TOGETHER, THEN OPENED WIDE FROM WRISTS.

STOP/APPLY BRAKES
ARMS RAISED UP, PALMS OUTWARD. RAPIDLY CLENCH FIST.

WING SWEEP 68°
ARMS FROM STRAIGHT OUT DROPPED TO SIDE.

COME AHEAD
HANDS AT EYE LEVEL EXECUTE BECKONING MOTIONS. RATE OF MOTION INDICATES DESIRED SPEED OF AIRCRAFT. FOR NIGHT OPERATION, WAVE WANDS SIDE TO SIDE.

WET START
SIGNAL PILOT BY PINCHING NOSE, THEN GIVING CUT ENGINE SIGNAL FOLLOWED BY WINDMILL ENGINE SIGNAL.

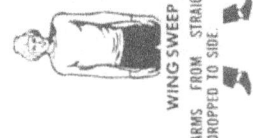

DECK/GROUND HANDLING SIGNALS

FLAPS UP
HANDS OPENED WIDE FROM WRIST, SUDDENLY CLOSED. ARMS IN CLOSE TO BODY.

WING SWEEP 50°
ARMS FROM STRAIGHT OUT SWEPT HALF WAY DOWN TO SIDE.

MANEUVER FLAPS DOWN
ARMS IN CLOSE TO BODY. HANDS FLAT TOGETHER THEN OPENED WIDE FROM WRISTS. FOLLOWED BY HAND HELD IN "X".

WING SWEEP 20°
ARMS IN HUGGING POSITION ACROSS THE CHEST THEN SWEPT OUT TO SIDES. ARMS HELD STRAIGHT OUT.

TRIM CHECK
ONE FINGER FOR RUDDER. TWO FINGERS FOR LONGITUDINAL. THREE FINGERS FOR LATERAL.

OVERSWEEP
ARMS, FROM STRAIGHT OUT SWEPT IN ACROSS THE CHEST HUGGING THE SHOULDERS.

RETRACT IFR PROBE
ARM EXTENDED TO SIDE, FIST BROUGHT TO CHEST AND CLENCHED.

INSTALL/REMOVE GROUND LOCKS
GRASP WRIST—REMOVING HAND IS REMOVING LOCKS, GRASPING WRIST IS INSTALLING LOCKS.

HANDS OFF CONTROLS
ARMS HELD UP OVER HEAD WITH PALMS FACING AIRCRAFT.

RELEASE BRAKES
ARMS RAISED UP. FISTS CLENCHED AND HELD IN SIMPLE "POLICEMAN'S STOP". FISTS OPEN TO SHOW PALMS.

REMOVE/INSTALL CHOCKS
PILOT MAKES SWEEPING MOTION OF FISTS WITH THUMBS EXTENDED OUTWARD. SIGNALMAN SWEEPS FISTS APART AT HIP LEVEL WITH THUMBS EXTENDED OUTWARD. INSTALL IS A THUMBS IN MOTION.

ANTI-SKID CHECK
FISTS HELD IN FRONT AND THEN OPENED. FOLLOWED BY HANDS HELD IN "X".

CLOSE SPEED BRAKES
ARMS EXTENDED AT WAIST. WRISTS TOGETHER. PALMS SPREAD. BRING PALMS TOGETHER RAPIDLY.

TURN LEFT
PULL DESIRED WING AROUND WITH REGULAR "COME AHEAD" AND POINT AT OPPOSITE BRAKE.

AIRCRAFT FIRE
DESCRIBE A LARGE FIGURE EIGHT WITH ONE HAND AND POINT TO THE FIRE AREA WITH THE OTHER HAND.

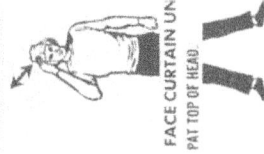

EMERGENCY SHUTDOWN
EMERGENCY STOP SIGNALS FOLLOWED BY CUT ENGINE.

EMERGENCY STOP
ARMS CROSSED ABOVE HEAD, FISTS CLENCHED.

TURNOVER OF COMMAND
BOTH HANDS POINTED AT NEXT SUCCEEDING TAXI SIGNALMAN.

SLOW DOWN
DOWNWARD PATTING MOTION, HANDS OUT AT WAIST LEVEL.

GROUND REFUELING INTERNAL TANKS
(NO EXTERNAL POWER)
CIRCULAR MOTION WITH THE PALM OF HAND TOWARD STOMACH (AS RUBBING STOMACH). FOLLOWED BY A DRINKING MOTION (THUMB TO MOUTH).

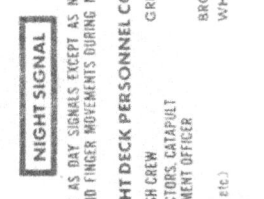

GROUND INTERCOM
CUP HANDS OVER EARS OR POINT WANDS TO EARS.

THROTTLE BACK
GRASP THUMB AND PULL BACK TO SIGNIFY REDUCTION OF THROTTLE SETTING.

FACE CURTAIN UNLOCK
PAT TOP OF HEAD.

NIGHT SIGNAL

NIGHT SIGNALS ARE THE SAME AS DAY SIGNALS EXCEPT AS NOTED. FLASHLIGHTS OR WANDS WILL SUBSTITUTE FOR HAND AND FINGER MOVEMENTS DURING NIGHT OPERATIONS.

CARRIER FLIGHT DECK PERSONNEL COLOR CODING

RED SHIRTS— ORDNANCE and CRASH CREW
YELLOW SHIRTS— PRI FLY, PLANE DIRECTORS, CATAPULT OFFICER, and ARRESTMENT OFFICER
BLUE SHIRTS— PLANE HANDLERS (Pushers, Chock Men, etc.)
PURPLE SHIRTS— FUEL HANDLING
GREEN SHIRTS— AIRCRAFT MAINTENANCE, CATAPULT CREW AND ARRESTMENT CREW
BROWN SHIRTS— PLANE CAPTAINS
WHITE SHIRTS— MEDICAL

NAVIGATION SYSTEM OVERVIEW

HUD
- BAROMETRIC ALTITUDE
- RADAR ALTITUDE
- VERTICAL SPEED
- MAGNETIC HEADING
- TACAN DEVIATION
- ROLL/PITCH

VDI
- COMMAND HEADING
- COMMAND ALTITUDE (D/L)
- COMMAND AIRSPEED (D/L)
- TACAN DEVIATION
- MAGNETIC HEADING
- ROLL/PITCH

HSD
- ADF BEARING
- TRUE AIRSPEED
- GROUNDSPEED
- TACAN BEARING
- TACAN RANGE
- DESTINATION RANGE
- TACAN DEVIATION
- COMMAND HEADING AND BEARING
- COMMAND HEADING
- COMMAND COURSE
- WIND SPEED AND DIRECTION
- MAGNETIC HEADING
- GROUND TRACK
- GROUND SPEED

ECMD

TID
- LATITUDE AND LONGITUDE
- WIND SPEED AND DIRECTION
- MAGNETIC HEADING
- MAGNETIC VARIATION
- DESTINATION, RANGE, AND BEARING
- TIME TO GO
- TRUE HEADING
- TRUE AIRSPEED
- GROUND TRACK
- GROUNDSPEED
- CAT #

- OWN AIRCRAFT
- HOME BASE
- DEPARTURE POINT
- WAYPOINT

VDIG CONVERTER

MDIG PROCESSOR

DISPLAYS CONTROL PANEL

DISPLAY MODE

DATA

NAVIGATION MODE

CSDC NAVIGATION COMPUTER

ALIGNMENT AND NAVIGATION COMPUTATIONS

WCS COMPUTER

DATA ENTRY READOUT

DATA READOUT
- LAT/LONG
- SPD
- CRS
- RGE
- BRG
- ALT
- HDG

DATA ENTRY
- DEST LAT/LONG
- CAT #
- CAT ANGLE
- MAG VAR
- CV SPD/HDG
- A/C HDG
- UPDATE
- WIND SPD/DIR

DATA

TACAN — RANGE BEARING

ADF — BEARING

AHRS — MAGNETIC HEADING

IMU — ROLL AND PITCH VELOCITIES

CADC — PRESSURE ALTITUDE AND TRUE AIRSPEED

RADAR ALTIMETER

DATA LINK — SINS ALIGNMENT

BEARING DISTANCE HEADING INDICATOR
- RANGE BEARING
- BEARING
- MAGNETIC HEADING

ALTIMETER

INDEX

INDEX (CONT)

INDEX (CONT)

INDEX (CONT)

INDEX (CONT)

INDEX (CONT)

INDEX (CONT)

INDEX (CONT)

INDEX (CONT)

INDEX (CONT)

INDEX (CONT)

INDEX (CONT)

INDEX (CONT)

INDEX (CONT)

INDEX (CONT)

INDEX (CONT)

INDEX (CONT)

INDEX (CONT)

INDEX (CONT)

INDEX (CONT)

INDEX (CONT)

INDEX (CONT)

INDEX (CONT)

INDEX (CONT)

INDEX (CONT)

☆ U.S. GOVERNMENT PRINTING OFFICE: 1982-517-007/327

THIS PAGE INTENTIONALY LEFT BLANK.